IB DIPLOMA PROGRAMME

Physics

Course Companion

Tim Kirk
Neil Hodgson

OXFORD
UNIVERSITY PRESS

IN COOPERATION WITH ib

OXFORD
UNIVERSITY PRESS

Great Clarendon Street, Oxford OX2 6DP

Oxford University Press is a department of the University of Oxford. It furthers the University's objective of excellence in research, scholarship, and education by publishing worldwide in

Oxford New York

Auckland Cape Town Dar es Salaam Hong Kong Karachi Kuala Lumpur Madrid Melbourne Mexico City Nairobi New Delhi Shanghai Taipei Toronto

With offices in

Argentina Austria Brazil Chile Czech Republic France Greece Guatemala Hungary Italy Japan Poland Portugal Singapore South Korea Switzerland Thailand Turkey Ukraine Vietnam

First published 2007

British Library Cataloguing in Publication Data

Data available

ISBN-13: 978-099151448

10 9 8 7 6 5 4 3 2

Printed in Great Britain by Bell and Bain Ltd, Glasgow

Paper used in the production of this book is a natural, recyclable product made from wood grown in sustainable forests. The manufacturing process conforms to the environmental regulations of the country of origin.

Acknowledgements

We are grateful to the following for permission to reproduce copyright material:

Carbon Dioxide Information Analysis Center (CDIAC) for extract from J-M Barnola, D Raynaud, C Lorius, and N I Barakov: 'Historical CO2 record from the Vostok ice core' in *Trends: A Compendium of Data on Global Change* (CDIAC, 2003).
Greenpeace for extract from website: www.greenpeace.org/international/campaigns/
Independent News and Media for James Lovelock: 'Nuclear Power is the only green solution', *The Independent*, 24.5.2006, copyright © The Independent 2006.
Intergovernmental Panel on Climate Change (IPCC) for extracts from *Climate Change 2007: The Physical Science Basis*, (IPCC Fourth Assessment Report, 2007)
Leukemia Research Foundation for extract from the LRF report after Royal Society of Edinburgh 'Symposium on Leukemia Clusters', 1994, and press statement by Cathy Gilman, Chief Executive LRF, in response to article by Nicholas Cecil in the *Evening Standard* 20.4.2007.
Richard S Lindzen for extract from 'The Press Gets it Wrong: Our report doesn't support the Kyoto Treaty', published in *The Wall Street Journal*, Opinion Journal, 11.6.2001
Organisation for Economic Co-operation and Development and the International Energy Agency (OECD/IEA) for extracts from 'China's Electrification Success Story' and other data from *World Economic Outlook 2002*, copyright © OECD/IEA 2002.
Penguin Books Ltd and **Random House**, Inc for John Updike: 'Cosmic Gall' from *Collected Poems 1953-1993* (Knopf/Hamish Hamilton 1993, Penguin 1995), copyright © John Updike 1993.
Solo Syndication for Nicholas Cecil: 'Power lines link to cancer in new alert', *Evening Standard*, 20.4.2007.
Music Sales Limited for extract from the lyrics of 'London Calling', The Clash, 1979, words and music by Joe Strummer, Mick Jones, Paul Simonon and Topper Headon, copyright © 1979 Nineden Ltd/Universal Music Publishing Ltd. All rights reserved. International copyright secured.

p80 Handbook for Acoustic Ecology, B. Truax. ed, Cambridge Street Publishing, CD-ROM edition, 1999; **pp82-3** www.geo.uib.no; **p97** From Salters Horners Advanced Physics AS Level Student Book by the University of York. Reprinted by permission of Harcourt Education; **p212** Physics World; **p219,223(table), 227** Energy Information Administration; **p222(figure and table)** BP Statistical Review of World Energy; **p228** World Energy Outlook, 2002 © OECD/IEA, 2002, Figure 13.8: "Number of People Without Electricity 1970-2030"; **p241(left)** Marland, G., T.A. Boden, and R. J. Andres. 2006. Global, Regional, and National CO2 Emissions. In Trends: A Compendium of Data on Global Change. Carbon Dioxide Information Analysis Center, Oak Ridge National Laboratory, U.S. Department of Energy, Oak Ridge, Tenn., U.S.A.; **p241(right)** K.M. Lugina, P.Ya. Groisman, K.Ya. Vinnikov, V.V. Koknaeva, and N.A. Speranskaya, 2006. In Trends Online: A Compendium of Data on Global Change. Carbon Dioxide Information Analysis Center, Oak Ridge National Laboratory, U.S. Department of Energy, Oak Ridge, Tennessee, U.S.A.; **p242(middle)** Petit, J.R., D. Raynaud, C. Lorius, J. Jouzel, G. Delaygue, N.I. Barkov, and V.M. Kotlyakov. 2000. Historical isotopic temperature record from the Vostok ice core; **p242(bottom)** Barnola, J.-M., D. Raynaud, C. Lorius, and N.I. Barkov. 2003. Historical CO2 record from the Vostok ice core; **p245(figure and table)** Climate Change 2007: The Physical Science Basis, Summary for Policymakers, Intergovermental Panel on Climate Change; **p258,259(table)** IAEA PRIS system; p282 STScI.

Oxford University Press would also like to thank Jonathan Allday for his permission to reproduce material.

We have tried to trace and contact all copyright holders before publication. If notified the publishers will be pleased to rectify any errors or omissions at the earliest opportunity.

Photo Credits

p7 (right) Carsten Reisinger/Shutterstock; **p7 (left)** Black Brent /pacific stock/ Photolibrary; **p7 (centre)** Science Photolibrary/VSC; **p21** Sha Tin College technicians; **p25 (top)** Dimitri Iundt/TempSport/Corbis; **p25 (centre)** REUTERS/Kim Kyung Hoon; **p25 (bottom)** Stefan Matzke/NewSport/Corbis; **p44 (left)** Ben smith/Shutterstock; **p44 (centre)** Walter Matheson; **p44 (right)** Hugo Maes/Shutterstock; **p63** Steve Pyke/Getty Images; **p72** Nonstock/Dinodia; **p81** Spencer Grant/Photo Researchers, Inc/Photolibrary; **p84** Nikolay Titov/Shutterstock; **p88** Dr. Larpent/G.R.E.H.G.E.P./Science Photo Library; **p91 (bottom)** Lourens Smak/Alamy; **p92** David B Fleetham/Oxford Scientific Films/ Photolibrary; **p93** Bill Brennan/Pacific Stock/Photolibrary; **p104 (top)** Nicola Vernizzi/ Shutterstock; **p129 (top left)** Craig Hill/Shutterstock; **p129 (top right)** Romanchuck Dimitry/Shutterstock; **p129 (bottom left)** Kropachev Nikolay/Shutterstock; **p136 (top)** Richard Sargeant/Shutterstock; **p138 (left)** Andrew Lambert Photography/Science Photo Library/Photolibrary; **p138 (right)** Helene Rogers/Alamy; **p141 (top)** Andrew Lambert Photography/Science Photo Library/Photolibrary; **p142 (top)** Peter Menzel/ Science Photo Library; **p142 (centre bottom)** John Barry de Nicola/Shutterstock; **p142 (bottom)** Eric Gevaert/Shutterstock; **p143 (top)** Kerry Hayes/Twentieth Century Fox/Bureau L.A. Collection/Corbis; **p143 (bottom left)** Elena Elisseeva/Shutterstock & NASA; **p143 (bottom right)** NASA; **p183 (bottom)** Andrew Lambert Photography/ Science Photo Library; **p184 (top left)** Science Photo Library: **p184 (top right)** Popperfoto/Alamy; **p184 (bottom left)** Prof. Peter Fowler/Science Photo Library; **p184 (bottom centre)** American Institute of Physics/Science Photo Library; **p185 (top)** A. Barrington Brown/Science Photo Library; **p185 (bottom)** Science Photo Library; **p193** Physics Today Collection/American Institute of Physics/Science Photo Library; **p196** Carl Anderson/Science Photo Library; **p198 (top)** to be confirmed; **p198 (bottom)** Hank Morgan/Science Photo Library/Photolibrary; **p213** Sha Tin College technicians; **p225** Sha Tin College technicians; **p244** Greenpeace / Jeremy Sutton-Hibbert; **p251 (top)** Kimberly White/REUTERS; **p275 (bottom)** Dr. Tony Brain/Science Photo Library; **p278 (top)** NASA; **p278 (bottom)** Sheila Terry/Science Photo Library; **p279** Andrew Lambert Photography/Science Photo Library; **p282 (top)** NASA; **p317** Carl Anderson/Science Photo Library

Meet the authors

Tim Kirk is a senior examiner and a former deputy chief examiner for IB and was a member of the team who wrote the syllabus for the 2007 physics course. He is an experienced workshop leader and has taught for over twenty years in a wide range of UK and international schools. He is currently teaching at the Godolphin and Latymer School in London.

Neil Hodgson is an assistant examiner in physics and an experienced author. He was a member of the team that wrote the syllabus for the 2007 physics course. For the last ten years he has taught at Sha Tin College in Hong Kong and is currently Head of Science.

Course Companion definition

The IB Diploma Programme Course Companions are resource materials designed to provide students with extra support through their two-year course of study. These books will help students gain an understanding of what is expected from the study of an IB Diploma Programme subject.

The Course Companions reflect the philosophy and approach of the IB Diploma Programme and present content in a way that illustrates the purpose and aims of the IB. They encourage a deep understanding of each subject by making connections to wider issues and providing opportunities for critical thinking.

These Course Companions, therefore, may or may not contain all of the curriculum content required in each IB Diploma Programme subject, and so are not designed to be complete and prescriptive textbooks. Each book will try to ensure that areas of curriculum that are unique to the IB or to a new course revision are thoroughly covered. These books mirror the IB philosophy of viewing the curriculum in terms of a whole-course approach; the use of a wide range of resources; international-mindedness; the IB learner profile and the IB Diploma Programme core requirements; theory of knowledge; the extended essay; and creativity, action, service (CAS).

In addition, the Course Companions provide advice and guidance on the specific course assessment requirements and also on academic honesty protocol.

IB mission statement

The International Baccalaureate aims to develop inquiring, knowledgeable, and caring young people who help to create a better and more peaceful world through intercultural understanding and respect.

To this end the IB works with schools, governments, and international organizations to develop challenging programmes of international education and rigorous assessment.

These programmes encourage students across the world to become active, compassionate, and lifelong learners who understand that other people, with their differences, can also be right.

The IB learner profile

The International Baccalaureate aims to develop internationally minded people who, recognizing their common humanity and shared guardianship of the planet, help to create a better and more peaceful world. IB learners strive to be:

Inquirers They develop their natural curiosity. They acquire the skills necessary to conduct inquiry and research and show independence in learning. They actively enjoy learning and this love of learning will be sustained throughout their lives.

Knowledgeable They explore concepts, ideas, and issues that have local and global significance. In so doing, they acquire in-depth knowledge and develop understanding across a broad and balanced range of disciplines.

Thinkers They exercise initiative in applying thinking skills critically and creatively to recognize and approach complex problems, and make reasoned, ethical decisions.

Communicators They understand and express ideas and information confidently and creatively in more than one language and in a variety of modes of communication. They work effectively and willingly in collaboration with others.

Principled They act with integrity and honesty, with a strong sense of fairness, justice, and respect for the dignity of the individual, groups, and communities. They take responsibility for their own actions and the consequences that accompany them.

Open-minded They understand and appreciate their own cultures and personal histories, and are open to the perspectives, values, and traditions of other individuals and communities. They are accustomed to seeking and evaluating a range of points of view, and are willing to grow from the experience.

Caring They show empathy, compassion, and respect towards the needs and feelings of others. They have a personal commitment to service, and act to make a positive difference to the lives of others and to the environment.

Risk-takers They approach unfamiliar situations and uncertainty with courage and forethought, and have the independence of spirit to explore new roles, ideas, and strategies. They are brave and articulate in defending their beliefs.

Balanced They understand the importance of intellectual, physical, and emotional balance to achieve personal well-being for themselves and others.

Reflective They give thoughtful consideration to their own learning and experience. They are able to assess and understand their strengths and limitations in order to support their learning and personal development.

A note on academic honesty

It is of vital importance to acknowledge and appropriately credit the owners of information when that information is used in your work. After all, owners of ideas (intellectual property) have property rights. To have an authentic piece of work, it must be based on your individual and original ideas with the work of others fully acknowledged. Therefore, all assignments, written or oral, completed for assessment must use your own language and expression. Where sources are used or referred to, whether in the form of direct quotation or paraphrase, such sources must be appropriately acknowledged through the use of footnotes and bibliographies.

Footnotes (placed at the bottom of a page) or endnotes (placed at the end of a document) are to be provided when you quote or paraphrase from another document, or closely summarize the information provided in another document. You do not need to provide a footnote for information that is part of a "body of knowledge". That is, definitions do not need to be footnoted as they are part of the assumed knowledge.

Bibliographies should include a formal list of the resources that you used in your work. "Formal" means that you should use one of the several accepted forms of presentation. This usually involves separating the resources that you use into different categories (e.g. books, magazines, newspaper articles, Internet-based resources, CDs, and works of art) and providing full information as to how a reader or viewer of your work can find the same information. A bibliography is compulsory in the extended essay.

What constitutes malpractice?

Malpractice is behaviour that results in, or may result in, you or any student gaining an unfair advantage in one or more assessment component. Malpractice includes plagiarism and collusion.

Plagiarism is defined as the representation of the ideas or work of another person as your own. The following are some of the ways to avoid plagiarism:

- words and ideas of another person to support one's arguments must be acknowledged
- passages that are quoted verbatim must be enclosed within quotation marks and acknowledged
- CD-ROMs, email messages, web sites on the Internet, and any other electronic media must be treated in the same way as books and journals
- the sources of all photographs, maps, illustrations, computer programs, data, graphs, audio-visual, and similar material must be acknowledged if they are not your own work
- copying works of art, whether music, film, dance, theatre arts, or visual arts, and where the creative use of a part of a work takes place, must be acknowledged.

Collusion is defined as supporting malpractice by another student. This includes:

- allowing your work to be copied or submitted for assessment by another student
- duplicating work for different assessment components and/or diploma requirements.

Contents

1	Learning in physics	7
2	Experimental measurements and internal assessment	10
3	Kinematics	25
4	Forces and motion	32
5	Energy, work, and power	44
6	Projectiles	52
7	Simple harmonic motion	55
8	Wave	78
9	Wave phenomena	94
10	Thermal physics	109
11	Thermodynamics	118
12	Electric currents	127
13	Three different non-contact forces	142
14	Field and potential	158
15	Electromagnetic induction	166
16	Atomic and nuclear physics	180
17	Quantum physics	207
18	Global energy sources	218
19	Solar radiation and the greenhouse effect	231
20	Climate change	241
21	Technologies to reduce climate change	248
22	Renewable energy	253
23	Nuclear power	258
24	Digital technology	270
25	The extended essay	283
26	Exam techniques	288
27	Option E: Astrophysics	294
28	Option F: Communications	307
29	Option J: Particle physics	317
	Answers	343
	Index	346

Introduction

The aim of this book is to encourage and support IB students who are studying physics as part of the IB Diploma Programme. Success in the final examinations does, of course, depend on gaining knowledge and understanding of physics but by the end of the course you should have gained much more than just an ability to answer examination questions. The IB learner profile provides a useful checklist of some of the wider skills you should be aiming to develop.

The goal of physics is no less than an attempt to understand the natural world in its entirety. Although it should be obvious that an IB physics course alone will not achieve this, the insights that physics can provide into the mechanisms that operate in the world around us can be both breathtakingly simple and hugely complex at the same time. If nothing else, we hope that after studying IB physics you gain a love, fascination and respect for the universe in which we live.

There are many excellent physics books available. In writing this course companion we have deliberately focused on some of the wider aspects of subject and attempted to emphasize how the subject links with other subjects within the IB Diploma Programme model.

It has not been possible to include material on all the possible options. We have chosen to focus on two of the new options that have been introduced in the syllabus for the first time – particle physics and communications, and also to include some material on one of the more popular options from the previous syllabus – astrophysics.

Over the years we have both been fortunate enough to teach fabulous students and we would like to thank those keen and enthusiastic students (you know who you are) for making teaching and learning so rewarding and fun.

Tim Kirk and Neil Hodgson
September 2007

Individual thanks

Neil Hodgson: Writing this book has been an extremely challenging process and it has only been possible with the support and best wishes of my colleagues at Sha Tin College, in particular the physics team. Paul Drew read some of the material, giving valuable advice. The team of technicians provided valuable support by photographing experiments and recording primary data. Of course the support and encouragement of my fantastic family; Meher and children, Raoul and Tanya has been invaluable and I would like to thank them for overlooking all my failings as a father and husband during this time. Then, there are the wonderful students at Sha Tin College who provide a purpose and inspiration for this book and I would like to thank them also for their interest, enthusiasm and for making teaching such a worthwhile profession.

Tim Kirk: This book was written while I had the privilege of teaching talented and motivated students at the Godolphin and Latymer School in Hammersmith, and I would like to start by thanking the students there. I would also like to give my heartfelt thanks to my family, friends and colleagues for allowing me the space and time to undertake this project. Without their individual support and understanding, the task would have been impossible. Above all, however, I thank Betsan for her support, patience and encouragement, which was above and beyond the call of duty.

1 Learning in physics

The nature of physics

Physics is an attempt to understand the universe in which we live. Richard Feynman, winner of the Nobel Prize in Physics in 1965, ranks as one of the world's greatest theoretical physicists of recent times. This is how he described physics:

Physics is the most fundamental and all-inclusive of the sciences, and has had a profound effect on all scientific development. In fact, physics is the present-day equivalent of what used to be called natural philosophy, *from which most of our modern sciences arose. Students of many fields find themselves studying physics because of the basic role it plays in all phenomena. (Feynman et al. 1965: 1)*

Ernest Rutherford, another Nobel Prize winner (for chemistry, in 1908), stressed the importance of physics to the progress of science in a more direct way:

All Science is either physics or stamp collecting. (quoted in Birks 1962: 108)

The same sentiment is echoed by Robert Bunsen (of Bunsen burner fame):

Ein chemiker der kein Physiker ist, ist gar nichts—A chemist who is not a physicist is nothing at all. (quoted in Partington 1961: 282)

Perhaps the most succinct description of the physicist's role comes from another Nobel Prize winner, Steven Weinberg. In his Nobel lecture in 1979 he said:

Our job in physics is to see things simply, to understand a great many complicated phenomena in a unified way, in terms of a few simple principles.

But building an understanding of the principles is only the beginning. The astronomer Martin Rees gave the following analogy:

The physicist is like someone who's watching people playing chess and, after watching a few games, he may have worked out what the moves in the game are. But understanding the rules is just a trivial preliminary on the long route from being a novice to being a grand master. So even if we understand all the laws of physics, then exploring their consequences in the everyday world where complex structures can exist is a far more daunting task, and that's an inexhaustible one I'm sure. (quoted in Wolpert and Richards 1988: 37)

The search for knowledge and understanding is a fundamental part of the human condition, but neither physics nor philosophy has found enough definitive answers for us to be able to stop asking questions. There are so many fundamental puzzles still to solve, and this is part of the thrill and excitement of physics.

The following list (taken from postings on the *Cosmic Variance* website) summarizes some of the reasons given by students for studying physics:

- To gain a deeper understanding of the fundamental principles that govern our universe and everything in it, while at the same time picking up a broad and eminently useful skillset: the ability to analyse and deconstruct problems, to effectively communicate solutions.
- You learn to see what is important.
- Because you want to succeed in (choose one): business, law, medicine, education, engineering, politics or research in physics.
- There are lots of people in the world who can read and write well. There are far fewer who can think clearly. The world needs more of the latter.
- The thrill of being on the brink of discovery is second only to being madly in love.
- You get to play with cooler, more expensive toys than your friends.
- It is fascinating and fun at the same time.

Approaches used in this book

In addition to introducing much of the material needed for IB Diploma Programme physics (and thus supporting the acquisition of knowledge and improving the ability to be a thinker), the following approaches are used in this book to try to support your learning in physics:

- **Higher level:** These sections contain material studied by higher level (HL) candidates, and are not examined for standard level (SL). However, they will provide useful background information and an extension of ideas to help SL students gain a deeper understanding.
- **The bigger picture:** These sections contain challenging material that goes beyond the curriculum (not just in breadth but in more depth). The aim is to improve real understanding of the issues, and to encourage inquiry.
- **Science tools:** These sections introduce and reinforce the practical techniques and manipulative internal assessment skills that form part of the skill of inquiry.
- **Working with data:** These questions foster and encourage the development of practical skills – particularly DCP and CE (see pages 19 and 23).
- **Rediscovering physics:** The practical situations introduced in these sections provide ideas for experimental work where the outcomes can be predicted from the topics being studied.

- **Investigating physics:** The practical situations introduced in these sections provide ideas for experimental work where outcomes could be unknown.
- **Physics issues:** These sections focus on wider issues in science, science and technology, and internationalism. They address some of the wider aspects of the learner profile, involving physics in the real world, risk, implications, CAS, etc.
- **Thinking about science:** These sections focus on aspects of the scientific method, and highlight opportunities for links to be made with aspects of your studies in theory of knowledge.
- **Questions:** The best way to test, reinforce, and stretch your understanding of physics is to attempt to answer questions. These have been added throughout the text and at the end of chapters.
- **Inquiry:** These sections present short topics for you to research yourself using information readily available in textbooks or the Internet. Clear instructions or a series of questions are given to guide the inquiry. The aim is to promote independent learning.
- **Data-based questions:** These encourage learning from the analysis of information, which may involve new information not on the syllabus (see "The bigger picture" above.)
- **Mathematical physics:** Mathematics is a very important tool in physics. These sections highlight useful approaches or necessary skills. Graphical display calculators can be used to support physics learning.

IB Diploma Programme physics can be studied without having to use the mathematical tool called calculus. However, many students learn about calculus in their mathematics courses. If you find the use of calculus notation hard, whenever a new quantity is introduced you should focus instead on the word equations being used.

References

Birks, J.B. 1962. *Rutherford at Manchester*. London. Heywood & Co.

Feynman, R., Leighton, R.B., and Sands, M. 1965. *The Feynman Lectures on Physics*, Vol. 3. Reading, MA. Addison-Wesley.

Partington, J.R. 1961. *A History of Chemistry*, Vol. 4. London. Macmillan.

Wolpert, L. and Richards, A. (eds) 1988. *A Passion for Science*. Oxford. Oxford University Press.

2 Experimental measurements and internal assessment

Estimating the very large and the very small

Physics concerns itself with the study of all natural phenomena, and so the range of possible measurements associated with the study of physics is huge. Everyday objects are, of course, studied in physics, but we also consider much smaller and much larger objects. The range that we consider goes from the very small (the tiny constituents of an atom, for example) up to the very large: individual galaxies, clusters of galaxies, or even the universe itself.

Some things cannot be measured directly. For example, we can estimate the size of a galaxy and the size of a nucleus, but we cannot measure either directly.

Scientific notation (and approximations in terms of orders of magnitude) allows the same basic unit to be used to measure any given quantity (such as mass, length, or time), no matter how large or small. It is important to get a "feel" for the range of the numbers used. You can use the following exercises to estimate the range of possible measurements in the three fundamental quantities of mass, length, and time.

?

For each of the following lists:

(a) Rearrange the list in terms of increasing order of magnitude.
(b) Write down a guess for the approximate value for each quantity in terms of the unit used for that measurement.
(c) Do some research to find more accurate values. Check the reliability of your data by comparing your values with other students' estimations.
(d) Add at least five other measurements of your own choosing that cover the full range of the quantity.
(e) Design a poster to illustrate the range of possible measurements for one of the fundamental quantities.

1 **Mass in kilograms:** a person, the Sun, an electron, the observable universe, a car, a grain of rice, a blood cell, a proton, an atom, the Earth, the Milky Way galaxy.
2 **Length in metres:** a person, the distance to the Sun, the radius of a proton, the thickness of a piece of paper, the radius of the Earth, the height of the tallest building, the furthest distance travelled by mankind (distance from the Earth to the Moon), a light year, the radius of the observable universe, a blood cell, sub-nuclear particles, the diameter of a hydrogen atom.
3 **Time in seconds:** an average human lifespan, the time needed for light to cross a nucleus, an average human reaction time, the time taken for a camera's flash, the age of the universe, the expected life of the Sun, the time needed for light to cross a room, the age of the Earth, the time taken for light to come from the Sun to Earth, your age, the time period for visible light.

?

1 In each case state the ratio of the two quantities as differences of orders of magnitude.
(a) the diameter of the hydrogen atom compared with its nucleus [1]
(b) the distance from the Earth to the Moon compared with the distance to the Sun [1]
(c) the expected life of the Sun compared with an average human lifespan [1]
(d) the age of the universe compared with the expected life of the Sun [1]
(e) the mass of a car compared with the mass of an atom [1]
(f) the mass of the Earth compared with the mass of a grain of rice. [1]

2 The diameter of the nucleus of a hydrogen atom is of the order of
A 10^{-8} m B 10^{-15} m
C 10^{-23} m D 10^{-30} m

3 The number of heartbeats of a person at rest in one hour, to the nearest order of magnitude, is
A 10^1 B 10^2
C 10^3 D 10^5

4 The diameter of a proton is of the order of magnitude of
A 10^{-12} m B 10^{-15} m
C 10^{-18} m D 10^{-21} m

5 Which of the following gives the approximate ratio of the separation of the molecules in water and in steam at atmospheric pressure?

	Water : Steam
A	1 : 1
B	1 : 10
C	1 : 100
D	1 : 1000

6 The order of magnitude of the weight of an apple is
A 10^{-4} N B 10^{-2} N
C 1 N D 10^2 N

Working with data: SI units

Physics is an experimental science. The process of experiment relies upon accurate measurement. To make a measurement, we need to define the units to be used. Many of the units used in science will already be familiar to you (the second, the metre, or the kilogram, for example), but others (such as the henry, or the weber) are not as well known. Many non-standard units are widely used in everyday measurements, but they are best avoided in physics. Some examples of common country-specific units that are widely used are calories, pounds, and gallons.

In order to rationalize the units being used worldwide, the international system of units (*Le Système international d'unités* or the SI system) was developed. It is based around measurement of seven fundamental quantities, with a base unit accurately defined for each one: see Table 1.

Table 1 The SI base units

Quantity	Unit name	Unit symbol
mass	kilogram	kg
length	metre	m
time	second	s
electric current	ampere	A
amount of substance	mole	mol
thermodynamic temperature	kelvin	K
luminous intensity	candela	cd

It would be clumsy to only use the base units in all the combinations that are necessary, so **derived units** can be defined from the base units. It is sometimes possible to use the units of given quantities to predict how different factors may be related.

Table 2 gives the units of many of the quantities that you will meet in your studies. In all cases, the SI derived unit can be worked out from the word equation. Be careful how you use this table, because the equations given refer to particular situations. Whenever you come across a quantity that is new to you, check this list to see if it is mentioned. If it is not here, then add it!

Table 2 Derived units

Quantity	Symbol	Word equation	SI derived unit	SI base unit	Alternative unit
area	A	length × breadth	–	m^2	–
volume	V	length × breadth × height	–	m^3	–
density	ρ (rho)	mass/volume	–	$kg\ m^{-3}$	–
velocity (speed)	v, u	displacement/time (distance/time)	–	$m\ s^{-1}$	–
acceleration	a	change of velocity/time	–	$m\ s^{-2}$	–
concentration		amount of matter/volume	–	$mol\ m^{-3}$	–
momentum	p	mass × velocity	–	$kg\ m\ s^{-1}$	N s
force	F	mass × acceleration (change of momentum/time)	newton (N)	$kg\ m\ s^{-2}$	
pressure (stress)	P	force/area	pascal (Pa)	$kg\ m^{-1}\ s^{-2}$	$N\ m^{-2}$
frequency	f	number of cycles/time	hertz (Hz)	s^{-1}	–
energy (work) (quantity of heat)	Q, W, E	force × distance ($\frac{1}{2}mv^2$ or mgh)	joule (J)	$kg\ m^2\ s^{-2}$	N m
power	P	rate of using energy (rate of doing work)	watt (W)	$kg\ m^2\ s^{-3}$	$J\ s^{-1}$
charge	Q	current × time	coulomb (C)	A s	–
potential difference (p.d.) (voltage) (e.m.f.)	V	energy/charge	volt (V)	$kg\ m^2\ s^{-3}\ A^{-1}$	W/A
resistance	R	voltage/current	ohm (Ω)	$kg\ m^2\ s^{-3}\ A^{-2}$	V/A
magnetic flux	ϕ (phi)	e.m.f. × time (magnetic field × area)	weber (Wb)	$kg\ m^2\ s^{-2}\ A^{-1}$	V s

Table 2 (continued)

Quantity	Symbol	Word equation	SI derived unit	SI base unit	Alternative unit
magnetic flux density (magnetic field strength)	B	force/(current × length) (flux/area)	tesla (T)	kg s^{-2} A^{-1}	Wb m^{-2}
moment of a force	T	force × perpendicular distance	N m	kg m^2 s^{-2}	–
heat capacity (entropy)	C (S)	energy/temperature rise	J K^{-1}	kg m^2 s^{-2} K^{-1}	–
specific heat capacity	c	heat capacity/mass	J kg^{-1} K^{-1}	m^2 s^{-2} K^{-1}	–
electric field strength	E	force/unit charge	V m^{-1}	kg m.s^{-3} A^{-1}	NC^{-1}
radioactivity	A	count rate	becquerel (Bq)	s^{-1}	–
exposure	X	charge/mass	C kg^{-1}	kg^{-1} s A	–
absorbed dose of radiation	D	energy/mass	gray (Gy)	m^2 s^{-2}	J kg^{-1}
dose equivalent (radiation)	H	absorbed dose × factor	sievert (Sv)	m^2 s^{-2}	J kg^{-1}
intensity	I	power received/area	W m^{-2}	kg s^{-3}	–
angle	θ (theta)	arc length/radius	(simple ratio)	–	radians (rad) *
angular velocity	ω (omega)	angle/time	–	s^{-1}	rad s^{-1}
albedo (reflectivity)	α (alpha)	power absorbed/incident power	(simple ratio)	–	–

*Technically the unit for angle, the radian, is known as a supplementary unit.

Thinking about science: uncertainties

All measurements in science suffer from uncertainty. No matter how hard we try to control all the variables in any situation, some level of experimental error is inevitable. This does not mean all experimental work is flawed, but it *does* mean that we need to consider the uncertainties whenever data are properly analysed.

A complete analysis of experimental data is very time-consuming, and involves a range of statistical techniques that go beyond the IB syllabus. There are simple statistical tests that allow researchers to calculate the likelihood that a given set of experimental readings are significantly different from what might be expected as the results of chance alone. These mathematical tests (such as the t-test and the χ^2-test) are often straightforward to apply to data, but should not be used unless one understands their limitations.

Central to many statistical theories is the concept of a **normal distribution**. Consider, for example, a machine designed to manufacture identical metal rods (perhaps for use as clamp stands in physics laboratories). In order to check the operation of the machine, the diameters of all the rods are measured. A perfect machine would make each rod exactly the same, but a real machine will produce rods that are all slightly different from one another. The variation will be random, with the distribution shown in Figure 1, which plots the number of rods against the measured diameter. The area under the curve between any two diameters represents the number of rods in that range of diameters.

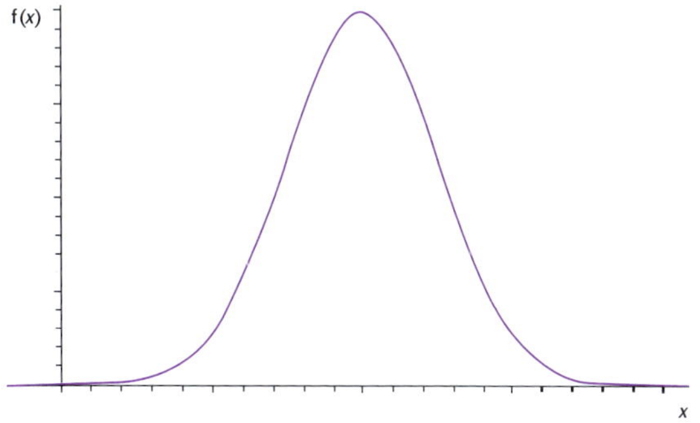

Figure 1 Normal distribution

This distribution is very common in practical situations. For example, if we measure the height of a large number of adults, the results will fit a normal distribution curve. Any normal distribution is characterized by two numbers: the **mean** ($\bar{x}$) and the **standard deviation, SD** (σ).

In a normal distribution:

- 68% of samples fall between ±1 SD.
- 95% of samples fall between ±2 SD
- 99.7% of samples fall between ±3 SD.

See Figure 2.

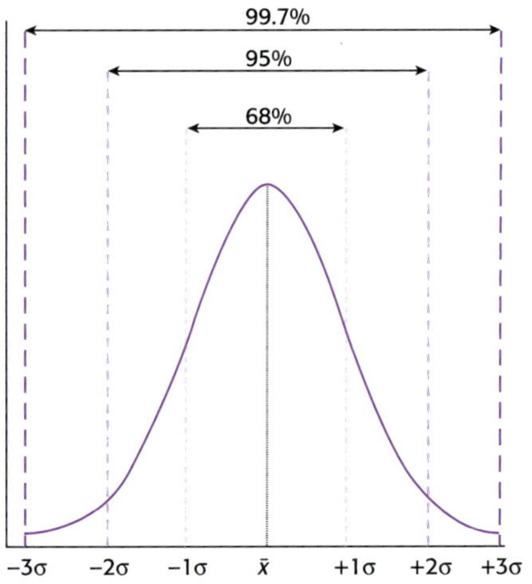

Figure 2 The normal distribution in more detail

We can expect a normal distribution to be produced whenever a given reading is repeated many times and there is the possibility of a **random error** on each reading. Repeating readings will tend to reduce the effect of random errors, because their effect will be "averaged out" over a large number of repeats.

Example of random errors include:

- the readability of a scale (parallax errors in particular: not reading the scale exactly "head on" – sometimes too large, sometimes too small)
- the experimenter's reaction time (if an experiment involves timing the beginning and end of a time interval using a stopwatch, then the recorded time will not be the same as the actual time)
- a meter (with a moving pointer) randomly reading too high or too low as a result of the meter needle having a tendency to "stick".

A full statistical analysis of repeated values would involve calculating the mean and the standard deviation of the recorded data. For IB physics it is usually sufficient to work with uncertainties assessed in terms of a limit of confidence that can be assigned to any given measurement. For example, most students would feel that the confidence limit when using a ruler to measure the length of this book correctly would be about ±1 mm.

It should be obvious from the graph in Figure 2 that a confidence limit is approximately equal to three standard deviations. Ideally, all experimental readings should be repeated several times, and the range of values obtained can be directly used to assign a confidence limit. However, the best practice is to assess practical uncertainties from first principles. See Chapter 7, page 59, for an example of how to do this.

We can use these confidence limits to evaluate the results of an experiment. You should already know that the gravitational field strength, g, at the surface of the Earth is approximately 10 N kg^{-1}. Suppose we develop a theory that predicts that g is exactly 10 N kg^{-1}. We devise an experiment to test this theory, and assess the resulting uncertainty. The result of our experiment is $g = 9.2 \pm 1.0$ N kg^{-1}. This value (9.2) is very different from our prediction (10), but the uncertainty range means that the theory could still be correct, because the possible range for g goes from 8.2 up to 10.2 N kg^{-1}.

We now devise a more **precise** experiment – one with a smaller associated uncertainty in the end result. This time the experimental result is $g = 9.85 \pm 0.06$ N kg^{-1}. This second result is closer to our predicted value, but the uncertainty range means that our theory and the experimental result cannot both be correct. So, provided we have not made any errors in the experiment, we have discredited the theory.

Good precision in readings does not necessarily mean the measurements are without error. For example, a meter might have a **zero error**: the zero mark on the scale does not correspond to the position of the needle when the quantity is zero – all readings are in error by the same amount. Simply repeating readings and then taking the average cannot resolve such **systematic errors**. An experiment with small systematic errors is an **accurate** experiment.

Mathematical physics: dealing with uncertainties

We have highlighted the need to assess the uncertainties associated with any experimental reading. Now we look at the techniques we can use to take these uncertainties into consideration.

Propagation of uncertainties through a calculation

Some simple experiments allow a quantity to be calculated directly from a set of practical measurements. Each measurement will have an associated uncertainty. The most complete way to assess the overall uncertainty of the final result is to do the calculation three times: first using the measured values, and then using the maximum or minimum values to give the maximum and minimum possible answers.

For example, measurements of the length and diameter of a piece of copper wire allow us to calculate the volume of copper in the wire. Say we record the following measurements:

length l = 234 mm ± 2 mm
diameter d = 1.86 mm ± 0.05 mm

The **absolute uncertainty** in the length measurement is ±2 mm, and in the diameter is ±0.05 mm.

So the possible maximum and minimum values of length and diameter are:

l_{max} = 236 mm, l_{min} = 232 mm

d_{max} = 1.91 mm, d_{min} = 1.81 mm

The equation for the volume V of copper is

$$V = \frac{\pi d^2}{4} l$$

So the measured volume, and the possible maximum and minimum values, are

$$V = \frac{\pi(1.86)^2}{4} \times 234\ \text{mm}^3 = 635.816\ \text{mm}^3$$

$$V_{max} = \frac{\pi(1.91)^2}{4} \times 236\ \text{mm}^3 = 676.190\ \text{mm}^3$$

$$V_{min} = \frac{\pi(1.81)^2}{4} \times 232\ \text{mm}^3 = 596.946\ \text{mm}^3$$

Therefore $V_{max} = V + 40.374$ mm³, and $V_{min} = V - 38.870$ mm³, which means that the uncertainty in V is approximately ±40 mm²: that is,

V = 636 ± 40 mm³

This procedure will always work, but can prove very time-consuming. We can use a short cut for quantities that have to be multiplied, divided, or raised to a power. In these situations it helps if we calculate the fractional uncertainty or the percentage uncertainty in each measurement. So, for example, the **fractional uncertainty** in our length measurement is $\pm\frac{2}{234} = \pm 0.0085$, and the **percentage uncertainty** is therefore ±0.85%.

A good way to estimate the overall percentage uncertainty in a quantity is to add all the individual percentage uncertainties in the values we use to calculate it, as follows:

Percentage uncertainty in length $l = \frac{2}{234} \times 100 = \pm 0.9\%$

Percentage uncertainty in diameter $d = \frac{0.05}{1.86} \times 100 = \pm 2.7\%$

Therefore percentage uncertainty in $d^2 = 2 \times (\pm 2.7\%) = \pm 5.4\%$

Therefore total uncertainty in volume = $(\pm 5.4\%) + (\pm 0.9\%) = \pm 6.3\%$

6.3% of 636 mm^2 = 40.068 mm^3

That is, $V = 636 \pm 40\ mm^3$

Note that, when using percentage uncertainties:

- This technique works only with quantities that have to be multiplied or divided. If they have to be added or subtracted, then the maximum and minimum values need to be calculated directly.
- Whether the numbers are multiplied or divided, we still *add* the percentage errors.
- If a reading has to be raised to a power n in calculating the result, the percentage uncertainty in the result is n times the percentage uncertainty in the reading. So, above, the uncertainty in d^2 was twice the uncertainty in d. Similarly, if a reading has to be cubed, the overall percentage uncertainty is three times the uncertainty in the original reading, and if we have to take the square root of a reading (that is, raise it to the power ½), the overall percentage uncertainty is half the value of the original reading.

Graphical representations of uncertainties

A graph visually represents the relationship between two variables. The line of best fit represents the trend shown by the results. In general there will be an uncertainty associated with each point on the graph. We can use **error bars** to show these uncertainties on the graph. For example, the graph in Figure 3 shows the variation of acceleration a with resultant force F.

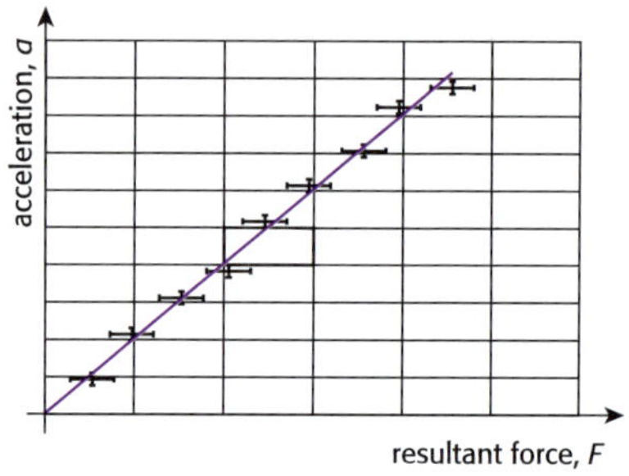

Figure 3 Error bars show the uncertainties in each reading

These results seem to represent a linear relationship. For this to be the case, the error bars for every point must agree with the best-fit line, as shown in Figure 3.

In many situations, the gradient of a straight-line graph can be used to calculate a constant related to the situation. In the example above the gradient (= acceleration/resultant force) turns out to be the inverse of the mass of the object involved. The range of possible gradients supported by the data is calculated by considering the maximum and minimum gradient lines that can be drawn.

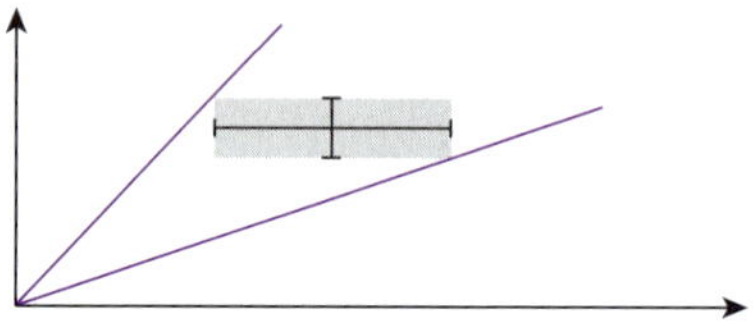

Figure 4 The shaded rectangle shows the range of possible values represented by the two error bars

When error bars exist in both the x and y directions, the confidence limits are represented by the rectangle of values that is defined by the two error bars (Figure 4).

So, for example, in our acceleration/force case, the maximum and minimum gradient lines would be as shown in Figure 5. We can determine these gradients, and so we can calculate the confidence limits for the gradient.

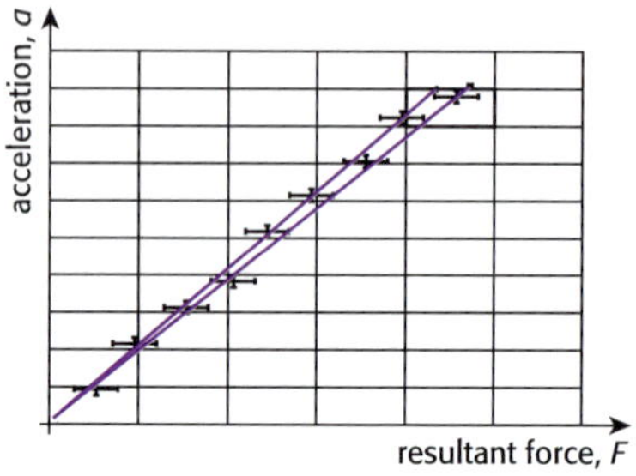

Figure 5 The maximum and minimum gradients of a line passing through the origin

It can be very time-consuming to plot individual error bars for every point. In practice, we often need only add error bars to the first and last data points. And if we know that the graph goes through the origin (because the two quantities are thought to be directly proportional) then it is often sufficient to consider just the point representing the largest values.

Science tools: general practical skills

Whether you are studying physics at standard or at higher level, 24% of your final marks in the Diploma Programme are awarded for your practical skills. Three quarters of these marks, or 18% of the Diploma, are assessed directly from the practical reports that you submit for marking. So it is important to keep up to date with your practical work, and to learn from earlier mistakes.

The marks for the practical (otherwise known as the internal assessment or IA) are awarded directly by your teacher throughout the course, and strict criteria are published to ensure that the same standards are applied worldwide. A sample of work from every IB school is sent off to check this. Remember: the final marks are awarded as a result of the skills that you can *demonstrate*. For example, you may know how to record raw data, but if you forget to do this properly in a piece of work that is being assessed then it cannot gain full marks.

The following sections focus on each of the five skills that will be assessed, but some general points apply to all of these skills.

- Each skill is given a mark out of 6.
- The first three skills (Design, D; Data collection and processing, DC; and Conclusion and evaluation, CE) are assessed from the written work that you submit.
- Your best two marks for D, DCP, and CE are the ones that count: each of these skills is thus worth a maximum of 12 marks. $3 \times 12 = 36$
- The Manipulative skills (MS) are **summative**. This means they are assessed once only, at the end of the course. The maximum MS mark is 6.
- The Personal skills (PS) are assessed once during the group 4 project. The maximum PS mark is 6.
- The maximum total is out of 48 ($36 + 6 + 6 = 48$).
- Each skill is assessed in terms of three separate aspects. The overall mark in each aspect is decided on the extent to which your work matches the criteria for that aspect. The match can be "complete" (2 marks), "partial" (1 mark), or "not at all" (0 marks).
- The assessment of these skills applies to an individual student and not to a team. So although practicals can be done with groups of students working together and helping one another, you need to be individually responsible for any tasks that are being assessed.

The following sections look at these practical skills in more detail.

Design (D)

Designing an appropriate experimental technique takes practice, and many students find this skill difficult to master.

All experiments must involve observations. Usually the observations will be quantitative measurements of a given property, such as mass, length, volume, or time. In any experiment, the value of the measured property depends on several variables. For example, the pressure of a sample of gas depends on at least four different variables:

- the volume of the gas
- the temperature of the gas
- the mass of the gas
- the particular gas used.

It is no use changing each of these variables at random and hoping that some pattern will be obvious from the results. We need to discover how each one affects the pressure. For the experiment to be a **fair test**, it needs to be designed so that we can vary only one at a time.

The variable that we choose to control (the one that is *manipulated*) is called the **independent variable**: for example, the volume of the gas.

The variable that we *measure* is called the **dependent variable**: for example, the pressure of the gas.

The variables that are held *constant* throughout the experiment (so that only one thing is affecting the dependent variable) are called the **control variables**: for example, the temperature of the gas, its mass, and the particular gas used.

Your experimental report needs to identify explicitly the dependent variable (measured), the independent variable (manipulated), and the control variables (constants).

Usually, the raw data from an experiment must include four things:

- corresponding values for the independent and dependent variables
- a wide range of values for the independent variable
- repeated readings whenever possible
- readings of all the control variables, when possible, to check that they have been kept constant during the experiment.

The three assessable aspects for the design skill are:

1 defining the problem and selecting variables;
2 controlling variables; and
3 developing a method for collection of data.

Aspect 1: defining the problem and selecting variables

This depends on the context in which the task is set, but for you to demonstrate your ability to define the problem, your teacher can at best give only a general outline of a possible situation to investigate. It is up to you to identify the variables and choose which ones you are going to measure, manipulate, or keep constant. Possible general outlines might be:

- Investigate some physical property of a bouncing ball.
- Investigate the deflection of a cantilever (it is allowable for the teacher to define the dependent variable).

It is up to you to develop this general outline into a specific investigation. If you do not manage to focus the question any further than this, then you have not made progress.

Aspect 2: controlling variables

Your laboratory report must explain your experimental design or method. You need to ensure that the independent variable will be manipulated appropriately, and you must mention how all the other variables are going to be controlled or monitored.

Aspect 3: developing a method for collection of data

The data you collect must be relevant to the problem, and sufficient for you to establish reliable conclusions. This depends on the context, but whenever repeats are easily possible, you must ensure that these take place. If every reading was repeated three times, it would be relatively easy to assess the uncertainty levels, and this technique would also allow large random errors to be immediately identified and checked.

You should aim to have at least ten different values of the independent variable. You may be able to reach valid conclusions on the basis of fewer readings, but it is always best to take as many readings as possible in the time available. If you are plotting average results on a graph, the absolute minimum numbers of readings needed to identify a linear trend is five, but ten would be better.

You also need to consider the data range. In some situations it makes sense to start the experiment by arranging for the dependent variable to take its largest reasonable value and its smallest reasonable value. You can then choose subsequent readings to cover the available range evenly between these two limits.

Data collection and processing (DCP)

This skill is at the core of any practical work. The data need to be recorded accurately, precisely, and in a way that is not ambiguous. Once the data have been recorded, the resulting information usually has to be processed in order for any conclusions to be drawn. Throughout this process it is important to ensure that uncertainties are considered. They need to be assessed and recorded during the data collection, and they need to be taken into account whenever data are processed.

A very useful approach to data analysis is a graph plot, but sometimes the choice of variables to plot is not straightforward. By convention, it is usual to plot the independent variable on the x-axis and the dependent variable on the y-axis. In this context time is often considered to be the independent variable, and is therefore plotted on the x-axis.

A line can be added to show the general trend represented by the data points. This **best-fit line** should either be a single smooth curve, or a straight line drawn with a ruler. Never "join the dots": the trend line is an attempt to highlight the general relationship suggested by the data. The graphs in Figure 6 show possible trend lines for different plots of the variation of the dependent variable D with the independent variable I.

Each of the graphs A to H in Figure 6 represents a common mathematical relationship that occurs in physics between a dependent and an independent variable. The graphs are discussed below. Before you read these descriptions, in each case try to describe the mathematical relationship that is suggested by the shape of the graph.

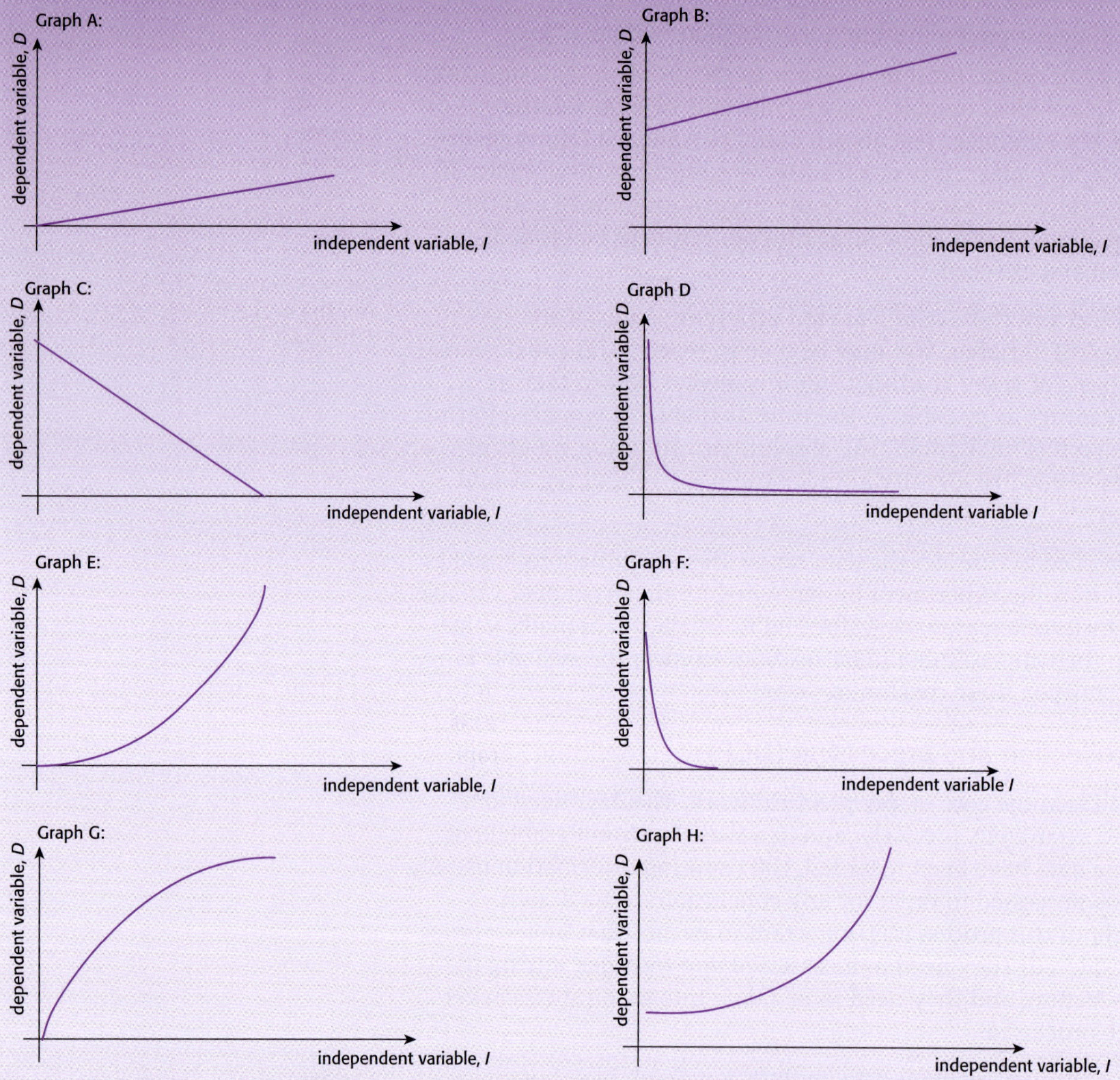

Figure 6 Common shapes for graphs

- *Graph A* This graph shows two variables that are **proportional**, because the line is straight *and* it goes through the origin. A graph with only one of these properties does not show proportionality. The phrase "directly proportional" means the same as "proportional". If the independent variable is doubled, then the dependent variable will also double. There is no such thing as "indirectly proportional". The general mathematical formula for a straight-line graph is $y = mx + c$. In this situation the y-intercept (c) is zero, because the line goes through the origin. Thus the equation of this line is $D = \text{constant} \times I$.

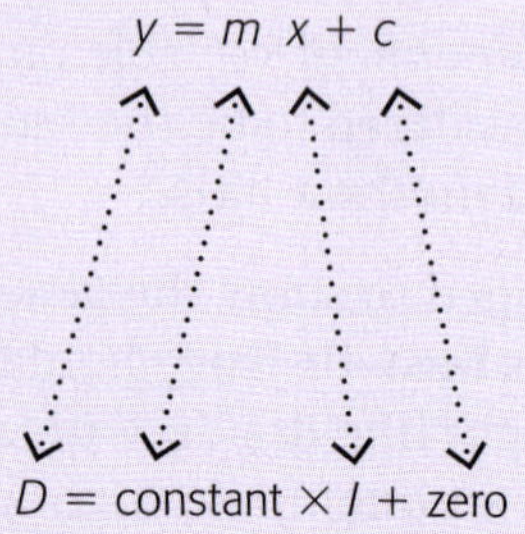

- *Graph B* This graph shows two variables that are *not* proportional because the line is straight *but* it does not go through the origin. There is a linear relationship between the variables. The equation of this line involves two different constants, and is $D = m\ 3\ I + c$.

- *Graph C* This graph shows two variables that are **inversely related**: as one increases, the other one decreases. They are not "inversely proportional", because one does not halve when the other one doubles. These variables could also be described as linearly related. This straight line graph has a negative gradient: $D = c - m \times I$.
- *Graph D* This graph shows two variables that are **inversely proportional**: as one doubles, the other one halves. Note that the graph does not touch either axis. D is proportional to the inverse of I and vice versa: $D = k \times 1/I$, where k is a constant.
- *Graph E* In this graph the y-value is proportional to the square of the x-value. Mathematically the relationship is $D = k \times I^2$, where k is a constant.
- *Graph F* The graph hits the y-axis but does not cross the x-axis. This is an exponential decrease. The mathematical relationship is given by $D = D_0e^{-kI}$: see page 233 for further information.
- *Graph G* In this graph the y-value is proportional to the square root of the x-value. Mathematically the relationship is $D = k\sqrt{I}$ or $I = k' \times D^2$, where k and k' are constants.
- *Graph H* The graph hits the y-axis and the slope continues to increase as x increases. This is an exponential increase. The mathematical relationship is given by $D = D_0e^{kI}$: see page 233 for an example.

It is useful to plot an appropriate straight-line graph if possible. For example, the time t taken for an object to fall, from rest, a distance s as a result of gravity is given by the equation

$$s = \frac{1}{2}gt^2$$

where g is the acceleration due to gravity. Readings can be taken of the time taken to fall and the distance fallen. A straight plot with s on the y-axis and t on the x-axis would look like graph E. It would be more useful to plot s on the y-axis and t^2 on the x-axis. A comparison with the general formula for a straight-line graph shows that this plot would be a straight line. The gradient of the line m would be equal to ½g, and the intercept on the y-axis, c, would be zero: that is, the line would go through the origin

$$\begin{array}{ccc} s = & \frac{1}{2}g\,t^2 & + \textit{zero} \\ \updownarrow & \updownarrow & \updownarrow \\ y = & m\,x & + \; c \end{array}$$

1 An experiment is set up to record the time taken by a ball bearing to fall a known distance (Figure 7). The ball bearing is initially held a measured distance above a surface by an electromagnet. An electronic timer records the time taken between the electromagnet being switched off and the ball bearing hitting the surface. The results are recorded in Table 3.
The experimenter assesses the uncertainties as follows:
- Uncertainty in time is ±0.005 s owing to residual magnetism in the ball bearing.
- Uncertainty in distance is ±0.5 cm.

Analyse these data in order to estimate the acceleration of free fall, g. Include an estimate of the uncertainty in your answer.

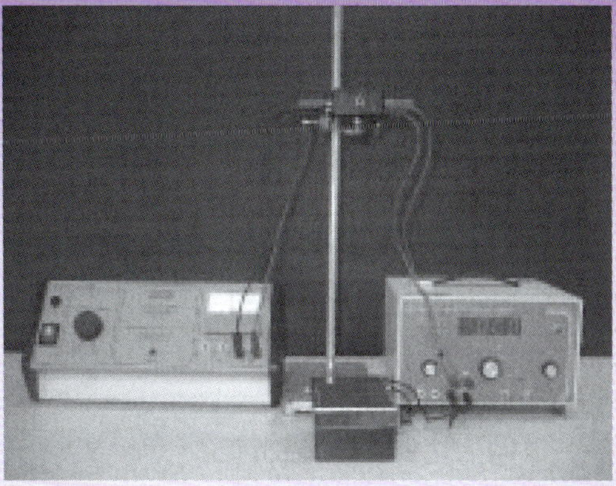

Figure 7 Apparatus to measure the acceleration due to gravity by freefall

2 The focal length f of a concave mirror, the object distance u, and the corresponding image distance v are related by the equation

$$\frac{1}{u}+\frac{1}{v}=\frac{2}{f}$$

Using the data below, draw a suitable graph in order to find the focal length of the mirror.

u / mm	204	250	345	455	714
v / mm	1000	526	333	270	222

Table 3 Experimental data for question 1

Distance, D/cm	Time for free fall, T / seconds			
	T_1	T_2	T_3	T (average)
30.0	0.270	0.269	0.269	0.269
40.0	0.305	0.305	0.304	0.305
50.0	0.333	0.334	0.333	0.333
60.0	0.366	0.365	0.366	0.366
70.0	0.392	0.392	0.392	0.392
80.0	0.413	0.413	0.412	0.413
90.0	0.438	0.438	0.439	0.438
100.0	0.458	0.459	0.457	0.458

Aspect 1: recording raw data

Most of the time you will be working with quantitative data. Whenever such data are recorded it is important to make clear the precise quantity that is being measured, the units it is being measured in, and the uncertainty range for the measurement. It is good practice to ensure that all this information is clearly in the headings for a data table. The example below shows a possible table designed for an investigation into the I–V electrical characteristics of a bulb.

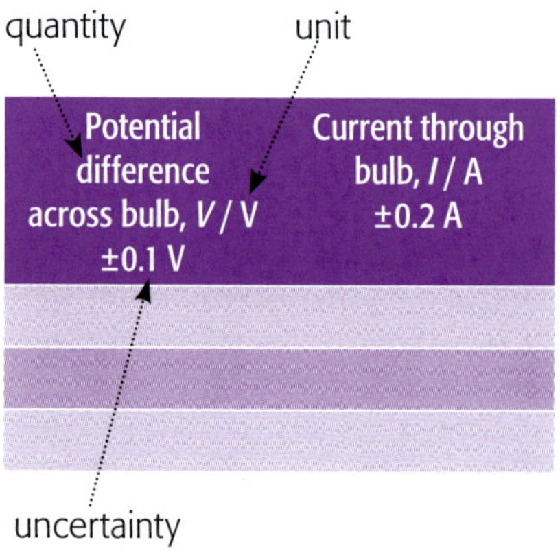

You should discuss in your report your reasons for choosing to assign any particular level of uncertainty, and the number of significant digits you use to record the data should be consistent with the stated uncertainty. In the above table it makes sense for there to be a potential difference reading of, say, 4.0 V. It would not make sense to record the reading either as 4.000 (too many significant digits) or as 4 (too few). If the uncertainty is the same throughout an experiment, then the number of decimal places used should always be the same. Average values should also be stated to an appropriate number of significant digits.

Aspect 2: processing raw data

To gain full marks for this aspect you need to be able to process the raw data that you recorded in some mathematical way. Often experiments involve plotting a graph, but a simple direct plot of the variables does not involve any processing. Graphs are ways of presenting data. If the data do not need processing before producing a straight-line graph, at the very least you need to choose to calculate the average of repeated readings, plot these average points, identify a suitable best-fit line, and calculate the gradient and/or the intercept.

Aspect 3: presenting processed data

You need to present the results of any processed data clearly and unambiguously, and you must take uncertainties into consideration by, for example, using the techniques shown on pages 13–16. Graphs need to have: appropriate scales; labelled axes with units; and accurately plotted data points with a suitable best-fit line or curve. Specifically, you need to:

- include uncertainty bars where significant
- explain where uncertainties are not significant
- draw lines of minimum and maximum gradients
- determine the uncertainty in the best straight-line gradient.

Conclusion and evaluation (CE)

The purpose of most physics experiments is either to measure a physical quantity or to investigate the relationship between two (or more) variables. A good overall conclusion and evaluation of an experiment does three things:

1 It summarises the findings of the experiment, and discusses these findings in the context of the assessed uncertainty limits.
2 In the light of what has been discovered, it critically analyses the experimental procedure to see whether the procedure contains any weaknesses or limitations.
3 It proposes sensible improvements to be made if the experiment is to be repeated in the future. These improvements should aim to correct any weaknesses or limitations already identified.

Aspect 1: concluding

The conclusion of an experiment should be statements of trends or patterns revealed by the data obtained rather than just a restatement of any predicted outcome. Once again, you need to take uncertainties into consideration when justifying any conclusions, and it is good practice always to discuss whether the main sources of error were systematic or random in nature.

If the aim of your experiment is to measure a physical quantity that is already known, you should compare your experimental value with literature values. If you have correctly assessed the uncertainty associated with your experimental procedure, these two values should agree within experimental error. You should fully reference any value that you quote from the literature.

Aspect 2: evaluating procedure(s)

You should comment on the design and method of the investigation as well as on the quality of the data. This should not just be a list of possible weaknesses; it should include some discussion of how significant particular weaknesses turn out to be. It is good practice always to discuss the precision and accuracy of the measurements taken, and to comment on the processes, use of equipment, and management of time.

?

1 The diagram below shows the position of the meniscus of the mercury in a mercury-in-glass thermometer.

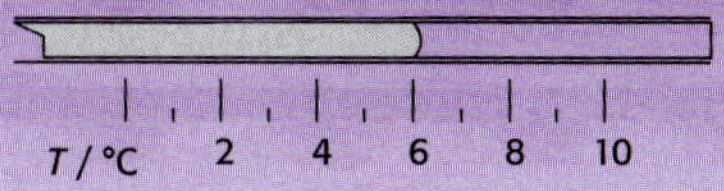

Which of the following best expresses the indicated temperature with its uncertainty?
A (6.0 ± 0.5)°C
B (6.1 ± 0.1)°C
C (6.2 ± 0.2)°C
D (6.2 ± 0.5)°C [1]

2 The time period T of oscillation of a mass m suspended from a vertical spring is given by the expression

$$T = 2\pi\sqrt{\frac{m}{k}}$$

where k is a constant.

Which **one** of the following plots will give rise to a straight-line graph?
A T^2 against m
B $\sqrt{T}$ against $\sqrt{m}$
C T against m
D $\sqrt{T}$ against m [1]

3 An ammeter has a zero offset error. This fault will affect
A neither the precision nor the accuracy of the readings.
B only the precision of the readings.
C only the accuracy of the readings.
D both the precision and the accuracy of the readings. [1]

Aspect 3: improving the investigation

The suggestions for improvement should arise naturally from the weaknesses and limitations that you have identified when evaluating the procedures used, and they should be realistic and appropriate. If, for example, the analysis of uncertainties in a particular experiment assigned ±1% as a result of human reaction time and ±15% to readability issues for the other measurements, then it would not be appropriate to simply suggest using more precise timing equipment.

Manipulative skill (MS)

By the end of your course you should have gained wide experience of different experimental procedures and approaches. Your teacher will submit a mark to the IB that reflects how far you are able to follow instructions accurately and precisely, carry out a wide range of techniques, and appreciate the importance of approaching all practical work in a safe way.

Personal skills (PS)

It is a requirement of the course for all IB Diploma Programme physics students to work together and be involved in a collaborative project across the science disciplines, known as the **group 4 project**. The theme and mechanics of the project will vary between schools and colleges worldwide. The aims are:

- to encourage an understanding of the relationships between scientific disciplines and the overarching nature of the scientific method
- to raise awareness of the moral, ethical, social, economic, and environmental implications of using science and technology
- to develop and apply information and communication technology skills in the study of science
- to develop an appreciation of the possibilities and limitations associated with science and scientists
- to develop an ability to analyse, evaluate, and synthesize scientific information.

The aim of every team involved in the group 4 project is to make scientific progress, but in this context the process is more important than the actual experimental outcomes.

The personal skills mark assesses your self-motivation and perseverance, your ability to work with other scientists as part of a team, and the extent to which you are aware of your own strengths and weaknesses and are able to take responsibility for your own learning.

3 Kinematics

Kinematics is the study of objects in motion. This chapter focuses on the accurate and precise measurement of the quantities that can be used to describe the way in which an object moves. Subsequent chapters will analyse these quantities in more detail. The principles involved can be applied in a wide range of situations, from the motion of individual subatomic particles up to complex extended objects such as stars or even galaxies.

Most everyday objects in motion undergo a complex mixture of different types of motion, all occurring together. For example, consider an athlete attempting the high jump. The athlete's overall motion through the air can be considered as a mixture of:

- The translational motion of the **centre of mass** of the athlete in each of three independent directions represented by the *x*, *y*, and *z*-axes. For these axes to be independent of one another they need to be mutually perpendicular, i.e. at 90° to one another, but we are free to choose their direction and/or orientation to fit the specific situation.
- The rotational motion of the athlete around each of three independent axes of rotation that pass through the centre of mass. Again, for these axes to be independent of one another, they need to be mutually perpendicular. It is often convenient to choose the axes of rotation to be the same as the axes used to describe the translational motion, but this is not always necessary.

We shall start by considering the simplest possible situations, with no rotational motion, and translational motion in only one direction, e.g. an object sliding along a surface in a straight line. We can then extend the ideas to motion in two dimensions, e.g. an object sliding along a surface, and going around a corner (circular motion) or the motion of an object thrown through the air. Three-dimensional motion and rotation about more than one axis are extensions of the same ideas.

To express an overall translational motion in terms of separate motions in independent directions, we need to consider not only the size (i.e. the **magnitude**) of the various quantities, but also their **direction**. Quantities that have both magnitude and direction are called **vector** quantities, whereas quantities that have only magnitude are called **scalar** quantities. Vector and scalar quantities need to be treated in different ways – see Chapter 4, page 32.

IB Diploma Programme physics limits the quantitative study of motion to either translational motion or constant rotational motion (circular motion) but not both at the same time. The mathematics of accelerating rotational motion is not included.

Figure 1 Athletes controlling motion

?

1 A bicycle is being ridden in a straight line. Identify the separate translational and rotational motions that fully describe the overall motion of one of the bicycle wheels.
2 A bicycle is being ridden around a corner. The bicycle leans over as it turns to the left. Identify the separate translational and rotational motions that fully describe the overall motion of one of the bicycle wheels.

Key terms

We need three quantities to describe an object's motion fully in any given situation: its **displacement**, its **velocity**, and its **acceleration**. These are all vector quantities. Two other related scalar quantities are often also used: the object's **distance** and its **speed**. For each of these five quantities it is important to understand the difference between an **average** value over a time interval and the **instantaneous** value at a particular time. Formulae exist that relate these quantities together, but they only apply in certain circumstances.

Many of the equations given below use calculus notation to define the quantities precisely. If your mathematics is not yet up to using this notation, it is sufficient to focus on the word equations and the associated units. If this is the first time you have encountered the definitions, it is best to start with the simplest situation of all: motion in only one direction. You should revisit these definitions once you have studied vector mathematics (see page 32).

Displacement, *s*

This quantity is the vector equivalent of distance. The displacement $\boldsymbol{s}$ of a particle is the length and direction of the line drawn to the particle from some fixed point (called the **origin**). In SI units the magnitude of the displacement is measured in metres, centimetres, kilometres, etc. For example, the displacement of my place of work from my home is 5.4 km, northeast (Figure 2).

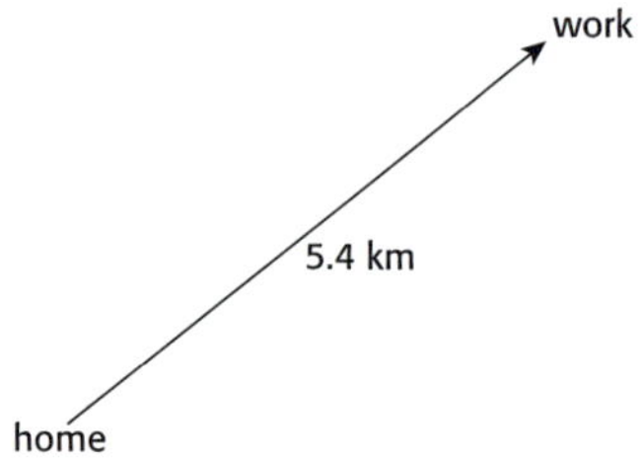

Figure 2

Velocity, *v*

This quantity is the vector equivalent of the scalar quantity, speed. A moving particle's displacement changes with time. In a time interval Δt, the change in displacement (found by subtracting the initial displacement from the final displacement using vector mathematics) is $\Delta \boldsymbol{s}$. The **average velocity** $\boldsymbol{v}_{av}$ during an interval of time Δt is defined as follows:

$$\boldsymbol{v}_{av} = \frac{\text{change in displacement}}{\text{time taken}} \qquad \boldsymbol{v}_{av} = \frac{\Delta \boldsymbol{v}}{\Delta t}$$

The units of average velocity are m s^{-1}. This quantity is different from **instantaneous velocity**, $\boldsymbol{v}$, which is the value calculated using the above equation when the change of displacement and the time interval become vanishingly small. This is called the rate of change of displacement. To signify that $\Delta \boldsymbol{s}$ and Δt are very small they are written as $\delta \boldsymbol{s}$ and δt.

$$\boldsymbol{v} = \frac{\text{small change in displacement}}{\text{small time taken}} = \text{rate of change of displacement}$$

$$\boldsymbol{v} = \lim_{\delta t \to 0}\left(\frac{\delta s}{\delta t}\right) = \frac{\mathrm{d}s}{\mathrm{d}t}$$

Speed

Speed is a scalar quantity. It is also measured in m s^{-1}. The average speed is not necessarily equal to the magnitude of the average velocity. The definition is

$$\text{average speed} = \frac{\text{distance travelled along actual path}}{\text{time taken}} = \text{rate of change of distance}$$

The instantaneous speed is the value calculated by the above equation when the time interval becomes vanishingly small, and is always equal to the magnitude of the instantaneous velocity.

Acceleration, *a*

This is a vector quantity related to the change in a moving particle's velocity with time. In a time interval Δt, the change in velocity is $\Delta \boldsymbol{v}$. The **average acceleration** $\boldsymbol{a}_{av}$ is defined as follows:

$$\boldsymbol{a}_{av} = \frac{\text{change in velocity}}{\text{time taken}} \qquad \boldsymbol{a}_{av} = \frac{\Delta \boldsymbol{v}}{\Delta t}$$

The units of the magnitude of average acceleration are m s^{-2}. This quantity is different from **instantaneous acceleration**, $\boldsymbol{a}$, which is the value calculated using the above equation when the change of velocity and the time interval become vanishingly small. This is called the rate of change of velocity.

$$\boldsymbol{a} = \frac{\text{small change in velocity}}{\text{small time taken}} = \text{rate of change of velocity}$$

$$\boldsymbol{a} = \lim_{\delta t \to 0} \left(\frac{\delta \boldsymbol{v}}{\delta t} \right) = \frac{d\boldsymbol{v}}{dt}$$

An important experimental observation is that the acceleration of an object falling in a vacuum near the Earth's surface is independent of its mass. This is called the acceleration of free fall, g. The average value for g at sea level is 9.81 m s^{-2}, or approximately 10 m s^{-2}.

Constant-acceleration equations

Provided an object moves with constant acceleration, we can manipulate the definitions of average velocity and average acceleration to create a set of five equations. Each of these equations relates four of the five relevant quantities that we would need to fully describe the object's motion:

initial displacement = zcro (by definition)
final displacement = $\boldsymbol{s}$
initial velocity = $\boldsymbol{u}$
final velocity = $\boldsymbol{v}$
acceleration = $\boldsymbol{a}$
time taken = t

?

This question is about linear motion.

A police car P is stationary by the side of a road. A car S, exceeding the speed limit, passes the police car P at a constant speed of 18 m s^{-1}. The police car P sets off to catch car S just as car S passes the police car P. Car P accelerates at 4.5 m s^{-2} for a time of 6.0 s and then continues at constant speed. Car P takes a time *t* seconds to draw level with car S.

(a) (i) State an expression, in terms of *t*, for the distance car S travels in *t* seconds. [1]

(ii) Calculate the distance travelled by the police car P during the first 6.0 seconds of its motion. [1]

(iii) Calculate the speed of the police car P after it has completed its acceleration. [1]

(iv) State an expression, in terms of *t*, for the distance travelled by the police car P during the time that it is travelling at constant speed. [1]

(b) Using your answers to (a), determine the total time *t* taken for the police car P to draw level with car S. [2]

In any given situation, if we know any three of these quantities, then we can calculate the other two.

$$v = u + at \qquad s = \frac{u+v}{2}t$$

$$v^2 = u^2 + 2as \qquad s = ut + \frac{1}{2}at^2$$

$$s = vt - \frac{1}{2}at^2$$

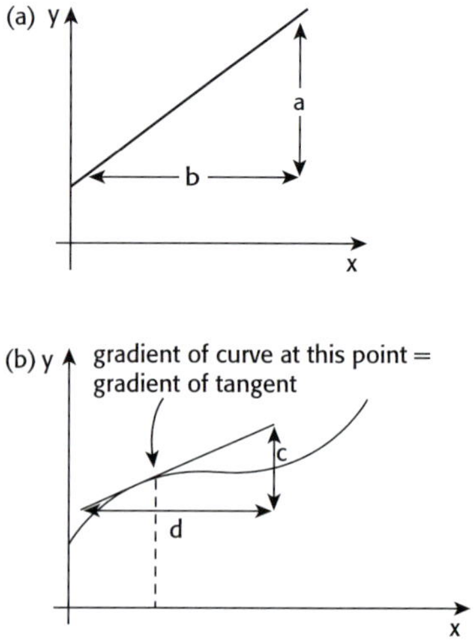

Figure 3 (a) gradient of straight line $= \frac{a}{b}$ (b) gradient of tangent $= \frac{c}{d}$

Relative velocity

Consider two points A and B, moving with velocities v_A and v_B, respectively. The velocity of A relative to B, v_{AB} (otherwise known as the relative velocity of A with respect to B), is the velocity that A appears to have according to an observer who is moving with B. This is calculated by subtracting the velocity of B from the velocity of A:

$$v_{AB} = v_A - v_B \qquad v_{BA} = v_B - v_A$$

Graphical work

The definitions of displacement, instantaneous velocity, and instantaneous acceleration correspond to either gradients or areas found from displacement–time graphs or velocity–time graphs.

The gradient is the ratio of the change in the y-axis measurement to the change in the x-axis measurement ("rise over run"). A straight-line graph can be analysed directly but a curve may be best analysed by calculating the gradient of the tangent to the line (Figure 3b).

The area under the graph takes into account the units used on the axes. We can calculate simple shapes mathematically, but to determine the area under complicated curves we need to count the number of squares on graph paper (Figure 4).

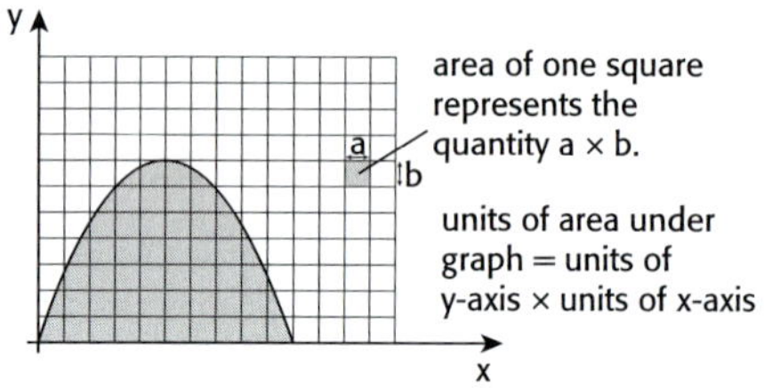

Figure 4 The area under a curve

Displacement–time graphs

For these graphs the slope of the graph at any point represents the instantaneous velocity, and the slope of the line joining any two points on the graph represents the average velocity over the chosen time interval (Figure 5). The area under the graph does not represent any physical quantity.

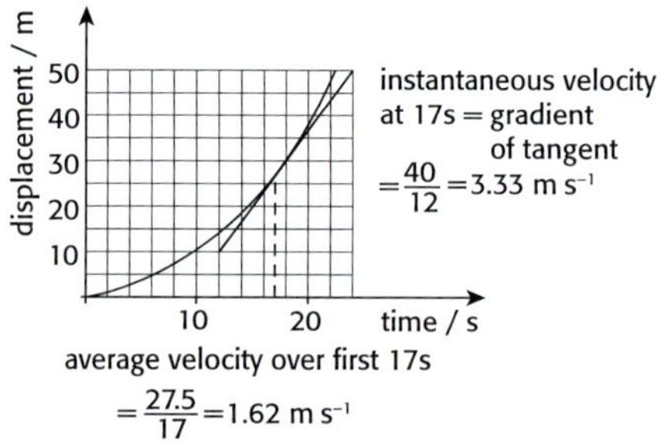

Figure 5 Calculating average and instantaneous velocity

Velocity–time graphs

For these graphs the slope of the graph at any point represents the instantaneous acceleration, and the slope of the line joining any two points on the graph represents the average acceleration over the chosen time interval. The area under the graph – between the line and the x-axis – represents the displacement that takes place over the chosen time interval (Figure 6). It is possible for the area to represent a negative number if either the displacement or the velocity is negative.

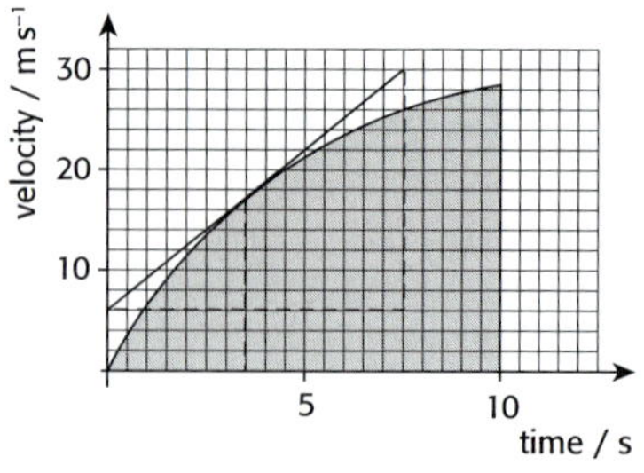

Figure 6 Displacement over a time interval

Acceleration–time graphs

For these graphs the slope of the graph at any point represents the rate of change of acceleration, and the slope of the line joining any two points on the graph represents the average rate of change of acceleration over the chosen time interval. The area under the graph represents the change in velocity that takes place over the chosen time interval (Figure 7).

Graphical display calculators often have powerful built-in tools for determining the gradient of the tangent to a curve or the area under the curve. See page 67 for an example of this for simple harmonic motion.

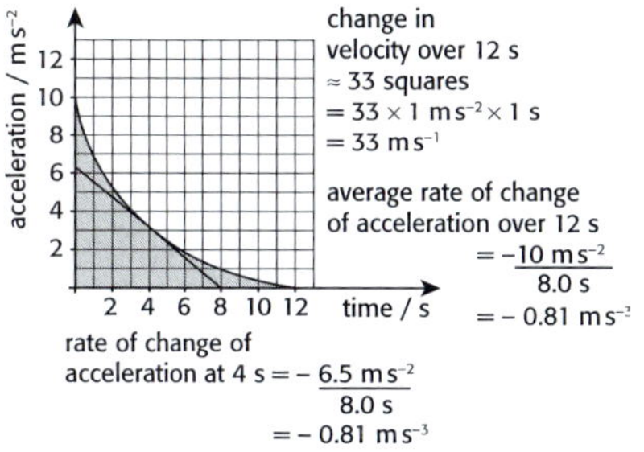

Figure 7 Calculations from an acceleration – time graph

Mathematical physics: modelling falling under gravity

Predicting the time taken for an object to free-fall a known distance under gravity involves a straightforward application of the constant-acceleration equations. This prediction assumes that, throughout the fall, air resistance can be taken as negligible.

For example, the time taken for an object to free-fall 10 m would be calculated as follows:

initial velocity $\boldsymbol{u}$ = zero
final velocity $\boldsymbol{v}$ = ?
constant acceleration $\boldsymbol{a} = 9.81\ \text{m s}^{-2}$
time taken t = ?
final displacement $\boldsymbol{s}$ = 10 m

$$s = ut + \frac{1}{2}at^2$$

$$\therefore t = \sqrt{\frac{2s}{a}} = \sqrt{\frac{2 \times 10}{9.81}} = 1.43\,\text{s}$$

If, however, the air resistance has a significant effect, the problem is much harder to solve mathematically, even though the principles are still straightforward. Air resistance acts to oppose the motion through the air. At any given instant of time this results in a reduction of the object's acceleration. The amount of reduction can be taken to be proportional to the speed of the object.

air resistance ∝ speed of object

∴ reduction in acceleration = constant × speed of object

The net result is that the object's acceleration is constantly changing, and this affects the speed of the object, which in turn affects the new value of air resistance.

A powerful mathematical technique for problem-solving in situations like this is to create a model of the changes that are taking place with time. By making the assumption that accelerations will not vary significantly over small time intervals, repeated approximate calculations can be used to predict accurately the variation over significant periods of time. Spreadsheets on computers or graphics calculators allow for a large number of calculations to be performed in a very short time. This is an example of an **iterative technique**, where the results of one calculation are fed into the next one and so on by following a simple rule.

The spreadsheet in Figure 8 shows a possible layout that could be used to check that the process works if friction is negligible. The object starts with a positive displacement, which represents a height measured upwards from the surface of the Earth, and a zero initial velocity. This means the acceleration will be negative. The minus sign represents the fact that the acceleration is downwards.

	A	B	C	D	E	F	G	H	I	J
1	**calculation number**	**Time**	**Displacement**	**Initial velocity**	**Change in velocity**	**Final velocity**	**Average velocity**	**acceleration due to gravity**	−9.80	m s⁻²
2	0	0.00	10.00	0.00	−0.10	−0.10	−0.05	**time increment**	0.01	s
3	1	0.01	9.99951	−0.10	−0.10	−0.20	−0.15	**initial displacement**	10.00	m
4	2	0.02	9.99804	−0.20	−0.10	−0.29	−0.25	**initial velocity**	0.00	m s⁻¹
5	3	0.03	9.99559	−0.29	−0.10	−0.39	−0.34			
6	4	0.04	9.99216	−0.39	−0.10	−0.49	−0.44			
7	5	0.05	9.98775	−0.49	−0.10	−0.59	−0.54			
8	6	0.06	9.98236	−0.59	−0.10	−0.69	−0.64			
9	7	0.07	9.97599	−0.69	−0.10	−0.78	−0.74			
10	8	0.08	9.96864	−0.78	−0.10	−0.88	−0.83			
11	9	0.09	9.96031	−0.88	−0.10	−0.98	−0.93			
12	10	0.10	9.951	−0.98	−0.10	−1.08	−1.03			
13	11	0.11	9.94071	−1.08	−0.10	−1.18	−1.13			
14	12	0.12	9.92944	−1.18	−0.10	−1.27	−1.23			
15	13	0.13	9.91719	−1.27	−0.10	−1.37	−1.32			

Figure 8 A spreadsheet to model acceleration due to gravity using an iterative method

The following points need to be understood:

- All the constants used in the calculation are in column I. Once we have created the spreadsheet, this allows us to change the variables easily and see the effect this has on the final answer. These cells need to be referenced in a direct way (e.g. $I$2).
- Column A is a record of the number of repeated calculations taking place.
- Column B uses the chosen interval size (the number in I2) to work out the time at the end of this interval, i.e. B3 = B2 + $I$2
- Column D is the velocity at the beginning of the time interval: D3 = F2
- Column E uses the acceleration and the time interval to work out the change in velocity during the time interval: E2 = $I$1*$I$2
- Column F works out the velocity at the end of the time interval: F2 = D2 + E2
- Column G works out the average velocity during the time interval: G2 = (D2 + F2)/2
- Column C works out the new overall displacement using the displacement at the beginning of the time interval, the average velocity during the time interval: C3 = C2 + G2*$I$2

This series of calculations is then repeated for the next time interval by "filling down" the equations. The software automatically works out the new formula for each line from that in the line above.

With the values chosen to fit the above situation, the calculations show that the object would be expected to hit the ground (the displacement goes negative if it is below the surface of the Earth) between 1.42 and 1.43 s after release (Figure 9). The value predicted from the constant acceleration equations is shown in cell I144 for comparison.

A few minor modifications allow the effect of friction to be modelled: see Figure 10.

- Another fixed constant, the friction constant, has been added to cell I5.

	A	B	C	D	E	F	G	H	I	J
1	calculation number	Time	Displacement	Initial velocity	Change in velocity	Final velocity	Average velocity	acceleration due to gravity	−9.80	m s⁻²
142	140	1.40	0.396	−13.72	−0.10	−13.82	−13.77			
143	141	1.41	0.25831	−13.82	−0.10	−13.92	−13.87			
144	142	1.42	0.11964	−13.92	−0.10	−14.01	−13.97	time to hit ground from rest	1.42857143	
145	143	1.43	−0.02001	−14.01	−0.10	−14.11	−14.06			
146	144	1.44	−0.16064	−14.11	−0.10	−14.21	−14.16			
147	145	1.45	−0.30225	−14.21	−0.10	−14.31	−14.26			
148	146	1.46	−0.44484	−14.31	−0.10	−14.41	−14.36			
149	147	1.47	−0.58841	−14.41	−0.10	−14.50	−14.46			
150	148	1.48	−0.73296	−14.50	−0.10	−14.60	−14.55			
151	149	1.49	−0.87849	−14.60	−0.10	−14.70	−14.65			
152	150	1.50	−1.025	−14.70	−0.10	−14.80	−14.75			
153	151	1.51	−1.17249	−14.80	−0.10	−14.90	−14.85			
154	152	1.52	−1.32096	−14.90	−0.10	−14.99	−14.95			
155	153	1.53	−1.47041	−14.99	−0.10	−15.09	−15.04			

Figure 9 Calculation of displacement is negative when the object hits the ground

	A	B	C	D	E	F	G	H	I	J
1	calculation number	Time	Displacement	Initial velocity	Change in velocity	Final velocity	Average velocity	acceleration due to gravity	−9.80	m s⁻²
2	0	0.00	10.00	0.00	−0.10	−0.10	−0.05	time increment	0.01	s
3	1	0.01	9.99951	−0.10	−0.10	−0.20	−0.15	Initial displacement	10.00	m
4	2	0.02	9.99804245	−0.20	−0.10	−0.29	−0.24	Initial velocity	0.00	m s⁻¹
5	3	0.03	9.995602238	−0.29	−0.10	−0.39	−0.34	Friction constant	0.50	
6	4	0.04	9.992194227	−0.39	−0.10	−0.49	−0.44			
7	5	0.05	9.987823255	−0.49	−0.10	−0.58	−0.53			
8	6	0.06	9.982494139	−0.58	−0.10	−0.68	−0.63			
9	7	0.07	9.976211668	−0.68	−0.09	−0.77	−0.72			
10	8	0.08	9.96898061	−0.77	−0.09	−0.86	−0.82			
11	9	0.09	9.960805707	−0.86	−0.09	−0.96	−0.91			
12	10	0.10	9.951691679	−0.96	−0.09	−1.05	−1.00			
13	11	0.11	9.94164322	−1.05	−0.09	−1.14	−1.10			
14	12	0.12	9.930665004	−1.14	−0.09	−1.24	−1.19			

Figure 10 Spreadsheet to include the effects of air friction

- The change in velocity that takes place over the time interval has been modified. The acceleration is not just the acceleration due to gravity. It needs to be reduced by an amount equal to the product of the friction constant and the initial velocity at the start of the time interval. E2 = ($I$1-$I$5*(D2))*$I$2.

In this example, the effect is that the time taken to hit the ground has been increased to approximately 1.62 s (Figure 11).

This process gives a good estimate of the time taken, because the chosen time interval is sufficiently small. The equations in column E effectively assume that the frictional force is constant over the chosen time interval. Once the data have been calculated, it is a simple matter to get the computer or a GDC to plot graphs showing the variation with time of either displacement or velocity.

1 Create a spreadsheet that models an object falling under gravity that is affected by air resistance and plots the variations of displacement and velocity.
(a) Explain the shapes of the curves when the value of friction is chosen
(i) to be significant, and
(ii) when it is not.
(b) Use your answers to (a) to explain the term "terminal velocity".
(c) Explore the effect on the predictions of the model of
(i) increasing the chosen time interval, and
(ii) decreasing the chosen time interval.
(d) Explore the effect of changing the other fixed constants; explain the physical relevance of any change of number.

	A	B	C	D	E	F	G	H	I	J
1	**calculation number**	**Time**	**Displacement**	**Initial velocity**	**Change in velocity**	**Final velocity**	**Average velocity**	**acceleration due to gravity**	−9.80	m s^{-2}
158	156	1.56	0.636443763	−10.63	−0.04	−10.68	−10.66			
159	157	1.57	0.529891545	−10.68	−0.04	−10.72	−10.70			
160	158	1.58	0.422892087	−10.72	−0.04	−10.77	−10.74			
161	159	1.59	0.315447626	−10.77	−0.04	−10.81	−10.79			
162	160	1.60	0.207560388	−10.81	−0.04	−10.85	−10.83			
163	161	1.61	0.099232586	−10.85	−0.04	−10.90	−10.88			
164	162	1.62	0.009533577	−10.90	−0.04	−10.94	−10.92			
165	163	1.63	−0.118735909	−10.94	−0.04	−10.99	−10.96			
166	164	1.64	−0.228372229	−10.99	−0.04	−11.03	−11.01			
167	165	1.65	−0.338440368	−11.03	−0.04	−11.07	−11.05			
168	166	1.66	1.448938166	−11.07	−0.04	−11.11	−11.09			
169	167	1.67	1.559863475	−11.11	−0.04	−11.16	−11.14			
170	168	1.68	1.671214158	−11.16	−0.04	−11.20	−11.18			

Figure 11 Air friction increases the time taken to hit the ground

4 Forces and motion

This chapter investigates how forces affect motion. Before you study it, make sure that you have read the earlier material on displacement, velocity, acceleration, and the links between them.

The block in Figure 1 is resting on a table. Experience suggests a force is needed to get the block moving along the surface, and that if the force were removed, the block would stop moving. From these observations, the hypothesis that a constant force causes a constant velocity might seem reasonable. However, this is not the case.

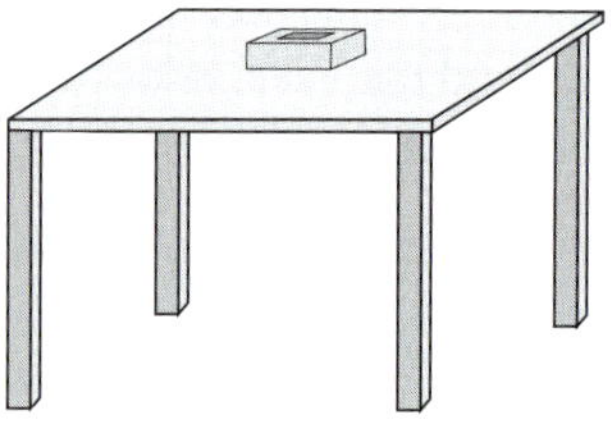

Figure 1

Careful observation shows that the block does not come to rest as soon as a force is removed, but that it takes time (and distance). Rather than reject the above hypothesis, some have attempted to modify it. Before the work of Isaac Newton in the 17th century, the accepted model was that the effects of a push continued after the actual event and took time to "run out" (Figure 2). However, this view arose because not all of the forces were being considered. In reality, there are *four* forces acting on the block as it moves along a flat surface: the forward push, friction from the surface, the pull of the Earth's gravity, and the vertical reaction force up from the surface.

Figure 2 The diagram is from an ancient manuscript and shows the predicted path of a cannon ball after being fired. It is close to being a straight line until the force "runs out" when the cannon ball falls down nearly vertically. Isaac Newton's work caused this prediction to be altered…

When studying motion experimentally, there are two forces that cannot be removed: gravity and friction. However, it is possible to compensate for each of these forces, or set up situations where their effects can be minimized. For example, if an object is dropped so that it falls freely, then, briefly, there is only the force of gravity acting on it. Unless it is moving very fast or is very light, the effects of air resistance can be ignored. Under these circumstances, experimental measurements show that the resulting motion is uniform acceleration. Resultant force causes acceleration.

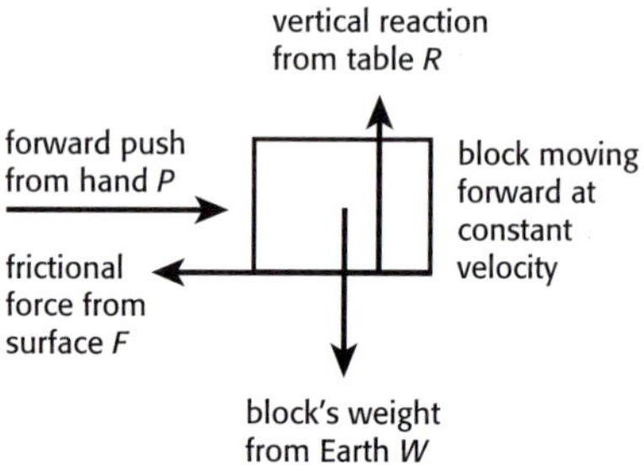

Later in the chapter, Newton's laws of motion will be introduced. However, before looking at these, one needs to understand how forces can combine, and why it is important to allow for their direction.

Mathematical physics: vectors and scalars

All quantities are either **vectors** or **scalars**. Vectors (such as force, velocity, and acceleration) have direction as well as magnitude (size). A change in either magnitude or direction, or both, means that the vector has changed. Scalars (such as mass, density, and energy) just have magnitude. There is no direction to take account of.

Velocity and speed are related, but different. Velocity is a vector, while speed is a scalar. In other words, if you say that a velocity is 10 m s^{-1}, you also have to give a direction (to the right, for example). However, if the speed is 10 m s^{-1}, no direction is implied. If a motorcycle is travelling around a circular track at 10 m s^{-1}, its speed is constant, but its velocity is not, because the direction is changing. And if the velocity is not constant, then the motorcycle has acceleration. Having acceleration does not necessarily mean getting faster or slower.

Mathematically, scalars and vectors need to be handled in different ways.

Table 1

Scalars	Vectors
Volume	Velocity
Density	Acceleration
Speed	Force
Concentration	Displacement
Frequency	Momentum
Energy / work	
Power	
Distance	
Mass	

Scalar addition Scalar addition is the everyday mathematics of numbers: a mass of 3 kg and a mass of 4 kg together give a total mass of 7 kg.

Vector addition A 3 N force and a 4 N force do not necessarily create a total force of 7 N. The maximum that they can create is 7 N, but the minimum is 1 N, and any value between 1 N and 7 N could be possible depending on the directions of the forces. The force produced by the combination is called the **resultant force**.

To add two vectors, one can use the **parallelogram rule**. First, the vectors are represented by arrows, with the length of each arrow being in proportion to the magnitude of the vector. Next, a parallelogram is completed, as in the example below. The diagonal gives the direction and magnitude of the resultant.

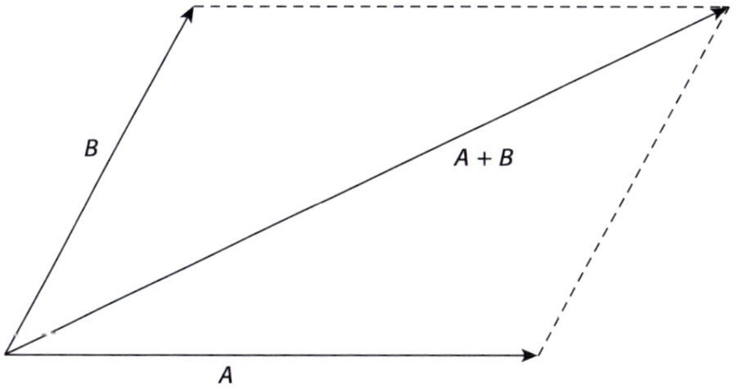

Figure 3 Parallelogram of vectors

If you have more than two vectors to add, the parallelogram rule can be applied with different pairs in turn. Alternatively, the arrows representing the different vectors can be drawn one after another with the tail of one arrow starting from the tip of the proceeding arrow. The arrow joining the tail of the first arrow to the tip of the last arrow represents the resultant vector.

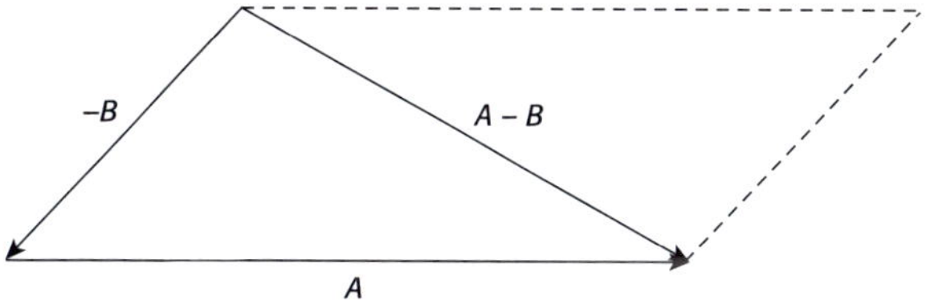

Figure 4 Addition of vectors

Vector subtraction A negative sign in front of a vector does not alter its magnitude, but means that the direction is reversed. The subtraction of two vectors just involves the addition of a negative vector, as shown below. The technique could be used, for example, to calculate the relative velocity between two moving objects.

Vector trigonometry The above techniques involve scale diagrams. However, it is often possible, and certainly more accurate, to use trigonometry to calculate the resultant. This is of course much simpler if right-angled triangles are involved.

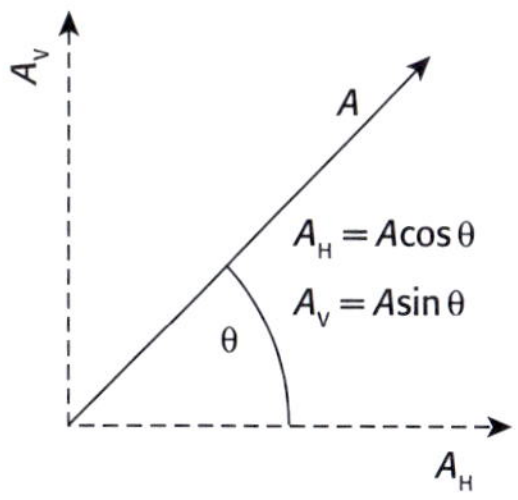

Figure 5 Decomposition of vectors

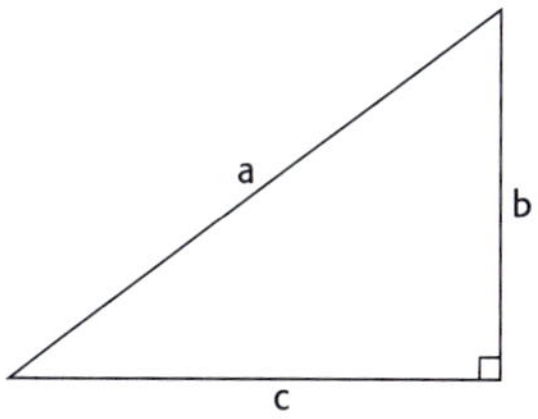

Figure 6 Pythagoras's theorem

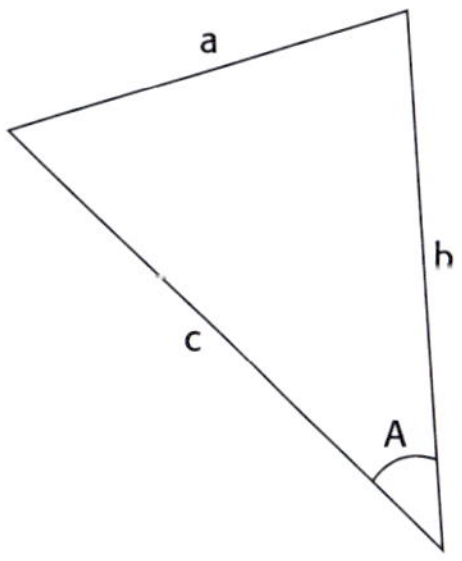

Figure 7 The cosine rule for any triangle is $a^2 = b^2 + c^2 - 2bc \cos A$

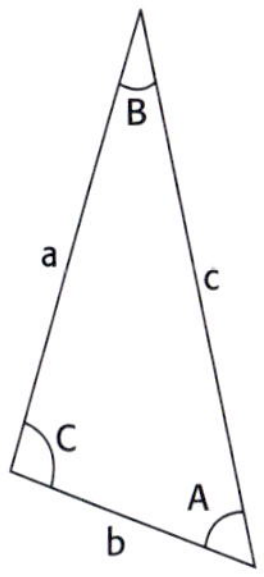

Figure 8 The sine rule for any triangle is $\frac{a}{\sin A} = \frac{b}{\sin B} = \frac{c}{\sin C}$

As the block is moving at constant velocity,

$T = f$

$R = W$

Example 2 An object in a lift. In this case, the object has a mass of 8 kg, and is accelerating upwards at 0.5 m s^{-2}.

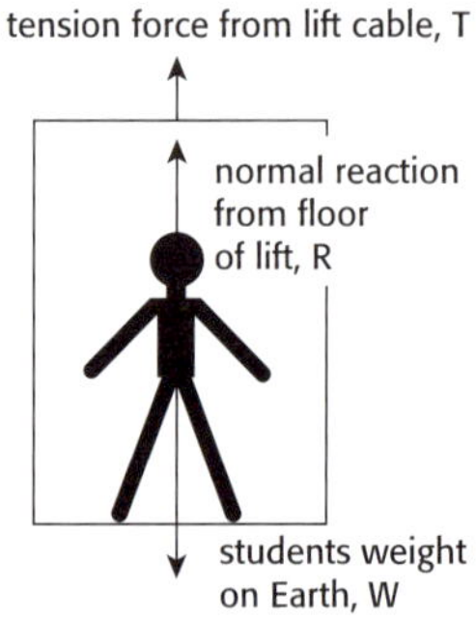

Figure 15 The forces on an object in a lift

Verifying Newton's second law

This section follows the design of an experiment that aims to verify Newton's second law. Although the experiment can provide evidence to support Newton's second law, a detailed analysis of the results shows that the procedure has been based on some false assumptions. Can you spot the mistake as it is introduced?

The experiment involves the following variables: resultant force, mass, and acceleration. This suggests two possible approaches:

First, the force could be chosen as the independent variable, the acceleration as the dependent variable, and the mass as the controlled variable, which is held constant. The prediction is that the acceleration should be proportional to the force.

Second, the mass could be the independent variable, the acceleration the dependent variable (as before), and the force as the controlled variable, which is held constant. The prediction is that the acceleration should be inversely proportional to the mass.

A possible setup for the experiment is shown in Figure 16.

In the first approach, the force is changed by altering the weights hanging from the string. The acceleration of the trolley is recorded electronically using motion sensors. The values of the force and the acceleration are recorded, then the experiment repeated with different forces.

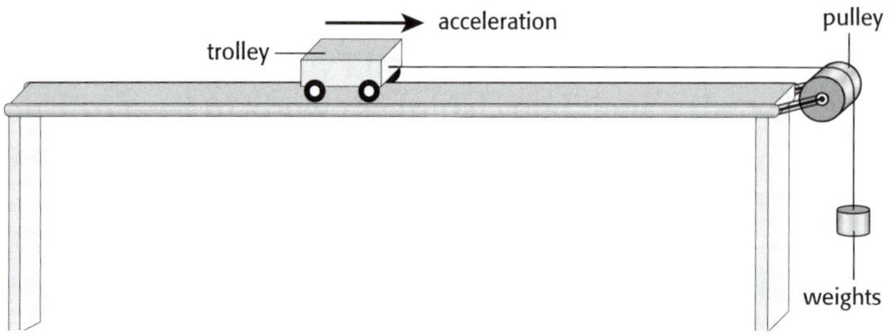

Figure 16 Possible experimental setup to verify Newton's second law

1 What shape of graph would you expect if the measured acceleration and the force were proportional? Explain your answer. [2]

2 The actual results are shown in Figure 17 with uncertainty limits shown for the force values.

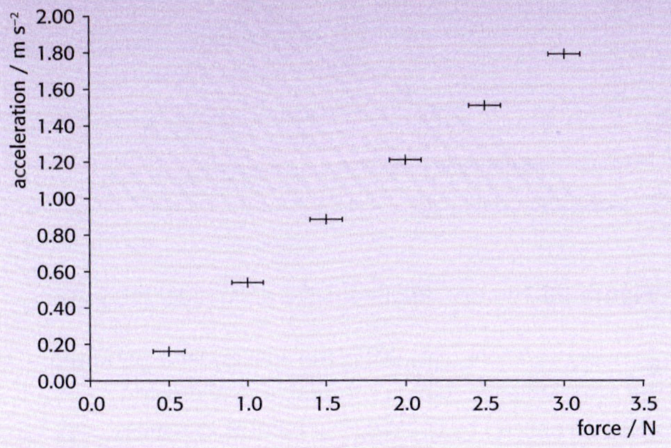

Figure 17

(a) Add the best-fit line for this data. [2]
(b) Use the graph to explain what is meant by a *systematic* error. [2]
(c) Estimate the value of the frictional force acting on the trolley. [1]
(d) Estimate the mass of the trolley. [2]

[Total 7 marks]

3 Careful analysis of the graph shows that the last two points are not quite in the same straight line as the first four points. The data is checked by taking further readings. This shows that the weight is *not* proportional to the acceleration as shown in Figure 18.

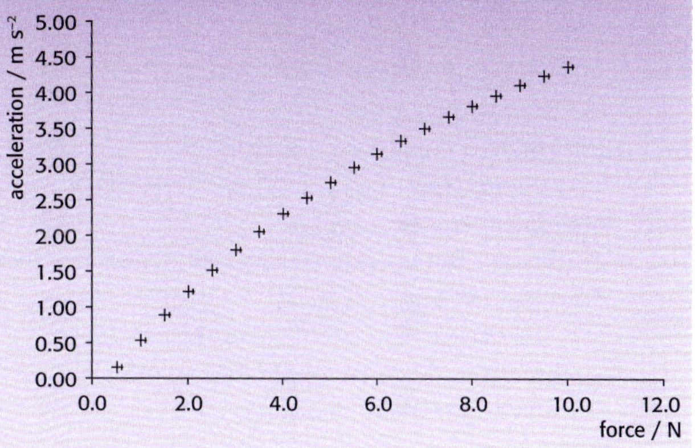

Figure 18

(a) Add the best-fit line for this data. [2]
(b) (i) Draw free-body force diagrams for the trolley **and** for the weight. [4]
(ii) Use the diagrams to explain why both the trolley and the weights accelerate together. [4]
(c) The value of the weights is continually increased. The acceleration of the trolley would eventually reach a maximum value. Explain why this is so and estimate its value. [3]
(d) How could the procedure be modified to improve the experiment? [4]

[Total 17 marks]

Mathematical physics: force and momentum

Linear momentum, $\boldsymbol{p}$, is a vector quantity defined by the following equation:

$$\boldsymbol{p} = m\,\boldsymbol{v}$$

m is the mass of the object, measured in kilograms
$\boldsymbol{v}$ is the velocity of the objects, measured in m s^{-1}
$\boldsymbol{p}$ is the momentum of the body, measured in kg m s^{-1}
Since 1 N = 1 kg m s^{-2}, an alternative unit for momentum is N s (these two units are equivalent).

The following definition provides a more fundamental way of stating Newton's second law:

The rate of change of momentum of a body is proportional to the resultant force acting.

$$F \propto \frac{d}{dt}(m\boldsymbol{v})$$

In SI units, the definition of Newton's second law means that the resultant force and the rate of change of momentum are equal.

$$F = \frac{d}{dt}(m\boldsymbol{v})$$

F is the resultant force, measured in newtons
$m\boldsymbol{v}$ is the momentum of the object, measured in kg m s^{-1}
$\frac{d}{dt}(m\boldsymbol{v})$ is the rate of change of momentum in kg m s^{-2} or N

The above equation relates the instantaneous value of force to the instantaneous rate of change of momentum. Over a time interval, Δt, the average force $\boldsymbol{F}_{average}$ is calculated using:

$$F_{average} = \frac{\Delta p}{\Delta t}$$

1 A 1600 kg car is travelling at 30 m s⁻¹. Calculate the average force on the car in each of the following ways that it can be brought to rest:
   (a) Gently applying the brakes, taking 20 s to stop.
   (b) An emergency stop taking 2.5 s.
   (c) Hitting a tree, taking 0.3 s to stop.

2 A 60 g tennis ball moving at 20 m s⁻¹ hits a wall horizontally and bounces back along the same path at 18 m s⁻¹. The ball is in contact with the wall for 0.5 s. Calculate:
   (a) The change in momentum for the ball. (*Be careful – the answer is not 0.12 kg m s⁻¹*)
   (b) The average force that acts on the ball.

Impulse

The previous equation can be rearranged to give this:

$$\Delta p = F_{average} \times \Delta t$$

The product of a force and the time for which it acts is called the impulse. The above equation shows that it is equal to the change of momentum. Impulse is a vector quantity.

Analysing a graph

In many practical situations, forces vary significantly over time. If one can record the actual variation of the force, the above relationship allows the overall change in momentum to be calculated.

The graph in Figure 19 represents a force that varies with time. It is reasonable to assume that over a very short time interval, δt, the instantaneous value of the force, F, does not significantly change. The impulse given to the object during this short time interval is the product of F and δt, represented in Figure 19 by the shaded area of the rectangle.

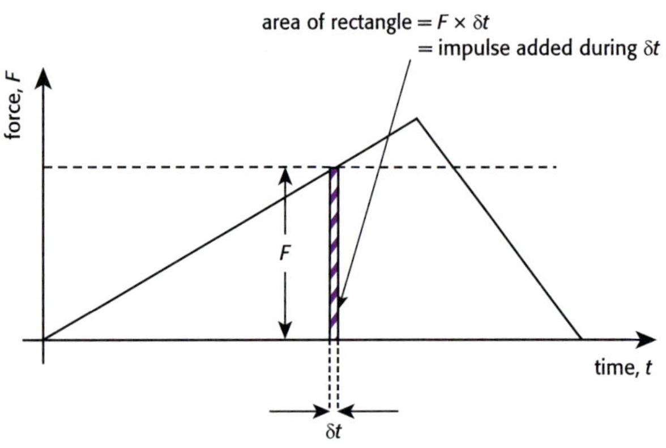

Figure 19 Incremental increase in impulse

The total impulse is the sum of all the small impulses, added over the time for which the force acts. Mathematically, the way of adding up these impulses is by calculating the area under the graph.

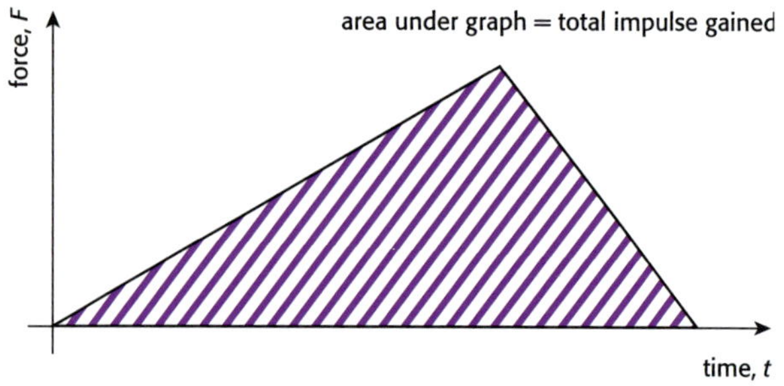

Figure 20 The total impulse is the area under the graph

In many situations, calculating the area under a graph can be achieved either with some simple mathematics or (less accurately) by counting the number of squares and calculating what one square represents. If the variation can be summarized in a mathematical equation, then the area under the graph can be calculated using a process called integration. The mathematics of integration is not required for IB physics, but it is studied in the SL and HL mathematics courses.

$$\text{change in momentum} = \text{impulse} = \int$$

3 During a tennis serve, a 60g tennis ball is hit with a tennis racket. The initial horizontal velocity of the ball is zero and Figure 21 represents the variation with time of the force on the ball.

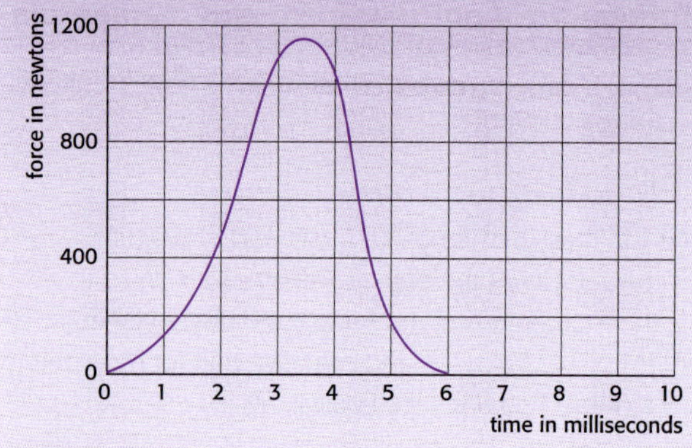

Figure 21

Calculate:
   (a) The area under the graph
   (b) The impulse given to the ball
   (c) The final velocity of the ball.

Conservation of momentum

When two objects collide, the momentum of each one changes. From Newton's third law, the forces on the two objects are equal but opposite. They also act for the same time, so the two impulses are equal but opposite. Therefore the change in momentum of one object must be equal but opposite to that of the other. Put another way, the total momentum of the objects is the same after the collision as before. This is an example of the **law of conservation of momentum**:

The overall momentum of an isolated system of interacting objects must be constant.

Here, an "isolated system" means one without any external forces acting on it. It is rare for any system to be completely isolated. However, being a vector, momentum, like force, can be resolved into components. So, if there is no component of a force in one direction, the component of the momentum *in that direction* is constant.

The law of conservation of momentum can sometimes provide a mathematical shortcut to help analyse complex situations. If you know that the total momentum of a system does not change, then you can ignore the internal interactions within it. The law of conservation of energy can be useful for similar reasons.

4 Research a formal mathematical derivation of the law of conservation of momentum. Explain the relevance of the following to the derivation:
   (a) Newton's second law
   (b) Newton's third law.

5 A 8 kg truck moving at 5 m s^{-1} collides with a stationary 12 kg truck. The collision takes 0.2 s. Afterwards, the two trucks are joined together and move off with a common velocity *v*. Calculate:
   (a) the common velocity *v* after the collision
   (b) the impulse given to each truck during the collision
   (c) the average force acting on each truck during the collision.

Circular motion

Can an object accelerate without getting any faster? The answer is yes, if it is moving in a circle. But before considering this, you need to know how angles and angular rotation are measured.

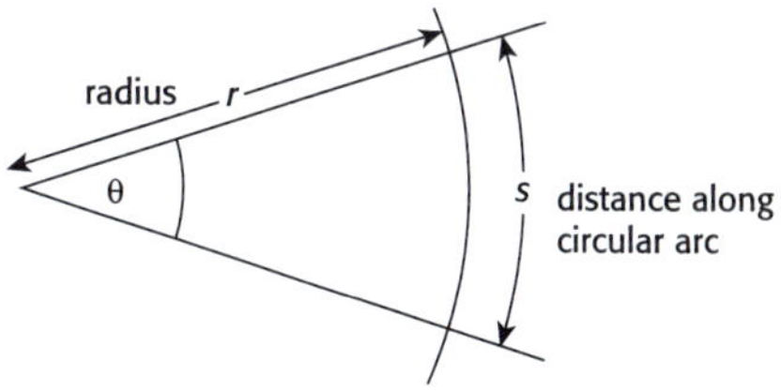

Figure 22 Calculating an angle in radians

Radian measure

Angles can of course be measured in degrees but, in studying rotation, the radian is a more useful measure.

The diagram in Figure 22 shows an angle θ.

The value of the angle in radians is found like this:

$$\theta = \frac{s}{r}$$

One radian (symbol rad) is approximately 60 degrees.

For a full circle (360°), $s = 2\pi r$, so θ in this case is 2π radians.

As $360° = 2\pi$ rad, it follows that $180° = \pi$ rad, and $90° = \pi/2$ rad

In each of the above examples, it isn't strictly necessary to include the unit (rad), because the value s/r is a number, with no dimensions.

For small angles (less than about 0.1 rad, or 5°), the arc and two radii in the above diagram form a shape that approximates to a triangle. Therefore the following relationship applies:

$$\sin\theta \approx \tan\theta \approx \theta$$

Table 2 Conversion between degrees and radians

Angle / degrees	Angle / radians
0	0.00
5	0.09
45	$0.79 = \frac{\pi}{4}$
60	1.05
90	$1.57 = \frac{\pi}{2}$
180	$3.14 = \pi$
270	$4.71 = \frac{3\pi}{2}$
360	$6.28 = 2\pi$

Angular velocity

In Figure 23, the object is moving round a circle at constant speed. Put another way, it has uniform circular motion.

The velocity is continuously changing direction. However, at any instant, it is always at right angles to the radius at that position. As the object rotates, this radius makes a different angle with the axes, as in Figure 24.

In a given time interval, the increase in angle is always the same. In other words, there is a constant rate of rotation. This rate called the **angular velocity** is given the symbol ω, and it is defined as:

$$\omega_{av} = \frac{\text{change in angle turned}}{\text{time taken}}$$

$$\omega_{av} = \frac{\Delta\theta}{\Delta t}$$

The definition above gives the *average* angular velocity. The following equations express the angular velocity at any instant:

$$\omega = \frac{\text{small change in angle}}{\text{small time taken}} = \text{rate of change of angle}$$

$$\omega = \lim_{\delta t \to 0}\left(\frac{\delta\theta}{\delta t}\right) = \frac{d\theta}{dt}$$

Angular velocity is measured in radians per second: for example, 2 rad s^{-1}. However, as the radian is dimensionless, this could be expressed as 2 s^{-1}.

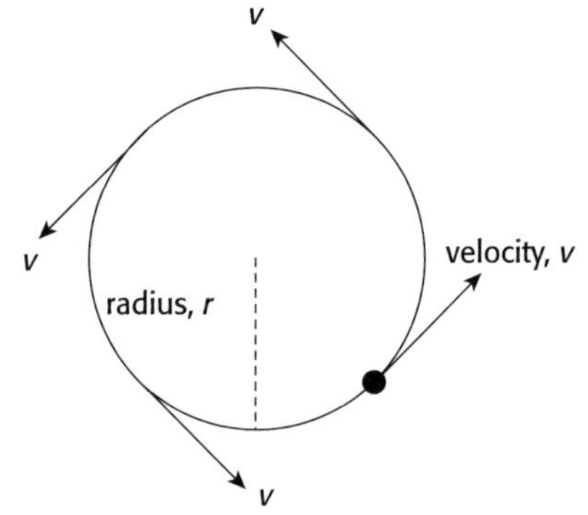

Figure 23 Uniform circular motion

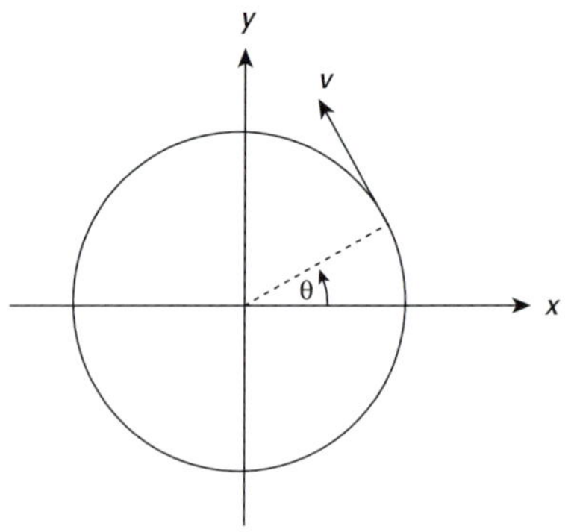

Figure 24

Linking speed and angular velocity

Look again at the diagram showing the arc of a circle with the angle θ (Figure 24).

Since $\theta = \frac{s}{r}$, it follows that $s = r\,\theta$

So: $v = \frac{s}{t} = \frac{\theta r}{t} = \omega r$

If the circular motion is uniform, there is no difference between the instantaneous and average values for angular velocity.

Period

This is the time taken for the object to complete one full circle. In a time period T, the object will have rotated through an angle of 2π radians. As the angular velocity is found by dividing the angle by the time:

$$\omega = \frac{2\pi}{T} \quad \text{or} \quad T = \frac{2\pi}{\omega}$$

Centripetal acceleration

Velocity is a vector quantity. So, if its direction is changing, it is not constant. And if it is not constant, the object must have acceleration (even though its speed remains steady). The full analysis of the vector situation is shown below, but the direction of the acceleration can be deduced from the change in velocity that takes place as an object moves around the circle between two points, say A and B (Figure 25).

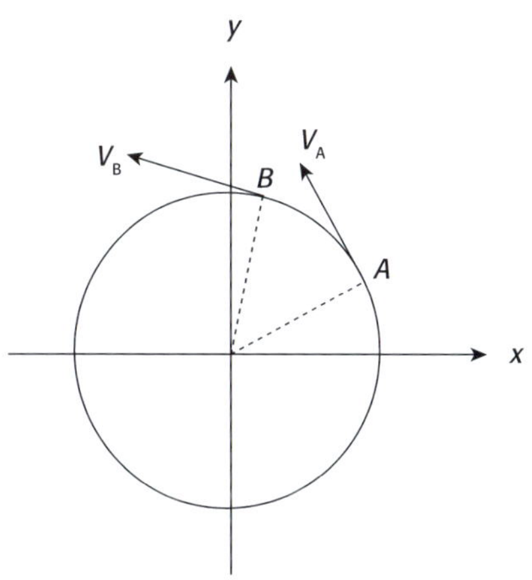

Figure 25 Calculating acceleration as a change in direction of velocity

The vector diagram shows that the average acceleration between A and B is directed towards the centre of the circle (Figure 26). This is also true of the instantaneous acceleration.

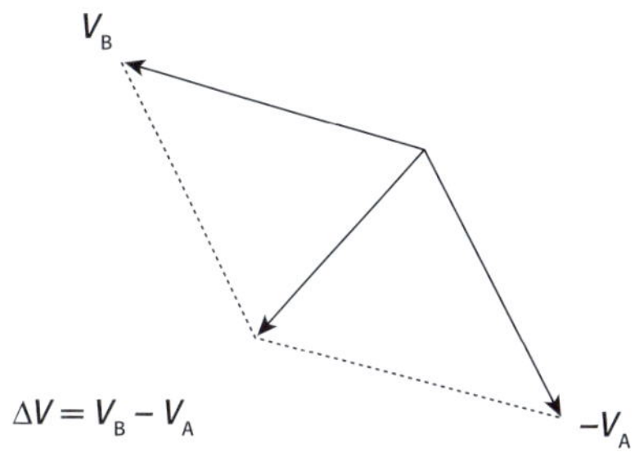

Figure 26 Using vectors to calculate acceleration

The acceleration of an object that is in uniform circular motion is called the **centripetal acceleration**. A more detailed analysis allows one to calculate the instantaneous magnitude of this centripetal acceleration. The result is as follows:

$$a_{\text{centripetal}} = \frac{v^2}{r}$$

This equation can be stated in several different equivalent forms:

$$a_{\text{centripetal}} = r\omega^2 = \frac{4\pi^2 r}{T^2}$$

Centripetal force

Newton's second law tells us that, as the object is accelerating, there must be a resultant force on it. This force is called the **centripetal force** (CPF). As $F = ma$:

$$\text{CPF} = \frac{mv^2}{r} = mr\omega^2 = m\frac{4\pi^2 r}{T^2}$$

The centripetal acceleration is towards the centre of the circle, so the centripetal force must also act in this direction. However, it is important to realize that CPF is not a new or additional force. Something must provide it. Without it, the object would travel in a straight line and not in a circle. Table 3 shows some examples of CPF providers:

Table 3

Situation	Origin of the centripetal force
Mass on the end of a string being whirled around in a horizontal circle	The string exerts a tension force on the mass which provides the necessary CPF
The Earth in orbit around the Sun	The gravitational force of attraction exerted by the Sun on the Earth provides the necessary CPF
A car turning around a corner	The sideways frictional force exerted by the road on the tyres provides the necessary CPF
An electron in a hydrogen atom	The electrostatic attraction exerted by the nucleus on the electron provides the necessary CPF

5 Energy, work, and power

In physics the three physical quantities of **work**, **energy**, and **power** have precise mathematical definitions.

The concepts of work and energy provide a useful set of tools that allow us to analyse complicated situations. For example, you should already be able to estimate the final velocity v of a 0.8 kg block as it slides from rest down a 30° slope from a height of 1.8 m to ground level (Figure 1).

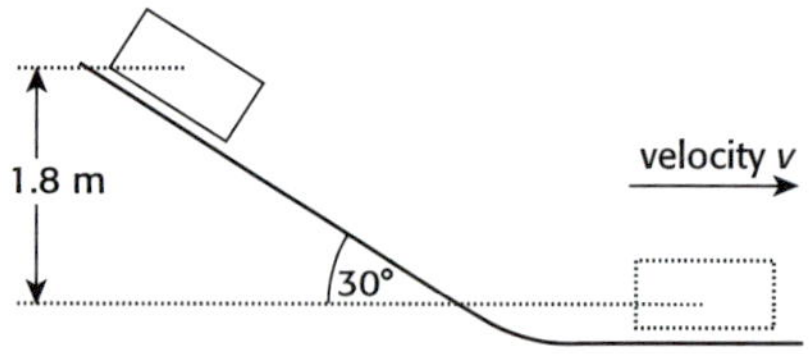

Figure 1 A block slides down a slope

The necessary stages in your calculation are as follows:

1 assume that friction is negligible
2 draw a free-body diagram for the block
3 resolve the forces in the directions (a) down the slope and (b) perpendicular to the slope
4 calculate the resultant force down the slope
5 use $F = ma$ to calculate the acceleration of the block down the slope
6 use trigonometry to calculate the distance down the slope
7 use the constant-acceleration equations to calculate the final velocity after going down the slope.

You should now have calculated that the block is moving down the slope at 6.0 m s^{-1}.

Note: This approach considers only the motion down the slope. We have not yet considered the curved section at the end. The forces during this section must be accelerating the block (it certainly changes direction), and so we might expect it to also change its speed. But if we assume that the effect on its speed will be negligible, then we have a prediction for its final velocity.

We can calculate the same final answer much more quickly using the energy concepts introduced below.

Work and the conservation of energy

Work and energy are related concepts. In any given situation, the work done measures the amount of energy that has been transferred.

Work

In physics, doing work involves moving a force through a distance. Although work is a scalar quantity, force is a vector quantity, and we need to consider the directions when calculating the work done. The simplest situation is when the force and the displacement are in the same direction.

work done, $W = Fs$

In general, however, the force and the displacement are not always in the same direction (Figure 2).

We define the work done with the equation

work done, $W = Fs\cos\theta$

By introducing the factor $\cos\theta$, we are effectively considering the component of the resultant force in the direction of the displacement.

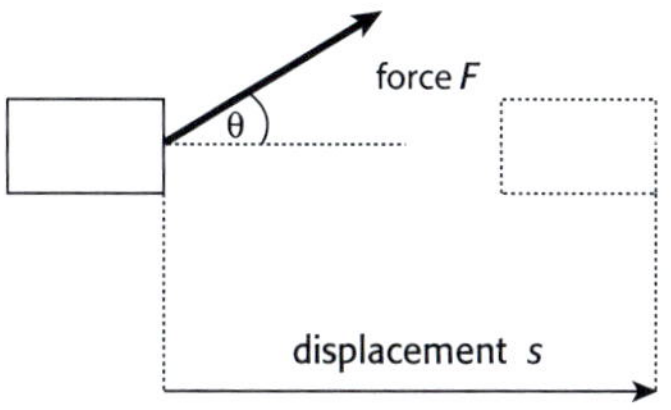

Figure 2 Force and displacement are not in the same direction

Energy

Energy is the ability to do work. Energy and work are both scalar quantities; they are measured in joules. For example, if we give an object 5 kJ of energy, it can do 5000 J of work. How do we give it the energy? The fact that it has gained 5000 J of energy means that something else must have done work on it. The object doing the work lost 5000 J of energy, whereas the object receiving the energy had work done on it. Doing work means that energy is being interchanged.

Many different forms of energy have been identified: see the next section and page 192 for a discussion on the nature of energy. Two extremely important forms that we can analyse quantitatively are **kinetic energy** (KE) and **gravitational potential energy** (gPE). KE is the energy that an object has as a result of its motion, and gPE is the energy that an object has as a result of its position in a gravitational field (i.e. its height).

Kinetic energy

Consider the situation of motion in a straight line when a constant resultant force F causes an object of mass m to accelerate from initial velocity u to final velocity v (Figure 3).

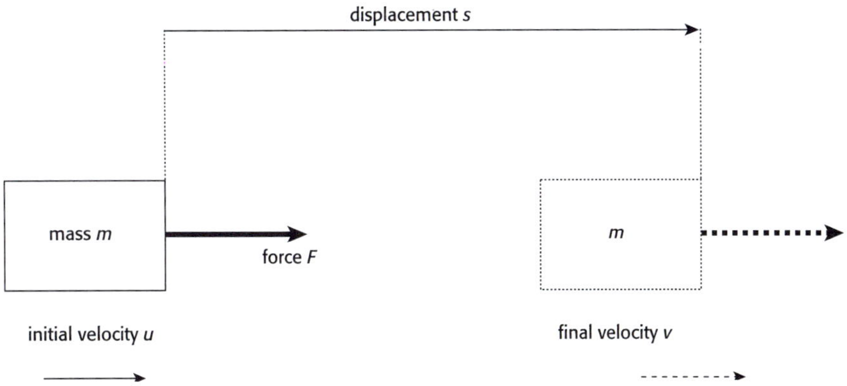

Figure 3 Work done considering change in KE

1 The diagram below shows the variation with displacement x of the force F acting on an object in the direction of the displacement.

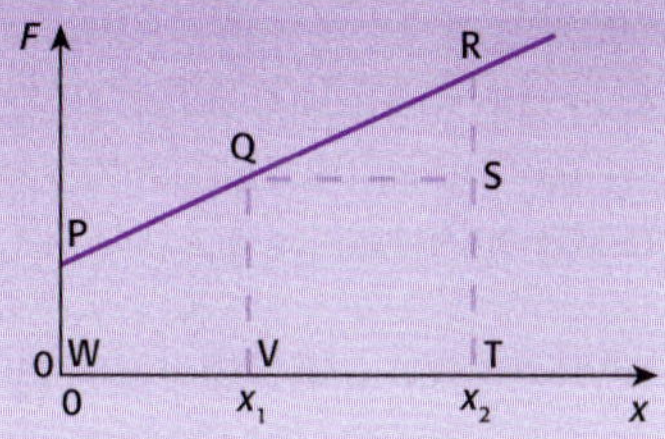

Which area represents the work done by the force when the displacement changes from x_1 to x_2?

A QRS C WPQV

B WPRT D VQRT

2 A stone of mass m is attached to a string and moves round in a horizontal circle of radius R at constant speed V. The work done by the pull of the string on the stone in one complete revolution is

A zero C $\frac{2\pi mV^2}{R}$

B $2\pi mV^2$ D $\frac{2\pi mV}{R}$

3 Which **one** of the following is a true statement about energy?

A Energy is destroyed due to frictional forces.

B Energy is a measure of the ability to do work.

C More energy is available when there is a larger power.

D Energy and power both measure the same quantity.

The work done is the product of the force and the displacement:

$$W = Fs$$

The force and the mass are related to the acceleration, a:

$$F = ma \quad \text{so} \quad W = mas \qquad (1)$$

The acceleration and the displacement can be related to the initial and final velocities:

$$v^2 = u^2 + 2as \quad \text{so} \quad 2as = v^2 - u^2 \qquad (2)$$

Combining equations (1) and (2):

$$W = \tfrac{1}{2}mv^2 - \tfrac{1}{2}mu^2$$

The work done equals the change of the quantity

$$\tfrac{1}{2}\text{mass} \times (\text{velocity})^2$$

The change in kinetic energy is the work done on the mass, so we have derived an equation for the kinetic energy of a moving object:

$$\text{KE} = mv^2$$

where
KE is the kinetic energy of the object, measured in joules (J)
m is the mass of the object, measured in kilograms (kg)
v is the velocity of the object, measured in metres per second (m s^{-1})

Gravitational potential energy

Consider the situation of an object of mass m being lifted (in a constant gravitational field) a vertical height h above its original position by a force F.

The work done, W, is calculated by

$$W = Fh$$

The force needed to lift the mass is equal to its weight:

$$F = mg$$

so

$$W = mgh$$

This work has increased the object's gravitational potential energy

$$\text{gPE} = mgh$$

where
gPE is the gain in gravitational potential energy of the object, measured in joules (J)
m is the mass of the object, measured in kilograms (kg)
g is the magnitude of the gravitational field strength, measured in newtons per kilogram (≈ 10 N kg^{-1} on the surface of the Earth)
h is the increase in vertical height, measured in metres (m)

Note that this equation only works when the gravitational field is constant. If an object moves a significant distance away from a planet, then the force due to gravity will reduce, and this equation does not apply. In this case, the calculation will involve either graphical work or the use of calculus.

?

1 An object of mass m_1 has a kinetic energy K_1. Another object of mass m_2 has a kinetic energy K_2. If the momentum of both objects is the same, the ratio $\frac{K_1}{K_2}$ is equal to

A $\frac{m_2}{m_1}$ C $\sqrt{\frac{m_2}{m_1}}$

B $\frac{m_1}{m_2}$ D $\sqrt{\frac{m_1}{m_2}}$

?

Calculate the KE of the following objects:
(a) An athlete of mass 72 kg running at 8 m s^{-1}
(b) A tennis ball of mass 60 g moving at 30 m s^{-1}
(c) A car of mass 1600 kg moving at 20 m s^{-1}
(d) A bullet of mass 150 g moving at 250 m s^{-1}

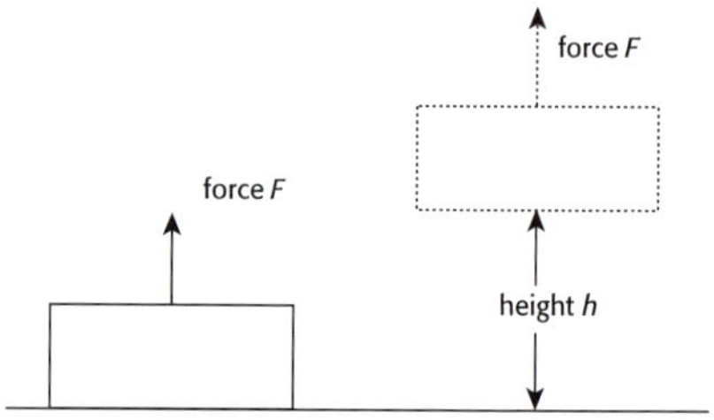

Figure 4 Increasing an object's gPE

?

Calculate the gPE of the following objects on Earth:
(a) A diver of mass 72 kg climbing to a diving board 8 m high
(b) A tennis ball of mass 60 g being hit 30 m into the air
(c) A car of mass 1600 kg driving up a 20 m hill
(d) A bullet of mass 150 g fired 250 m into the air

The conservation of energy

When work is done, energy is interchanged between objects. The amount of energy possessed by any given object can change, but the total amount of energy shared among all the objects involved must remain the same. This principle is known as the **law of conservation of energy**: "energy is neither created nor destroyed, it just changes form" or "the total energy of the universe is constant".

We can use this principle to analyse complex situations without having to predict every physical change that may take place. For example, we began this chapter by considering a mass sliding down a slope (see Figure 1 on p. xx).

If we assume that frictional forces can be neglected, then by analysing the forces involved we can predict the final velocity.

The block starts with gPE, and this is converted into KE. If friction is negligible, then the final KE must equal the initial gPE. This provides a quick method for calculating the final velocity, v.

final KE = initial gPE

$$\frac{1}{2}mv^2 = mgh$$

$$\therefore v = \sqrt{2gh} = \sqrt{2 \times 10 \times 1.8} = \sqrt{36} = 6.0 \text{ m s}^{-1}$$

Note that this analysis predicts that the final speed of the block is independent of:

- the mass of the block
- the angle of the slope
- the shape of the final curve.

The same prediction applies whatever route the block takes from its start to the finish. For example, in the absence of friction, the block in Figure 5 would also end up travelling at 6 m s^{-1}.

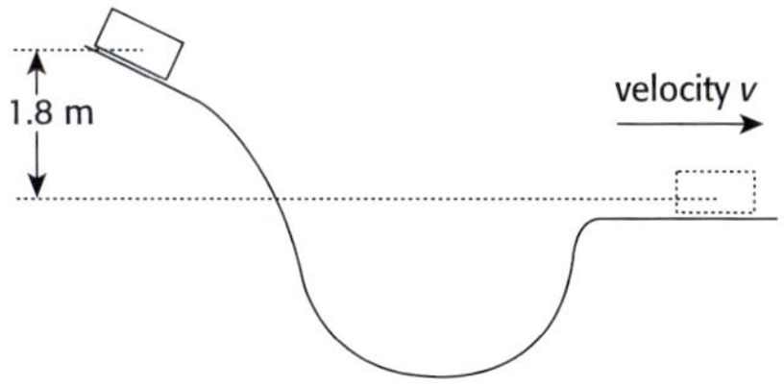

Figure 5 The final velocity only depends on the change in height of the object

Energy transformations

In the previous section we introduced quantitative equations for two common forms of energy, KE and gPE. There are many different energy forms that we can usefully consider, but ultimately we can classify all forms of energy as either kinetic (if associated with movement) or potential (all other forms).

The various forms of energy include:

- kinetic energy
- potential energy
  - gravitational
  - electrostatic/electric
  - elastic
  - chemical
  - magnetic
- thermal energy
- radiant energy
- light (and the rest of the electromagnetic spectrum) energy
- sound energy
- nuclear energy.

?

1 A boy is sitting on a swing of length 2 m. The swing is pulled away from its resting position in the middle until it makes an angle of 30° with the vertical. The swing is then released.
   (a) How fast will the boy be moving when he passes through the middle?
   (b) How far will the boy travel beyond the middle on the other side?

2 The Internet can provide lots of examples of simple physics (Newton's laws and energy considerations) being used to analyse complex situations. One very good example is Kung Fu Science, which follows a physics PhD student, Michelle, as she attempts to break three 2 cm thick pieces of pine board using her bare hands (http://www.kungfuscience.org/site.asp). Do some research for similar sites and create a list of your favourites.

If energy appears to be lost (or gained) in any particular situation, then rather than abandon the law of conservation of energy we need to try and identify a different form of energy. The nature of energy is discussed further on page 192.

Note that in the list above we included "thermal energy" rather than "heat energy". Many scientists use the phrase "heat energy" or even just "heat" to refer to thermal energy. Technically "heat" (or better, "heating") is a verb so using the word "heat" to represent a form of energy can lead to some confusion and it is better to avoid these terms if at all possible.

The quantitative relationships between many of these energy forms are discussed throughout the IB physics course. At this stage, it is sufficient to note the general principle that many everyday practical situations involve much of the initial energy in a system ending up as thermal energy shared among a large number of molecules. This includes all situations in which frictional forces act. The energy is said to be **degraded** (see page 220). If thermal energy is released to the surroundings and cannot be recovered, it is said to be **dissipated**. Thermal energy concepts are discussed in more detail in Chapter 10 (page 109) and Chapter 18 (page 218).

Collisions and explosions

In any collision or explosion, the total energy involved must be conserved. In addition, the conservation of momentum must also apply to an isolated system. These two principles do not provide enough information for us to predict the final outcome in all situations – this depends on the context. By focusing on one particular type of energy (the kinetic energy) we can categorize different types of collision and explosion, and compare similar situations.

- **Elastic collisions** are those in which the KE is conserved. Everyday collisions are not elastic, because a significant portion of the KE is transferred into thermal energy and sound energy. Perfectly elastic collisions can take place only at the atomic scale.
- **Inelastic collisions** are those in which the total KE is decreased as a result of the collision. Most everyday collisions are inelastic. A **completely inelastic** collision is one in which the objects stick together after the collision.
- **Explosive collisions** are those in which the total KE is increased as a result of the collision. A source of energy (e.g. chemical potential energy or elastic energy) is needed to provide the "extra" KE.

?

1 An 8 kg truck moving at 5 m s^{-1} collides with a stationary 12 kg truck. The collision is completely inelastic, and the two trucks join together and move off with a common velocity v. Calculate:
   (a) the common velocity v after the collision
   (b) the energy "lost" to the surroundings.

2 (mathematically harder) An 8 kg truck moving at 5 m s^{-1} collides with a stationary 12 kg truck. The collision is perfectly elastic. Calculate the velocity of each truck after the collision.

Examples of energy transformations

If we can identify the energy transformations that are taking place in any situation it can often help us analyse the changes that are taking place. However, remember that changes do not take place as a *result* of the energy changes. A ball does not roll down a hill *because* gPE is transferred into KE. It rolls down the hill because of the gravitational force of attraction on it.

There must always be a physical reason for the changes to take place; there are many things that are energetically possible but could not actually occur. For example, it should be straightforward to identify the energy changes that take place when a glass is accidentally dropped. The main energy changes are gPE transforming into KE, which transforms into heat and sound.

From an energy point of view, the reverse process (the fragments coming together to form a whole glass that jumps back on to the table; is not ruled out – but we know it never occurs. See page 125 for an explanation.

?

Outline the principle energy conversions taking place in the following situations. In each case you need to identify the energy *transformations* that are occurring; do not merely list the types of energy involved. Many of the situations involve more than one energy transformation. One possible approach would be for each student in a class to take responsibility for analysing different situations and then report back to the rest of the class to see whether they agree. In addition to the situations listed below, describe a situation of your own choice.

(a) A torch that runs on batteries is used to shine light.
(b) A doorbell rings when a button is pushed.
(c) A burglar alarm sounds when a laser beam is broken.
(d) A lift carries passengers to the top floor of a skyscraper.
(e) In order to live, you need to eat.
(f) A pendulum swings back and forth.
(g) Listening to music on a personal hi-fi device of your choice.
(i) A firework rocket goes into the air and explodes.
(j) Listening to the radio.
(k) A voyage to the Moon.
(l) A hydroelectric dam used to generate electrical energy.
(m) An electromagnet used to pick up and sort metallic waste.
(n) An electric hob used for cooking.
(o) A microwave cooking a potato.
(p) A parachute jump.
(q) An aircraft taking off.
(r) A pole-vaulter jumping over the bar.
(s) A gun being fired.
(t) A hair dryer.
(u) A mechanical watch or clock.
(v) A coal-fired power station.
(w) A car being driven at constant velocity.
(x) A solar-powered calculator.
(y) Playing a piano.
(z) A "hole in one" shot at golf.

Power and efficiency

Two useful quantities related to the study of any energy changes are the **power** and the **efficiency**.

Power

Power is the rate at which energy is transferred. This is the same thing as the rate at which work is done. The average power P_{av} over a time interval Δt is defined in terms of the energy transferred ΔW as follows:

$$P_{av} = \frac{\text{energy transferred}}{\text{time taken}} \qquad P_{av} = \frac{\Delta W}{\Delta t}$$

or

power = resultant force × velocity

The units of average power are J s^{-1}, and a new unit, the watt (W), is defined:

$$1\,\text{watt} = \frac{1\,\text{joule}}{1\,\text{second}}$$

?

1 A man of mass 75 kg climbs 8 m up a vertical rope in 20 s. Calculate his average power.

2 A car maintains a constant velocity of 25 m s^{-1} against a friction force of 2 kN. How much power is the car's engine developing?

Efficiency

The efficiency of any process is a measure of the proportion of energy that is usefully transferred. What does "usefully" mean? That depends on the context. For example, a light bulb is designed to give out light energy, but it will also dissipate thermal energy to the surroundings at the same time. So in this context the useful energy is the light energy, and the thermal energy is the wasted energy. An electric heater, however, is designed to produce thermal energy and, when it is operating, very little of the energy it produces will be in any other form.

$$\text{efficiency} = \frac{\text{useful work done}}{\text{total energy transferred}} = \frac{\text{useful energy transferred}}{\text{total energy transferred}}$$

We can also express this definition in terms of power:

$$\text{efficiency} = \frac{\text{useful power}}{\text{total power used}}$$

Efficiency is a ratio, and has no units. It is often expressed as a percentage.

?

A 60 W light bulb is 15% efficient. Calculate the amount of energy dissipated into heat during 8 hours.

The efficiency of different fuels is a crucial concept that runs through the whole issue of fuel use. See Chapter 18 (page 241) for further details.

Work done with non-constant force

The basic defining equation for work done, on page 47, assumed that the force was constant. In many practical situations forces vary significantly with distance. For example, in order to increase the length of a spring, we need to increase the applied force. As a spring is stretched, the varying amount of work done is stored as elastic energy in the spring.

The extension, x, is defined as the increase in length compared with the original length. **Hooke's law** applies to springs, and states that the extension is proportional to the force, up to a certain limit. Stretching a spring involves using a varying force.

The equation of the straight line in Figure 7 is given by

$$F = kx$$

where

F is the force acting on the spring, measured in newtons (N)
x is the extension of the spring, measured in metres (m)
k is a constant, called the **spring constant**, measured in newtons per metre (N m^{-1})

The varying force means that we cannot just multiply the final force and the extension to calculate the work done. Once again we can use graphical techniques (or the mathematics of integration) to analyse the situation. The graph in Figure 8 represents a force that varies with distance. It is reasonable to assume that, over a very small distance δx the value of the force, F, does not change significantly. The work done over this small distance is the product of F and δx, which is represented by the shaded area of the rectangle.

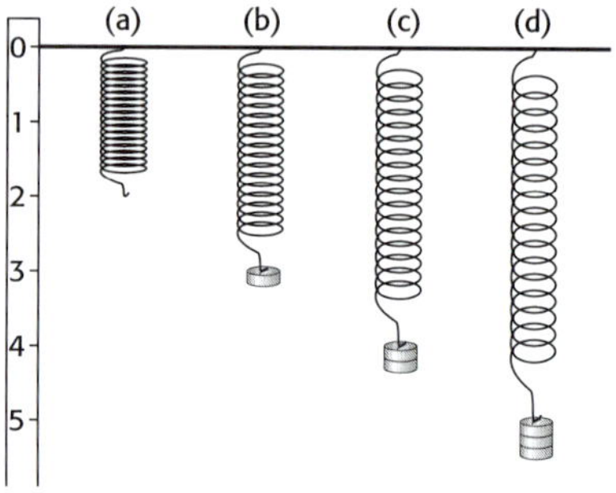

Figure 6 A spring loaded with various masses

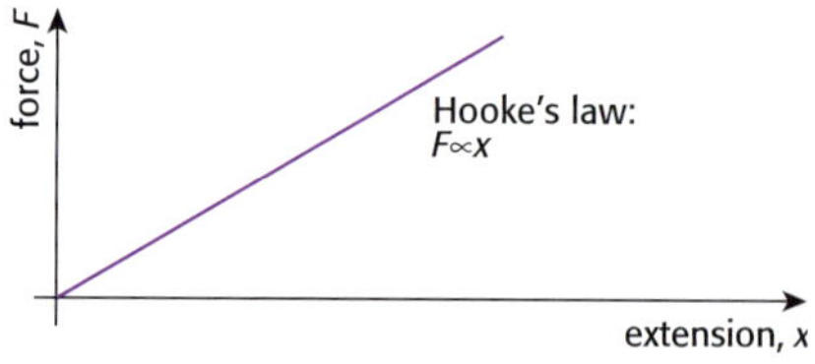

Figure 7 Hooke's Law. Force is proportional to extension

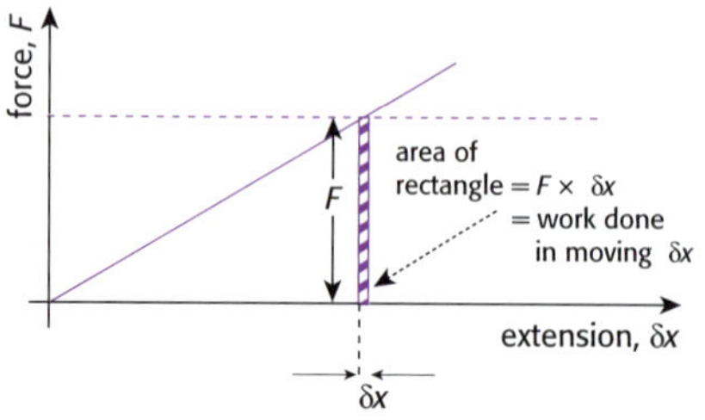

Figure 8 Work done in extending a spring by a small distance

The total work done that results from the varying force is the sum of all the extra small amounts of work done, for the whole distance over which the force acts. This sum is calculated as the area under the graph (Figure 9).

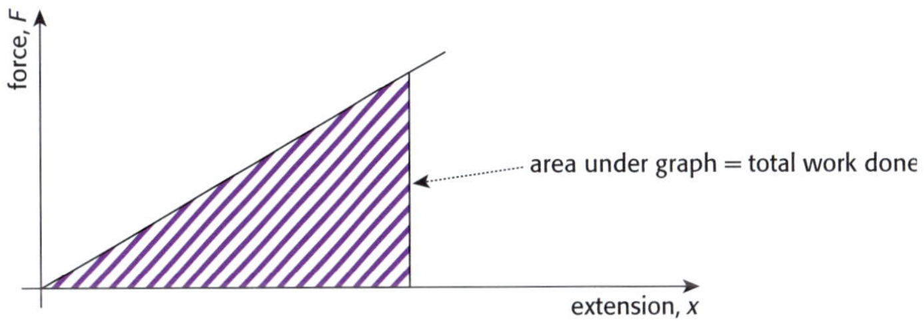

Figure 9 Work done is the area under the graph

$$\text{total work done} = \frac{1}{2}\,Fx = \frac{1}{2}\,kx^2$$

This question analyses the spring shown earlier in Figure 6. The masses used on the spring are 100 g ± 5 g. Take readings from the figure to see whether Hooke's law is verified within experimental error. Use your graph to estimate:

(a) the spring constant
(b) the uncertainty range for your value for the spring constant
(c) the elastic energy stored in any of the photos showing a stretched spring.

Investigating physics: force–extension graphs

In the previous section we assumed that, for any given value of applied force, there was only one possible value for the extension. This is often the case for springs, but there are situations where this is not the case. Careful measurement can reveal a difference between loading and unloading.

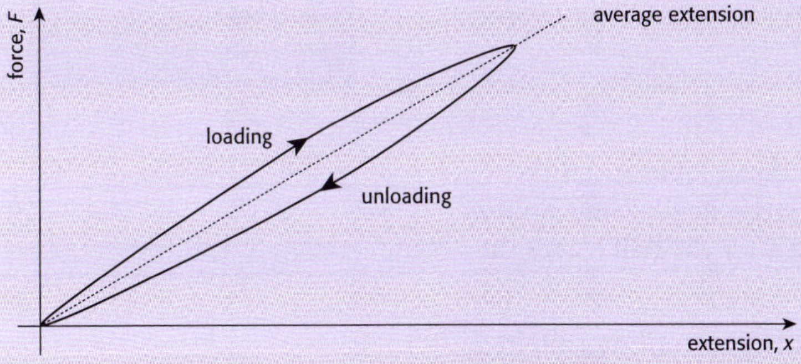

Figure 10 Loading and unloading can give different extensions

The area under the upper curve in Figure 10 represents the work done on the material during loading; the area under the lower curve represents the work done by the material during unloading. The area between the two lines represents the energy stored in the material after one cycle of loading and unloading.

Possible situations to investigate include:

- The extension of a copper wire. What does the shape of the graph of force versus extension tell you about the nature of the forces between copper atoms?
- The extension of a rubber band. What can you deduce about the nature of the forces between molecules of rubber?
- The breaking load for human hair. You could study the effects of different shampoos and/or conditioners.
- The extension of spaghetti and/or noodles to find the ideal cooking time.

Safety note

There are many possible situations and materials that we can investigate under tension. Any material under tension is potentially dangerous. If the material were to break, some of the energy stored would be transferred into KE of the free end: so safety glasses should be worn at all times.

6 Projectiles

HL This chapter focuses on the motion of projectiles in a gravitational field. Although the chapter has been written for HL candidates, SL candidates could study the material as part of the "wider picture".

Going in more than one direction

The effect that any vector quantity has in a particular direction can be quantified by resolving the vector into components. At an angle of θ, a vector V has an effect $V \cos \theta$, as shown in Figure 1.

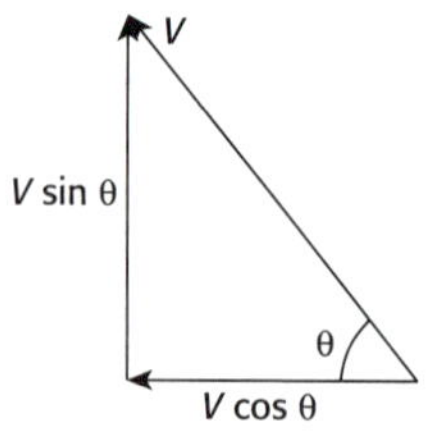

Figure 1 Resolving a vector into components

When θ is 90°, the component at angle θ will be $V \cos(90°) = 0$ (Figure 2). So vectors that are at right angles to one another are completely independent. This means that, for example, horizontal motion and vertical motion can be analysed independently.

Figure 2 The component at 90° is zero

Which **one** of the following is a true statement concerning the vertical component of the velocity and the acceleration of a projectile when it is at its maximum height? (*The acceleration of free-fall is g.*)

	Vertical component of velocity	Acceleration
A	maximum	zero
B	maximum	g
C	zero	zero
D	zero	g

Consider a ball that rolls at constant velocity along a table and then falls over the edge, as shown in Figure 3. Friction is negligible. Can the shape of the path (called a **parabola**) followed after the ball leaves the table be predicted?

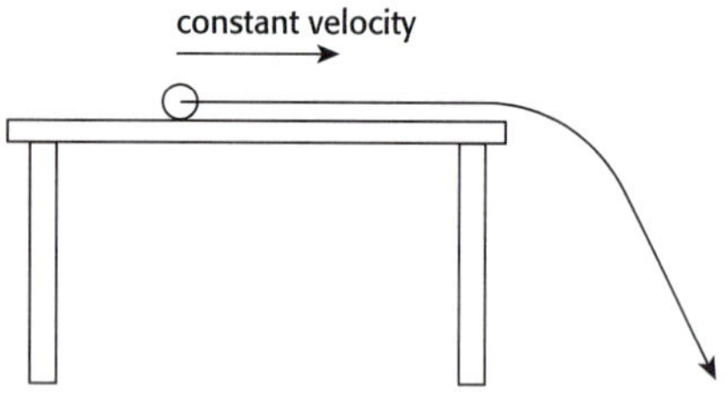

Figure 3

To simplify the analysis, we can first consider the overall motion in the horizontal direction, and then in the vertical direction. When the ball leaves the table, the only thing that changes is that the upward reaction force from the table no longer exists.

When the ball was moving along the table, the resultant horizontal force was zero, and the ball moved equal distances in equal times. When it leaves the table, the resultant horizontal force is still zero, so the ball continues to travel equal horizontal distances in equal times (Figure 4).

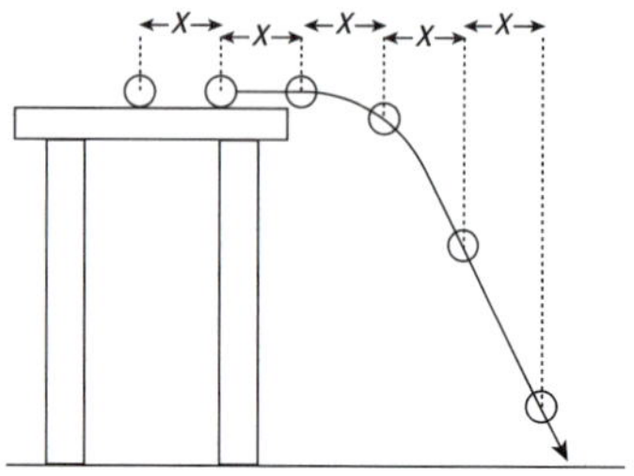

Figure 4 Horizontal motion is constant velocity

A resultant force acts on the ball in the vertical direction when it leaves the table. This means that the ball must accelerate in the vertical direction (Figure 5). The value of the vertical acceleration is g, the free-fall acceleration due to gravity. The vertical distance travelled can be worked out using the constant acceleration formula.

The vertical motion is influenced only by the vertical gravitational force. The time taken for this ball to hit the floor is exactly the same as the time taken for a ball to be dropped from the table vertically down (Figure 6).

It can be helpful to set up the situation and test this statement. If a stationary ball is dropped from the height of the table when a rolling ball leaves the table (this might take a little practice), they will fall together and you will hear only one sound as they both hit the floor together.

Note that the above analysis is valid even if the initial horizontal velocity is extremely large. A bullet fired horizontally from a gun must also follow a parabolic path towards the Earth. The high initial horizontal velocity will mean that the extent to which the bullet falls vertically may be negligible over small distances, but the bullet will still take the same time to fall to the Earth as it takes a bullet dropped from the same height next to the gun.

In the absence of air resistance, the motion of all projectiles must be a constant horizontal velocity and a constant vertical acceleration down. The result of these two independent motions is a parabolic path. The precise shape of the path depends on the initial conditions of the projectile motion.

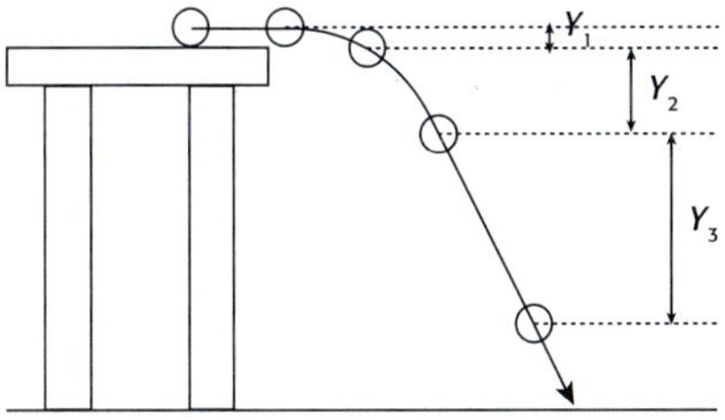

Figure 5 Vertical motion is constant acceleration

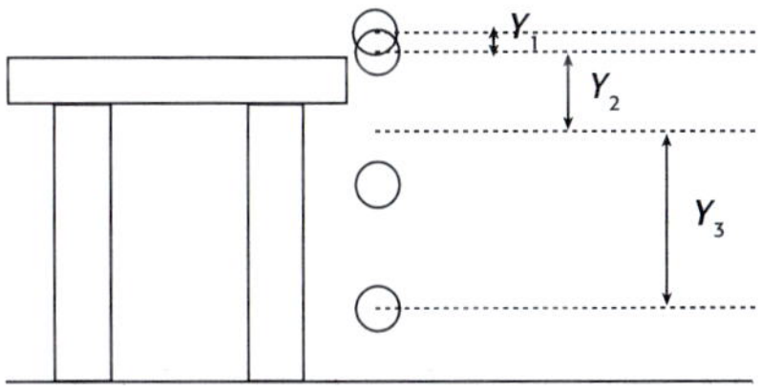

Figure 6 The time taken for vertical motion is independent of the horizontal motion

An object is thrown upwards at an angle of 30° to the horizontal from the top of a 10 m vertical cliff and lands in the sea. The initial velocity of the object is 8 m s^{-1}. See Figure 7.

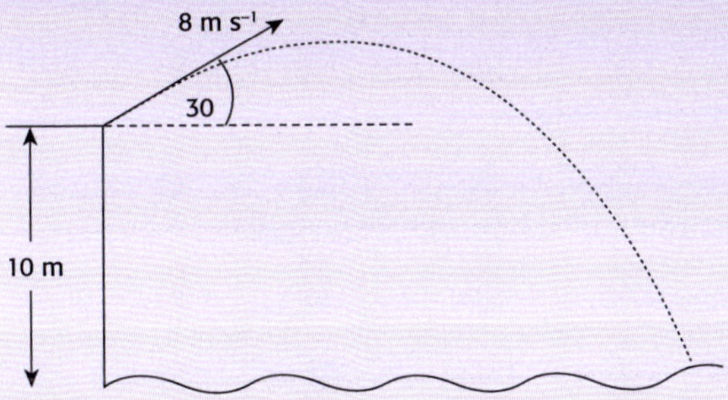

Figure 7

(a) Describe how the velocity of the object varies
   (i) vertically
   (ii) horizontally.
(b) Calculate the time taken
   (i) before the object gets to the highest point in its flight
   (ii) before the object hits the sea.
(c) Calculate the velocity of the object
   (i) at the highest point in its flight
   (ii) when it hits the sea.
(d) Calculate the displacement
   (i) vertically, at the highest point in its flight
   (ii) horizontally, when it hits the sea.

The effect of air resistance

In the above analysis we assumed that frictional effects were negligible. In reality, air resistance will cause any projectile motion to deviate from parabolic motion. Air resistance opposes motion. The magnitude and the direction of the resistive force will vary during an object's flight. The resistive force always opposes the motion, and the magnitude of this force can always be taken to be proportional to the speed.

A full analysis is complex, because the varying resistive force has components in both the horizontal and the vertical directions, but these can be analysed independently. The horizontal component of the resistive force will be proportional to the horizontal velocity, and the vertical component of the resistive force will be proportional to the vertical velocity. One way of modelling the situation would be to use spreadsheets.

Describe qualitatively the changes to the answers you have given to the above question that would result if resistive forces were not negligible.

Overall, the effect of air resistance will be to slow down the object. As the object does work against the resistive force, it will lose energy. See Figure 8.

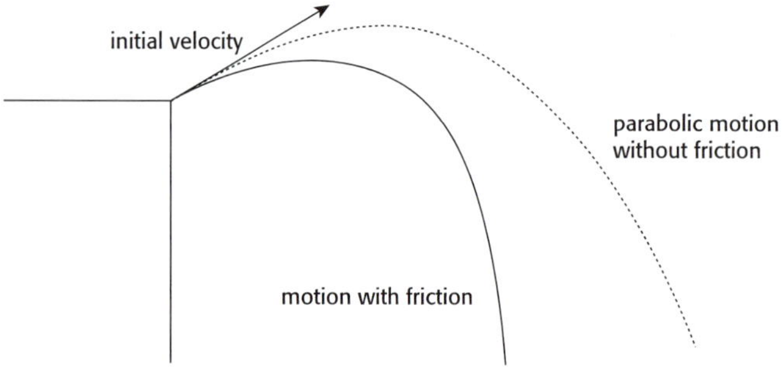

Figure 8 Effect of air resistance (friction) on parabolic motion

Data-based question: basketball shots

Digital video techniques can be used to capture the motion of any object. A good application is the analysis of basketball shots. Once the data has been captured on video, simple software can be used to trace the motion of the ball. The photograph in Figure 9 of the trajectory of a basketball shot was created using a free piece of software called "vidshell", but many others exist.

Figure 9 Parabolic path of a basket ball shot

The basket shot was recorded using a standard digital video camera, and the software was used to mark the position of the ball in each of the pictures, or frames, taken during its flight. Your task is to analyse the photographs below in order to calculate a value for the:

(a) initial velocity of the basket ball as it leaves the girl's hands

(b) acceleration due to gravity.

The following information is available:

- The basket ring is 10 ft (3.05 m) above the ground.
- The time between frames is 0.04 s.

At the time of writing vidshell was available from http://cripe03.rug.ac.be/Vidshell/Vidshell.htm, and useful video clips were available at:

http://www.didaktik.physik.uni-muenchen.de/materialien/inhalt_materialien/videosequenzen/index.html

The same procedure can be used to analyse any type of motion. Many pieces of software allow the user to scale the image, and can automatically calculate the instantaneous displacement, velocity, and acceleration from the captured images. This can be also useful in gaining a good understanding of simple harmonic motion (see page 55).

If you have access to the appropriate equipment, this procedure can allow very detailed investigations to be achieved in a reasonably short amount of time, and would form an excellent basis for an extended essay (see Chapter 25).

7 Simple harmonic motion

The term "oscillation" describes any vibration that goes "back and forth" without an overall resulting movement. It turns out that many real-world situations approximate to one particular type of oscillation, called Simple Harmonic Motion (or SHM for short). Before starting any of the sections in this chapter it would be useful to:

- review circular motion and how to measure angles in radians
- ensure that you understand and can precisely define what is meant by the basic terminology used when considering oscillations. A good definition stands by itself and does not need further explanation. The units that are used to measure a quantity should be clear from the definition. See if you can come up with correct definitions (and units) for the following terms:
  - *displacement*
  - *amplitude*
  - *frequency and angular frequency*
  - *time period*
  - *angular speed*
  - *phase difference.*
- produce a list of some practical examples of oscillations. As you work through this chapter, check to see if all your examples were appropriate
- research formulae for the time period of SHM, particularly a mass on the end of a spring and a pendulum.

Everyday SHM

An object undergoing SHM oscillates about a fixed point. This fixed point is the object's mean position, often called its equilibrium position because it is the point where the object would come to rest if no external forces were acting on it. If displaced away from this point, there must be a **restoring force** on the object. The motion is SHM if this restoring force, and hence the object's acceleration, is always proportional to its displacement from the mean position and always directed towards it.

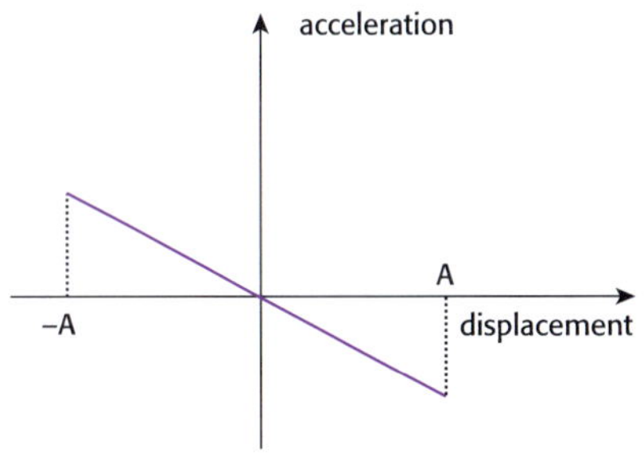

Figure 1 SHM: acceleration is proportional to −displacement

The equation for the resulting displacement oscillation involves a sine function and a graph showing the variation with time of displacement is described as **sinusoidal**:

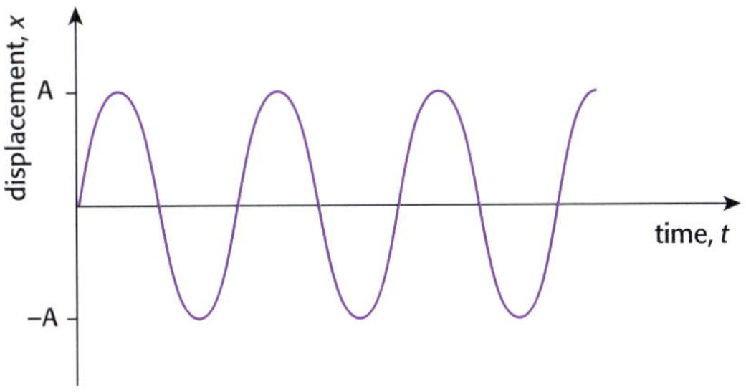

Figure 2

The following situations are some of the many day-to-day examples of SHM that are all around us. Some of these examples can also be considered to be examples of resonance (see page 76). None of these situations are perfect examples of SHM – they are merely approximations to SHM under certain conditions.

Table 1

Situation	Comment
Pendulum	Any suspended object moving from side to side can be considered to be a pendulum. Chandeliers in cathedrals can sometimes be seen to swing back and forth. According to legend, while listening to a Mass in the cathedral of Pisa in 1581, a 17-year-old student noticed something: the chandelier overhead was swaying in the breeze. Sometimes it was barely moving and at other times it was swinging in a wide arc but one complete swing always took the same amount of time (as measured by the student's heart beat). The student was called Galileo Galilei (1564–1642).
Bungee jumping	The fall from the platform is an acceleration due to gravity but once the elastic material has straightened out, the jumper is effectively a mass on the end of a spring and so oscillates vertically. ? Explain why only some parts of the motion can be considered to be SHM. Would it be possible to just bounce back up to the platform?
Diving board	Once the diver has left the board, it will oscillate up and down. The board is one example of a very common situation called a cantilever. It is quite straightforward to build a cantilever in the laboratory to study its motion.
Object bobbing in water	Any object floating in water can bob up and down. If the cross-section going into the water is approximately constant, the oscillations will be SHM. A test tube containing a small mass will float in water and bob up and down with SHM.
Musical instruments	All musical instruments make use of oscillations. For some of these instruments, the cause of the oscillation is obvious, for example, the triangle is hit and vibrates. For some others the musician produces the driving oscillation (e.g. a trumpeter's lips), the amplitude of which is reinforced by the oscillation of the air in the instrument.
Earthquakes	There are different types of waves associated with earthquakes, some of which are SHM waves. Much of the damage (and hence injuries) associated with earthquake disasters are a result of buildings collapsing because they have not been built to withstand the shaking of the ground. See page 81 for further details.

Thinking about science: comparison of two different SHM formulae

Consider three very simple oscillating systems. A mass on the end of a string (the simple pendulum) and a mass on a spring are common basic examples of Simple Harmonic Motion. The bouncing ball is not an example of SHM.

Mathematical analysis is such a powerful tool that these two apparently very different situations can be shown to have aspects of their motion in common. Even though the situations are so different, it is possible to use Newton's laws of motion to derive equations that accurately describe the position,velocity, and acceleration of these objects. Some of the assumptions involved in these derivations will be discussed later.

$$T = 2\pi\sqrt{\frac{m}{k}}$$

Time period of a mass on a spring

$$T = 2\pi\sqrt{\frac{l}{g}}$$

Time period of a pendulur

T time period of the oscillation
m mass of the object on the spring
k spring constant (force per unit extension for the spring)
l length of the pendulum
g gravitational field strength

Perhaps the biggest surprise is that these formulae predict that the time period for each oscillation depends only on two variables. For the spring, m and k fix the time period, T, whereas for the pendulum only l and g are involved. In neither case does the amplitude of the oscillation appear and so we can say that the equations predict that the time period is independent of the amplitude.

The variable g only appears in the equation for the pendulum and is not part of the equation for the mass on the end of the spring. This means that a change of gravitational field strength should not change the frequency of oscillation for a spring. In other words, if a mass on the end of a spring were to be taken to the Moon, the equation predicts that it would oscillate with the same frequency as it did on Earth.

The time period of the pendulum is predicted to be independent of the mass of the pendulum's bob. Do you believe that a 1 m pendulum with a 100 g mass on its end would swing with exactly the same time period as a 1 m pendulum that had a 5 kg mass?

The mass on the end of the spring and the simple pendulum will, for most practical situations, oscillate with a time period that can be correctly predicted by the mathematics of SHM. It is, however, impossible for either of these situations to be anything other than an approximation to SHM.

In both situations the SHM equation arises because we assume that the resultant restoring force on the mass is directly proportional to its displacement from the equilibrium position (see page 58). In mathematical notation:

$$F = -kx$$

where k is a constant of proportionality.

Mass on a spring

The derivation of the equation for the time period, T, for one oscillation

$$T = 2\pi\sqrt{\frac{m}{k}}$$

includes the following assumptions (doubtless there are others):

1 the mass of the spring is negligible compared to the mass of the load
2 friction is negligible
3 Hooke's law applies to the spring at all times
4 the gravitational field strength is constant
5 the fixed end of the spring cannot move.

Some of these assumptions (e.g. number 1) may not apply in particular setups but others (e.g. number 4) will always be reasonable assumptions.

Pendulum

The derivation of

$$T = 2\pi\sqrt{\frac{l}{g}}$$

makes the following assumptions:

1 the mass of the string is negligible compared to the mass of the load
2 friction is negligible
3 the angle of swing is small
4 the gravitational field strength is constant
5 the length of the pendulum is constant during the experiment.

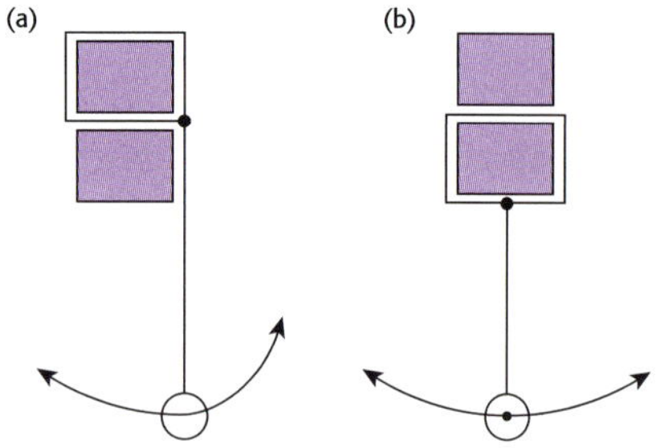

Figure 3 (a) Bad way of attaching a pendulum (b) good way of attaching a pendulum

Although you would probably agree that the fifth assumption is obvious, a surprisingly large number of students fail to think about how to attach the string to the support when setting up their own simple pendulum.

Physics is often about making simplifications in order to create models that aid understanding of real-world situations. A good model makes enough approximations to turn the situation into one that we can analyse, without making so many that we can no longer relate it to the real world.

If we try to apply a model to a situation where the simplifications no longer apply then, not surprisingly, the model will not make accurate predictions. Often a simple model can be adapted and improved to deal with any over-simplifications – see page 73 on damped SHM.

Inquiry: pendulum motion

Design and carry out an investigation to measure the effect that the angle of swing has on the time period of a pendulum. In this investigation, the **independent** variable is the angle of swing and the **dependent** variable is the time period of the pendulum. Other aspects such as the length of the pendulum and the gravitational field strength are the **controlled** variables.

Note: as far as the IB Diploma Programme is concerned, this task would **not** be suitable for assessment against the IB Internal Assessment criteria for the **design** skill as the variables have been given. The experiment can be used to assess DC or CE.

Investigating physics: using a pendulum with SHM to measure *g* and assess the uncertainty limits

In calculations involving the gravitational field strength at the surface of the Earth, g, different values can be used: $g = 10\ \text{N kg}^{-1}$ for rough calculations and $g = 9.81\ \text{N kg}^{-1}$ if we want to be more precise. What exactly is the value of g?

If the Earth were a perfectly uniform sphere with constant density, we would expect g to have the same value anywhere on the Earth. At different places around the world, however, we actually observe some variations. Height above sea level is one obvious factor affecting the value of g but local geological conditions and the effects of the Earth's spin at different latitudes can also be important.

? Non-mathematical communication

Mathematical proof is a very powerful technique but is not always useful when discussing ideas. This question asks you to justify, in two different ways, the following statement:

"A bouncing ball's motion is fundamentally different from the motion of a simple pendulum and of a mass oscillating vertically on a spring."

1 By mathematically analysing the forces acting on each of the oscillating masses, show what the final two situations have in common with each other and explain why the first situation is different.
2 Write a brief article justifying the above statement without the use of any mathematical equations or symbols.

The aim of this experiment is to find an experimental value for g and to compare your own value with other values calculated by students both in your school and, if possible, from around the world.
In this experiment you should fix the length of a simple pendulum, l (the **independent** variable) and measure the time period of the pendulum, T, (the **dependent** variable). These values can be substituted into the SHM formula $T = 2\pi\sqrt{\frac{l}{g}}$ in order to calculate g.

By using the SHM equation in the form

$$g = \frac{4\pi^2 l}{T^2}$$

you can find a more accurate value for g.
This approach will only work if all the assumptions made at the start are valid.

Uncertainty in measurements

All experimental measures are subject to some degree of uncertainty. This means that two students working in the same laboratory are unlikely to calculate exactly the same answer. Suppose one student calculated the value $g = 10.22$ N kg^{-1} whereas the other one got $g = 9.73$ N kg^{-1}. Which one is right? A straight comparison between two values is not going to allow you to decide, nor is a comparison with the accepted value. The student with $g = 9.73$ N kg^{-1} is closer to the accepted value than the other student but this does not mean that she is right.

The existence of uncertainties does not mean that you cannot do good science, but it does mean that when drawing conclusions you must calculate and discuss the uncertainty in your result. The aim is to assign a "confidence level" to any value that results from an experiment. Different experimental techniques result in different confidence levels but if two experiments are supposed to be measuring the same value, then they should agree within their confidence levels.

In this context, if two experimenters find different values of g for the same location on the surface of the Earth then, providing they have not made any mistakes, **new physics must have been discovered**. Unfortunately mistakes are possible (even for a physicist!).

Possible sources of mistakes include:

- reading the scales on the instruments wrongly
- the instruments not being correctly set to zero
- making mistakes in the mathematics
- making mistakes when copying data from one place to another.

For this reason, the scientific community would probably not reject a law just because of a single experiment. The experimental evidence needs to be consistently repeatable.

Assessing the uncertainty from first principles

Typical results could be as follows:
Length of pendulum = 42.1 cm
Time for 10 swings = 12.75 s
The calculation gives $g = 10.22$ N kg^{-1}.

The **uncertainty** for the length of the pendulum needs to take into account the readability of the metre rule used (typically ±0.5 mm) as well as the uncertainty resulting from the string not being perfectly flat, say another ±0.5 mm. The uncertainty in the location of the centre of mass of the pendulum bob, approximately ±2 mm, and the uncertainty in the location of fixing should not be ignored. A total figure of ±3 mm or more would not be unreasonable.

Overall the length uncertainty is thus ±0.3 cm in 42.1 cm = $\frac{0.3}{42.1}$ or ±0.7%.

Uncertainty in (l) = ± 0.7%.

The uncertainty of the timing due to the readability of the stopwatch is ±0.01 s. If the readability was the only source of uncertainty then the error would be an impressive ±0.01 in 12.75 s or ±0.08%. In reality, assessing when the pendulum bob has completed a full oscillation is not that easy and one should also take into account the reaction time of the experimenter. It is, however, possible to practice taking readings and accurately anticipate the end of a swing. An estimation of ± (a quarter of one time period) or 0.32 s would not seem unreasonable as an upper limit.

The uncertainly over time is thus ±0.32 s in 12.75 s or ±2.5%.

$$\begin{aligned} \text{Uncertainty in } (T) &= \pm 2.5\% \\ \therefore \text{ Uncertainty in } (T^2) &= 2 \times \text{Uncertainty in } (T) \\ &= \pm 5.0\% \end{aligned}$$

$$\begin{aligned} \therefore \text{ Uncertainty in } (g) &= \text{Uncertainty in } (T^2) + \text{Uncertainty in } (l) \\ &= \pm(5.0 + 0.7)\% \\ &= \pm 5.7\% \\ &\approx \pm 6\% \end{aligned}$$

6% of 10.22 is 0.61 so the appropriate result to quote for this experiment would be:

$$g = 10.2 \pm 0.6 \text{ N kg}^{-1}$$

The experimental value is now known to be somewhere between 9.6 N kg^{-1} and 10.8 N kg^{-1}. You could, of course, argue that some of the values given above are pessimistic and that the uncertainty is probably somewhat less than the calculated value.

If the other experimenter assessed the uncertainties to be of the same magnitude then her result would be quoted as $g = 9.7 \pm 0.6$ N kg^{-1} so these two students agree that g is somewhere between 9.6 N kg^{-1} and 10.3 N kg^{-1}.

What is the accepted uncertainty in the values normally quoted for g? The way in which we quote the value implies the uncertainty of the measurement. If we state that g = 10 N kg^{-1} this implies an uncertainty of ±1 N kg^{-1}. We are saying that the numerical value for g is 10, not 11 or 9. Writing $g = 9.81$ N kg^{-1} implies an uncertainty of ±0.01 N kg^{-1} (it is 9.81, not 9.82 or 9.80). When the accepted value is quoted to one **significant digit** we write $g = 10 \pm 1$ N kg^{-1} whereas the latter is quoted to three significant digits and thus implies g = 9.81 ± 0.01 N kg^{-1}.

These experiments both agree with the 10 ± 1 N kg^{-1} and with 9.81 ± 0.01 N kg^{-1} but do not worry if your experiment does not agree with the more accurate figure. This figure is the accepted *average* value of *g at sea level*. There is no particular reason for your local value to be the same as a global average but there is every reason to expect all experiments to get the same value in the same location.

Improving the accuracy

The accuracy of this experiment can be improved by both repeating readings for an individual value and by taking a range of different values of the time period for different lengths of pendulum.

Repetitions allow average values to be calculated but they also provide another quick way of assessing the uncertainties.

If, for example, the time for 10 swings was recorded as 12.75 s, 12.84 s, and 12.63 s, then the average value is 12.74. The maximum value was +0.10 s and the minimum value was −0.12 s, so it would be reasonable to conclude that our readings imply that the correct value is 12.74 ± 0.12 s. It should be noted that this method is not statistically precise and can only be used to provide an estimation of the actual uncertainty of the mean.

One way of analysing these results would be to use each reading to calculate a different value for g and then to average the result. This approach would work but a numerical average treats all results equally and does not take into account any assessment of whether an individual reading is subjected to a large or small amount of error. A better method would be to use each reading to construct a graph that allows you to mathematically identify the most precise link between the variables.

In this situation, many students think it helps to plot a graph showing the variation of the time period with length.

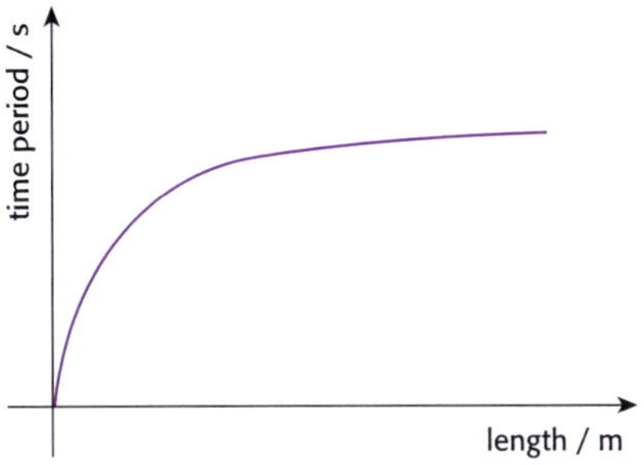

Figure 4 It is hard to see a precise relationship in a graph of this shape

As the best-fit line is a curve, it is not practical to use this plot to assess any formal relationship between time period and length. A better approach is to look at the hypothesised relationship, $T = 2\pi\sqrt{\frac{l}{g}}$, and choose to plot variables (or combinations of variables) in order to produce a straight-line graph.

$$T = 2\pi\sqrt{\frac{l}{g}} \qquad \therefore T^2 = \frac{4\pi^2 l}{g}$$

This equation can be compared with $y = mx + c$.
Thus if we plot T^2 on the y-axis and l on the x-axis, we will get a straight line going through the origin (intercept c = zero) with gradient, m, given by:

$$m = \frac{4\pi^2}{g} \qquad \therefore g = \frac{4\pi^2}{m}$$

In principle, error bars can be constructed for every point and a best-fit straight line can be added as shown on page 16. In practice it is often only necessary to construct error bars for one or two points (often the largest values and/or the point that seems furthest from the line).

Table 2

Summary of the comparison with $y = mx + c$		
dependent variable:	y	T^2
independent variable:	x	l
gradient:	m	$4\pi^2/g$
y-intercept:	c	0

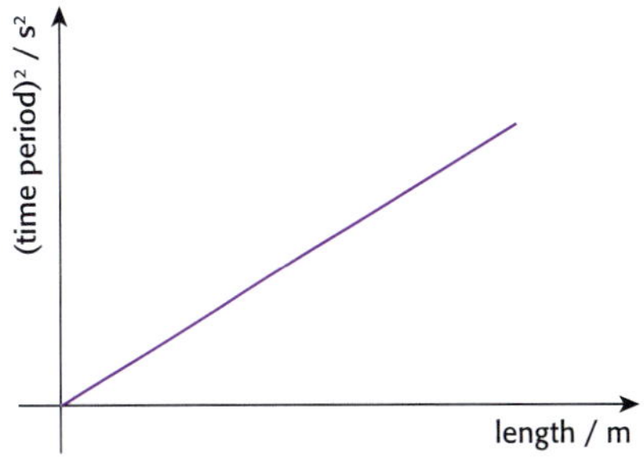

Figure 5 It is clearer from this graph that there is a square root relationship between the two variables

The value for g calculated from the gradient of this line has taken every reading into consideration by creating a visual trend (the straight line) that can be used to identify the actual mathematical link between time period and length.

$$g = \frac{4\pi^2}{\text{gradient of straight line}}$$

This final result is more accurate than any value arising from a single calculation. An assessment of the uncertainty can be achieved by calculating the maximum and minimum values for the gradient of all the straight-line graphs that can be drawn whilst still including every data point.

?

The time taken for a simple pendulum of varying length to undergo 20 oscillations was recorded and used to calculate the average time period for the pendulum as shown in Table 3. Analyse this data to calculate a value for g. The uncertainty in the time period is negligible.

Table 3 The uncertainty in the length measurement is estimated to be ±1 cm.

Length	Time for 20 oscillations				Period
l	T_1	T_2	T_3	T(average)	T
cm	second				
10	12.84	12.91	12.88	12.88	0.64
20	18.35	18.30	18.21	18.29	0.91
30	22.16	22.16	22.23	22.18	1.11
40	25.36	25.48	25.36	25.40	1.27
50	28.63	28.58	28.62	28.61	1.43
60	31.16	31.09	31.13	31.13	1.56
70	33.85	33.97	33.75	33.86	1.69
80	36.07	35.79	36.07	35.98	1.80
90	38.34	38.31	38.03	38.23	1.91
100	40.33	40.30	40.53	40.39	2.02

Inquiry: the scientific method

It is commonly accepted that it is impossible to prove a law experimentally. An experiment can only produce results that are consistent with the law; if this is the case, we say that the experiment **verifies** the law. However, experiments can disprove (or falsify) a law.

Your experiments and those performed by more experienced scientists are equally valid in this regard. Although highly sophisticated equipment has been designed to measure g with great accuracy, it is important to realize that a simple experiment performed by you in the laboratory is just as valid as a highly technical professional setup as long as the uncertainties are specified.

This aspect of the scientific method is sometimes referred to as "empirical falsifiability" and is often used to distinguish scientific theory from non-science. The 20th-century philosopher Karl Popper (1902–1994) is perhaps best known for rejecting the classical observationalist-inductivist account of scientific method and for advancing the principle of falsification in "The Logic of Scientific Discovery", 1934.

Research the terms underlined in the paragraph above so that you could explain them to a non-scientist.

Figure 6 Karl Popper (1902–1994)

Mathematical physics: the "driving" equation for SHM

If any object is oscillating with SHM, then the most complete and succinct way to describe the variation of its motion over time is to use a branch of mathematics called differential calculus. If you find mathematics challenging, fear not – you do not need to understand calculus to be able to do well at IB Diploma Programme physics. It is true, however, that if you choose to study physics after your IB Diploma, then it is extremely useful to have acquired a high level of mathematical skills. If your mathematics is up to the challenge, you should be using it for IB physics.

At time $t = 0$

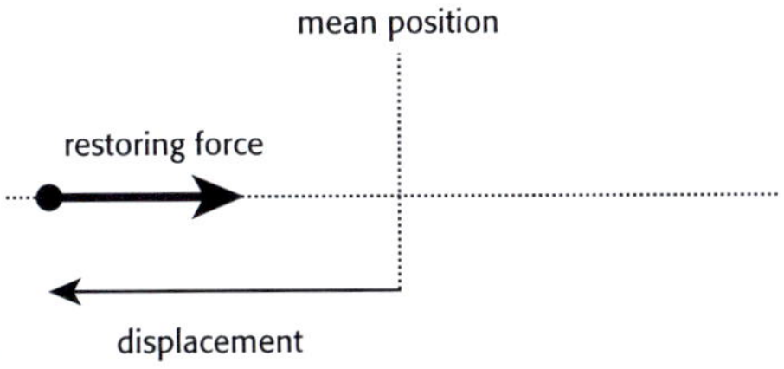

Figure 7

The great benefit of using mathematical analogies is that if a situation can be shown to be an example of SHM, then it is possible to "jump" straight to the known solution without necessarily needing to follow through all of the calculations from first principles.

Perhaps the simplest description of a physical situation that results in SHM is an object oscillating about a mean (average) position. As discussed on page 55, the forces on it are arranged so that the **resultant force** acting on it is always directed towards the mean, or equilibrium position (so it is called a *restoring force*) and is **proportional** to the object's **displacement** from the mean position. Remember: this means that when the displacement from the mean position doubles, the magnitude of the restoring force also doubles.

At a later time, $t = t'$

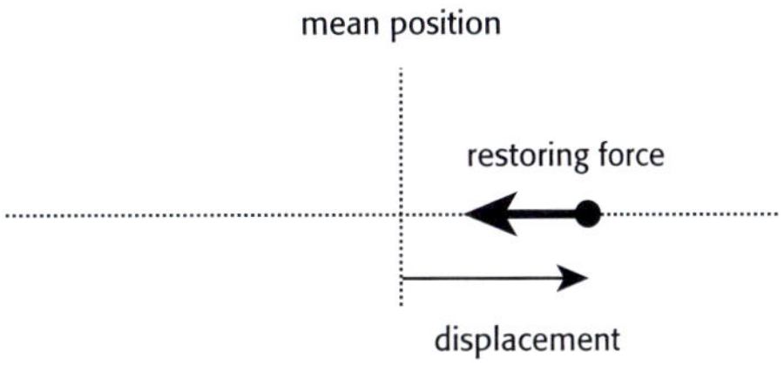

Figure 8

The resulting motion of the object is an oscillation about the mean position.

Mathematically this can be expressed as:

$$\text{resultant force} \propto -\text{displacement}$$

or $$F \propto -x$$

Therefore $$F = -k\,x \qquad (1)$$

Equation (1) defines SHM. The negative sign is important: it is there because the displacement vector and the resultant force acting on the object are always in opposite directions. k is a constant of proportionality and in the case of the mass on a spring, it turns out to be a measure of the force per unit extension.

From Newton's second law of motion we know that an object's acceleration is proportional to the resultant force acting on it:

$$F = ma$$

So we have $$ma = -kx$$

$$a = -\frac{k}{m}x \qquad (2)$$

The term k/m is a constant and is often written as ω^2, where ω is called the angular frequency. The reason for this is discussed on page 68.

Therefore $$a = -\omega^2 x$$

The acceleration is the rate of change of velocity and velocity is the rate of change of displacement. This allows equation (2) to be re-written as a second-order differential equation.

Using calculus to solve the "driving" equation for SHM

Acceleration is the rate of change of velocity (or the rate of change of the rate of change of displacement). Using calculus notation this can be expressed as:

$$\frac{d^2x}{dt^2} = -\frac{k}{m}x \text{ this is sometimes written as } \ddot{x} = -\frac{k}{m}x$$

Since k and m are constants for any given situation, we have an equation whose general form is:

$$\frac{d^2x}{dt^2} = -\text{constant} \times x$$

We must now find a solution to this second-order differential equation. Essentially we wish to identify a mathematical function that, if differentiated twice, yields the same function. In this situation this means deriving a formula for x in terms of t (and the other constants) that satisfies the condition above, and which can be used to work out the value of x at any given time t. The mathematical way of showing that x is some function of t is by writing $x = f(t)$. f is an unknown function and the variables are inside the brackets.

Solutions to the general equation

Luckily there exists a second-order differential equation of this general form that already has known solutions.

The equation $\frac{d^2x}{dt^2} = -\omega^2 x$ has a solution:

$$x = A\sin(\omega t) \qquad (3)$$

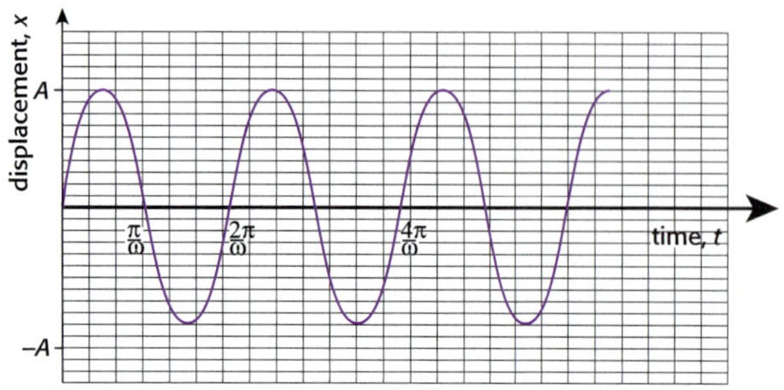

Figure 9 This general shape is called sinusoidal and the pattern repeats with a time period given by $T = \frac{2\pi}{\omega}$.

This equation describes an object that moves back and forth along the x-axis with a maximum displacement A. Figure 9 shows the motion graphically.

If you have studied the calculus of trigonometric functions, it is reasonably straightforward to prove the solution described by equation (3).

differentiate once to obtain $\frac{dx}{dt} = A\omega\cos(\omega t)$

differentiate again to obtain $\frac{d^2x}{dt^2} = -A\omega^2\sin(\omega t)$

Since $x = A\sin(\omega t)$, this final equation can be written as $\frac{d^2x}{dt^2} = -\omega^2 x$

If you have not studied this level of maths, you should still be able to understand SHM from graphs of the displacement, velocity, and acceleration.

How to apply the general solution

By comparing the equation of the general solution

$$\frac{d^2x}{dt^2} = -\omega^2 x$$

with the equation which describes SHM $\frac{d^2x}{dt^2} = -\frac{k}{m}x$

we can see that they become exactly the same equation if we write $\omega^2 = \frac{k}{m}$ or $\omega = \sqrt{\frac{k}{m}}$. As we found earlier, the first equation has a solution $x = A\sin(\omega t)$, so the SHM equation must have a solution given by:

$$x = A\sin\left(\sqrt{\frac{k}{m}} \times t\right).$$

The time period of SHM must also be given by $T = \frac{2\pi}{\omega}$ or $T = 2\pi\sqrt{\frac{m}{k}}$.

$$\text{Time period of SHM} = 2\pi\sqrt{\frac{\text{oscillating mass}}{\text{restoring force per unit displacement}}}$$

Complete solutions

You may already realize that the general solution does not exactly match the example situation at the start of this section. In that case, the displacement of the object was at its maximum when we started timing the motion (x = maximum when $t = 0$). The solution above has zero displacement when $t = 0$. Has something gone wrong?

The equation that we have found is only one of a general set of solutions possible. You can see from the way the graph regularly repeats itself that the place where we chose to define $t = 0$ was an arbitrary one. There are an infinite number of other similar solutions possible, providing that they are of the same shape as this graph. These other solutions are, in effect, the same shape graph moved one way or another along the time axis. For example, $x = A\cos(\omega t)$ must also be a solution to the SHM equation because this has the same shape as $x = A\sin(\omega t)$ but is shifted along the time axis by a quarter of the period of the graph.

One way of writing the very general solution is $x = A\cos(\omega t + \phi)$. The variable ϕ is dependent on the initial conditions and is known as the **phase angle**, or in other words, the point where we start timing the motion. So changing its value simply has the effect of moving the graph along the time axis. The equation for the time period of SHM is, of course, unaffected by a shift along the time axis and thus does not need to be modified.

Thinking about science: the links between SHM and circular motion

Any motion involving oscillations can be analysed from first principles. If the mathematical description of the oscillations can be rearranged and expressed in the form $a = -\text{constant} \times x$ (often written as $a = -\omega^2 x$), then the oscillation under consideration must be SHM.

Once we know this to be the case we know that the general solution must take the form $x = A\cos(\omega t + \phi)$. But what do the variables, A, ω, and ϕ mean, and how do we find their values?

The numerical value for the angular frequency, ω, comes from the equations used in the analysis of the original situation. It is related to the frequency of the oscillation:

$$\omega = 2\pi f$$

The values for A (the amplitude) and ϕ (the phase) also need to be deduced for the situation being considered and it would help to understand what they physically represent.

We can exploit the fact that the mathematics of circular motion is very similar to the mathematics of SHM. This allows us to use our understanding of the physics of circular motion to help build a mathematical interpretation for what ω, A, and ϕ represent in the equations of SHM. As an example, two very different situations are compared below.

In the left-hand column, we are considering an object that is going around in a circle *with constant angular speed*, and in the right-hand column we are considering a mass on the end of a rigid spring. The equations are exactly equivalent but remember that some of the symbols mean slightly different things in the two situations.

Table 4

Circular motion	Symbol	SHM of mass on spring
The time elapsed from the starting conditions	t	The time elapsed from the starting conditions
The component of the displacement in the x-direction	x	The displacement from the mean position
The radius of the circle	A	The maximum displacement (the amplitude)
The angular speed (must be constant)	ω	A constant defined by $\omega = \sqrt{\frac{k}{m}}$
The initial angle when $t = 0$	ϕ	A constant fixed by initial displacement when $t = 0$

Circular motion

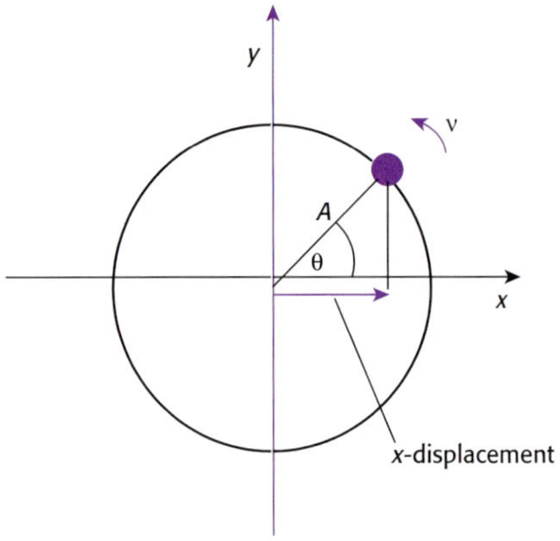

Figure 10 A mass going around in a horizontal circle of radius A at constant speed v

In this situation, the mass is moving in two dimensions.

The time period of the motion can be calculated from the time it takes to complete the circle:

$$T = \frac{2\pi A}{v}$$

Another way of expressing this is to use the concept of the angular velocity of the mass, ω. (see page 65).

$$T = \frac{2\pi}{\omega}$$

If we resolve the motion of the mass into the x and y directions, the x-component of the mass's displacement from the centre is given by:

$$x = A\cos(\theta)$$

Another way of expressing this is:

$$x = A\cos(\omega t)$$

In this example the mass was assumed to have started its motion on the x-axis. We could have started timing from anywhere so the general formula for the x-component of displacement is:

$$x = A\cos(\omega t + \phi)$$

The angle ϕ depends on where the mass is when $t = 0$.

SHM of mass on a stiff spring

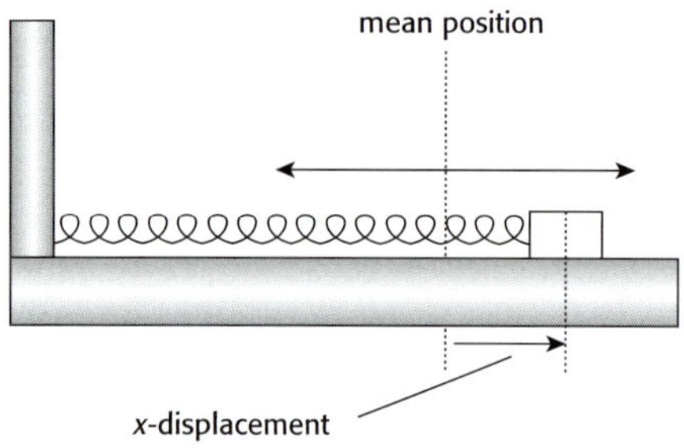

Figure 11 A mass on a frictionless surface attached to a stiff spring undergoing horizontal SHM

In this situation, the mass is moving in one dimension.

The time period of the motion can be calculated from the SHM equation:

$$T = 2\pi\sqrt{\frac{m}{k}}$$

Another way of expressing this is to substitute using $\omega = \sqrt{\frac{k}{m}}$ (see page 65).

$$T = \frac{2\pi}{\omega}$$

Since the motion is known to be SHM, the formula for its displacement is:

$$x = A\cos(\omega t)$$

In this example the mass was assumed to have started its motion at the far right-hand side. We could have started timing from anywhere so the general formula for displacement is:

$$x = A\cos(\omega t + \phi)$$

The angle ϕ depends on where the mass is when $t = 0$.

In both situations x represents the displacement in the x-direction and if we concentrate on the motion of the mass moving in a circle projected onto this axis only, it is identical to Simple Harmonic Motion. In other words, there is no difference at all between the two situations as far as the x-direction is concerned. The equation of the mass on the end of the spring is exactly the same as the equation of the mass moving in a circle so long as we only consider the motion in the x-direction and ignore any extra motion in the y-direction. If you look at the shadow of an object moving in a circle, it will look like mass oscillating on a spring.

Since the two motions are exactly equivalent, any algebraic manipulation that applies to circular motion must also apply to SHM. The only difference between the motion of two objects is that the one moving in a circle is also moving in the y-direction at the same time.

By applying Pythagoras's theorem to the circular motion situation you should be able to show that the x-component of the object's velocity is given by

$$v = \pm\omega\sqrt{A^2 - x^2}.$$

If this works for circular motion, it must also work for SHM.

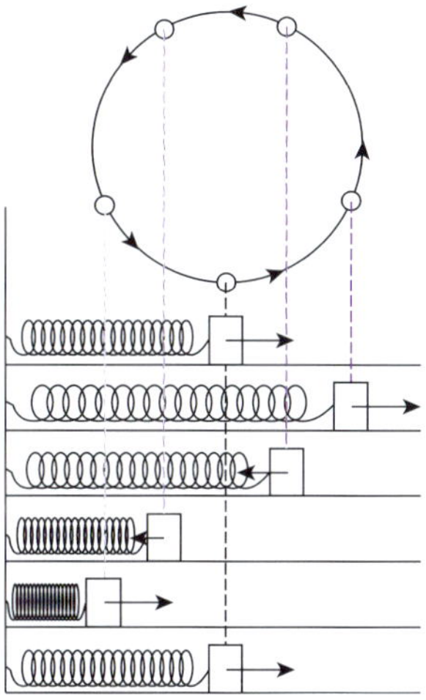

Figure 12

Using a graphic display calculator to analyse SHM

A graphic display calculator (GDC) can be used for modelling, data logging or equation analysis.

Modelling

A graphic display calculator, personal computer, or any device with spreadsheets can be used to model SHM using the iterative method introduced on page 29. An outline of the general procedure for modelling the motion of an oscillating mass is given below. In this method we divide the motion into intervals, with each interval representing a fixed period of time (for example 0.1 s or 1 s, depending on the overall time period).

- At any instant of time, the resultant force acting on the oscillating mass is related to the displacement of the mass ($F = -kx$).
- This instantaneous force can be used to calculate the instantaneous acceleration ($F = ma$).
- The change in velocity can be calculated by assuming that the acceleration remains approximately constant over the time interval ($\Delta v = a\Delta t$).
- The initial velocity and the final velocity over the time interval can be used to calculate the average velocity $\left(v_{\text{average}} = \dfrac{u+v}{2}\right)$.
- The average velocity over the time interval can be used to calculate the change in displacement ($\Delta x = v_{\text{average}}\Delta t$).
- The change in displacement can be used to calculate the new displacement at the end of the time interval.
- The whole process can be repeated with the new displacement.

The procedure for using a GDC is outlined here.

1 On the spreadsheet, enter appropriate values for all of the constants:
   - oscillating mass
   - restoring force per unit displacement (spring constant)
   - time interval
   - initial displacement
   - initial velocity.

2 Add headings for the following columns on the spreadsheet:
   - total time elapsed
   - displacement at beginning of time interval
   - instantaneous force
   - instantaneous acceleration
   - initial velocity
   - change in velocity
   - final velocity
   - average velocity
   - change in displacement.

3 Add equations to calculate these values and fill the equations down the sheet.

4 Use the graphing ability of the GDC to display graphs that show the variation with time of:
   - displacement
   - velocity
   - acceleration.

5 The spreadsheet can be modified to answer the following questions:
   - What happens when the restoring force is assumed to be proportional to x^2 rather than x?
   - What happens when frictional forces act on the oscillating mass?

Data logging

If you have position data sensors available to connect to a graphic display calculator, it should be very simple to experimentally log data that records the variation with time of the displacement of a mass undergoing SHM.

A setup such as the one in Figure 13 should not only yield a plot of the variation of displacement with time, but from it you will be able to calculate and plot the variation of velocity and the variation of acceleration with time. There are, however, many adjustments that you will have to make before the graphs plotted by the calculator are useful.

- The position sensor measures the total distance from the sensor to the oscillating mass. Work out the mean position of the mass from the graph.
- Subtract the mean position of the mass from each reading. The displacement graph should then be sinusoidal with a mean value of zero.
- The displacement graph can be used to work out the amplitude, period, and phase shift of the oscillation.
- Use these three pieces of data and the calculator to define your best-fit line through the data. You know the graph should turn out to be sinusoidal so use this option and switch, if you have been working in degrees, to radians.
- The calculator will also be able to work out and plot the variation of the velocity of the oscillating mass. Try to predict the shape of the graph before you look at the answer. You will probably have to rescale the axes in order to view the velocity graph properly.
- At what position do you predict the velocity of the mass will be a maximum? Where is the mass when its velocity is zero?
- Find a mathematical equation for the variation of the velocity with time.

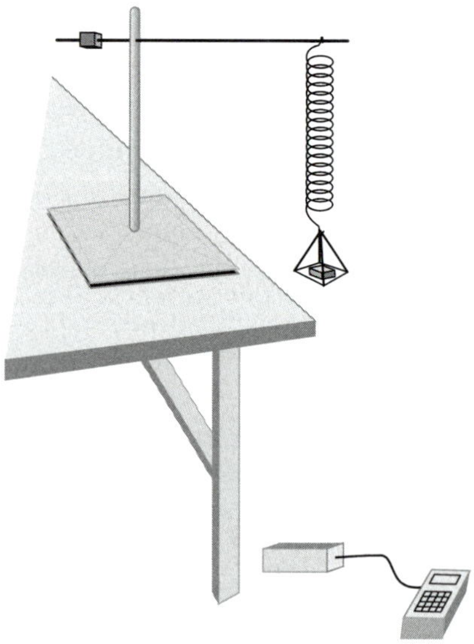

Figure 13 Possible set-up of a position sensor to record SHM

- Finally, the variation in acceleration can also be displayed. What relationship do you expect to see between the displacement graph and the acceleration graph?
- If you know the mass of the oscillating object, you can use the equations that you have discovered to calculate the value for the spring constant.

Equation analysis

A graphic display calculator is an extremely powerful tool for analysing experimental data, but it is also possible to gain a good understanding of SHM by using it to manipulate the basic equations. Try the following exercise on your graphic display calculator.

1. Plot a graph of $y = A\sin(\omega x)$ with $A = 2$ and $\omega = 0.05$. The x range can go from 100 to –100 and the calculator needs to be set to work in radians (see page 13).
2. The GDC should display several complete oscillations of a sinusoidally varying graph. The y value should vary between 2 and –2.
3. Move the curser around the graph and use the values displayed to measure the time period. Does this value agree with the value predicted using $T = \frac{2\pi}{\omega}$?
4. What difference does it make to your graph if you add in a phase angle ϕ to the formula? Try different values of ϕ: π, $\pi/2$, $\pi/4$, and so on.
5. Use the calculator to plot a tangent to your displacement graph for a value of time that you choose. The equation of the tangent can also be displayed.
6. Since it is a straight line, the equation of the tangent will be in the form $y = mx + c$. The value of m is the gradient of the tangent.
7. We are plotting a graph of displacement against time, so the gradient of the tangent to any point on this graph will equal the instantaneous gradient, and therefore the velocity at this time. Select several different positions for plotting the tangent and record the different velocities that these represent.
8. Where is the velocity at its maximum and where is it zero?
9. It should be possible to get the calculator to plot a second graph to represent the variation of the velocity with time. In the mathematical language of calculus, we want to graph the derivative of the first function. The notation for doing this calculation varies among the different makes of calculator. Texas Instruments use the following notation for their TI-83 calculator for graphing the derivative: *nDeriv(function,X,X)*. In this case the second graph to be plotted would be Y2 = *nDeriv*(2 × sin(0.05 × *X*),*X*,*X*).
10. Once the graphs have been rescaled and you can identify which plot represents velocity and which plot represents displacement, you should check to see if you understand the shapes that have been plotted. What is the velocity when the displacement is zero? What is the velocity when the displacement is at its maximum?
11. The derivative function can be used again to display the acceleration. What is the relationship between the acceleration graph and the displacement graph?

Investigating energy in SHM using a graphic display calculator

The aim of this section is to predict how the energy of the oscillating mass varies during the cycle. If you do not have access to a GDC, the graphs can still be plotted (it will just take longer). We will consider the variation of the KE and PE for an object moving with a displacement given by:

$$\text{displacement} = A \sin(\omega t)$$

Kinetic energy

The motion of the mass is described by SHM so we know that its velocity can be represented by:

$$\text{velocity} = A\omega \cos(\omega t)$$

A mass, m, moving at a velocity, v, has a kinetic energy given by $\text{KE} = \frac{1}{2}mv^2$. So if we substitute the SHM equation for velocity into this equation we get:

$$\text{KE} = \frac{1}{2}m(A\omega \cos(\omega t))^2$$

or

$$\text{KE} = \frac{1}{2}mA^2\omega^2 \cos^2(\omega t)$$

In order to visualize this variation (a "cos-squared" variation), try plotting the following graphs on your GDC.

1 Plot graphs of the displacement $y = A \sin(\omega x)$ and velocity $\frac{dy}{dt} = A\omega \cos(\omega t)$ with $A = 2$, $\omega = 0.8$. The x range can go from –10 to +10 in steps of 0.01 and the calculator must be set to work in radians. You should be able to see several complete oscillations.

2 Now plot the KE equation $\text{KE} = \frac{1}{2}mA^2\omega^2 \cos^2(\omega t)$ using 0.7 as the value for mass.

3 Your GDC should display the KE variation shown below:

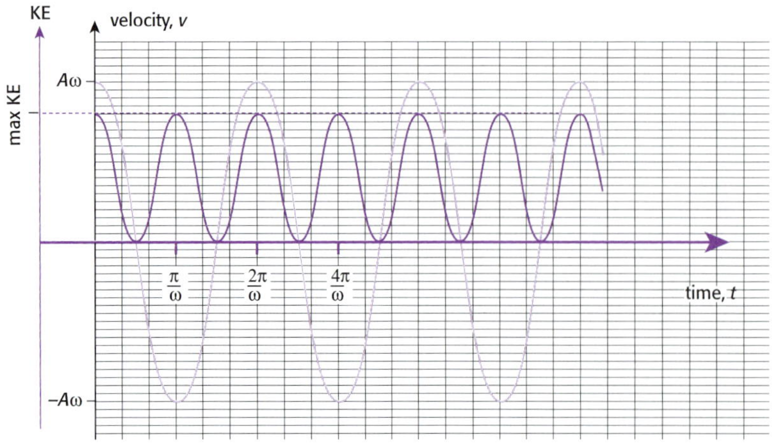

Figure 14 Variation of KE and its relationship with velocity in SHM

Potential energy

In this section we will analyse several different graphs. Remember that as we consider different graphs on the GDC the y-axis represents different variables.

As the mass is moving, the change in PE must always be equal to the total work done. We know that if a force F is moved through a distance x, then the work done can be calculated from:

$$\text{work done} = \text{force} \times \text{distance}$$

However, in this situation we cannot simply multiply the value of the force by the value of the displacement because the magnitude of the force on the object is constantly changing. At any given displacement, the magnitude of the force acting on the object depends on x:

$$F(x) = k\,x$$

This means that the extra little bit of work done in moving a small distance δx is given by:

$$\text{extra work done} = F(x) \times \delta x$$

This is equivalent to using a graph and working out the area of the section of the graph between the line and the x-axis.

The total work done between any two points (often called the lower limit and the upper limit) is the addition of all the "extra little bits" of work done during the motion. This work done is equal to the extra potential energy stored when moving between these two points. The way of adding up all the extra little bits of work done is to calculate the area under the variation of force with distance graph. The mathematical technique for working out the area under a graph is called integration and is written like this:

$$\text{PE stored} = \text{total work done} = \int F(x)\,dx$$

Since the magnitude of $F(x) = k\,x$ we can deduce that the equation for the PE of a mass moving with SHM is:

$$\text{PE} = k \int x\,dx$$

If you have studied calculus you might recognize the solution to this equation:

$$\text{PE} = \frac{1}{2}kx^2$$

We can now complete this substitution, and you can also verify this formula using the GDC. As before, the calculator must be set to work in radians and we will use the same constants as before: $A = 2$, $\omega = 0.8$, and $m = 0.7$.

- Use the GDC to plot a graph of the variation of force with displacement. The graph to plot is $y = kx$ or $y = m\omega^2 x$ ($\omega^2 = k/m$ – see page 64)
- Use the "calculate" function to work out the area under the graph between the origin and a range of different values of x.

Thinking about science

Consider the following points.

- Since the velocity has been squared, the kinetic energy is always positive.
- The axes of Figure 14 have a different scale to the ones that you have plotted on the GDC, but shape of the graphs must be the same.
- Do the maximum and minimum values for the KE agree with the variation of the velocity?

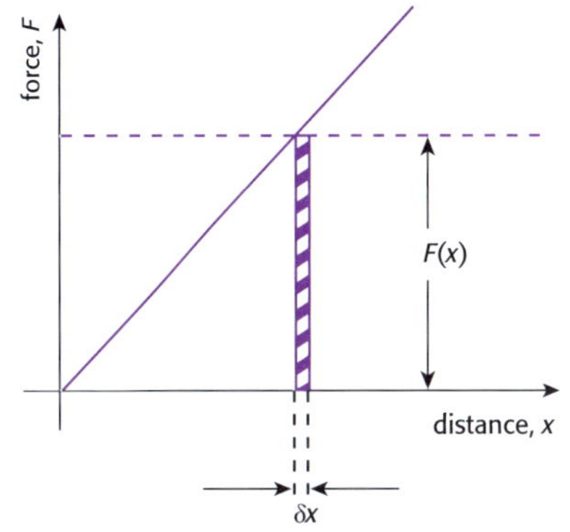

Figure 15 Extra work done is the area under the graph

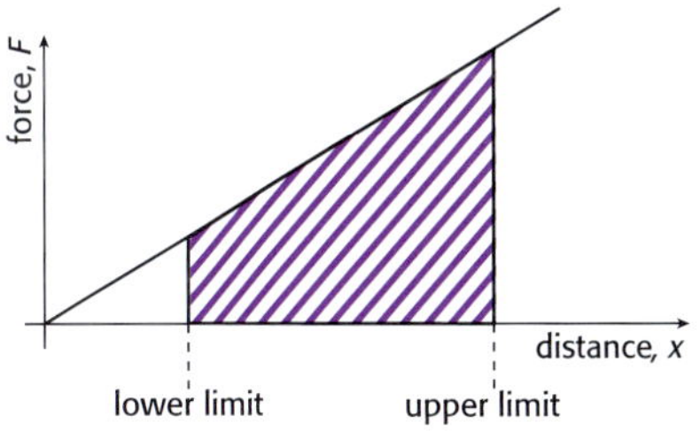

Figure 16 Work done is the area under the graph

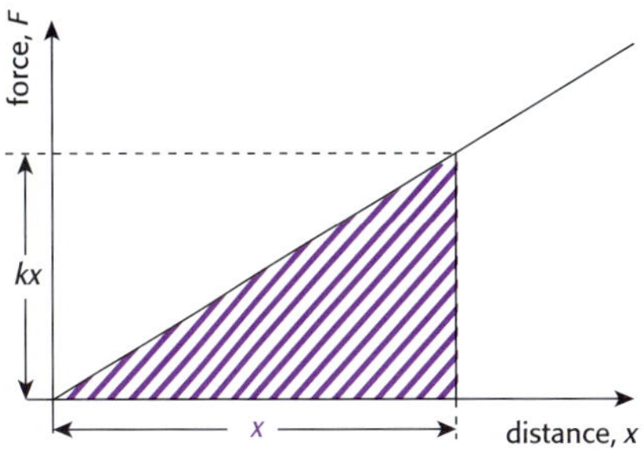

Figure 17 PE gained is the area under the graph

- The calculator can probably also display a graph of the integral of $y = kx$. The Texas instruments notation for this is *fnInt(function,X,0,X)*, so the equation to enter is Y2 = fnInt($m\omega^2x$, *X,0,X*).
- Check the shape of this graph. Is it of the form $Y_2 = \frac{1}{2}kx^2$?

  Compare the individual values that you found earlier with the values on this new graph. They should be equal.

The two equations that we need to combine are:

$$PE = \frac{1}{2}kx^2 \text{ and displacement } x = A\sin(\omega t)$$

- Use the GDC to plot the graph $y = \frac{1}{2}m\omega^2(A\sin(\omega t))^2$.
- Is the shape of this graph the one that you expected?

As the mass is moving under SHM we know that its velocity can be represented by:

$$\text{velocity, } \frac{dx}{dt} = A\omega\cos(\omega t)$$

- A mass, *m*, moving with a velocity, *v*, has a KE given by $KE = \frac{1}{2}mv^2$.

  So if we substitute the first equation for the velocity into the second equation for KE, this results in the equation $KE = \frac{1}{2}mA^2\omega^2\cos^2(\omega t)$. Use the GDC to plot this graph.
- Is the shape of this graph the one that you expected? How does this compare with the PE graph?

Total energy

The above two sections have derived formulae for the KE and PE of an object doing SHM. The formulae are very similar:

$$PE = \frac{1}{2}mA^2\omega^2\sin^2(\omega t)$$

$$KE = \frac{1}{2}mA^2\omega^2\cos^2(\omega t)$$

The total energy at any given time is just the addition of PE and KE. Using your graphic display calculator add these two functions together and comment on your answer.

?

Show algebraically why the result is:

$$\text{Total energy} = \frac{1}{2}mA^2\omega^2$$

The bigger picture: Fourier analysis

The creation and propagation of a sound wave involves oscillations. So far we have only considered SHM but most sound waves (apart from when you whistle) are not sinusoidal – they involve more complicated oscillations. The shape of sound waves made by a musical instrument or everyday speech seem so far away from being perfectly sinusoidal that it is tempting to think that the mathematics of SHM will not be of any use in these situations.

Figure 18 Some different musical instruments

Starting with a simple example does help to understand the more complicated ones. Complicated oscillations can be shown to be the sum of lots of SHM oscillations of different frequencies and different amplitudes. You might like to revise the ideas of superposition of waves before reading more of this section.

In the language of music a single note being played by one instrument is described as having a fundamental frequency (the main note that you hear – also called the first harmonic) along with other frequencies, called harmonics, which give this note its particular characteristic.

If a violin and a flute both played the same note, the two individual sound waves would have the same fundamental frequency but each instrument would also have different harmonic frequencies of different amplitudes. It is the "additional" harmonic frequencies that make a violin sound different to a flute. You can recognise different people's voices from their own unique combination of harmonics. Musicians refer to the sound of a note as its *quality*, or *timbre*.

It is possible to analyse a complicated sound wave and work out what component frequencies (or harmonics) exist. The mathematics of frequency analysis is quite complicated but the result will always be a series of sine and cosine functions of different amplitudes. For many waveforms the (sometimes infinite) series has already been calculated and is called a Fourier series after the French physicist and mathematician Joseph Fourier.

Equipment designed to perform real-time frequency analysis of sounds is generally very expensive and only widely used in research laboratories and sound recording studios. These days, however, software is available which allows sound files held on a computer to be analysed and some can perform the analysis directly from a microphone input. There are even versions that are available free on the Internet.

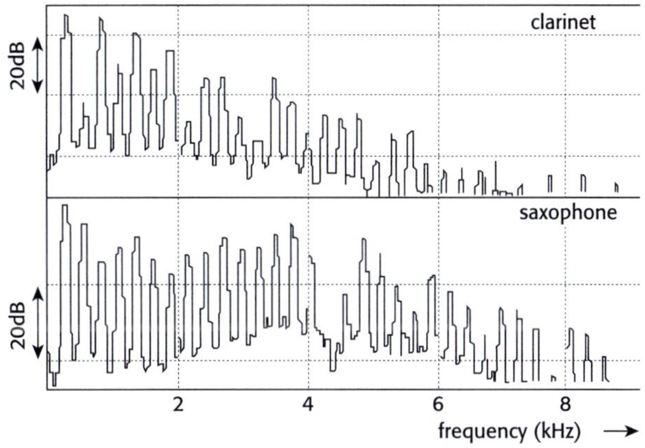

Figure 19 The same note played by a clarinet and a saxophone

If a sound wave can be resolved into its component frequencies, is it possible to create different sounds by adding together different component frequencies? Of course, the answer is yes and this is the basic principle behind the instrument called a synthesizer. The sounds of different instruments are created by combining different sine waves. The Internet is, once again, a good place to locate demonstrations of how Fourier-Synthesis can work (see http://www.phy.ntnu.edu.tw/ntnujava/viewtopic.php?t=33 and also http://www.zeitnitz.de/Christian/Scope/Scope_en.html)

The first four terms for the Fourier series for a square wave are:

$$y = A\sin(x) + \frac{A}{3}\sin(3x) + \frac{A}{5}\sin(5x) + \frac{A}{7}\sin(7x)$$

?

Use a GDC to plot this function and to explore to what extent the addition of extra terms alters the shape of the resulting wave.

Damping

Until now we have been considering an SHM system in isolation and we have not considered possible energy losses from the system. But energy loss does happen whenever friction acts on the moving mass (slowing the object down) and is called energy dissipation or **damping**.

When an object is oscillating with SHM its total energy is constant and depends on three things: the object's mass, the initial amplitude of the oscillation, and the angular frequency of the oscillation (which is proportional to its frequency). Note that the total energy is not proportional to its amplitude but proportional to the square of the amplitude. This is an important relationship and it not only applies to mechanical objects undergoing SHM but also to any oscillation that can be described using the SHM formulae and in many other areas of physics (for example, wave–particle duality on page 211). The reasons are shown below.

The total energy of an object moving with SHM can be calculated by adding its potential energy and its kinetic energy together. Mathematically:

$$\text{PE} = \frac{1}{2}mA^2\omega^2\sin^2(\omega t)$$

$$\text{KE} = \frac{1}{2}mA^2\omega^2\cos^2(\omega t)$$

$$\therefore \text{total energy} = \frac{1}{2}mA^2\omega^2 \qquad (\text{since } \sin^2\omega t + \cos^2\omega t = 1)$$

$$\therefore \text{total energy} \propto A^2$$

If we add a resistive (or damping) force to the Simple Harmonic Motion the general "driving" equation for the oscillation will change. The damping (frictional) force usually depends on the velocity with which the object is moving so we can write:

$$\text{Damping force} \propto -\text{velocity}$$

or

$$\text{damping force, } F = -k_{\text{damping}}\frac{\mathrm{d}x}{\mathrm{d}t}$$

The negative sign is there because the force resists the motion so it must be acting in the opposite direction to the velocity. This situation results in the following differential equation to solve:

$$\frac{\mathrm{d}^2x}{\mathrm{d}t^2} = \frac{-\left(k_{\text{spring}}\,x + k_{\text{damping}}\dfrac{\mathrm{d}x}{\mathrm{d}t}\right)}{m} \qquad (4)$$

where k_{spring} and k_{damping} are different constants and are particular to the practical setup under consideration. Solving this is well beyond the scope of the IB Diploma Programme physics course, However, it is possible to predict the effect of light damping on an object by considering the energy equations again.

The energy of an object moving under SHM is proportional to the (amplitude) ^{2}. If work is being done by the system against a resistive force, then the amplitude of the oscillation must fall. Since the time period of the oscillation is independent of the amplitude, the time period, the frequency of the oscillation must be unaffected. Figure 20 shows the variation of displacement with time for **lightly damped SHM**.

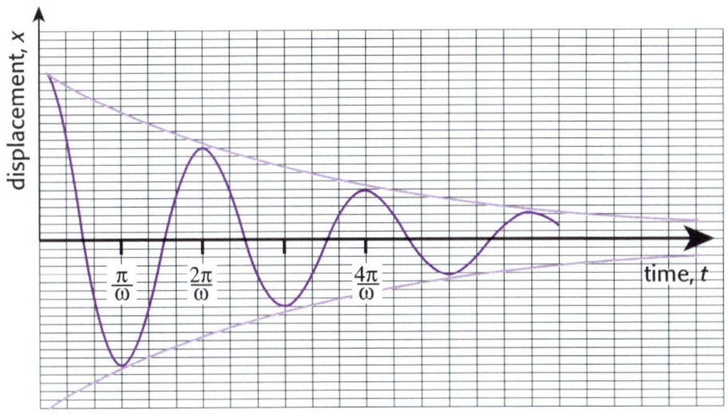

Figure 20 Lightly damped SHM

Use readings from the graph to show that the amplitude is decreasing exponentially.

The amplitude of the oscillations can be shown to decrease exponentially with time as the system loses energy by doing work against the resistive force.

An exponential decrease of the amplitude only happens if the damping is "light" – that is to say that the frictional force removes only a small portion of the total available energy during one oscillation, e.g. a pendulum moving through air. The air resistance acting on the pendulum bob would cause the amplitude of the oscillations to decrease over time but the time period remains constant.

If, however, the pendulum were to be set up not in air but in viscous oil, the damping force would be so great that oscillations could not take place. In this situation, called **heavy damping** (or even "heavy over-damping"), a pendulum displaced and released would just slowly make its way back to the mean position (Figure 21).

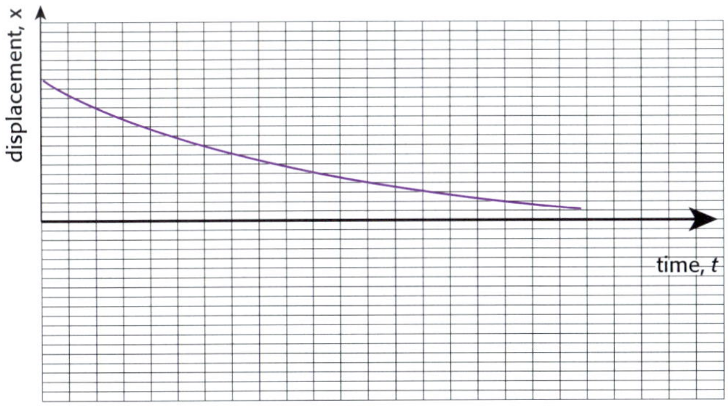

Figure 21 Heavily damped SHM

There are many engineering situations where neither light damping nor heavy damping is desirable. An example of one such situation is the suspension fitted to a car. In all cars, the passengers travel in a compartment that is joined to the wheels of a car by heavy-duty springs.

One of the aims of these springs is to give the driver and passengers a comfortable ride (another is to improve road-holding). If the damping is too light then whenever the car goes over a bump the car will start oscillating up and down with SHM giving a rather unpleasant ride. If the damping is too heavy then the car behaves as though there was no damping at all and the driver feels every bump.

It is possible to arrange for a system to be "critically damped". This describes an arrangement where the damping is somewhere between light and heavy so as to make the oscillations "die out" in the shortest time possible. When this happens, the system does not "overshoot" before returning to zero displacement (see Figure 22).

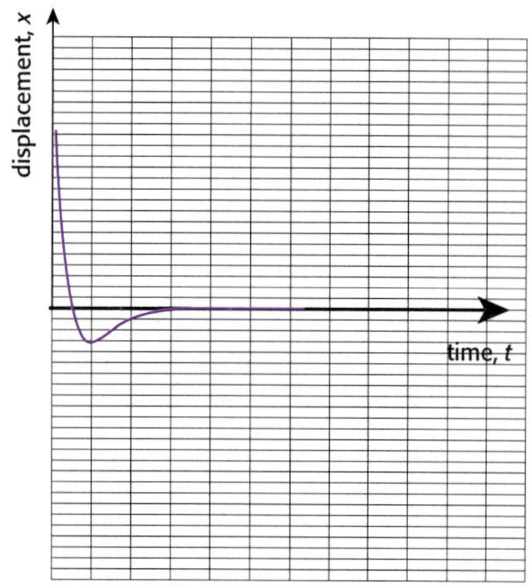

Figure 22 Critical damping

It is possible to use the iterative method outlined on page 29 to investigate damping. The equation for overall acceleration (equation 4) can be inserted into the program, generating new values for velocity and displacement. By changing the coefficient for damping, light, critical, and over-damping can be investigated.

1. Another example of a situation that is critically damped is an electrical meter with a moving pointer. Whenever such a meter is connected to take a reading the needle will, if properly damped, "overshoot" the correct reading slightly and then arrive back at the correct value in the shortest possible time. What would happen if the meter's needle was too lightly or too heavily damped?
2. Go back and look again at your list of practical examples of SHM. Decide whether no damping, light damping, heavy damping, or critical damping would be most beneficial in each situation. For example, a pendulum will continue to swing for longer if the damping is light.

Resonance

Damping removes energy from a system. It is also possible to put energy into the system by having an external force pushing the mass forward, called a **driving force**.

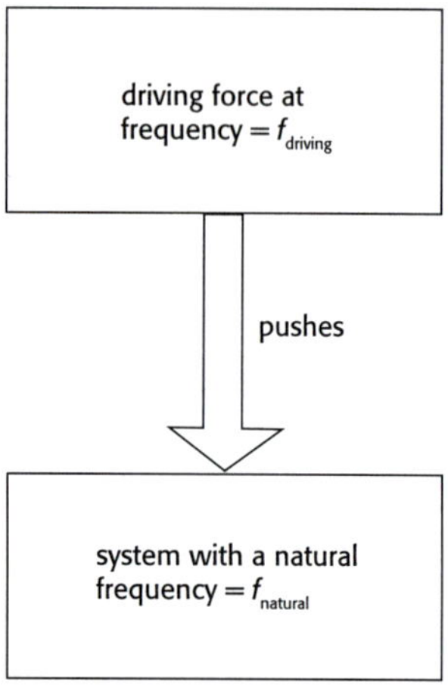

This driving force varies in size and direction so in order to properly derive the solution it is necessary to set up the equation of motion for the object. Once again the full derivation is beyond the scope of the IB Diploma Programme physics course but we can use energy arguments to understand these ideas.

Note that we are now considering two different oscillations and you should ensure that you understand the difference between them. Firstly we consider the oscillations that the system would do if there was no driving force, this is the **natural oscillation** of the system and has a frequency, $f_{natural}$ or f_N, given by the SHM equations. We also consider the driving oscillations with the frequency $f_{driving}$ or f_D.

The driving force can be sinusoidal, defined by the following equation:

$$\text{driving force} = F_D \cos \omega_D t$$

This means the equation for the system's motion is:

$$m\frac{d^2x}{dt^2} = -kx + F_D \cos \omega_D t$$

The solution to this equation is still SHM with the system oscillating at the driving frequency, $f_{driving}$. The amplitude of the resulting oscillation depends on how close the driving frequency is to the **natural frequency**. A strange thing happens as the driving frequency approaches the natural frequency. The amplitude of the oscillations increases rapidly as the frequencies become nearly equal

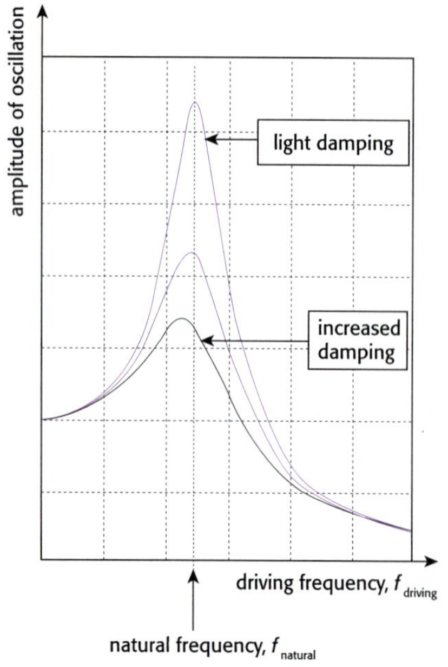

Figure 23 Variation of amplitude with driving frequency

and the only amplitude-limiting factor is the amount of damping present in the system. This phenomenon is called resonance.

At resonance the driving force is doing work, in other words, putting energy into the oscillating system. As energy goes into the system, its amplitude increases. It stops growing in amplitude when the energy put into the system in one cycle is equal to the energy lost in one cycle due to damping. Resonance is an extremely important phenomenon and occurs in a very wide range of situations. Sometimes resonance is something to be avoided and in extreme cases can cause a lot of damage, whereas in other situations resonance is relied upon to produce a desired effect.

Inquiry: examples of resonance

The following are all examples of resonance. For each situation, with a small amount of research, you should be able to identify the system providing the driving frequency and the system that has the natural frequency and thus outline the physics of the situation. In addition, you should be able to decide if the resonance condition is something that should be avoided or is something desirable.

1. A removal van's combustion engine is controlled by the accelerator pedal. With the engine out of gear, the engine can be made to idle at different frequencies. At one particular frequency, one of the wing mirrors on the van can be seen to shake violently.
2. A note being played on a clarinet.
3. A microwave oven used to cook a meal.
4. When singing in the shower, an IB Diploma Programme student notices that one particular note sounds louder to him than all the others even though he is not singing that note more loudly.
5. A radio can be tuned so that one particular radio station is heard.
6. An opera singer claims to be able to break a wine glass by loudly singing a note of a particular frequency.
7. If the keys on a piano are pushed down gently enough it is possible to avoid playing any notes. With the keys held down, if any loud noise happens in the room (e.g. somebody shouting), then some of the notes held down will start to sound.

8 Waves

This chapter builds on the ideas introduced in Chapter 7 and brings together two different, but related, areas of physics: oscillations and waves. Before starting, review the work you have done in Chapter 7 and to try to write down the general definitions/characteristics that are common to all waves.

When imagining a mechanical wave, try to imagine ripples on a pond.

There are some properties that are common to all types of waves, whether they turn out to be mechanical waves, electromagnetic waves, or something else altogether. These properties are that any wave can be shown to be able to undergo **reflection**, **refraction**, **diffraction**, and **interference**. All of these phenomena are discussed later in this chapter. It is the latter two (diffraction and interference) that are often of particular interest to physicists because they are phenomena that are unique to wave motion. A tennis ball "reflects" off a wall when it bounces and its motion can be described assuming it is a point particle. It is possible to hypothesize an explanation for refraction in terms of particles but there is not yet a successful particle model for diffraction or interference. Sometimes it is not obvious whether a phenomenon is best described by wave motion or a stream of particles.

Light and sound can be shown to diffract and interfere. This means that waves of some sort must be involved in a proper explanation of how the energy is transmitted from source to receiver. In fact, an absolutely complete description of both light and sound needs to involve wave *and* particle properties (see page 207 for more details).

Amplitude, intensity, and energy in waves

All waves transmit energy from one place to another. This is achieved *without any overall transfer of mass* using oscillations. In many cases these oscillations can be shown to involve SHM.

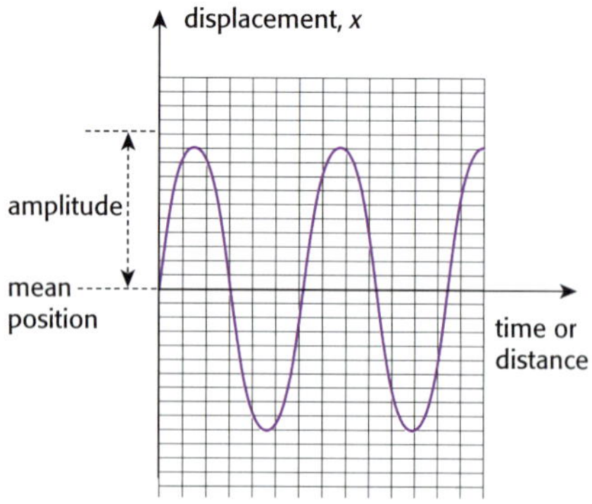

Figure 1 The amplitude of a wave motion

The amplitude of a wave is not the same as its intensity. Amplitude is defined as the maximum variation from the mean value (or rest position) for the oscillation concerned.

The precise definition of intensity can vary depending on the context in which it is being considered, but in general it is defined as the power received per unit surface area.

$$\text{intensity} = \frac{\text{power recieved}}{\text{area of reciever}}$$

the unit of intensity is: W m^{-2}

The total energy of an object doing SHM is proportional to $(\text{amplitude})^2$ and the intensity of a wave is also proportional to its $(\text{amplitude})^2$.

When energy is received from a continuous source of waves, the total energy absorbed is proportional to two variables: the area of the receiver and the time for which the wave was received.

Waves always spread out from their sources so the amplitude and the intensity of any wave must decrease with distance from the source. If the source is isotropic (this means that the source emits uniformly in all directions), then the decrease in intensity turns out to be what is called an "inverse square relationship". As you move away from the source, the total wave energy is spread over a greater area so the intensity (and the amplitude) must be reduced. This is illustrated in Figure 2.

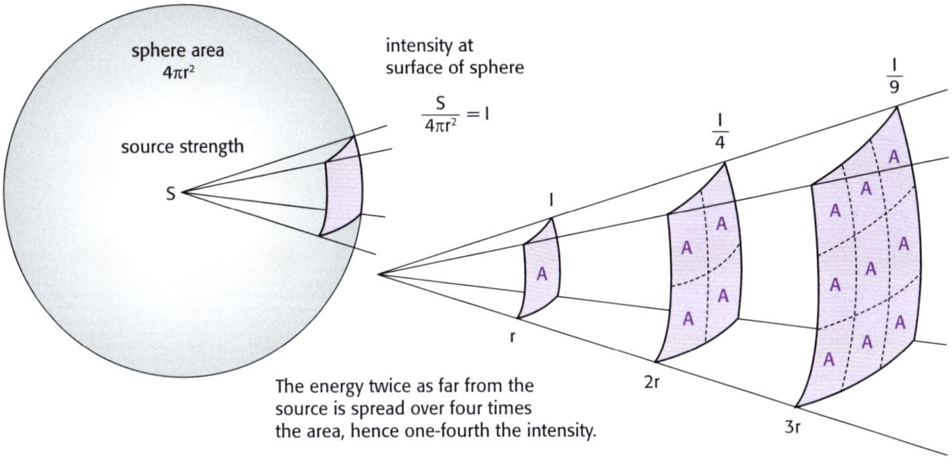

Figure 2 Inverse square relationship

Sound

The human ear is able to detect waves over a very large range of frequencies and amplitudes. The frequency of a sound is related to the pitch that we hear – a low frequency sound is a low note. The range of human hearing is from approximately 20 Hz to 20 kHz. Increasing the amplitude of the sound increases the intensity received by our ears, and hence the loudness that we hear.

Amplitude, intensity, and loudness are all different quantities. Strictly speaking, the amplitude of a sound wave travelling in air would be just the maximum displacement of the air molecules as a result of the sound. This piece of information is hard to measure and not particularly useful. Sound engineers often refer to the amplitude of the electrical signal produced by a microphone or electrical system that is used to process the sound.

The human ear is capable of responding to intensities from as low as 1×10^{-12} W m^{-2} up to about 1×10^{3} W m^{-2}. Different intensities are compared on the decibel scale, which is logarithmic. This means that if a sound intensity is doubled, then doubled again, and so on, the listener experiences, very approximately, the same regular increases in the sensation of loudness.

The apparent loudness of a sound to a given listener depends on many factors and is also very subjective. Increasing the intensity of a sound whilst keeping other variables fixed will make the sound seem louder. The ear does not, however, respond equally to different

Table 1 The decibel scale is logarithmic

Source	Intensity/W m^{-2}	Intensity/dB
Threshold of hearing	1×10^{-12}	0
Rustling leaves	1×10^{-11}	10
Whisper	1×10^{-10}	20
Normal conversation	1×10^{-6}	60
Busy street traffic	1×10^{-5}	70
Vacuum cleaner	1×10^{-4}	80
Large orchestra	6.3×10^{-3}	98
MP3 player at maximum level	1×10^{-2}	100
Front rows of a rock concert	1×10^{-1}	110
Threshold of pain	1×10^{1}	130
Military jet takeoff	1×10^{2}	140
Instant perforation of eardrum	1×10^{4}	160

frequencies. Age is another factor that affects the human ear's response to a sound. Exploring variations in perceptions of loudness could form the basis for an interesting extended essay.

Another possible theme for an extended essay could be an investigation into some aspect of unwanted sound, called noise. Noise pollution is a growing problem in our towns and cities and has been shown to have a significant detrimental effect on both physical and mental health. Technological solutions can prevent some of these effects but much depends on the particular situation being considered. For example, a very low amplitude noise would be disturbing in an examination hall whereas a much louder noise would be ignored in a busy restaurant.

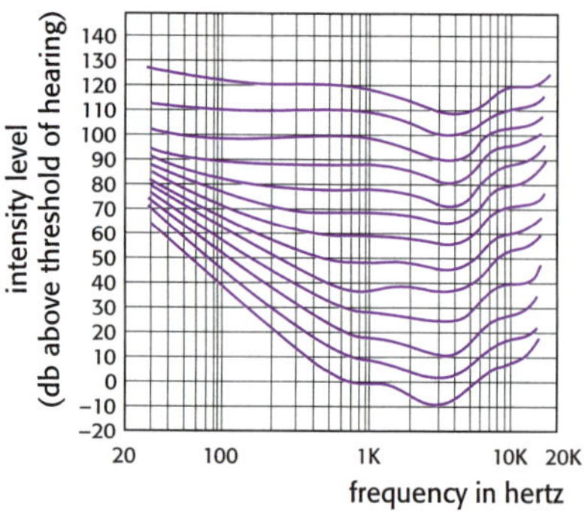

Figure 3 Equal loudness curves (based on Fletcher and Munson)

Light

Light is an electromagnetic wave (see Figure 10 on page 85). All electromagnetic waves travel at the same speed in a vacuum, 3×10^8 m s^{-1}, but they travel at different speeds in different media. This produces a phenomenon called **dispersion**. The speed at which light propagates through glass depends on its frequency and this is the reason why a prism splits white light into its different component frequencies.

We perceive different frequencies of electromagnetic radiation as different colours of light. The human eye is able to respond to frequencies ranging from about 8×10^{14} Hz to 4×10^{14} Hz (corresponding to a wavelength range of 400 nm to 750 nm). The perception of colour is extremely complicated as two frequencies arriving at the same time are not perceived to be a mixture of the two colours: we just see it as one colour. For example, red light and green light added together is seen as yellow light. If all frequencies of visible light arrive together at the same time, this is usually perceived as white light (depending on the amplitudes of the individual frequencies).

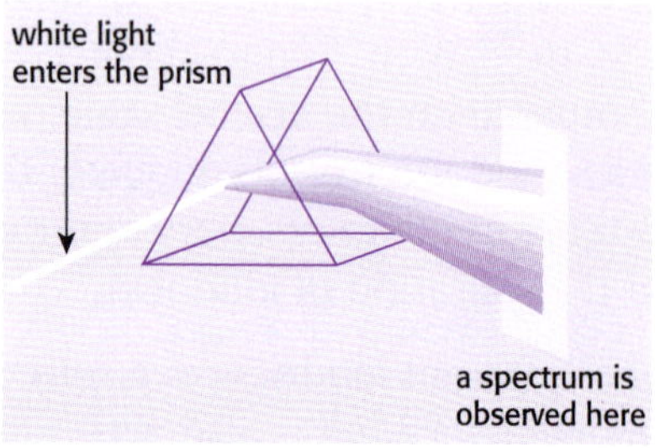

Figure 4 Prism splitting white light into colours

Do not confuse this additive property of light with the mixing of coloured paints. A surface appears coloured because it absorbs most of the frequencies contained in white light. The frequencies that are not absorbed are reflected off the surface and arrive at your eyes. If red paint and green paint are mixed together, we would not get yellow but something more like black.

Light is an electromagnetic wave so if we are precise, the amplitude of light is measured in terms of the maximum electromagnetic field variation. In practical situations the intensity is a much more useful piece of information. Increasing the amplitude of light increases its intensity and this is seen as increasing the brightness – another subjective measurement. Different eyes have different perceptions of brightness.

Earthquakes

The energy involved in earthquake oscillations is very large indeed. As before, the total energy transmitted by the wave is proportional to the square of the amplitude. If the amplitude of one earthquake wave is double that of another, then the energy associated with it is four times more than the energy associated with the smaller earthquake.

The size of earthquakes can be measured on the Richter scale, which was invented by Charles F. Richter in 1934. The Richter magnitude is calculated from a measurement of the amplitude of the largest seismic wave recorded for the earthquake. It is also called the local magnitude scale.

The scale is logarithmic so, for example, a magnitude 5 earthquake would result in ten times the level of "ground shaking" as a magnitude 4 earthquake. The calculation of the total amount of energy transmitted by any given earthquake is complex but we can say that an increase of one point on the Richter scale is roughly equivalent to the release of 32 times as much energy.

Figure 5 Earthquake damage

Table 2 Describing earthquakes

Descriptor	Richter magnitudes	Earthquake effects	Frequency of occurrence
Micro	Less than 2.0	Microearthquakes, not felt.	About 8,000 per day
Very minor	2.0–2.9	Generally not felt, but recorded.	About 1,000 per day
Minor	3.0–3.9	Often felt, but rarely causes damage.	49,000 per year (estimated)
Light	4.0–4.9	Noticeable shaking of indoor items, rattling noises. Significant damage unlikely.	6,200 per year (estimated)
Moderate	5.0–5.9	Can cause major damage to poorly constructed buildings over small regions. At most slight damage to well-designed buildings.	800 per year
Strong	6.0–6.9	Can be destructive in areas up to about 100 miles across in populated areas.	120 per year
Major	7.0–7.9	Can cause serious damage over larger areas.	18 per year
Great	8.0–8.9	Can cause serious damage in areas several hundred miles across.	1 per year
Rare great	9.0 or greater	Devastating in areas several thousand miles across.	1 per 20 years

Earthquakes: longitudinal and transverse waves

There are many different types of earthquake waves. The location of the disturbance that causes the earthquake (called the **hypocentre** or **focus**) is below the surface of the Earth. The point on the surface above the hypocentre is called the **epicentre**. The hypocentre is sometimes very deep beneath the epicentre. For example, on 9th June 1994 an 8.2 magnitude earthquake took place beneath Bolivia. The hypocenter was calculated to be 670 km below the surface yet oscillations were recorded in seismic stations around the world. So how was the energy transferred? There are two fundamental types of wave: S waves and P waves.

P waves are an example of **longitudinal waves**. In this type of wave, the oscillations involved are in the same direction as that of energy transfer. Areas of high pressure (**compressions**) and low pressure (**rarefactions**) travel alternately through the medium. Other examples of longitudinal waves include sound waves and one type of wave along a stretched spring.

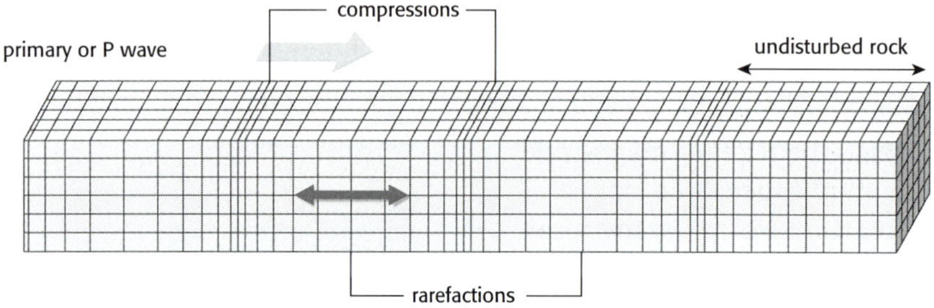

Figure 6 Primary or P waves

Remember that longitudinal waves are NOT just "waves where the oscillations are from left to right". The oscillations are along the direction of energy transfer, whatever that direction might be.

S waves are examples of the other fundamental type of wave: **transverse waves**.

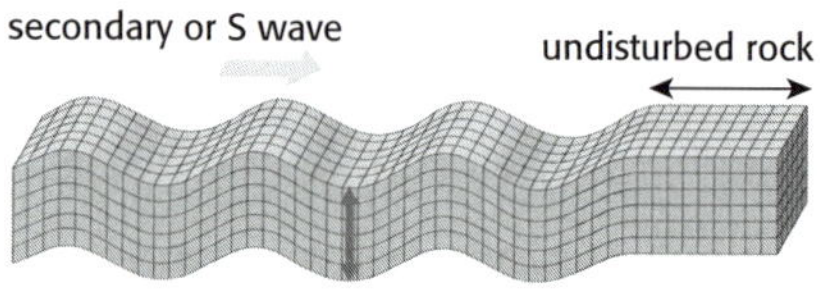

Figure 7 Secondary or S waves

In transverse waves, the oscillations are perpendicular to the direction of transfer of energy. A pattern of peaks and troughs travel through the medium. There are many other examples of transverse waves: one type of waves on a spring, waves on the surface of a pond, and the whole electromagnetic spectrum of waves: visible light, IR, UV, radio and TV broadcasting, gamma rays, microwaves.

The wave equation

All waves (including, of course, transverse and longitudinal) have a particular wavelength λ, frequency, f and velocity, v. These quantities are related by the **wave equation**:

$$v = f\lambda$$

The wave equation can be derived from the basic wave definitions and is applicable to every wave type, including earthquake waves. It cannot, however, be used to predict the speed of a particular wave from first principles, as this depends on the situation being considered.

?

Derive the wave equation from the basic wave definitions asked for on page 55.

S and P waves travel through the Earth at different speeds from one another and thus they take different times to arrive at any given point on Earth. Geophysicists use this fact to locate the likely focus of an earthquake from seismic recordings taken from around the world. In general, the P waves travel faster and are thus called **primary waves** because they will arrive at a seismic station first. The S waves, or **secondary waves**, will arrive later. P waves are also sometimes called "pressure (or compression) waves" and S waves are sometimes called "shake waves".

A final class of earthquake waves that travel even slower than S or P waves are called surface waves. As the name suggests, these waves do not travel through the main bulk of the Earth, but they only move over its surface. The speed of these types of wave depends on their frequency and they tend to spread out as they travel away from the epicentre of the earthquake. Two examples of possible surface waves are **Love waves** and **Rayleigh waves**:

Love waves are another example of a transverse wave whereas Rayleigh waves are more complicated and can be considered to be a mixture of transverse and longitudinal.

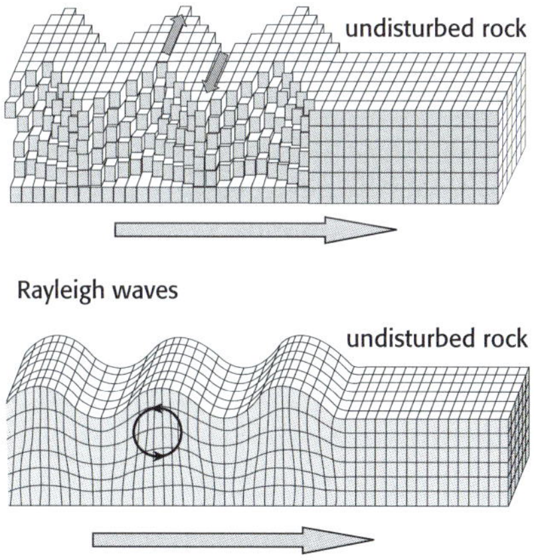

Figure 8 Love waves and Rayleigh waves

Thinking about science: an example of the political use of science

In October 2006 the government of North Korea announced that it had successfully detonated a test nuclear bomb. The earthquake wave signature of the explosion was analysed around the world in order to attempt to verify this claim but initially there was widespread disagreement.

Within a few hours, the Russian government announced that it was "100% certain" that a nuclear test had been carried out and measured its size to be equivalent to an explosion of between five to fifteen kilotons of TNT. South Korea, France, and the US, however, all announced that the measurement suggested the test was less than one kiloton. This could imply that the North Korean test had not been successful or that they had engineered a large conventional explosion and chosen to announce that it was nuclear in nature.

A couple of days later, the US government announced that it had acquired further evidence (traces of radioactive gas in the air near the site of the alleged nuclear test) supporting the claim, but they stressed that more tests were needed to reach a conclusion.

Before this announcement, South Korean and Chinese scientists said that they had detected no evidence of radioactivity in their own air, soil, and rainwater tests. The UN Security Council passed a resolution imposing sanctions on North Korea.

There are, of course, many possible procedures that could be used to check if a nuclear explosion had taken place. Earthquake wave analysis and testing for radioactive substances near the alleged site are just two possible methods. The following questions can be asked:

- Are the two methods of inquiry mentioned above the most reliable methods to use? Can you think of any other possible methods?
- In situations like these, do you think the scientific method can be relied upon to discover the truth behind the political claims or is science being used to make a political point?
- How could a citizen of one of these countries decide what to believe?
- What evidence would you need to convince you either way in this case?

Physics issues: risk assessment and EM waves

Every day we are exposed to risks. Many people would agree that some risks are worth taking, in fact, we sometimes tolerate these "extra" risks because we wish to enjoy all the technological advantages of living in the 21st century. For example, whenever you cross the road there is a risk of being involved in an accident. Most people would not argue for banning all vehicles from our roads in order to save the number of lives that are lost as a result of the occasional accidents that do happen, even though more than 1.2 million people worldwide die in road traffic crashes every year.

Figure 9 Is this an acceptable level of risk?

Perceived risks are often balanced against perceived benefits in this way. For example many people enjoy skiing but there are many accidents associated with this sport. Incredibly, given the often verified direct link between smoking and the increased risk of suffering from many diseases including cancer, there are still thousands of people who choose to ignore the risks associated with smoking.

Often it is not possible to avoid potential risks. As a result of modern communication systems, for example, we are all now exposed to much more electromagnetic radiation than would have occurred only 10 years ago. Some people worry about the potential risks of using mobile phones or living near the associated aerials. So what are the risks and are they risks worth taking? Can the level of risk be quantified or calculated?

The formal study of risk assessment and its management is beyond the scope of the IB Diploma Programme physics course but it is important to understand the sort of checks that companies and governments need to undertake whenever a project is proposed. They need to weigh up the possible consequences of an accident verses the potential benefits that the project brings. In simple terms a risk can be calculated by multiplying the probability of a particular event or accident occurring by some agreed measure of the consequences of the event.

$$\text{risk} = \text{probability} \times \text{consequence}$$

The **probability** is a number between 1 and 0. If, for example, the probability that a new hydroelectric dam would fail was assessed to be 0.0001 year^{-1}, this would mean that on average it would only fail once in every ten thousand years. Most people would accept that this risk is very small. However, if there were several thousand of the same type of dam in existence around the world, then you might expect an accident every few years. Given the potential consequences, is this an acceptable level of risk?

The **consequence** is a measure of what happens if the event under consideration does take place. In the example above, the failure of a dam would be measured in human lives as well as financial and environmental costs.

Most alleged dangers related to the use of mobile phones are associated with the electromagnetic radiation used. Are all types of electromagnetic radiations as harmful as each other?

All the different types of electromagnetic radiation are the same type of wave – they are a transverse oscillation of electric and magnetic fields and they all propagate at the same speed in a vacuum: 3.0×10^8 m s^{-1}. They do, however, have very different properties as a result of their very different wavelengths.

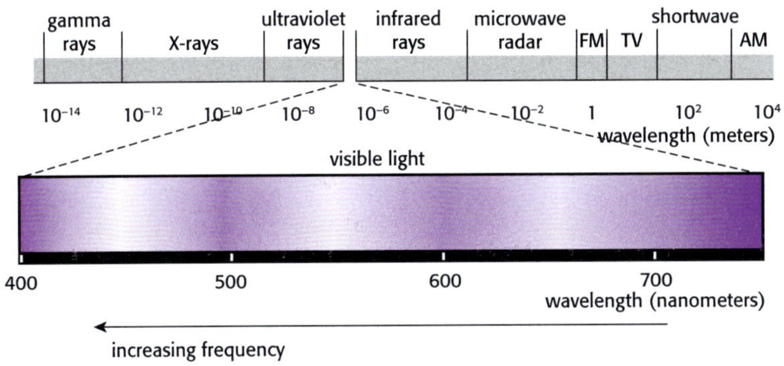

Figure 10 The electromagnetic spectrum

One aspect of the transmission of energy by EM radiation is that it can be modelled as being emitted in "packets" or **photons** of energy. These packets of energy are also called **quanta**. The evidence for this quantum phenomenon is explored more on page 207. The energy of each wave packet is proportional to the frequency of radiation being considered and each photon has an energy given by the following equation:

$$\text{energy of photon, } E = h \times f \qquad (1)$$

The constant h is called Plank's constant and is equal to 6.63×10^{-34} J s

The alleged dangers of mobile phones are largely attributed to the ionizing effects of the radiation they use (microwave radiation) on human tissues. However, equation 1 can be used to show that the photon energies of microwaves are not large enough to be able to cause these ionizations. It is true, however, that some types of electromagnetic radiation, for example gamma waves, can produce these effects.

The experimental evidence available, at the time of writing, does not provide any robust statistical evidence for a definite link between the widespread use of mobile phones and, for example, increased incidence of brain tumours. The World Health Organization coordinates assessments of the health risks. These are in-depth critical reviews conducted through independent, scientific peer-review groups aimed at evaluating the evidence from around the world. Their current conclusions can be downloaded from www.who.int/.

For a further discussion of the risks associated with a possibly related situation (living near electrical power lines) see page 177.

?

1 How many photons of visible light are being emitted by a 60W light bulb every second? Use an average value of the wavelength of visible light, say $\lambda = 600$ nm.
2 The ionization energy of hydrogen is 2.18×10^{-18} J. Calculate the frequency of a photon just able to achieve this ionization.
3 Use your answer to question 2 to categorize the different parts of the EM spectrum into those that you would expect to be able to cause ionizations and those that would not be expected to achieve this.
4 Imagine that there is a proposal to install a mobile phone aerial on the roof of your school in return for a small annual fee from the company. Discuss this proposal and create as long a list as possible of the risks associated with this project – what could go wrong and what would be the consequences? What level of risks do you find acceptable? In your opinion what conditions need to be met for the cost–benefit analysis of this project to be favourable?

Data-based question: graphs of wave motion

Consider the graph (Figure 11) representing the oscillations involved during a particular type of wave motion. Try to write down brief answers to the questions that follow without reading ahead to the solutions.

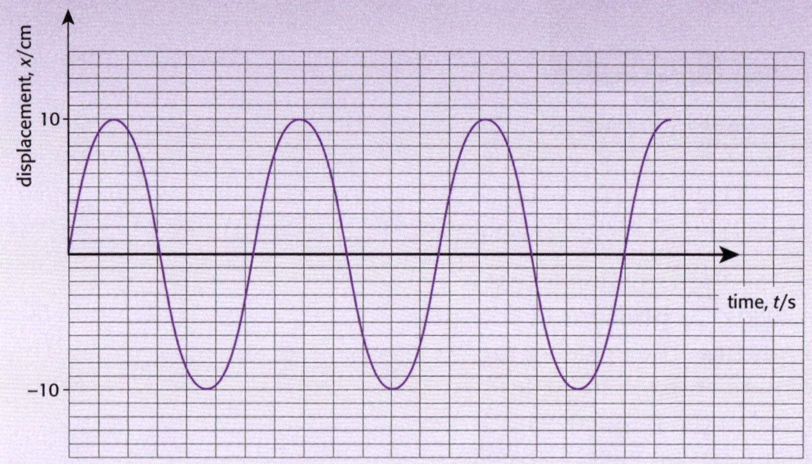

Figure 11 Variation with time of the displacement in a wave

1 Is the wave represented longitudinal or transverse?
2 What is the wavelength of the wave?
3 The time period of the oscillations is 8 seconds. How fast is the wave travelling?

Answers

It is important to understand what exactly is being represented. Under the pressure of an examination situation, many candidates attempt to analyse a graph without reading what is being plotted on each axis!

In Figure 12, the *y*-axis is labelled displacement and the *x*-axis, time. Thus the graph represents the variation of displacement with time. In the context of a wave motion, this must be how the displacement **of one particular particle** associated with the wave motion varies as time progresses. It would, of course, be possible to plot other, similar, graphs showing how the different particles oscillate as time progresses but in this graph we have chosen to focus on one particular point.

Now we can discuss the questions posed on the previous page.

1 The diagram is a graphical representation and not a physical picture of the waves. You don't have enough information to answer the question; the direction of the displacement is not specified and the graph could equally apply to a transverse wave or a longitudinal wave. Without knowing the direction of energy transfer, you cannot answer this question.

2 Since the graph is only providing information about how one particular particle oscillates with time, we are not being given any information about the separation between different particles along the wave. Another way of thinking about this is that the *x*-axis is labelled time and thus cannot be used as a measure of wavelength. Any measurements from the *y*-axis relate to the displacement of an individual particle but this is not the same as the length occupied by one complete oscillation along the wave. The graph cannot be used to provide the information requested.

3 Once again, the graph does not provide enough information to calculate the required quantity. It is possible to calculate the **average speed** of the particle during one quarter (or any complete multiple of a quarter) of an oscillation, as shown below. Remember the speed at which any particle moves during wave motion constantly changes (the graph for speed is a cosine wave).

time taken to complete one complete oscillation = 8 s
from the graph,
distance gone in one quarter oscillation = 10 cm
time taken to complete one quarter oscillation = 2 s

$$\text{speed} = \frac{\text{distance}}{\text{time}} = \frac{10}{2} = 5\,\text{m}\,\text{s}^{-1}$$

During the first quarter oscillation, it started at its maximum speed and slowed down to zero. 5 m s^{-1} thus represents the particle's average speed. This is a completely different quantity to the speed at which the wave pattern is propagated through space so we have not actually answered the question.

The graph provided in Figure 11 is very different to a graph which plots the displacement of **all the points** associated with the wave motion **at one particular time**. An example of this type of graph is shown in FIgure 13:

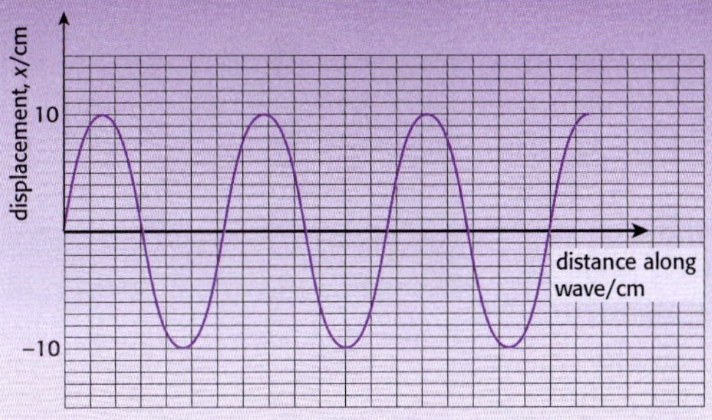

Figure 12 Variation of displacement with distance along a wave

It is potentially very confusing for the two graphs to have very similar shapes, but if the axes are carefully labelled then the difference between the information represented in each should be clear.

To summarize, the first graph (Figure 12) showed the variation **with time** of (the displacement **of one point** along the wave). Different points would need different graphs.

The second graph shows one "snap shot" of the different variations **with distance along the wave** of (the displacement of all the points along the wave). Different times would need different graphs.

?

1 The paragraph above states that with each approach, further graphs would be needed to fully describe the wave motion. For each approach draw another similar graph to represent either:
   (i) the variation of displacement for a different point, or
   (ii) the variation of displacement of all the points at a later time.
2 Annotate your graphs and explain how the following quantities can be calculated from these graphs:
   - amplitude
   - frequency
   - period
   - phase difference
   - wavelength
   - wave speed.

The endoscope: an example of reflection

In this section we will discuss one example of reflection. There are many more and after reading this section you might like to identify and research other examples.

In general, whenever a wave crosses a boundary between two different media it is both partially transmitted and partially reflected. Under particular conditions (see page 88) it is possible for a wave that is travelling in a dense medium to undergo **total internal reflection** (TIR). This means that the entire wave energy is reflected and no wave energy is transmitted into the neighbouring material.

A light ray can be continuously totally internally reflected along a glass fibre, called a fibre optic cable.

Fibre optics are thin strands of optically transparent material that use the principle of total internal reflection to transmit light waves along the path of the fibre with very little loss (see Figure 13). Refer to Chapter 27 for a discussion of some more applications of fibre optic cables. A common application of TIR is in a medical device called an endoscope (Figure 14). This device allows doctors to see inside the body of patient in a way that does not involve surgery.

An endoscope consists of two bundles of fibre optic cables which can be introduced into the body. One of the bundles of fibre optics transmits light from an external source to the point of interest

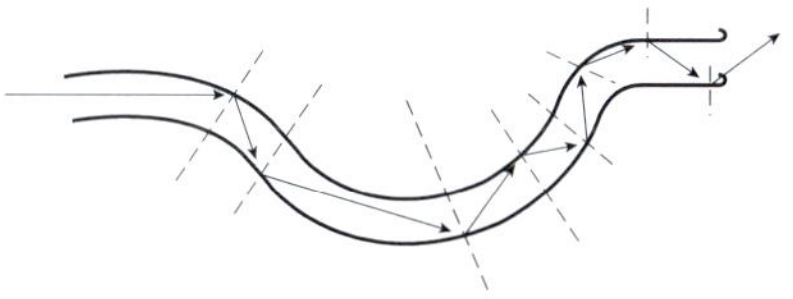

Figure 13 Total internal reflection along a fibre optic

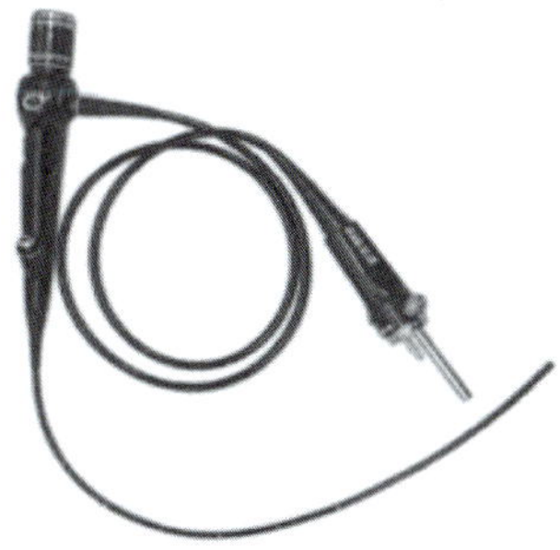

Figure 14 An endoscope

and another bundle is used to bring the reflected light back to an external imaging device, effectively providing a view inside the patient. In Figure 15 you can see the image of a polyp, a benign precancerous growth projecting from the inner lining of the colon.

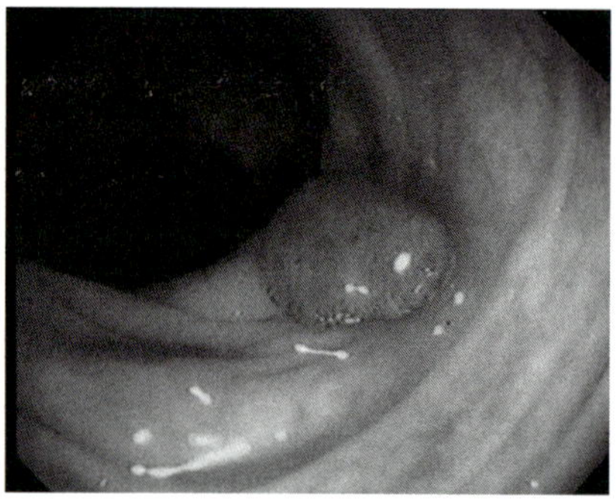

Figure 15 A polyp in the colon

Refraction

As we stated earlier in this chapter, when a wave meets the boundary between two different media, some of the wave is reflected back and some of the wave is transmitted on. The portion that is reflected obeys the law of reflection (angle of incidence = angle of reflection) but the direction of the refracted ray depends on the speeds of the wave in each medium. In the example in Figure 16, the speed of the wave in medium 1 is greater than the speed of the wave in medium 2.

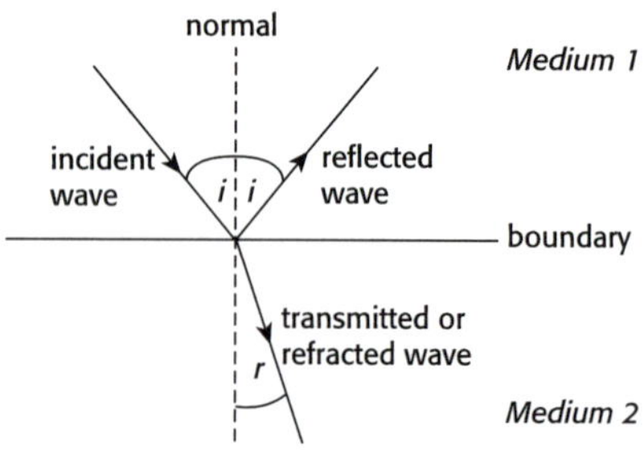

Figure 16 Reflection and refraction

Experiments show that in all situations:

$$\frac{\sin i}{\sin r} = \text{constant} = \frac{\text{speed of wave in medium 1}}{\text{speed of wave in medium 2}} \qquad (2)$$

This equation is known as Snell's law.

The ratio expressed in equation 2 defines a constant called the **refractive index**. Note that the general wave equation ($v = f\lambda$) must still apply as the wave crosses the boundary between the two media. As it crosses the boundary, its wavelength changes in proportion to the change in speed. The frequency of the wave must remain unchanged because the number of waves leaving medium 1 must equal the number of waves entering medium 2, or in other words, the law of conservation of energy must apply.

If the wave is travelling from a medium where its speed is slow into one where its speed is faster, the above situation is essentially reversed. However, there exists a critical angle above which it is impossible for the wave to still be transmitted into the new medium. At exactly the critical angle, the wave is refracted along the boundary, and at any larger angles of incidence all of the wave's energy is reflected (total internal reflection).

This description of refraction only applies whenever a wave makes the abrupt change from one medium to another. If, however, a wave travels through a region where its speed gradually changes then the path of the wave turns out to be a curve. You can imagine the wave passing through thin layers and changing its speed by a small amount each time is passes into a new layer (Figure 17).

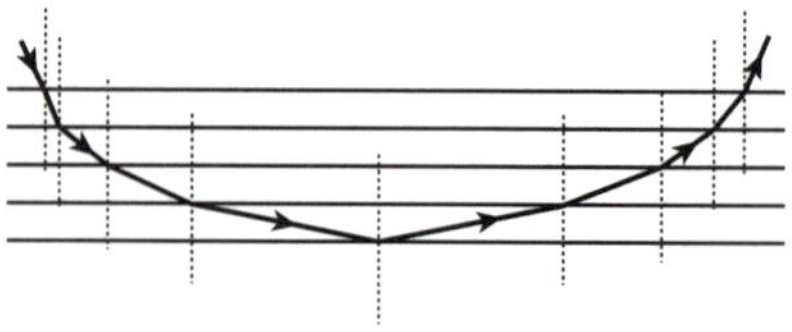

Figure 17 A curved ray path resulting from gradually changing densities

Earthquake waves: an example of refraction

This section outlines one specific example of refraction; after reading it you might like to identify and research some more.

Imagine a huge scale model of the Earth that is 13 m high.

Figure 18 A model Earth, 13 m high

> ***Thinking about science: discussion***
>
> Given the relatively tiny section of the Earth that the human race is able to explore, how are we able to describe, with a great deal of confidence, the internal structure of the Earth?

On this scale, Mount Everest would be approximately 9 mm high and the deepest diamond mine drilled in South Africa would only be about 3.5 mm deep.

It has been known for a long time that the density of the centre of the Earth is greater than the density at its surface. In the 17th century, Newton was able to use his theory of universal gravity to show that the average density of the Earth was approximately twice the density of the rocks at the surface. If we assume that the density of the material that makes up the Earth increases uniformly with depth, this would result in an increase in the speed of earthquake waves with depth, so the path that they take would be predicted to be curved as shown in Figure 19.

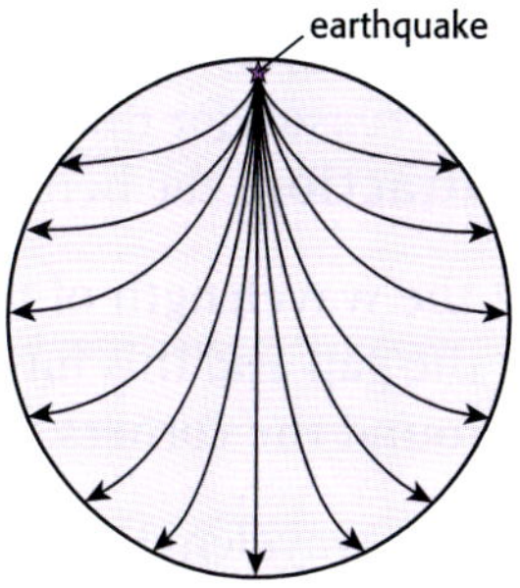

Figure 19 Path of earthquake waves if the density of the Earth varied uniformly

However, what we actually observe depends on whether we are considering P waves or S waves. We will consider these situations separately.

P waves

The P waves do show the expected patterns of refraction but there is a shadow zone where no P waves are observed. This can be explained if we hypothesize a central core region to the Earth where the density (and other properties) of the material suddenly changes.

More information about the nature of this core region comes from the study of the behaviour of S waves.

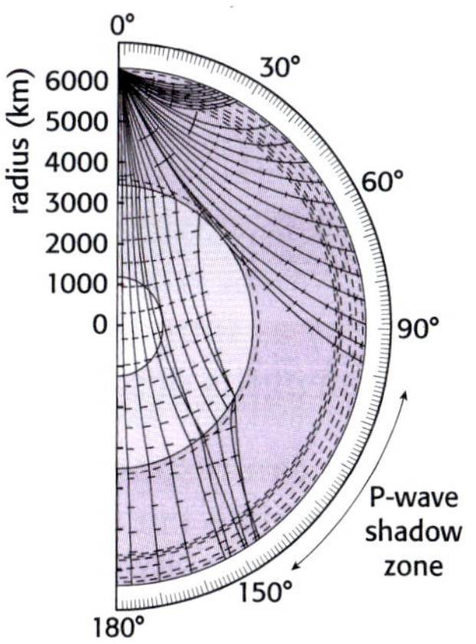

Figure 20 P wave shadow zone

S waves

There is a total S wave shadow zone observed on the opposite side of the Earth to the hypocentre, where no S waves ever arrive.

We can develop an explanation for this shadow zone by considering the way S waves are propagated. They are transverse waves and thus can easily pass through any solid. Transverse mechanical waves are, however, unable to pass through the body of a liquid or a gas. It is possible to move a block of liquid "sideways" without causing a corresponding sideways movement of the adjacent liquid. When trying to visualize the situation, it is important to remember that we are considering the bulk movement of a section of a liquid that is within the body of the liquid. This is very different to the movement of the surface of a liquid when transverse waves will be propagated as a result of the surface tension forces.

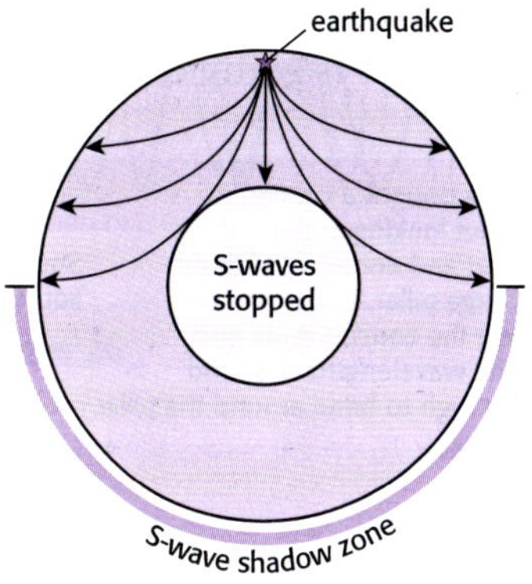

Figure 21 S wave shadow zone

waves with nodes at either end. Figure 4 shows some of the possible standing waves on a string that is fixed at both ends.

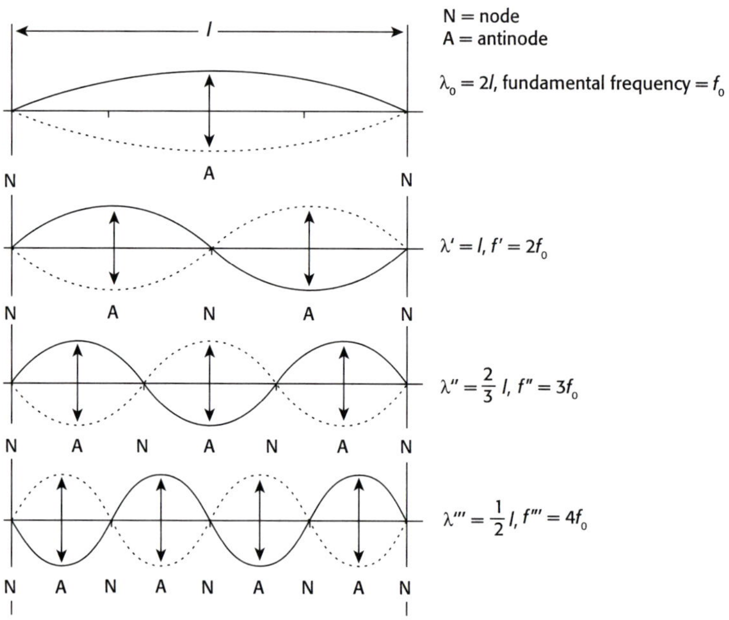

Figure 4

The standing wave with the lowest possible frequency is called the **fundamental** or **first harmonic** and the other possible frequencies are called **harmonics** (also sometimes called **overtones**). The difference between the sound of, for example, a guitar and a violin which are both playing the same note, is a result of the relative amplitude of the different harmonics that are produced by the instrument. This different sound, even though the instruments are playing the same note, is called the **quality** or **timbre**.

Many wind instruments use a similar principle, involving longitudinal standing waves produced in a column of air. The boundary condition at each end of the air column depends on the type of reflection taking place. At a closed end the air is unable to move and there must be a node. At an open end some of the original travelling wave is transmitted out of the instrument and some is reflected back in.

Although we often say that there is an antinode at the open end, in practice, the exact position of the antinode is slightly different to the physical end of the instrument. This can be ignored in many situations but a more accurate approach would be to assign a small end correction c to the length of the instrument l and define the effective length as $(l + c)$.

Note that the nodes referred to above are all displacement node. This means that they mark the position where the displacement does not vary. Similarly, antinodes mark the positions where the displacement shows maximum variation. This is not the same as a pressure node – the point where the pressure does not vary with time.

?

The fundamental frequency of a pipe closed at one end is f. A pipe of the same length but open at both ends has a fundamental frequency (first harmonic) of

A $\frac{1}{2}f$ C $2f$

B f D $4f$

Inquiry: musical instruments from around the world

Investigate a range of different musical instruments from around the world. The different tasks below outline some possible lines of inquiry. Before starting this inquiry, decide which instruments you will investigate, their country/countries of origin and the method you will use to present your findings.

For each instrument chosen you could:

A outline the method of production of the original disturbance that creates the sound

B identify the boundary conditions that apply to the instrument

C use the physical dimensions of the instrument to estimate a typical fundamental wavelength associated within the instrument.

D use your answer to C and the speed of the wave in the instrument to estimate a typical frequency of a note played by the instrument. If the original oscillations are in a column of air then the speed of a sound wave can be taken to be approximately equal to 330 m s^{-1}. If the original oscillations are in a string under tension, then the speed of the sound waves along the string can be calculated from speed, $v = \sqrt{\frac{T}{\mu}}$ where T is tension in the string in N and μ is the mass per unit length of the string in kg m^{-1}

E estimate the range of possible frequencies achievable by the instrument.

Possible practical extensions include:

- the measurement of the fundamental frequency of an individual instrument using a microphone and cathode-ray oscilloscope
- measuring the variation in a string of frequency of vibration with tension. The equation for the speed of a wave in a string under tension (introduced in D, above) can be used to predict the expected frequency
- a digital recording of an instrument can be analysed to discover the main component harmonics involved. This involves the use of a spectrum frequency analyser (see Fourier analysis in Chapter 7 page 72).

Rediscovering physics: the speed of sound

This experiment uses a recorder to calculate the speed of sound in air. A recorder is essentially an open pipe, which, when played in the normal way, develops a standing wave between the open end where you blow and the first open hole below the mouthpiece. For this experiment, take the length of the standing wave as between the tip of the mouthpiece and the far end of the first open hole (Figure 5).

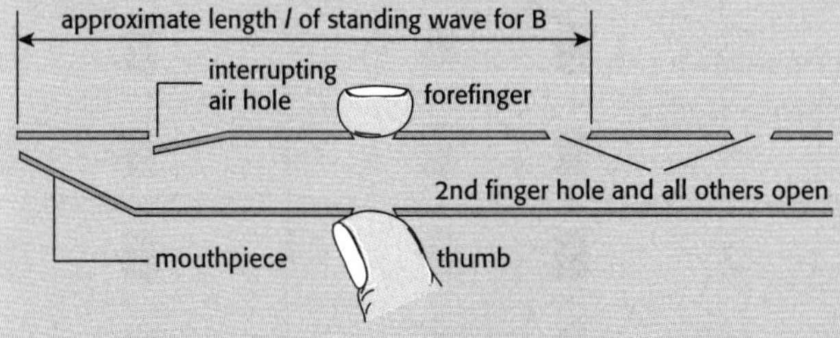

Figure 5 Cutaway side view showing how to play the note B

The wavelength $\lambda = 2l$ and thus the frequency of the note f is given by:

$$f = \frac{v}{2l}$$

where v is the speed of sound measured in m s^{-1}.

The method involved is to take l as the independent variable, measure f (the dependent variable) and analyse the data obtained to calculate v. Controlled variables include the temperature and pressure of the air in the room.

The method of measurement of frequency will depend on the equipment available. Possibilities include:

- using a microphone and a CRO to display the waveform. An analysis of the waveform against the time-calibrated axis allows the time period and hence the frequency of the fundamental to be assessed
- the sound produced can be recorded. A spectrum analyser (available to download from the Internet) can be used to identify all of the frequencies including the fundamental (the lowest frequency present with a relatively large amplitude).

Play the notes C, D, E, F, G, A, and B on a recorder. Figure 6 shows you how to do this.

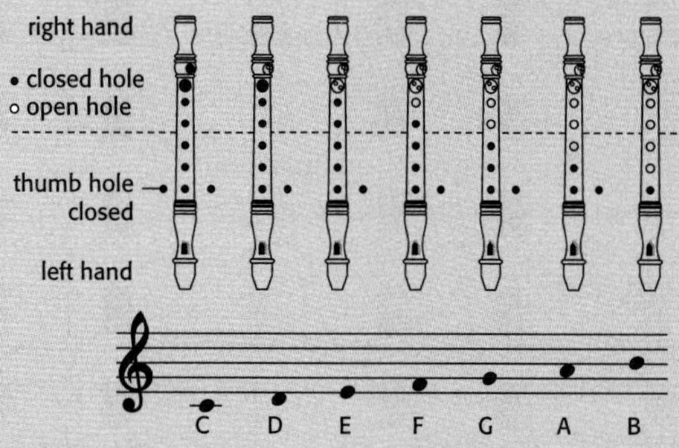

Figure 6 Notes on a recorder

For each note, identify and record the fundamental frequency. Analyse your data to estimate the speed of sound in air, taking care to use a suitable method of graphical analysis. Remember to include your uncertainties.

The bigger picture: standing and travelling waves

The equation for a standing wave can be derived from the equation for a travelling wave.

The equation for a wave travelling in the positive *x* direction (to the right) is:

$$y_1 = A\sin(\omega t - kx)$$

Where:

y_1 is the displacement, measured in m
A is the amplitude, measured in m
ω is the angular frequency, measured in radians s^{-1}
t is the time, measured in s
k is a phase constant, called the wave number, measured in radians m^{-1}
x is the distance along the wave, measured in m.

The time period, $T = \frac{2\pi}{\omega}$ and the wavelength, $\lambda = \frac{2\pi}{k}$.

A wave with an equal amplitude, angular frequency, and wave number travelling to the left is represented by:

$$y_2 = A\sin(\omega t - kx)$$

The equation for the standing wave is thus:

$$y = y_1 + y_2$$
$$y = A\sin(\omega t - kx) + A\sin(\omega t - kx)$$
$$y = 2A\cos(kx)\sin(\omega t)$$

This can be analysed as follows:

$$y = (\text{amplitude term})\cdot(\text{time-dependent term})$$

The amplitude term $(2A\cos(kx))$ depends on the position along the wave. The maximum value is $2A$ and the distance between the nodes is $\frac{\lambda}{2}$, or $\frac{\pi}{k}$.

The Doppler effect

The Doppler effect is the change in the measured frequency of a wave that results from the relative motion of a source and/or an observer relative to the medium in which the wave is propagated. An example of the Doppler effect is the change in pitch that is heard as a police car sounding its siren drives past.

Two things should be noted:

- the speed of the wave through the medium is not affected but the *received* frequency is altered
- although an analogous effect can be measured for both sound and light waves, no medium is necessary for the propagation of light and the derivations below do not apply (see page 100).

In the following discussion,
the velocity of the waves with respect to the medium is v
the velocity of the observer is u_o
the velocity of the source is u_s
the frequency emitted by the source is f_s
the frequency received by the observer is f_o.

Moving observer

An observer moving towards a wave source will receive more waves per unit time when compared with a stationary observer. This is shown in Figure 7.

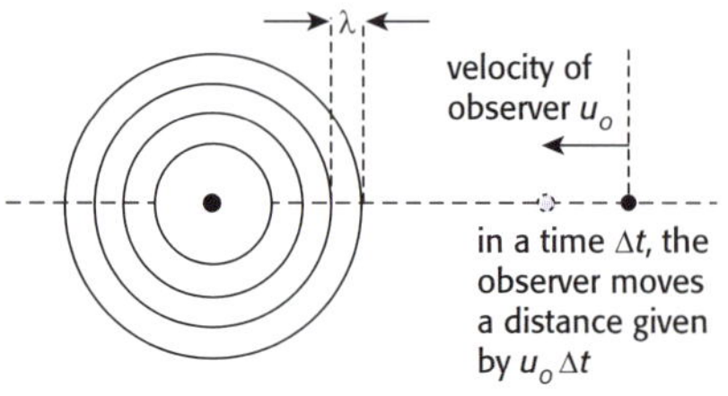

Figure 7 Doppler effect for a moving observer

In a time interval Δt,

The observed frequency $f_o = f_s + \left(\frac{\text{number of waves in a length } u_o\Delta t}{\Delta t}\right)$

$$f_o = f_s + \left(\frac{u_o}{\lambda}\right)$$

but $v = f_s\lambda$ so

$$f_o = f_s + \left(\frac{u_o f_s}{v}\right)$$

$$f_o = f_s\left(1 + \frac{u_o}{v}\right) = f_s\left(\frac{v + u_o}{v}\right)$$ [observer moving TOWARDS source]

If the observer is moving away from the source (with a velocity u_o which is less than v) then

$$f_o = f_s\left(1 - \frac{u_o}{v}\right) = f_s\left(\frac{v - u_o}{v}\right)$$ [observer moving AWAY from source]

Moving source

The source movement causes a change in the wave speed relative to the source. This means that the waves will "bunch" together in the forward direction and be spread apart in the reverse direction. This causes a change of wavelength for the observer.

$$f_o = f_s\left(\frac{1}{1 - \left(\frac{u_o}{v}\right)}\right) = f_s\left(\frac{v}{v - u_o}\right)$$ [source moving TOWARDS observer]

$$f_o = f_s\left(\frac{1}{1 + \left(\frac{u_o}{v}\right)}\right) = f_s\left(\frac{v}{v + u_o}\right)$$ [source moving AWAY from observer]

Overall effect

If the velocities of source or observer, v are small compared with the speed of the waves, c (i.e. $v \ll c$) all four equations can be reduced to one equation:

$$\frac{\Delta f}{f_s} \approx \frac{v}{c} \qquad (1)$$

where
Δf is the difference in frequency between source and observer, measured in Hz
f_s is the frequency of the source, measured in Hz
v is the relative velocity between source and observer, measured in m s^{-1}
c is the velocity of the wave, measured in m s^{-1}

?

A source of sound emits waves of wavelength λ, period T, and speed v when at rest. The source moving away from a stationary observer at speed V, the wavelength of the sound wave received by the observer is

A $\lambda + vT$ C $\lambda + VT$
B $\lambda - vT$ D $\lambda - VT$

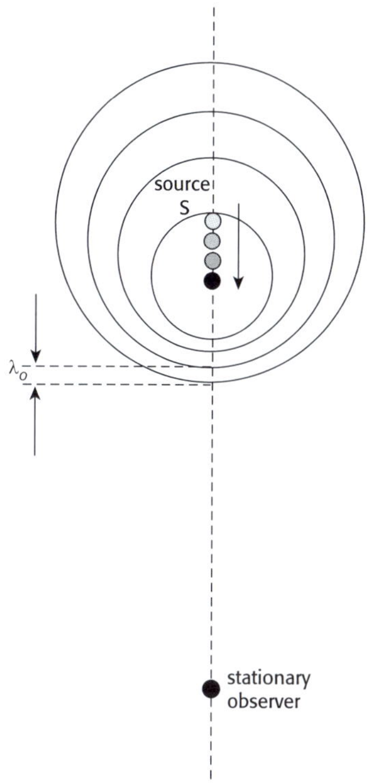

Figure 8 Doppler effect for a moving source

?

Analyse the situation shown in Figure 8 to derive the frequency relationships stated below.

?

Show that equation 1 applies for each of the above situations providing v is small compared with c.

Thinking about science: comparison between sound and EM waves

The frequency of received sound and EM radiation both depend on the relative motion of source and observer. Both are described by the Doppler effect but the underlying reasons for the change in frequency (and the associated equations) are not completely identical.

The derivations of full equations for the Doppler effect involve using a value for the speed of the wave relative to the speed of the medium *c*. Figure 10, for example, shows waves bunching together because the speed of the wave in the medium is fixed and the source is able to "catch up" with some of the waves that have already been emitted. This means that the relative speed of the waves as measured by the source would be reduced.

This derivation cannot apply to electromagnetic waves, however, as the waves do not travel through a medium. Indeed, a central assumption (**postulate**) of Einstein's theory of special relativity is that all inertial observers will measure exactly the same value for the speed of electromagnetic waves in a vacuum irrespective of their relative motions.

The full analysis for EM radiation results in an equation which predicts the change of frequency when there is a relative motion between source and observer. If the source is moving away from the observer the received radiation will be shifted to a lower frequency, a shift towards the red end of the spectrum, which is called a **redshift**. If the source and observer are moving towards one another, a blueshift is observed. If the relative motion is small compared with the speed of the EM waves in a vacuum, this formula reduces to the same approximation as in the case of sound waves:

$$\frac{\Delta f}{f_s} \approx \frac{v}{c}$$

Einstein's general theory of relativity predicts that gravitational fields will also affect the frequency of electromagnetic waves. When EM waves travel away from a massive object their frequency decreases in an effect called a **gravitational redshift**.

Physics issues: examples of the Doppler effect

We will now highlight some examples and uses of the Doppler effect. It might be useful to research some of them further.

Train going through a station

The sound emitted by a moving train's whistle is of constant frequency, but the sound received by a passenger standing on the platform will change. At any instant of time, it is the resolved component of the train's velocity towards the passenger that should be used to calculate the frequency received.

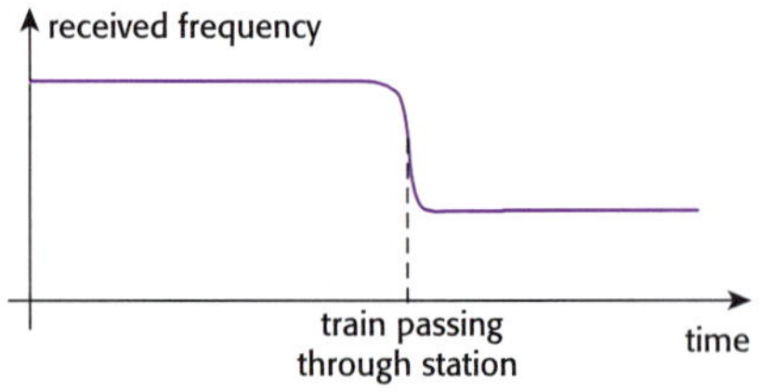

Figure 9 A passenger on a platform hears a change in pitch

Speed measurement for vehicles

In many countries the police use radar to measure the speed of moving vehicles to see if they are breaking the speed limit. A pulse of radio waves of known frequency is emitted, reflected off the moving car and received back at the source. The change between the frequency emitted and the frequency received is used to calculate the speed of the car.

There is a double Doppler effect in this situation. The frequency received by moving car will be higher than the emitted source (if the car is moving towards the radar) and the car also acts as a moving source when the signal is sent back to the receiver.

Blood-flow measurements

Doctors use an analogous setup to the speed measurement for vehicles in order to measure the blood flow rate around a body and identify circulation problems.

An ultrasound pulse is arranged to reflect off blood flowing around the body. The frequency that is received can be analysed to calculate the speed of blood flow. Measurements of the dimensions of the vessels are then used to directly calculate the volume flow rate.

Redshift

The light from the vast majority of stars is redshifted. When the light is analysed the frequencies represent the absorption spectrum (see page 185) of the elements contained in the star's outer layers. The measured frequencies are, however, not the same as the frequencies associated with particular elements as measured in the laboratory.

For a given star, all the received frequencies have been shifted by the same amount towards the red end of the visible spectrum, i.e. to lower frequencies. Different stars have different redshifts and the magnitude of the shift can be used to calculate the star's recession velocity. This provides evidence for the big bang model of the creation of the universe.

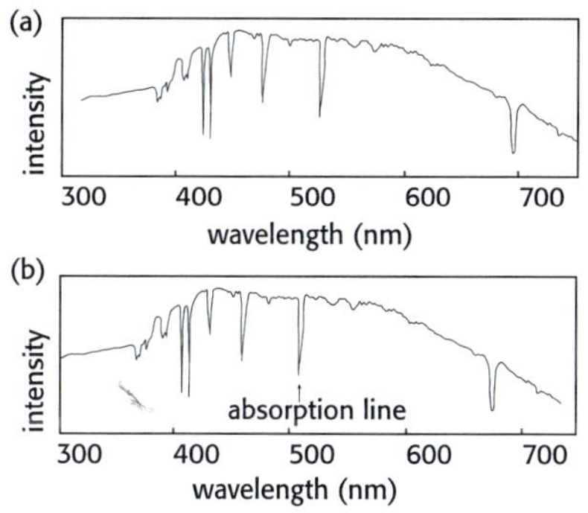

Figure 10 Red shift of a star's absorption spectrum

Rotating object

When a luminous object rotates, one side will be moving towards the observer and one side will be moving away. For example, the Sun is rotating on its own axis. When an observer looks at the Sun, (it is not safe to do this without special glasses so do not attempt this) the left-hand side is moving towards the observer and the right-hand side is moving away. This means that the absorption spectra for the left-hand side will be blue shifted and the right-hand side will be red shifted. This can be used to find the speed of the rotation of the star.

The same splitting of spectral lines can be used to identify binary stars where only one source of light can be seen. See page 297 in the astrophysics section.

Broadening of spectral lines

Absorption and emission spectra provide evidence for discrete atomic energy levels. Precise measurements show that each individual level is actually equivalent to a small but defined wavelength range.

The gas molecules that are emitting the light have different velocity components, when resolved along the line of observation. Light from different molecules will be subjected to Doppler shifts so there will be a general "Doppler broadening" of the discrete wavelengths – any given spectral line will broaden into a range of measured wavelengths. In any given spectrum, this amount of broadening must depend on the temperature of the substance emitting the light. A higher temperature means that the molecules have a wider distribution of kinetic energy and hence the spectral line is broader.

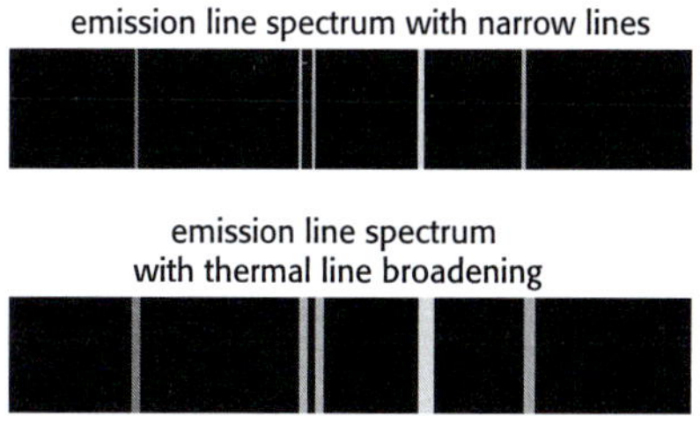

Figure 11 Doppler broadening

Diffraction

In Chapter 8 we introduced the concept of diffraction, that is, any wave that is constrained to pass through a narrow aperture will spread out. The extent of the spreading depends on the size of the aperture compared with the wavelength involved. An extremely narrow aperture behaves like a point source, with equal intensities being transmitted in all directions whereas a wide slit can result in no effective diffraction at all.

The mathematics of diffraction can be very complex and depends on the physical setup used. It is important to distinguish between **Fresnel** and **Fraunhofer** diffraction. Fresnel diffraction involves diverging rays (spherical wavefronts) arriving at the aperture whereas Fraunhofer diffraction involves parallel waves (plane wavefronts). Analysis of the former (Fresnel) is complex whereas the later (Fraunhofer) is more straightforward. Unfortunately, simple experimental setups often involve Fresnel diffraction.

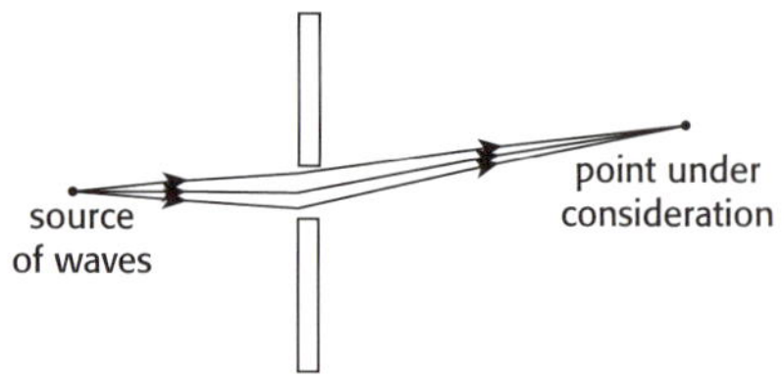

Figure 12a Fresnel diffraction

A wavefront may be considered to be constructed from an infinite number of point sources. The position of the next wavefront can be determined by considering the waves produced by each point source within the previous wavefront. This idea is known as **Huygens principle**. The overall diffraction pattern at an aperture of a given width b is the result of superposition of all the individual point sources in the wavefront across the width of the slit.

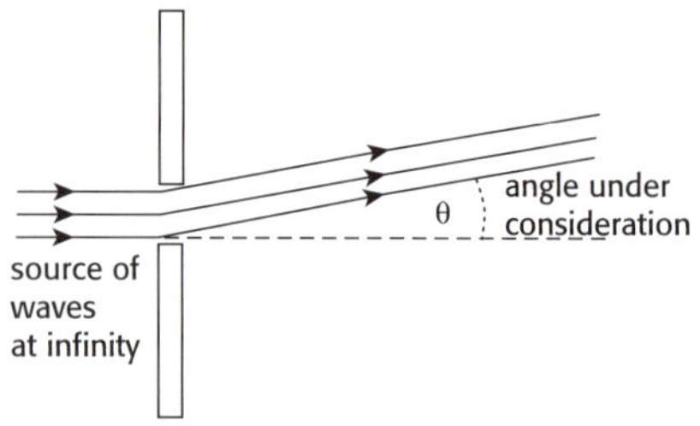

Figure 12b Fraunhofer diffraction

In order to calculate the variation with angle of the intensity of wave energy received after passing through an aperture, it is useful to consider the overall path difference across the slit. As the angle under consideration gets bigger, the path difference between the point sources across the aperture increases.

Consider a slit of width b.

$$\sin\theta = \frac{\text{path difference across aperture}}{\text{slit width, } b}$$

At a certain angle, the path difference across the slit will equal one whole wavelength. Waves from the point sources in one half section of the slit will all cancel with the corresponding point source in the other half.

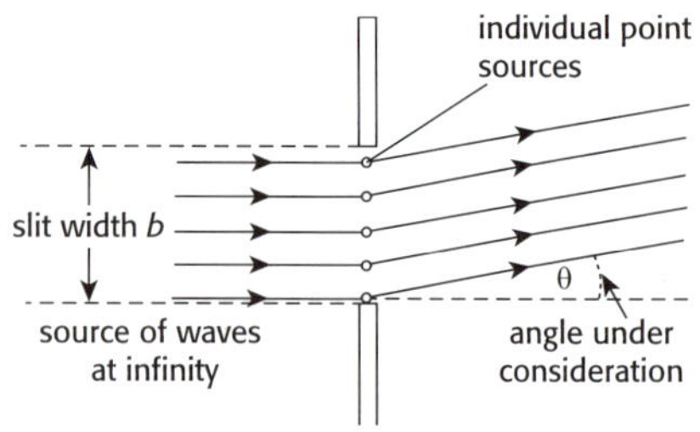

Figure 13 A single slit can be considered to contain many individual point sources

The result will be the first minimum. This takes place when

$$\text{path of difference} = b\sin\theta = \lambda$$

$$\sin\theta = \frac{\lambda}{b}$$

Provided θ is small, we can apply the small angle approximation.

$$\sin\theta \approx \theta = \frac{\lambda}{b}$$

θ is the angle between the straight through direction and the first minimum of the diffraction pattern measured in radians
λ is the wavelength, measured in m
b is the slit width, measured in m.

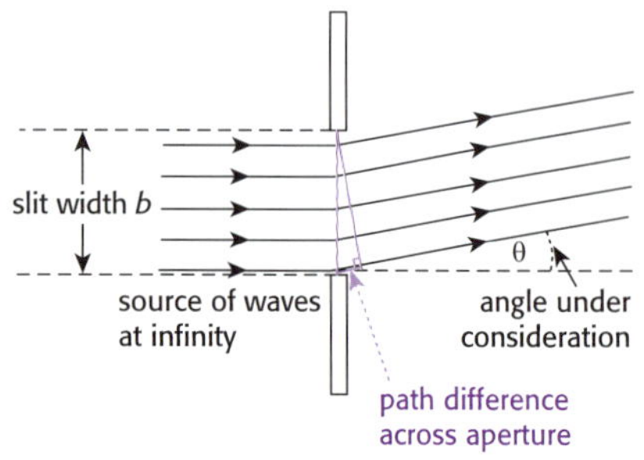

Figure 14 The path difference between individual point sources in a single slit

At some larger angles, the path difference across the slit will equal one and a half wavelengths. Waves from the point sources in two of the sections will cancel, leaving the amplitude reduced to a third of the maximum. This is the first maximum. The intensity at this angle (which is proportional to the square of the amplitude) will be one ninth of the initial value.

The next minimum occurs at $\theta = \frac{2\lambda}{b}$ and the next maximum is $\frac{1}{25}$ of the intensity of the initial value. The resulting diffraction pattern is shown in Figure 15:

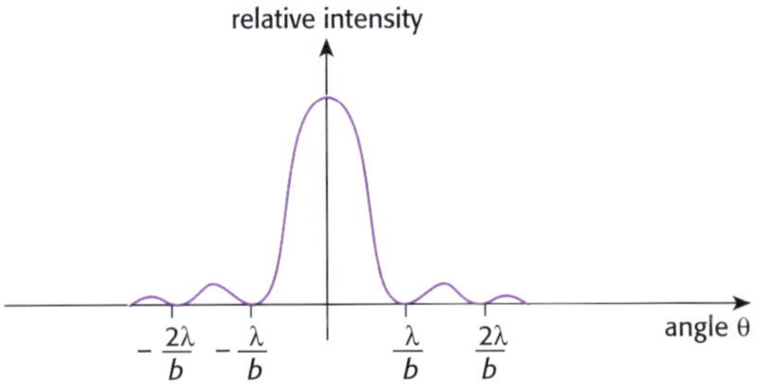

Figure 15 Single slit diffraction

Figure 16 below shows light from a laser (Helium-Neon) that has passed through a single slit of width $\frac{1}{16}$ mm and then projected onto a plastic ruler 3.0 m away from the slit. By taking measurements from the photograph, calculate the wavelength of laser light used. Estimate the uncertainty in your value.

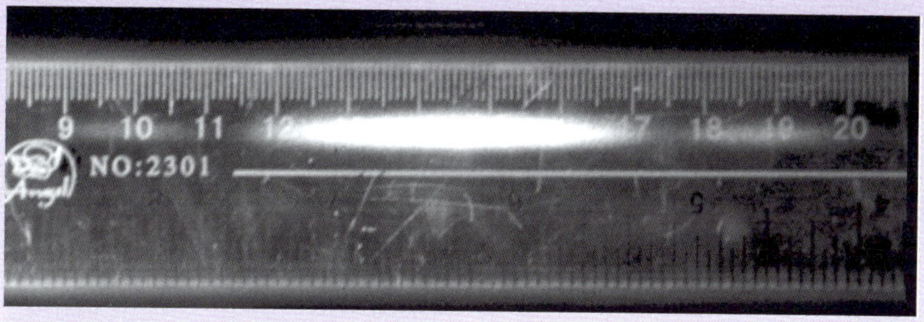

Figure 16

Rayleigh criterion

When two sources have a small angular separation, it may be difficult to resolve them as two separate objects, in which case they appear as a single source. For example, many of the "individual" stars that you can see at night actually turn out to be two stars in close orbit around one another (a binary star system). On top of this, a significant number of the "stars" are, in fact, galaxies (large collections of stars). In both cases your eyes are unable to resolve the individual sources because the angular separation between them is too small.

When light from two different sources is brought together, the principle of superposition determines the final image. When viewing two sources through an aperture, such as stars through a telescope, the overall result is the addition of individual diffraction patterns. Rayleigh's criterion (which can be practically verified) is the suggestion that:

Two sources are just distinguishable when the first minimum of the diffraction pattern of one of the sources falls on the central maximum of the diffraction pattern of the other source.

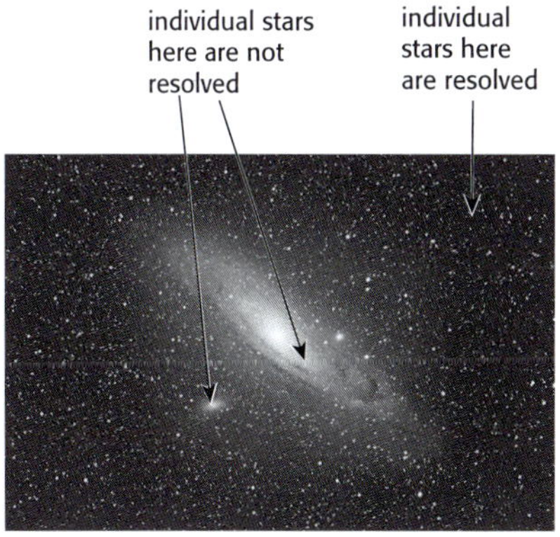

Figure 17 The Andromeda Galaxy

The angle at which two sources can just be resolved is fixed by the wavelength being used λ and the size of the aperture involved. If the aperture is a slit of width b then the minimum angle must be $\frac{\lambda}{b}$, as shown on page 104. More typically, the aperture is circular in which case the minimum angle is $1.22\frac{\lambda}{b}$, where 1.22 is a correcting factor for the slightly different geometry. This equation applies equally to all waves. For example, visible light is diffracted when it passes through the circular aperture of a telescope or even the pupil of your eye and radio waves are also diffracted when collected by a radio telescope. Increasing the collecting diameter of a telescope will decrease the angle between two sources that are just resolved.

1 A radio telescope is receiving radio waves of frequency 5.0 GHz and its dish size is 76 m. Two stars that are 100 light years away from Earth are known to be 0.05 lights year apart.
   (a) Will the radio telescope be able to resolve these two sources?
   (b) Two separate identical radio telescopes are situated 200 km apart. The signals that they receive can be combined to effectively create a large radio telescope that is 200 km in diameter. Will these be able to resolve the two stars?

2 Figure 19 shows laser light that has passed through a circular hole of width 0.24 mm and then projected onto a plastic ruler 3.0 m away from the slit. By taking measurements from the photograph, calculate the wavelength of laser light used. Estimate the uncertainty in your value.

3 Research one situation or device in which a resolution limit results from the diffraction of waves through an aperture and decide how significant this factor is compared with other factors. The particular example and the method of presenting the results of your research are up to you. In each case:
   - outline the situation and explain where diffraction takes place
   - state the wavelength or range of wavelengths involved
   - state the dimensions of the aperture involved
   - calculate the resolution limit according to the Rayleigh criterion
   - state the practical resolution limit.

   Situations that could be considered include:
   (a) the human eye
   (b) an animal's eye
   (c) a camera
   (d) a land-based astronomical telescope
   (e) the Hubble space telescope
   (f) an array of radio telescopes
   (g) a visible light microscope
   (h) an electron microscope.

Figure 18

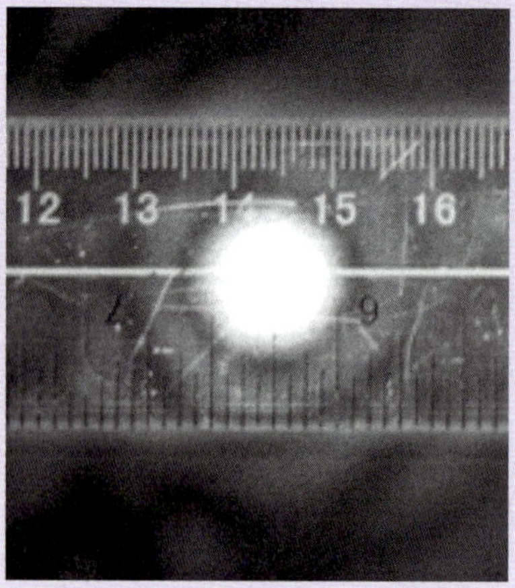

Figure 19

Polarization

Polarization is a property that only applies to transverse waves. Unlike longitudinal waves, where the direction of the oscillations is fixed by the direction of energy transfer, there are an infinite number of possible directions for the oscillations of transverse waves (all at 90° to the energy transfer direction). A wave that is plane polarized is one that only has one direction of oscillation.

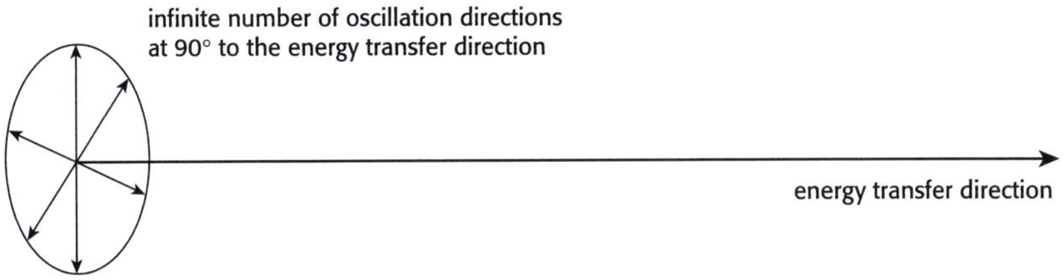

Figure 20 This wave is not polarized

In electromagnetic waves, for example, the plane of vibration is defined as the plane that contains the electric field. Some EM radiation is polarized as a result of its method of production. For example, the radio waves used in the broadcasting of terrestrial television signals are often polarized either horizontally or vertically. The plane of polarization is fixed by the chosen arrangement of the broadcasting aerials.

On the other hand, light from a standard light bulb is not polarized. This means that the oscillations of the electric field associated with the light take place in all directions. It is also possible for light to be partially polarized, which means that the electric field oscillations are more common in one direction than in all other directions. One final possibility is to add together two plane-polarized waves to create a wave whose plane of polarization varies. Depending on the phase of the two waves this is said to be **elliptically** or **circularly polarized**.

The simplest method of polarizing light is to use a sample of a substance called **Polaroid**. This man-made material consists of long-chain molecules that are aligned together. The long chains absorb light that has its electric field in the same direction as the chains and only allows through light with an electrical field at 90° to the chains:

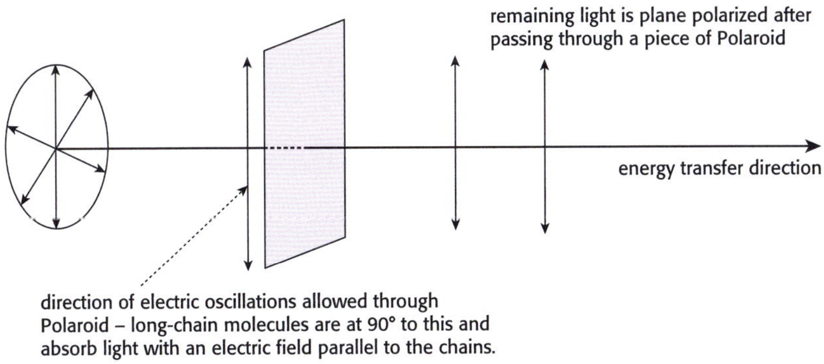

Figure 21 Polarizing a wave

Polaroid absorbs light and can be used to manufacture dark glasses. The advantages of using polaroid are discussed on page 113–14.

Explain why "crossed" polaroids (two samples of Polaroid at 90° to one another) absorb all the light in the region of their overlap whereas they transmit some light when parallel, as represented in Figure 22:

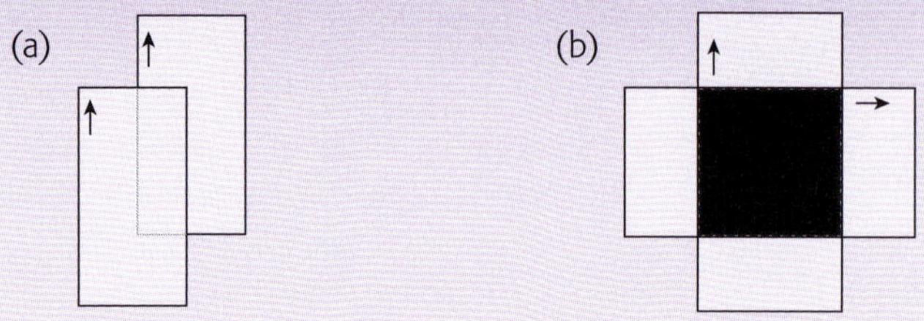

Figure 22

Some materials (e.g. sugar solutions) are **optically active**. This means that the plane of polarized light is rotated as it passes through the material. This rotation is a result of the interaction between the molecules in the material and the incident light. The degree of rotation for a given substance is measured using two polarizers (one is called an **analyser**) as shown in Figure 23.

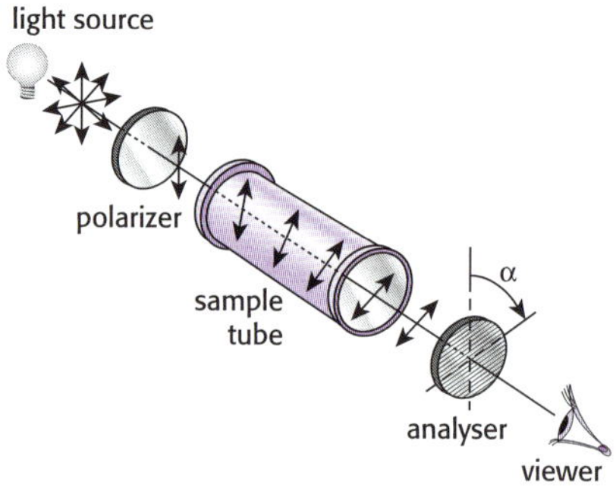

Figure 23 Measuring the optical activity of a sample

The polarizer and analyser are initially aligned without the sample tube being present. The introduction of the optically active sample causes a rotation of the plane of polarization of the light. The analyser is rotated through an angle α to find the maximum intensity of transmitted light. This apparatus is known as a **polarimeter**.

The angle through which the plane of polarization of the light is rotated depends on the length of the sample and its concentration. An example of a polarimeter can be easily constructed using a light source, a beaker, and two pieces of Polaroid. Once built, you can test it using sugar solutions of different concentrations.

1 Different lengths of a particular concentration of sugar solution were analysed in a polarimeter. Table 2 shows the variation with solution length of the angle of rotation of the plane of polarization.
   (a) Hypothesize the relationship between angle of rotation and solution length.
   (b) Plot a suitable graph to test your hypothesis.

2 Different concentrations of sugar solution were analysed in a polarimeter. Table 3 shows the variation with concentration of the angle of rotation of the plane of polarization. The lengths of each sample were the same.
   (a) Hypothesize the relationship between angle of rotation and solution concentration.
   (b) Plot a suitable graph to test your hypothesis.

Table 2

Angle of rotation/°	Length of sugar solution/cm
7	2.0
14	4.0
20	6.0
27	8.0
34	10.0

Table 3

Angle of rotation/°	concentration of sugar solution/g cm^{-3}
17	0.25
33	0.50
50	0.75

3 A bright source of light is viewed through two sheets of Polaroid whose preferred directions are initially parallel.
(a) Calculate the angle through which one sheet needs to be turned in order to reduce the amplitude of the observed electric field vibrations to half its original value.
(b) Calculate the effect this has on the transmitted intensity.
(c) Calculate the angle through which one sheet needs to be turned to reduce the transmitted intensity to half its original value.

Mathematical physics: polarization calculations

There are two mathematical laws associated with polarization – Malus's law and Brewster's law.

Malus's law

The intensity of plane-polarized EM radiation from the analyser depends on the intensity of the plane-polarized light I_0 and the angle between the direction allowed by the analyser and the electric field direction of the polarized light, θ. The analyser only allows one component direction of the electric field through.

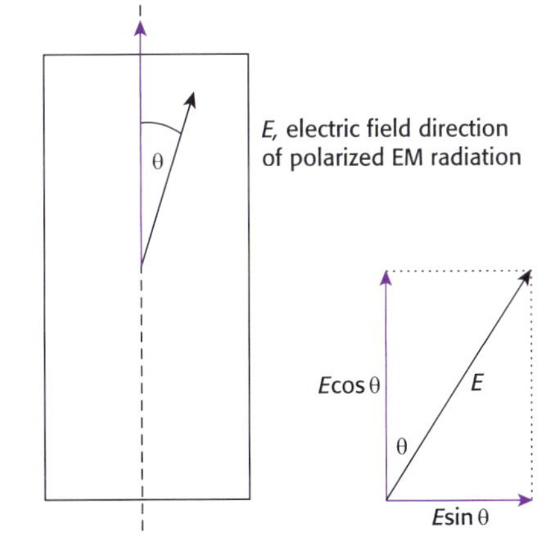

Figure 24 Geometry of transmitted component

The transmitted amplitude of the electric field is $E\cos\theta$. Since the intensity is proportional to the square of the amplitude, the transmitted intensity I is given by:

$$I = I_0 \cos^2\theta$$

Brewster's law

When light is reflected from a surface some light energy is reflected and some is refracted. The reflected light is predominately plane polarized in a direction that is parallel to the surface. If the transmitted ray and reflected ray are at right angles to one another, then the reflected ray is totally plane polarized, as represented below. The angle of incidence for this condition is called **Brewster's angle**.

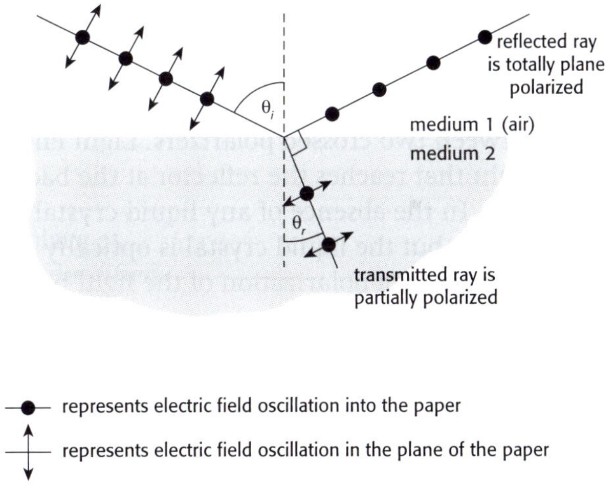

Figure 25 Polarization upon reflection

$$\theta_i + \theta_r = 90°$$

refractive index of medium 2 is given by:

$$n = \frac{\sin\theta_i}{\sin\theta_r} = \frac{\sin\theta_i}{\cos\theta_i} = \tan\theta_i$$

Physics issues: examples of polarization

Polaroid dark glasses

Reflections from all surfaces are predominately polarized in a plane that is parallel to the reflecting surface. In normal use (i.e. a person standing looking at a surface), Polaroid dark glasses are arranged

The zero points for the various scales are set at different temperatures. The sizes of the degrees on the Celsius and Fahrenheit scales are different, but to convert from a temperature T on the thermodynamic scale to one on the Celsius scale (θ), all we need do is subtract 273.15:

$$\theta/°\text{C} = (T - 273.15)/\text{K}$$

An object's temperature is used to predict the direction in which thermal energy naturally flows – that is, from hot to cold. When two objects are put in thermal contact, thermal energy will flow from the object with the higher temperature to the object with the lower temperature.

It can be very difficult to control the experimental uncertainties associated with experiments that involve heat and temperature. This is because any hot object will tend to lose thermal energy to its surroundings and cool down. Conversely, any cold object will tend to receive thermal energy from the surroundings and heat up.

Careful experiments show that, provided a state change does not take place, the temperature change of an object is related to the quantity of thermal energy transferred. For example, a sample of water requires 4200 J for its temperature to change from 22 °C to 23 °C with no loss of thermal energy. It will need the same amount of energy to change from, for example, 57 °C to 58 °C.

The amount of thermal energy Q needed to raise the temperature of a body through 1 K is called its **thermal capacity**, C:

$$Q = C\,\Delta T$$

where
Q is the thermal energy transferred to the object, measured in J
C is the thermal capacity (also called the total heat capacity), measured in J K^{-1}
ΔT is the temperature change of the object, measured in K (or Celsius degrees)

This equation applies so long as there is no change of state.

An object's thermal capacity depends on its material and its mass. The **specific heat capacity** c is the amount of thermal energy required to raise the temperature of 1 kg of a substance by 1 K. In this context, the word "specific" means "per unit mass".

$$Q = mc\,\Delta T$$

where
Q is the thermal energy transferred to the object, measured in J
m is the mass of the object, measured in kg
c is the specific heat capacity, measured in J kg^{-1} K^{-1}
ΔT is the temperature change of the object, measured in K (or Celsius degrees)

This equation applies so long as there is no change of state.

?

1 Explain the difference between a rate of loss of temperature and a rate of loss of thermal energy. State the units that would be used to measure each quantity.

2 The specific heat capacity of water is 4200 J kg^{-1} K^{-1}, and that of copper is 385 J kg^{-1} K^{-1}. Calculate:
   (a) the temperature change of 500 g of water given 20 kJ of energy
   (b) the energy released by 200 g of copper cooling down from 700 °C to 20 °C.
   (c) the final temperature of a mixture of 200 g of water at 70 °C and 300 g of water at 10 °C, assuming no energy is lost to the surroundings
   (d) the final temperature when a 200 g block of copper heated to 700 °C is placed in 500 g of water at 10 °C, assuming no energy is lost to the surroundings.
   (e) Explain why your answer to (d) is likely to be an overestimate of the final temperature.

3 A hole is drilled in an iron block and an electric heater is placed inside. The heater provides thermal energy at a constant rate of 600 W, and its temperature is recorded as it varies with time. The 800 g block increases in temperature and reaches a final maximum temperature. The specific heat capacity of iron is 480 J kg^{-1} K^{-1}. Calculate:
   (a) the initial rate of increase in temperature
   (b) final rate of loss of energy of the block.

The **molar heat capacity** can also be useful; see page 115 for an explanation of the mole.

There is no net flow of thermal energy in or out of an object at constant temperature. Typically the object is probably gaining and losing thermal energy all the time. The rate of energy gained equals the rate of energy lost.

Thermal energy transferred to the surroundings is often described as "lost" energy. This correctly conveys the fact that it would be difficult, from a practical point of view, to harness this energy to do useful work. The conservation of energy means that this energy cannot have been destroyed, but is just in a form (the thermal energy of the surroundings) that makes it hard to utilize. The energy is said to be **degraded** (see Chapter 18, page 221).

Microscopic description

Any substance is composed of particles, and these are in random motion whether the substance is a solid, liquid, or gas. Heating an object causes, on average, these particles to move faster. The total thermal energy that is transferred to an object when it warms is shared among all its particles as random kinetic energy (KE).

Note that there is a difference between the organized KE that a moving object's particles must possess (which equals the KE of the object as a whole) and the random thermal kinetic energy that the particles must possess (which is part of the internal energy of the object and is related to its temperature).

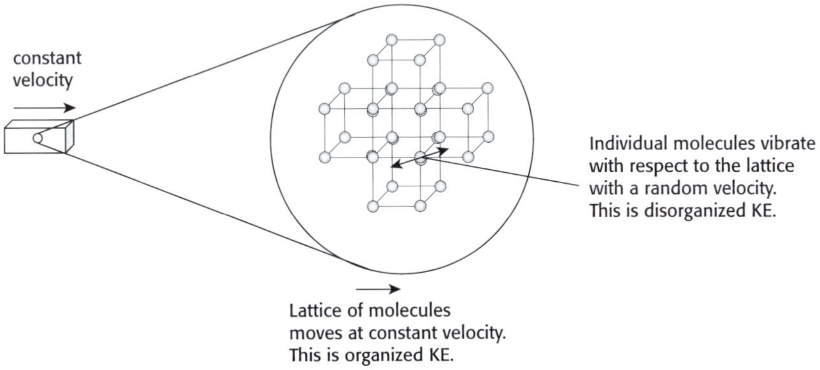

Figure 1 Organized and random KE.

Increasing a substance's temperature means that the average random KE of the particles has been increased. The temperature as measured on the thermodynamic scale is proportional to the average KE per particle.

There is no net flow of thermal energy in or out of an object at constant temperature, but individual molecules are interchanging energy all the time. When energy is "lost" to the surroundings this degraded energy is shared amongst all the molecules in the surroundings as disorganized KE. This makes it a difficult practical task to utilize.

?

1 A substance changes from solid to liquid at its normal melting temperature. What change, if any, occurs in the average kinetic energy and the average potential energy of its molecules?

	Average kinetic energy	Average potential energy
A	constant	constant
B	increases	constant
C	increases	decreases
D	constant	increases

2 The temperature of an ideal gas is reduced. Which **one** of the following statements is true?

A The molecules collide with the walls of the container less frequently.

B The molecules collide with each other more frequently.

C The time of contact between the molecules and the wall is reduced.

D The time of contact between molecules is increased.

?

A car accelerates from rest up to a constant velocity and then slows down again to rest. When moving forward all the molecules of the car had more KE than when the car was at rest. Does this mean that a moving car has a higher temperature than a stationary one?

Inquiry: convection, conduction, and radiation

So far we have considered only the effect on an object's temperature of gaining or losing thermal energy. There is another very different property: the ease (or not) with which thermal energy can be transferred between objects or flow through any given object.

There are three important mechanisms by which thermal energy can be transferred from one object to another:

- convection
- conduction
- radiation (particularly in the infrared part of the electromagnetic spectrum).

We shall take the study of radiation further when we consider the electromagnetic spectrum, black body radiation, and the greenhouse effect.

Your task is to research some detail relating to the mechanisms of one of the three processes listed above. You should agree with your teacher the way in which you are going to share your findings with the rest of your group. Possible inquiries include:

1 the application of convection, conduction, and radiation to the design of an astronaut's space suit
2 how gliders are able to stay airborne
3 the design of a hot water system in a house
4 the macroscopic explanation for convection
5 the microscopic explanation for conduction
6 the factors that affect the rate at which infrared radiation is emitted by a surface
7 the factors that affect the rate at which infrared radiation is absorbed by a surface
8 how a vacuum flask works
9 whether the inside of the International Space Station could be kept warm by convection
10 Newton's law of cooling
11 the factors that affect the rate at which thermal energy is conducted along a material
12 the difference between thermal conductors and insulators
13 the coefficient of thermal conduction (also known as the thermal conductivity)
14 how wetsuits/drysuits keep divers warm
15 how the Inuit keep warm in igloos
16 the cooling effect of wind towers.

Investigating physics: saving Jack

This work can be attempted individually or in small groups. The investigation can be used to assess Design (D), Data collection and processing (DC), and Conclusion and evaluation (CE).

Cast to your minds back to the tragic end of the film *Titanic*, when Rose is saved but Jack tragically dies of hypothermia because he is totally immersed in almost freezing water. (See youtube clip, http://www.youtube.com/watch?v=1AVXPFRvc2A accessed May 2007) Your task is to think of ways in which Jack could have been saved, and then try to model them in the laboratory. You need to design an experiment that:

- models Jack
- allows you to vary the conditions suitably
- enables you to take a suitable range of data for analysis.

Produce a plan with an equipment list. Discuss it with your teacher or technician before you carry out the investigation. Try to use some data-logging equipment if possible.

Your final report should be a full laboratory report, including:

- a prediction and outline of the theory involved
- experimental details
- data tables
- graphs
- conclusion and evaluation.

Table 1

Phase	Macroscopic	Microscopic	Mean molecular separation	Number per m^3	Volume of molecules relative to that of substance
solid	both volume and shape are approximately constant	molecules are held in position by bonds and vibrate about a fixed point in the lattice	≈ atomic radius	10^{28}	1
liquid	has variable shape but volume is approximately constant	molecules vibrate within ordered clusters but there are no permanent bonds and can move	≈ atomic radius	10^{28}	1
gas	has variable shape and volume	The intermolecular forces are only significant during collisions, molecules are in random motion	≈ 10 × atomic radius	10^{25}	10^{-3}

Source: adapted from P.M. Whelan and M.J. Hodgson, Essential Pre-University Physics, John Murray

Rediscovering physics: thermal experiments

This section outlines three experimental situations that can be investigated in order to reinforce your understanding of thermal physics. A safety point to note in all these situations is to take great care near hot objects. Before starting any experiment, you should always discuss an assessment of the risks involved with your teacher.

Calibrating an unmarked thermometer

A mercury-in-glass thermometer (or equivalent) capable of recording temperatures between 0 °C and 100 °C can be used to measure an unknown temperature (room temperature, for example), even if the scale is not clearly marked. Measurements of the length of the column of mercury can be recorded for three different situations: ice/water mix, steam/water mix, and the unknown temperature. The ice/water mix is known to be at 0 °C and the steam/water mix must be at 100 °C. The unknown temperature can be deduced from these measurements. The experiment could provide a good opportunity to practise the treatment of errors.

Measuring a substance's specific heat capacity

A simple electrical heating method can be devised to measure the specific heat capacity of a substance, provided the loss of thermal energy to the surroundings is negligible. The energy dissipated in a time t by an electrical heater with potential difference V across it and current I flowing through it, is calculated using

$$Q = ItV$$

This energy can be assumed to be all transferred to the substance under consideration. The mass of the substance is m and the recorded maximum temperature change is ΔT. The specific heat capacity c is given by

$$c = \frac{Q}{m\Delta T}$$

Again, the experiment could provide a good opportunity to practise the treatment of errors, and to suggest improvements to the technique in the light of the results obtained.

Further research can provide other detailed methods that:

- help correct for any loss of thermal energy
- involve a "method of mixtures" and use a known specific heat capacity to calculate an unknown value
- use a continuous-flow method for a liquid. The apparatus is quite complex to set up, but accurate results can be obtained.

Estimating the temperature of a Bunsen flame

A metal block of mass m_{metal} and specific heat capacity c_{metal} is heated directly in a Bunsen flame and is assumed to reach a steady temperature equal to that of the Bunsen flame, T_{hot}. The block is carefully and quickly transferred to a mass of cold water m_{water} of temperature T_{cold}, and the final combined temperature is T_{final} (Figure 2).

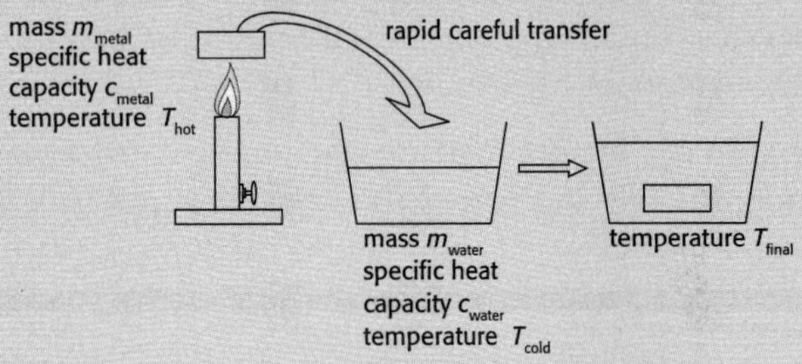

Figure 2 Experiment to estimate the temperature of a Bunsen flame.

Ignoring all heat transfers to the surroundings, the energy lost by the metal cooling down must equal the energy absorbed by the water warming up:

energy lost by metal cooling down = energy gained by water warming up

Therefore

$$m_{metal}c_{metal}(T_{hot} - T_{final}) = m_{water}c_{water}(T_{final} - T_{cold})$$

Substitute in all the values and rearrange for T_{hot}.

A better estimate can be achieved by devising methods to:

- correct for the thermal energy lost while the water is warming up
- allow for the energy involved in changing the temperature of the water container
- correct for the energy involved in the change of state of the small amount of water evaporated during the process (see later work).

Macroscopic and microscopic: phase change

When a substance changes phase, the potential energy of the particles changes as, on average, they move further apart.

Solids, liquids, and gases have different macroscopic properties corresponding to the different structures of the substances in molecular terms. Table 1 summarizes some aspects of these differences, and gives some typical values. A separate fourth phase, **plasma**, exists when a gas is heated to extremely high temperatures (e.g. 10^4 K or above), sufficient to ionize the atoms and molecules. A plasma consists of a mixture of ions and some neutral particles in random motion. The hydrogen and helium in the Sun are in the plasma state.

Macroscopic description

Whenever a substance changes state it does so at a fixed temperature, and there is energy associated with the change: this energy is called the **latent heat**. The amount of thermal energy involved depends on the type of substance, on its mass, and on the phase change involved. It can also be affected by other external factors, such as external air pressure.

The change of state from solid to liquid is called **fusion**, and the change from liquid to gas is called **vaporization**. Each of these changes involves energy being given to the substance. The reverse processes (liquefying and solidifying) must involve energy being released from the substance.

The **specific latent heat** for any phase transformation (at constant temperature) is defined as the latent heat per unit mass:

$$Q = mL$$

where
Q is the thermal energy transferred to the substance, measured in J
m is the mass of the substance, measured in kg
L is the specific latent heat, measured in J kg^{-1}

Microscopic description

The making or breaking of bonds involves energy. When bonds are broken, the potential energy of the molecules is increased, and this requires energy input. In some changes (for example vaporization) the change is also associated with work being done: when a liquid changes into a gas the volume greatly increases, so that during the expansion work must be done (pushing the atmosphere away), which adds to the increase in potential energy. When bonds are formed (solidification or liquefaction) then energy must be released.

It is important to note that when intermolecular bonds are made or broken, this happens independently of the KE of the individual molecules. When water boils, the temperature of the liquid and the temperature of the vapour is the same: 100 °C. The bonds between water molecules are being broken, but the average kinetic energy per molecule remains the same, and thus the molecules do not, on average, move faster.

Evaporation and boiling

There are two common processes by which liquids can turn into the vapour state: evaporation and boiling. Boiling takes place at one fixed temperature, and happens throughout the liquid concerned (bubbles appear throughout the body of the liquid). Evaporation, however, is the process by which the faster-moving molecules can escape from the surface of a liquid (Figure 3).

As the faster-moving molecules are the only ones that can escape, evaporation causes the average KE per molecule to decrease. In other

?

1. The latent heat of vaporization of water is 2300 kJ kg^{-1}. How long would it take a 2 kW electric kettle containing 800 g of boiling water to boil off all the water?
2. In order to maintain a constant body temperature, a sunbather needs to lose about 320 J of thermal energy to the surroundings every second through sweating. Estimate the amount of sweat evaporated from the skin of the sunbather every hour. (The specific latent heat of sweat is about 2300 kJ kg^{-1}.)

?

Some liquid is contained in a shallow dish that is open to the atmosphere. The rate of evaporation of the liquid does **not** depend on

A the temperature of the liquid.
B the temperature of the atmosphere.
C the depth of the liquid.
D the pressure of the atmosphere.

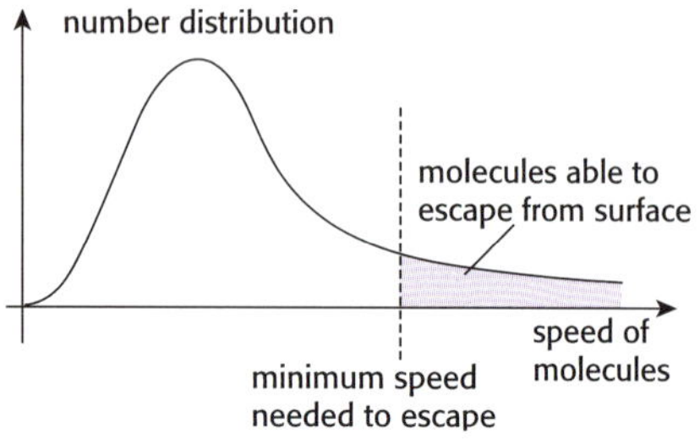

Figure 3 Standard energy distribution curve

words, the temperature reduces. Evaporation causes cooling, and is an additional method of transferring thermal energy.

Mathematical physics: the mole

A detailed analysis of the physics (or indeed the chemistry) behind the macroscopic behaviour of materials must involve knowing how many atoms or molecules exist in a given sample of a substance. The numbers involved are extremely large, so it helps to have a unit that deals with large numbers of particles. The amount of substance can then be measured in terms of a fixed number of atoms or molecules. The unit used for the amount of substance is the **mole**.

A mole of any substance will always contain the same number of particles, but the mass will vary depending on the particular substance considered. The mole can be thought of as a very large number. It is defined as the amount of substance that contains the same number of elementary units as there are atoms in 12 g of carbon-12. The number of particles in one mole is called the **Avogadro constant**, and is equal to 6.02×10^{23}.

The word "molar" just means "per mole", so the **molar mass** of any substance is the mass per mole. For example, 1 mole of water molecules (H_2O) has a mass of 18 g, whereas 1 mole of oxygen molecules (O_2) has a mass of 32 g. These masses can be worked out from the individual molar masses for the elements involved (the molar mass of the element hydrogen is 1 g, and for oxygen it is 16 g). Data tables often provide lists of the **relative molecular mass** (rmm) of different elements and molecules. This quantity does not have any units, and is sometimes (wrongly) also called the "molecular weight".

$$\text{relative molecular mass} = \frac{\text{mass of a molecule}}{\text{mass of } {}^{12}_{6}\text{C atom}} \times 12$$

so

$$\text{molar mass} = \frac{\text{relative molecular mass}}{1000}\ \text{kg mol}^{-1}$$

The chemical reaction between hydrogen and oxygen to form water provides an example of how the mole can be used to bring together the microscopic and the macroscopic points of view. A mixture of hydrogen and oxygen is stable at room temperature but will explode if ignited.

The chemical reaction is described by the following equation:

$$2H_2 + O_2 \rightarrow 2H_2O$$

The information represented by this equation is that, at the microscopic scale, two molecules of hydrogen (H_2) and one molecule of oxygen (O_2) combine to produce two molecules of water (H_2O). Since one mole of anything contains the same number of particles, this means that two moles of hydrogen will combine with one mole of oxygen to produce two moles of water. In other words, 4 g of hydrogen (rmm of hydrogen = 2) will react completely with 32 g of oxygen (rmm = 32) and produce 36 g of water.

There are two related pieces of information that can prove useful to those also studying chemistry. First, the charge on a mole of electrons ($= 9.6 \times 10^4$ C) is called the Faraday constant, and is used in calculations involving electrolysis. Second, a mole of any ideal gas at

STP (standard temperature and pressure: 273.15 K and atmospheric pressure) has a volume of 22.4×10^{-3} m^3.

Conversions between the mass of a substance and the number of atoms or molecules involved are relevant when we are considering the behaviour of an ideal gas in detail (see below) and the radioactive decay of nuclei (see page 189).

Ideal gas behaviour

The quantitative analysis of the behaviour of gases is studied in more detail by HL students (see page 118).

Compared with the temperature of solids and liquids, the temperature of a gas is more dependent on external factors. For a given sample of gas, three quantities are interdependent: the pressure p, the volume V, and the temperature T. Macroscopically, many gas experiments show that these quantities are related in approximately the same way, as summarized by the ideal gas equation of state (see page 120). From the macroscopic point of view, an **ideal gas** is one that follows the ideal gas equation for all values of pressure, volume, and temperature.

One of the triumphs of classical physics is that it is possible to predict the macroscopic behaviour of a gas by using Newton's laws to analyse the interactions taking place on the microscopic (atomic) scale. This analysis leads to the identification of temperature as a measure of the average random kinetic energy of the molecules of an ideal gas.

The **kinetic theory** views all matter as consisting of individual atoms and molecules. Certain microscopic assumptions need to be made in order to be able to deduce the macroscopic behaviour of a gas. For an ideal gas, these are as follows:

- The molecules are assumed to behave in an idealized way: that is, Newton's laws of mechanics apply to the individual molecules' motions.
- The intermolecular forces are assumed to be negligible (except during a collision).
- The molecules are assumed to be spherical and their volume negligible (compared with the volume occupied by the whole gas).
- The molecules are assumed to be in random motion.
- The collisions between molecules are assumed to be perfectly elastic.
- The time taken for a collision is assumed to be negligible.

Pressure: macroscopic and microscopic views

From the macroscopic point of view, a general definition of pressure is as follows:

$$P = \frac{F}{A}$$

where
F is the force exerted, measured in N
A is the normal area over which the force acts, measured in m^2
P is the pressure, measured in $N\ m^{-2}$ (or pascals, Pa)

$$1\ Pa = 1\ N\ m^{-2}$$

In the above equation for pressure, area is at right angles to the direction in which the force acts. If a force does not act at 90° to a surface, then the pressure is calculated using the component of the force that is at 90° to the surface.

Gas pressure can be understood by considering the large number of collisions that take place between the molecules of the gas and the walls of the container:

- When a gas molecule hits the walls of the container, it bounces off.
- The momentum of the gas molecule has changed during this collision, and Newton's second law applied to this situation means that there must have been a force on the molecule from the wall.
- Newton's third law applied to this situation means that there must have been a force on the wall from the molecule.
- Each time a molecule collides with the wall, there will be a small force from the molecule on the wall.
- In a given time, there will be a certain number of collisions.

- The average result of all of these individual molecular collision forces acting for a short time will be a constant force on the wall of the container.
- The value of the constant force on the wall divided by its area is the pressure that the gas exerts on the wall of the container.

A gas exerts a pressure on every surface of its container: this pressure depends on the mass of the gas, its volume, and its temperature. The macroscopic behaviour of an ideal gas can be understood in terms of the motion of the molecules of the gas. In each of the following scenarios, one variable has been changed and all the other ones are kept constant.

1 Ideal gases increase in pressure when more gas is introduced into the container. The increase in mass of the gas means that there are more gas molecules in the container, and therefore an increase in the number of collisions that take place in a given time. The force from each molecule remains the same, but an increased number of collisions in a given time means that the pressure increases.

2 Ideal gases increase in pressure when their volume decreases. The decrease in volume means that molecules hit a given area of the walls more often. The force from each molecule remains the same, but an increased number of collisions in a given time means that the pressure increases.

3 Ideal gases increase in pressure when their temperature increases. The increased temperature means the molecules are moving faster, and thus they hit the wall more often. The force from each molecule goes up on average and an increased number of collisions in a given time means that the pressure increases.

4 Ideal gases increase in temperature when their volume is decreased. As the volume is reduced, the walls of the container move inwards. The molecules are colliding with a moving wall and will, on average, speed up. Faster-moving molecules mean that the average kinetic energy per molecule has increased – that is, the temperature has increased.

1

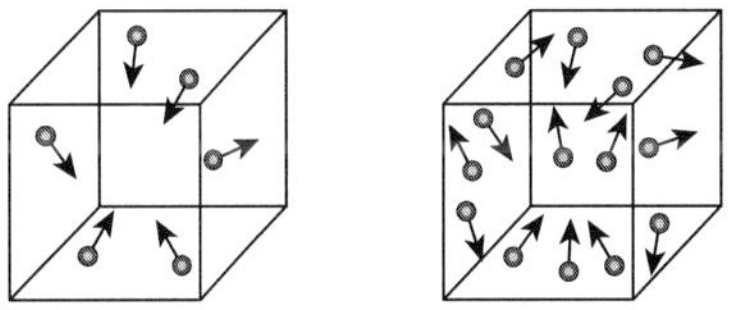

few molecules low pressure | many molecules high pressure

2

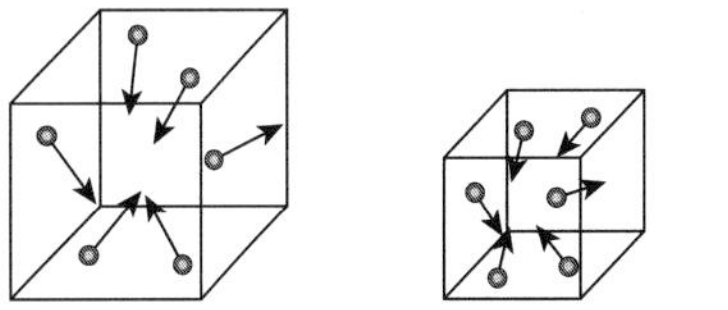

high volume low pressure | low volume high pressure

3

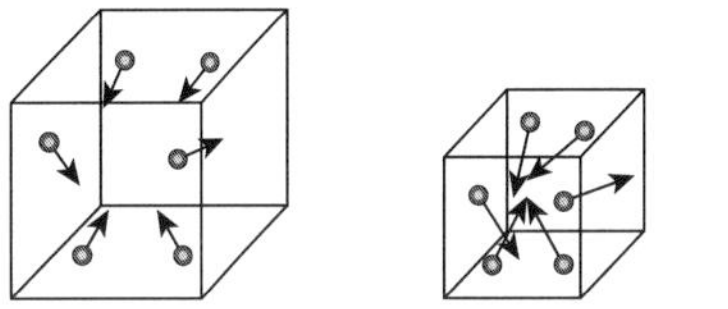

low temperature low pressure | high temperature high pressure

4

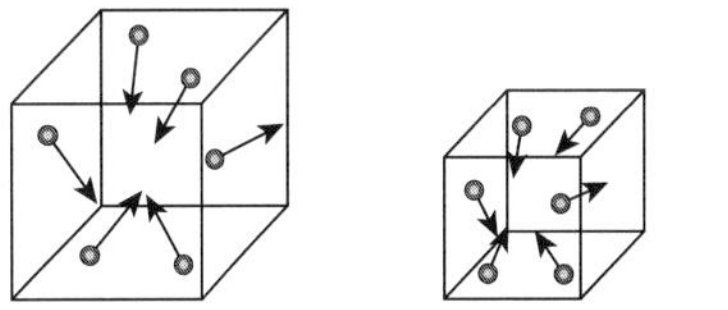

low temperature high volume | high temperature low volume

11 Thermodynamics

HL Before studying this chapter it would be worth reviewing the concepts introduced in Chapter 10. Although the material covered is all higher level material, standard level students would gain useful background information by studying this chapter.

Working with data: analysis of gases

Three interrelated quantities are needed to describe a fixed mass of gas: its pressure, its volume, and its temperature. In order to investigate how these quantities are related, three separate experiments are needed. In each experiment, one of the variables is kept constant (the controlled variable), one is altered (the independent variable), and the variation in the third variable (the dependent variable) is recorded.

The following three experiments provide outline experimental descriptions and data that should be analysed as practice for the DCP and CE skills.

Experiment 1: variation of volume with pressure at constant temperature

The apparatus illustrated in Figure 1 was used to subject a fixed mass of trapped dry air to different pressures. As the pressure increases, the oil moves up the graduated tube and the air occupies a smaller volume. The Bourdon gauge records the pressure.

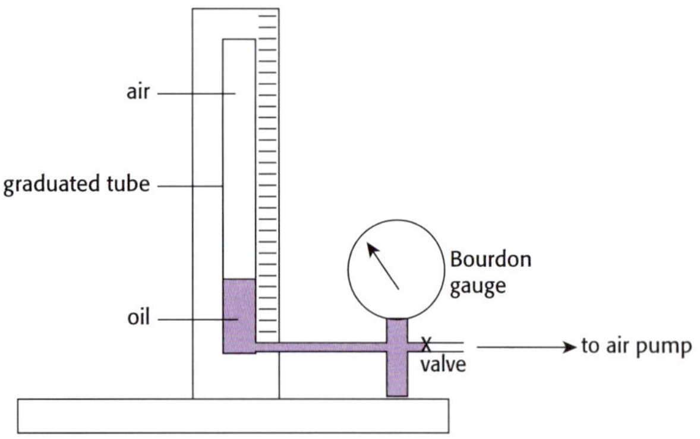

Figure 1 Apparatus to measure the variation of volume with pressure, at constant temperature

The following procedure was followed:

1 Switch on the pump and compress the air.

2 Stop the pump when the Bourdon gauge registers its maximum pressure.

3 Record the pressure on the gauge and the volume of air in the graduated tube.

4 Carefully release the valve to reduce the pressure slightly.

5 Record the new measurements.

6 Repeat steps 4 and 5 until the pressure is back to atmospheric pressure.

Sample data are listed in Table 1.

Table 1 Sample data

Volume of trapped air		Pressure, P / kN m^{-2} (kPa)
V_1 / cm^3	V_2 / cm^3	
34.0	34.0	120
29.0	29.0	140
25.5	25.5	160
22.5	22.5	180
20.0	20.0	200
18.0	18.0	220
16.5	16.5	240
15.5	15.0	260
14.0	14.0	280
13.0	13.0	300

Temp = 17.5 °C Temp = 18.0 °C

Analyse the data in Table 1 and come to a conclusion about how the volume of a gas is related to the pressure.

Experiment 2: variation of volume with temperature at constant pressure

The apparatus shown in Figure 2 was used to subject a fixed mass of trapped dry air to different temperatures.

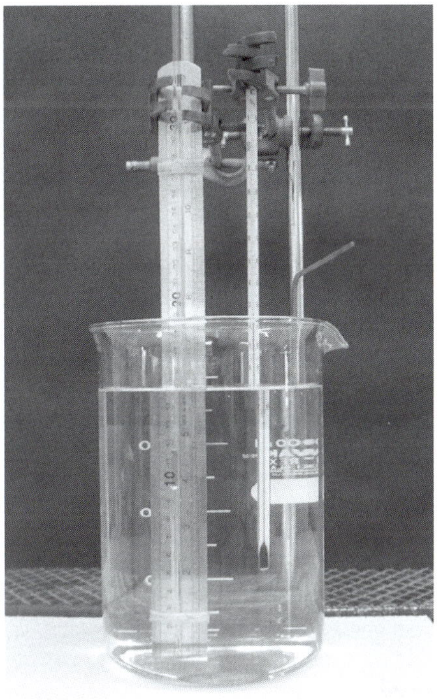

Figure 2 Apparatus to measure the variation of volume with temperature, at constant pressure

Sample data are listed in Table 2.

Table 2 Sample data

Temperature, T / °C	Length of trapped air	
	L_1 / 10^{-3} m	L_2 / 10^{-3} m
80	98.0	98.0
75	96.0	97.0
70	95.0	95.0
65	93.0	94.0
60	92.0	92.0
55	90.0	90.0
50	89.0	89.0
45	88.0	88.0
40	86.0	86.0
35	85.0	85.0
30	83.0	83.0

Estimate the uncertainties associated with the readings and then analyse the data in Table 2 and come to a conclusion about how the volume of a gas is related to the temperature.

Experiment 3: variation of pressure with temperature at constant volume

The apparatus shown in Figure 3 was used to subject a fixed mass of trapped dry air to different temperatures. Sample data are listed in Table 3.

Figure 3 Apparatus to measure the variation of pressure with temperature, at constant volume

Table 3 Sample data

Temperature, T / °C	Pressure, P / 10^5 Pa
82	1.175
73	1.150
66	1.125
57	1.100
49	1.075
40	1.050
32	1.025
25	1.000

Estimate the uncertainties associated with the readings and then analyse the data in Table 3 and come to a conclusion about how the pressure of a gas is related to the temperature.

Absolute scale of temperature

Note that in experiment 2 (or 3) it would be incorrect to state that volume (or pressure) is proportional to the temperature. It is, however, possible to continue the trend line (this process is known as **extrapolating**) to lower temperatures and redefine a new scale of temperature. For example, the graph in Figure 4 represents the variation with temperature of pressure.

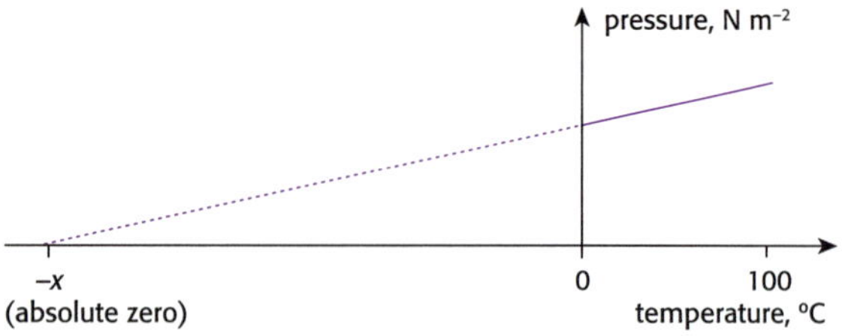

Figure 4 Variation of temperature with pressure.

The *x*-axis can be rescaled so that a new origin is defined to be at the point where the trend line meets the axis (Figure 5). This temperature is called **absolute zero**. For convenience, the size of one degree on the new scale is kept the same as on the Celsius scale. The new temperature scale is called the **absolute scale of temperature**, the **thermodynamic scale of temperature**, or the **kelvin scale**. The unit of this scale is kelvin (K).

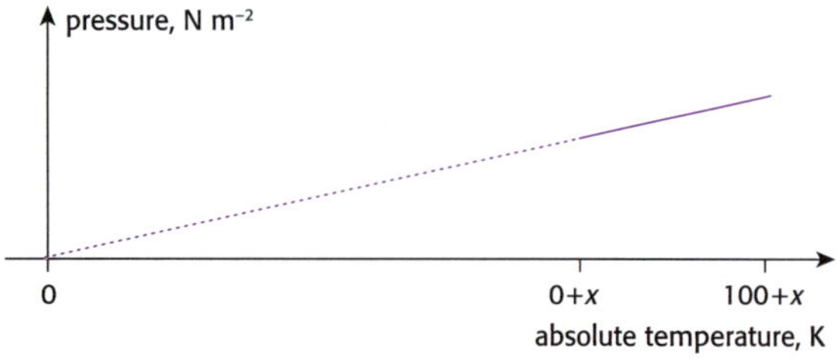

Figure 5 The pressure of a gas is proportional to its absolute temperature.

The graph in Figure 5 shows that the pressure of the gas is proportional to its absolute temperature.

Note that by extrapolating the graph we are assuming that the gas continues to vary in the same way at low temperatures as it did at higher temperatures.

Summary

The above three experiments point towards three general laws that can be used for many different gases (particularly at low pressures). For a fixed mass of gas, *M*, we have:

- *Boyle's law*: If T is constant, pV = constant
- *Charles's law*: If p is constant, $\frac{V}{T}$ = constant
- *Pressure law*: If V is constant, $\frac{P}{T}$ = constant

where

p is the pressure of the gas, measured in N m^{-2}
V is the volume of the gas, measured in m^3
T is the absolute (thermodynamic) temperature of the gas, measured in K

In each case the constant is proportional to the mass of the gas considered. These laws can be brought together into one general law for each gas:

$$\frac{pV}{T} = \text{constant}$$

Many different gases have approximately the same value of constant for one mole, so the equation can be rewritten as the ideal gas equation:

$$\frac{pV}{nT} = R \qquad \text{or} \qquad pV = nRT$$

where

p, V, and T are as defined above
n is the amount of matter in the gas, measured in moles
R is the universal molar gas constant, which has the value 8.31 J mol^{-1} K^{-1}

An **ideal gas** is one that obeys the above equation for all values of p, V, and T. The **state** of an ideal gas is specified by p, V, and T, and the above equation is sometimes called the **equation of state** for an ideal gas. This use of the word "state" should not be confused with the different states of solid, liquid, and gas. Remember that ideal gases cannot turn into liquids or solids.

The bigger picture: microscopic explanation for ideal gases

If we apply Newton's laws to the kinetic theory of gases (provided the ideal gas assumptions detailed on page 116 apply), it allows us to derive an equation that links the pressure of a gas to the molecular speed distribution. A full statistical analysis is beyond the scope of the IB Diploma Programme, but an understanding of the principal conclusions aids understanding.

Random interactions between the molecules of a gas will result in a distribution of different molecular speeds. A full analysis shows that this distribution, called a **Maxwell–Boltzmann** distribution, has the shape shown in Figure 6.

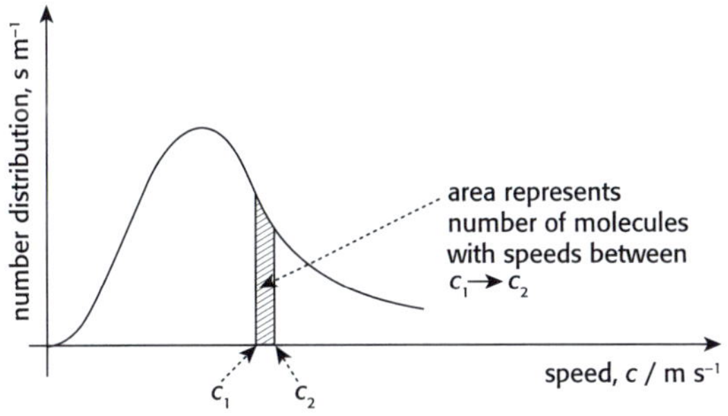

Figure 6 Maxwell–Boltzmann distribution

The area under the graph represents the number of molecules in the chosen speed range (from c_1 to c_2), hence the unit on the *y*-axis. When a molecule collides with one of the walls of its container, the force from the molecule on the wall will depend on the molecule's velocity. A full analysis shows that the force depends on the square of the velocity, c^2.

The individual forces from each molecule can all be added together and this leads to an expression for the overall pressure in terms of the average value of the squares of the different molecular velocities, $\overline{c^2}$.

Note that:

The bar above the c^2 represents the average (mean) value.

This average value is different to the square of the average value of the different molecular speeds, which would be represented by $(\bar{c})^2$.

The **r.m.s.** (root mean squared) value of the molecular speeds is a useful quantity, which is equal to $\sqrt{(\overline{c^2})}$. This is not the same as the average speed, but represents a "typical" speed of the gas molecules. If all the molecules had this speed, then the pressure of the gas would be the same as given by the random distribution of speeds.

R.m.s. values are important in many other areas of physics, notably the analysis of alternating voltages and currents.

Ten molecules have the following speeds c in km s^{-1}:
1.0, 2.0, 3.0, 5.0, 5.0, 5.5, 6.0, 7.0, 8.0, 10.0
Calculate:

(a) the average speed, $\bar{c}$
(b) the square of the average speed, $(\bar{c})^2$
(c) the average of speed squared, $\overline{c^2}$
(d) the square root of your answer to part (c), i.e. the r.m.s. speed.
(e) Compare and comment on your answers to parts (a) and (d).

A full analysis, applying Newton's laws to the random motion of molecules, predicts that the pressure of an ideal gas is given by

$$pV = \frac{1}{3}\rho\overline{c^2}$$

where

p is the pressure of the gas, measured in Nm^{-2}
ρ is the density of the gas, measured in kg m^{-3}
$\overline{c^2}$ is the average of the speed squared for the molecules, measured in m^2 s^{-2}

The gas contains N molecules, each of mass m. The definition of density means that this equation can be rewritten as

$$pV = \frac{1}{3}Nm\overline{c^2}$$

or

$$pV = \frac{2}{3}N\left(\frac{1}{2}m\overline{c^2}\right)$$

This means that the assumptions of kinetic theory predict that

$$pV \propto \frac{1}{2}m\overline{c^2}$$

where $\frac{1}{2}m\overline{c^2}$ is the average kinetic energy of a molecule. This can be compared with

$$pV \propto T$$

which is the ideal gas equation expressed in terms of the absolute scale of temperature. This shows that the absolute temperature of an ideal gas is proportional to the average translational KE of its molecules.

Mathematical physics: energy changes for ideal gases

There are two different processes by which energy can be transferred to (or from) a gas. Either work is done on a macroscopic scale by or on the gas, or thermal energy is transferred. The latter process is doing work on the microscopic scale.

When a quantity of thermal energy ΔQ is given to an ideal gas, one of three things can happen:

- It can raise the internal energy of the gas by an amount ΔU. The internal energy of the gas is equal to the translation KE of its molecules. The average KE of each molecule is proportional to the absolute temperature of the gas.
- It can enable the gas to do work ΔW. A positive value for ΔW means that the gas has done work. A negative value means that work has been done on the gas.
- Both processes may occur together.

The principle of the conservation of energy tells us that

$$\Delta Q = \Delta U + \Delta W$$

This equation is known as the **first law of thermodynamics**.

The various different ways by which the state of an ideal gas can be altered will have different effects on the internal energy of the gas and on the work done. The energy required per degree change in temperature will depend on the method. For example, the specific heat capacity at constant volume, c_V, will be less than the specific heat capacity at constant pressure, c_p. The gas does work when expanding at constant pressure, but does no work when the volume is constant. Four important changes are considered below.

Calculation of work done

Macroscopically, a gas does work whenever it expands; conversely, work is done on the gas when it is forced to contract. We can calculate the amount of work done by considering a gas expanding and pushing back a frictionless piston, as shown in Figure 7.

Provided the displacement is small, we can assume the pressure remains constant. Since $F = p\,A$, then

$$\delta W = (pA) \times \delta x$$

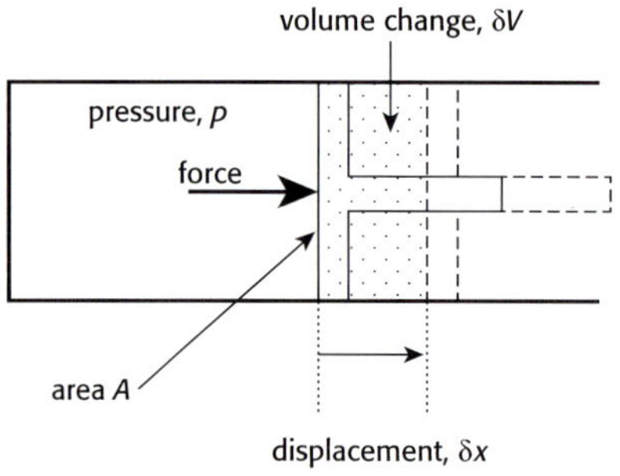

Figure 7 Work done by an expanding gas

Since $A\delta x = \delta V$, then

$$\delta W = p \times \delta V$$

This small amount of work done corresponds to the area under a p–V curve (Figure 8).

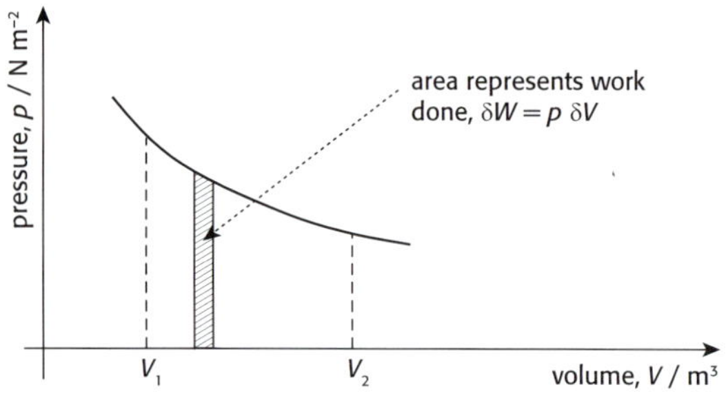

Figure 8 Work done during expansion

In general, the total work done by the gas is the area under the graph if the volume increases (i.e. from V_1 to V_2 in the example in Figure 8). Using calculus notation, $W = \int_{V_1}^{V_2} p \cdot dv$. If the volume decreases, the area represents the work done on the gas.

This principle can be applied to a gas that undergoes a series of state changes and returns back to its original state. The net work done on (or by) the gas is represented by the area contained on the p–V diagram, as shown in Figure 9.

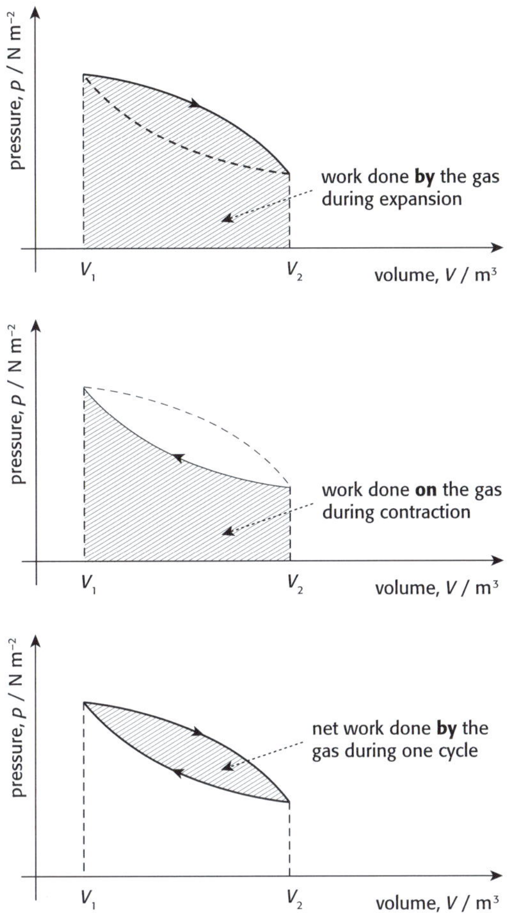

Figure 9 Net work done during one cycle

Isochoric (isovolumetric) change

This state change for a gas takes place at constant volume, so the work done must be zero (Figure 10).

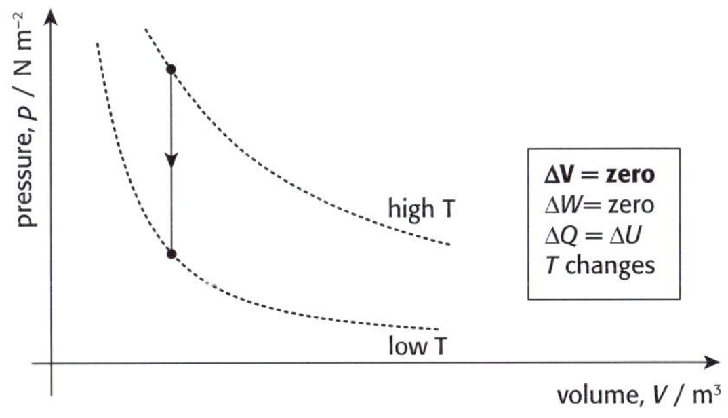

Figure 10 Isochoric (isovolumetric) change

Isobaric change

This state change takes place at constant pressure. The overall work done is easy to calculate (Figure 11).

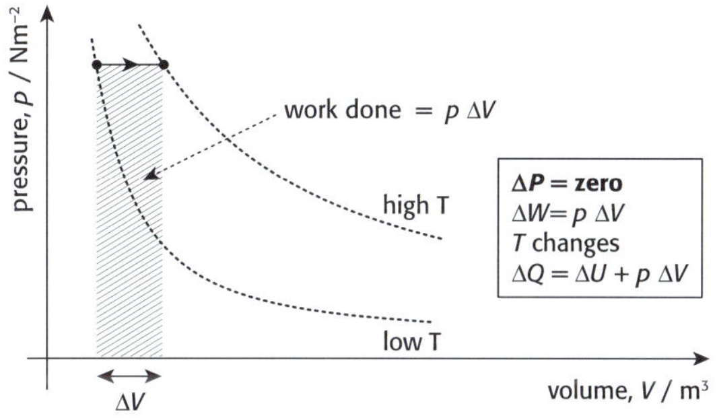

Figure 11 Isobaric change

Isothermal change

This state change takes place at constant temperature. The overall work done is calculated from the area under the curve (Figure 12).

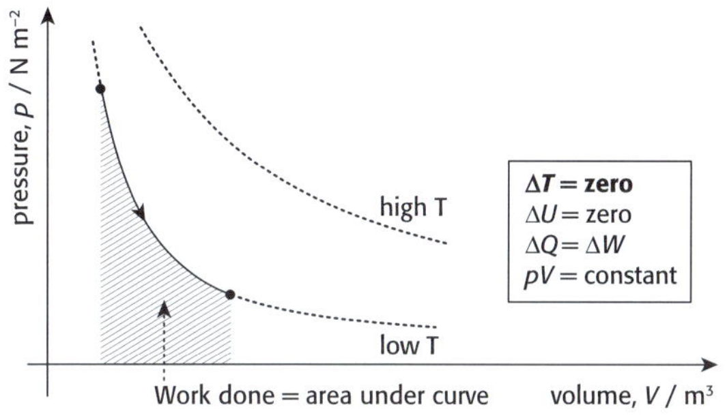

Figure 12 Isobaric change

Adiabatic change

This state change takes place without any transfer of thermal energy to or from the gas. The overall work done is calculated from the area under the curve (Figure 13).

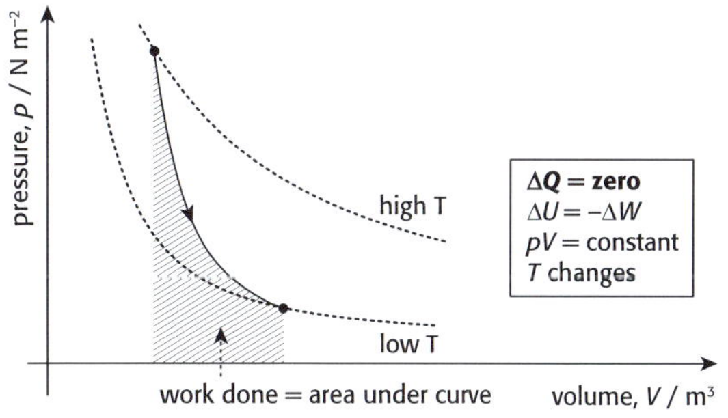

Figure 13 Adiabatic change

One technical point that should be noted is that all the changes that have been considered in this section are classified from a thermodynamic point of view as **reversible**. We have assumed that the universal gas equation applies all the time. This implies that the gas has never departed from **thermodynamic equilibrium** – that is, there have never been any temperature or pressure gradients *within* the gas that might cause the measured pressure, volume, or temperature to change as time passes.

In order to achieve reversibility in these processes, the practical implications are that all changes need to take place very slowly, so that the gas never significantly departs from equilibrium. In reality, perfectly reversible changes would take an infinite amount of time. Real, non-reversible, changes can be **dissipative** – that is, energy can be lost.

Heat engines to do work

One of the key inventions that lead to the western Industrial Revolution of the 18th century was the steam engine.

Given the importance of the steam engine to industry, any slight improvement in it could be worth a lot of money. Not surprisingly, much intellectual effort was expended on improving the efficiency of the design.

The first law of thermodynamics does not put a limit on the maximum possible efficiency of a heat engine. According to this law, it is possible to design systems that convert thermal energy into useful work completely, but they cannot continue to operate over long periods of time.

For example, an ideal gas being heated under isobaric conditions converts thermal energy into work with 100% efficiency. But the gas cannot freely expand for ever; there must be a physical limit imposed by the equipment used. For this reason an engine must involve a cycle of operations.

A steam engine is one example of a device called a **heat engine**: other examples include a car engine and a thermal power station. In all of these examples thermal energy is converted into useful work. Figure 14 analyses a generalized heat engine in terms of the energy transfers taking place.

To increase the efficiency of an engine, the amount of energy lost to the surroundings needs to be reduced to a minimum. In the 19th century a general principle called the **second law of thermodynamics** was identified. The following two statements can be shown to be equivalent:

"Thermal energy cannot spontaneously transfer from a region of low temperature to a region of high temperature. In order for this transfer to take place, work must be done."

"No heat engine that operates in a cycle can be 100% efficient."

An analysis of possible cycles demonstrated that the maximum possible efficiency would be achieved if the ideal gas followed the following process:

1 an isothermal expansion
2 an adiabatic expansion
3 an isothermal contraction
4 an adiabatic contraction back to the original state.

1 Explain why an ideal gas cannot be liquefied.
2 For each of the ideal gas assumptions listed on page 116, outline reasons why and how these assumptions may not apply to real gas molecules.

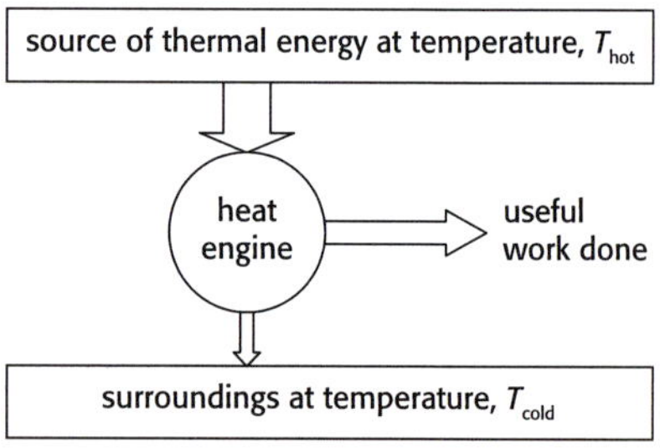

Figure 14 The basic principle of a heat engine

The above cycle is called a **Carnot cycle**, and an engine with the maximum possible efficiency is known as a **Carnot engine**. The efficiency of a Carnot engine can be calculated using the following relationship:

$$\text{Efficiency} = 1 - \left(\frac{T_{cold}}{T_{hot}}\right)$$

That is, the maximum possible efficiency of any engine is determined by its operating temperature and the temperature of the surroundings. This result has a profound effect on the design of all engines. See page 248 (Chapter 21) for more information.

Second law of thermodynamics and entropy

The previous section introduced the second law of thermodynamics, as expressed in the 18th century. Nowadays it is much more common to use a concept called **entropy**, S. The entropy of a system is a property that expresses the degree of disorder in the system. For example, the particles in a solid are held in an ordered state compared with the same particles in a gaseous state.

If all the air molecules in a room were collected in one corner of the room, this would be a very ordered state (and so have low entropy) compared with the "normal" arrangement of molecules spread throughout the room in random motion (in which they would have high entropy). In everyday situations, the total entropy of a system is extremely hard to calculate from first principles, but an entropy change can often be quantified.

When thermal energy ΔQ is given to an object at a temperature T the system becomes more disorganized. The increase in entropy can be calculated using:

$$\Delta S = \frac{\Delta Q}{T}$$

where
ΔQ is the thermal energy added to the system, measured in J
T is the absolute temperature of the system, measured in K
ΔS is the increase in the entropy of the system, measured in J K^{-1}

When a thermodynamic process results in a temperature change, we need to use calculus techniques in order to sum up the entropy changes taking place.

The concept of entropy allows another equivalent statement of the second law of thermodynamics:

In any process, the entropy of the universe must either stay the same or increase.

Macroscopically, the second law in this form is a powerful tool to predict thermodynamic changes. It should be noted that on the microscopic scale, there is no guiding control exerted on atoms and molecules making them behave in a particular way. They interact randomly and the changes that take place on the macroscopic scale are simply the ones that are more likely to happen. The macroscopic concept of entropy just quantifies microscopic statistical analysis.

?

The diagram below shows the relation between the pressure and the volume of the air in a diesel engine for one cycle of operation of the engine. During the cycle there are two adiabatic processes, an isochoric process, and an isobaric process.

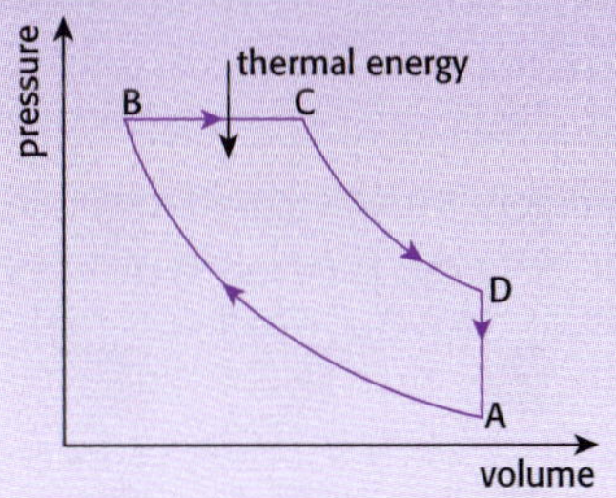

(a) Explain what is meant by
  (i) an adiabatic process [2]
  (ii) an isochoric process [1]
  (iii) an isobaric process [1]
(b) Identify, from the diagram, the following processes
  (i) adiabatic processes [1]
  (ii) isochoric process [1]
  (iii) isobaric process [1]

During process B → C thermal energy is absorbed.

The diesel engine has a total power output of 8.4 kW and an efficiency of 40%. The cycle of operation is repeated 40 times every second.

(c) State what quantity is represented on the diagram by the area ABCD. [1]
(d) Determine the value of the quantity that is represented by the area ABCD. [1]
(e) Determine the thermal energy absorbed during the process B → C. [2]

[Total 11 marks]

Example

In Chapter 5 (page 48-9) we introduced the example of an impossible situation (the broken bits of a glass spontaneously coming back together and the whole glass jumping back onto the table) that is not excluded by energy considerations. The explanation of this in terms of entropy could be as follows:

- Before falling to the ground, the glass was in an ordered state.
- As a result of falling to the ground, the pieces of glass are randomly spread around the floor, and the original PE of the glass is now shared with the surroundings: this is a more disordered state.
- The process of the glass falling to the ground and breaking into pieces is an entropy increase.
- The reverse process would be an entropy decrease and would violate the second law of thermodynamics.

?

When water freezes, the entropy change is $-22\ \mathrm{J\ K^{-1}\ mol^{-1}}$ and the latent heat of fusion is $6\ \mathrm{kJ\ mol^{-1}}$. Use these data to predict the temperature at which water freezes.

12 Electric currents

The analysis of an electrical circuit involves many related quantities. This chapter starts by focusing on the differences between the various quantities that are involved, and then introduces the techniques and approaches used when analysing a circuit.

Electrical quantities in brief

The main electrical quantities involved are as follows:

Charge

Charge is a fundamental property of objects. There are two types of charge: **positive** and **negative**. An object with equal amounts of positive and negative charge is **neutral**. Charge is usually measured in terms of the charge on one electron but in electric circuits there are a lot of charge carriers (electrons or ions) so we define a new unit called the **coulomb** (unit symbol C). There are 6.25×10^{18} electrons in one coulomb of negative charge. The letter Q is used to represent charge.

Current

Current is the rate at which charge flows. It is measured in amperes (unit symbol A), which are equivalent to coulombs per second. For historical reasons, current is conventionally considered as the flow of positive charge around a circuit. The letter I is used to represent current.

Potential difference (p.d.), voltage

Potential difference is the energy difference per charge (i.e. the number of joules per coulomb, or watts per ampere) between two points in a circuit. When measured in volts (unit symbol V) it is usually referred to as **voltage**. The letter V is used to represent potential difference, or voltage.

Resistance

Resistance is defined to be the ratio of potential difference to current. The unit is volts per ampere, which is given the name ohm (symbol Ω). As the name suggests, we can think of the resistance as a quantity that measures the resistance to the flow of charge around a circuit. A high resistance needs a high p.d. across it to make a current flow, whereas a low resistance needs a small p.d. across it to achieve the same current. The letter R represents resistance.

A current involves a charge moving around a circuit and gaining or losing energy. The circuit in Figure 1 illustrates the energy changes that are taking place for each coulomb. When energy is transferred from the moving charge, the energy is said to be **dissipated**.

The result is that the coulomb was given 12 joules in order to complete its journey around the circuit, and all of this has been dissipated $(6 + 2 + 4 = 12)$.

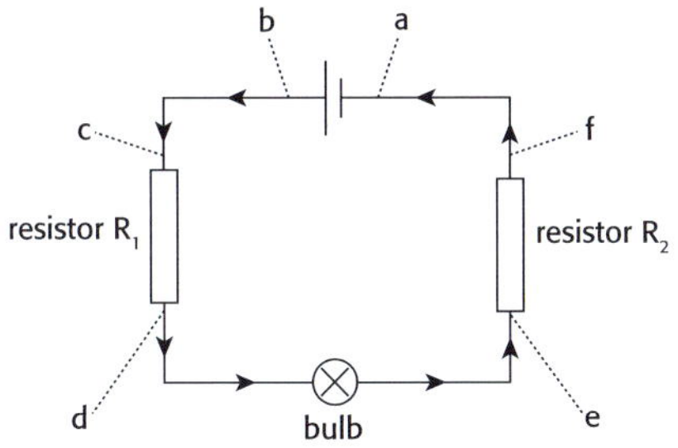

Figure 1

Table 1

a to b	The battery gives some electrical energy to any charge that moves through it. It is a 12 volt battery which means that every coulomb leaving the battery from the left-hand side has 12 joules more energy than when it went in on the right-hand side.
b to c	The wire is a very good conductor, which means that almost none of the energy is used up – a little must be, but it is so small we take it as zero.
c to d	The coulomb must have used up some of its energy moving through this region of higher resistance. Since p.d. measures the energy difference per coulomb we must be able to define a voltage across this component. Let's say it is 6 V in this case.
d to e	Now it goes through the bulb – again it uses up energy, so there must be a potential difference across the bulb. Let's say 2 V.
e to f	Once again more energy is used up. Again there is a p.d. across the resistor, this time 4 V.
f to a	Just like b to c, virtually no energy is used up. We pretend the wire is such a good conductor that no energy is used at all.

This is always true. A complete journey around a circuit exactly uses up the energy per coulomb.

You may ask: "What if it hasn't got enough energy to get all the way round? Does it stop after the first resistor?" Remember that the rate of flow of charge (i.e. the current) must be the same before and after the resistor; otherwise more coulombs would go into the resistor than would come out of it. This would mean a build-up of charge, which is impossible.

Where does the energy per coulomb go? To understand this, let's concentrate on one device. We will consider resistor R_2 (Figure 2).

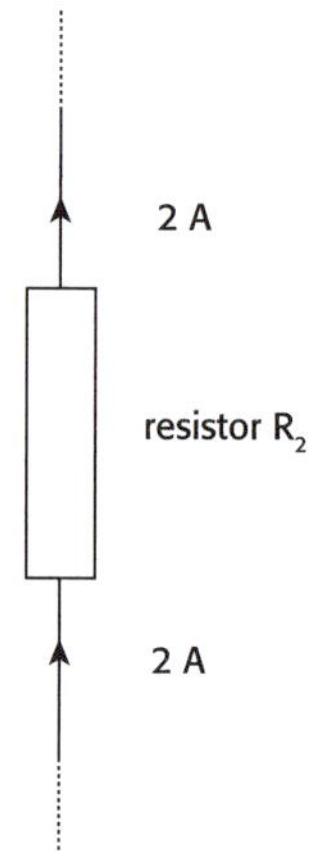

Figure 2

The energy used up per coulomb depends on the value of R_2, and the rate of flow of charge (the current) through it. Assume that in this example the current is 2 A: that is, 2 coulombs move through R_2 every second. We already know that this takes 4 V.

The total energy dissipated is 4 J for every coulomb, and we know that there are 2 C every second. So we use up 8 J every second. This make R_2 get hotter and hotter.

We now build a completely different circuit incorporating R_2: one that can put 4 A through it. This will take more energy per coulomb. R_2 now needs 8 V across it, because to double the current we need to double the p.d. Here 4 coulombs per second are each dissipating 8 J per coulomb, so the total rate at which energy is used up in this new circuit is 32 J/s.

In summary, when we set up a circuit we fix the *total* energy per coulomb (i.e. the total voltage) by the power supply we choose. The total p.d. it provides is exactly used up in the various devices around the circuit, and the p.d. across each one and the current through it are related.

Resistors are an easy group of devices to work out, because the p.d. across them and the current through them are proportional:

$$\text{p.d. across} = \text{current through} \times \text{resistance} \qquad (1)$$

Example 1: a series circuit

In a series circuit, the components are connected end-to-end (see Figure 3).

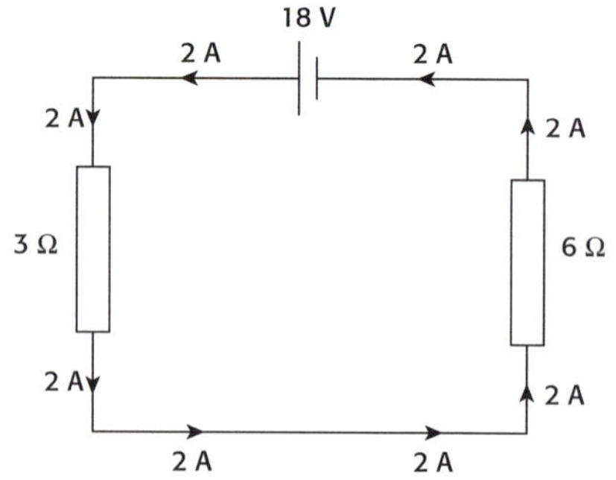

Figure 3

In the circuit in Figure 3, 18 V is put across two resistors in series, one of 3 Ω and the other 6 Ω. In this case, 2 A must flow through the circuit for the numbers to work:

$$\begin{aligned}\text{The p.d. across the } 3\ \Omega \text{ resistor} &= \text{current through} \times \text{resistance}\\ &= 2\text{ A} \times 3\ \Omega\\ &= 6\text{ V}\end{aligned}$$

$$\begin{aligned}\text{The p.d. across the } 6\ \Omega \text{ resistor} &= \text{current through} \times \text{resistance}\\ &= 2\text{ A} \times 6\ \Omega\\ &= 12\text{ V}\end{aligned}$$

$$\begin{aligned}\text{So the total p.d. across the two resistors} &= 6\text{ V} + 12\text{ V}\\ &= 18\text{ V}\end{aligned}$$

The numbers balance!

Example 2: a parallel circuit

Components in a circuit are said to be connected in parallel if their ends are connected to the same point in the circuit. In Figure 4, the three components (the battery and the two resistors) are connected in parallel.

In the circuit in Figure 4, the same p.d. (18 V) is put across both resistors by the battery.

The p.d. across the 3 Ω resistor = 6 A × 3 Ω
= 18 V

The p.d. across the 6 Ω resistor = 3 A × 6 Ω
= 18 V

So the total current that flows must equal 6 A + 3 A = 9 A

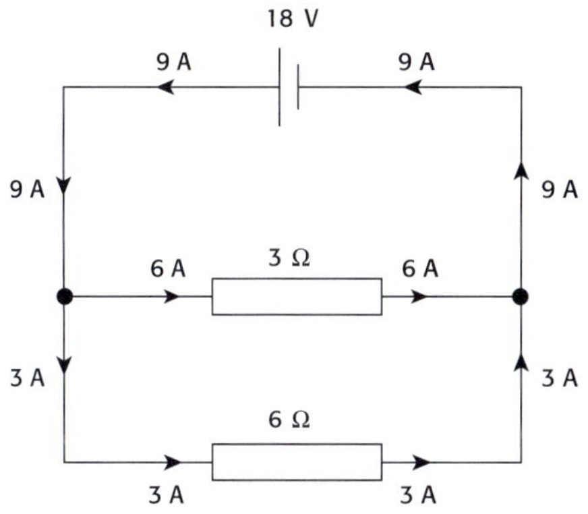

Figure 4

All you have to do is to apply equation 1 to resistors. Remember that you must use the p.d. across the resistor, the current through the resistor, and the value of the resistor all in the same equation. You must not use, for example, the p.d. due to the battery, the current through resistor 1, and the value of resistor 2 together, as they don't match.

Thinking about science: what is the real nature of charge?

Most people are very familiar with everyday electrostatic effects. For example, using a plastic comb may make your hair stand on end. Devices such as photocopiers and air ionizer/purifiers rely on the application of this effect. All electrostatic phenomena are based around charges that are, at least temporarily, at rest.

An electric current always involves charge carriers (usually electrons) moving around a circuit.

Whether we are switching on a light or a complicated piece of electronic equipment, a flow of charge must be involved. But what *is* charge?

As we have already said, there are two types of charge: **positive** and **negative**. They are opposites. If a positive charge and an equal amount of negative charge are in close proximity, the resulting object is uncharged or **neutral**. The SI unit of charge is the coulomb (C).

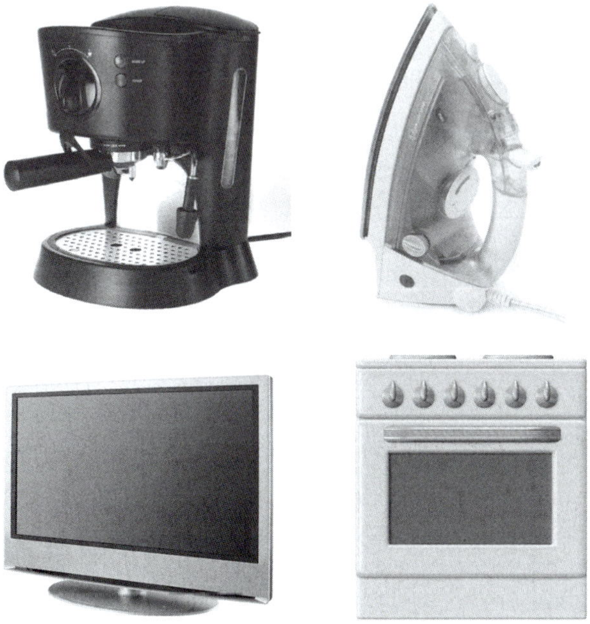

Figure 5

Some objects can become charged as a result of friction. Study of the charging process shows that the total amount of charge is always conserved. When charge is generated by rubbing two objects together, the positive charge gained by one object is balanced by the negative charge gained by the other object.

For example, if a neutral insulating rod is rubbed with a neutral cloth, and the rod gains a positive charge of $+5 \times 10^{-9}$ C, the cloth will gain an equal but opposite charge (-5×10^{-9} C).

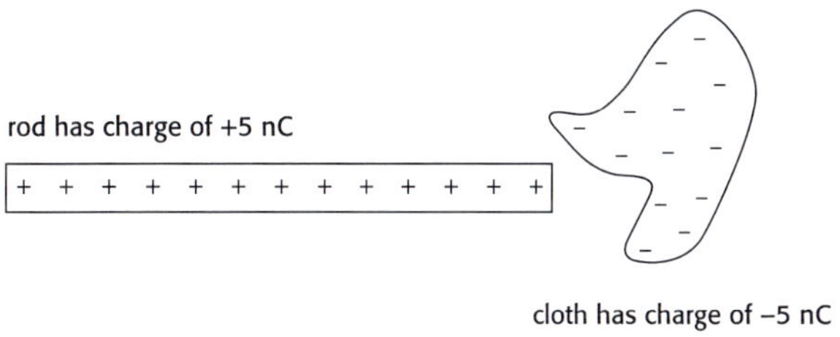

Figure 6

Charge is conserved in electric currents. In any given time, the number of coulombs that flow into a device must equal the number of coulombs that flow out. No charge can be lost; nor can the rate of flow of charge change.

Charge carriers can stay still or be moved about, but an object's charge is something completely different from its mass. Charge can be given to, or flow through, an object. What is happening when an everyday object becomes charged, or when a current flows?

Everyday matter can be thought of as being made up of atoms, or atoms combined into molecules. At this atomic scale we can identify which particles are charged and the type of charge that they carry. In this simple atomic model, everything is made up of fundamental particles – electrons, protons, and neutrons:

- electrons are negatively charged
- protons are positively charged
- neutrons are neutral.

All electrons have the same mass (9.11×10^{-31} kg) and the same amount of negative charge (1.6×10^{-19} C). This is the smallest observable quantity of charge. It is not possible to break an electron down into smaller parts, or remove or change its charge and leave its mass.

A proton has an exactly equal but opposite amount of charge to that of an electron, so a proton and an electron together (as in a hydrogen atom, for example) can be considered neutral. Protons are bound inside nuclei, so the movement of charge is often due to the movement of electrons. In the example in Figure 6 the rod lost electrons, so it has more protons than electrons and, overall, is positive. The cloth now has an excess of electrons and so overall is negative.

When an electric current flows along a copper wire the overall effect is that charge moves around the circuit. In this case the protons are fixed because the atoms are not free to move along the wire: the copper is solid, so the electrons carry the charge in a wire. But current does not always just involve the flow of electrons. Many liquids contain mixtures of atoms or molecules that have gained or lost electrons. These are called **ions**.

For example, a water molecule is a combination of atoms: two hydrogens and one oxygen – H_2O. If a soluble ionic impurity (e.g. table salt) is added to liquid water, the water molecules tend to split into hydroxide ions (OH^-) and hydrogen ions (H^+). When a current flows in this water, it is the ions that are moving, carrying the charge with them.

Semiconductors are another example of conduction in which a current is not just free-moving electrons (see the next section).

Conductors and insulators

A conductor is a substance that allows an electric current to flow through it. This process requires the input of energy but the better the conductor, the less energy that is used up for each coulomb that flows. An insulator is a substance that does not allow the flow of electric current because a great deal of energy would be required for a charge carrier to move.

A substance is classified as a conductor or an insulator according to how free the charge carriers are to move within its structure. In a solid, the only charge carriers that are available to move are the electrons. In insulators, the electrons are tightly bound to their own individual atoms, and are not free to move. In conductors, and particularly in metals, these electrons are **delocalized**: that is, they are free to transfer from atom to atom. At any given time every atom will be neutral, because it has equal numbers of electrons and protons, but the electrons are free to move: when an electron leaves an atom, a replacement is provided from the neighbouring atom.

How fast do the electrons in a normal current move around the circuit? You already know that electrons are moving around randomly inside the metal due to thermal motion. When a p.d. is applied the electrons still move essentially randomly but with an overall or "bulk" motion in the direction of the current. Typically, the speed of this "bulk motion" can be measured in mm s^{-1}.

Electrical definitions and approximations

Definition of the units for current and charge

We define currents in terms of the quantity of charge that is moving – current is the rate of flow of charge. It would seem sensible to define the unit of current in terms of the units of charge: that is, to define the ampere in terms of the coulomb. SI units actually do this the other way around. The ampere is defined in terms of an experimental quantity (the force between two long, current-carrying wires) because it can be measured easily. The conceptual link between current and charge is then used to define the coulomb: 1 coulomb is the charge that flows past any point in the circuit when a current of 1 amp flows for 1 second:

$$1 \text{ coulomb (C)} = 1 \text{ ampere second (A s)}$$

Resistance and resistivity, and their units

The relationship between resistance R, potential difference V, and current I is given by

$$R = \frac{V}{I}$$

This equation is the *definition* of the resistance R. Whatever conditions exist in a particular circuit, this relationship must be true, because it is how we define the resistance. This definition leads to the definition of a new unit, the ohm (symbol Ω). It implies that resistance is measured in volts per ampere, and we define

$$1 \text{ ohm } (\Omega) = 1 \frac{\text{volt (V)}}{\text{amp (A)}}$$

A change in temperature or external pressure can change an object's resistance. Each time we change the p.d. across a device, we should expect a different value for the resistance. In general, the resistance of any object is not just one fixed value but it has a range of values that depend on external circumstances, particularly the circuit to which it is connected and the particular values of current and p.d.

The different components in a circuit can be connected in series, parallel, or a combination of the two. Simple equations can be derived to calculate the total resistance of any given combination. The derivations of equations 2 and 3 involve applying the laws of conservation of charge and energy.

$$R_{\text{series}} = R_1 + R_2 + R_3 + \ldots \qquad (2)$$

$$\frac{1}{R_{\text{parallel}}} = \frac{1}{R_1} + \frac{1}{R_2} + \frac{1}{R_3} + \ldots \qquad (3)$$

Resistivity, ρ, is a useful quantity that allows you to find the resistance of a wire experimentally. The definition is given in terms of the resistance of the wire R, its length l, and the area of its cross section, A.

$$R = \rho \frac{l}{A} \qquad (4)$$

This definition means that the units of resistivity must be ohm metres (Ω m). Note that this is the ohm multiplied by the metre, NOT the ohm divided by the metre ("ohms per meter"). Resistivity is related to the microscopic properties of the material.

?

Using equation 4, show that the units of resistivity must be ohm metres.

Perfect electrical meters

The two most common electrical measurements made in circuits are measurements of current and voltage. The general names for the devices that achieve these measurements are ammeters and voltmeters, respectively. When analysing real circuits, the properties of the individual meters need to be taken into consideration.

A perfect measuring device should measure and not do anything else. Thus when we imagine connecting a perfect meter into a circuit we assume that it does not affect the circuit in any way. Suppose we wish to measure the current and the potential difference shown in Figure 7.

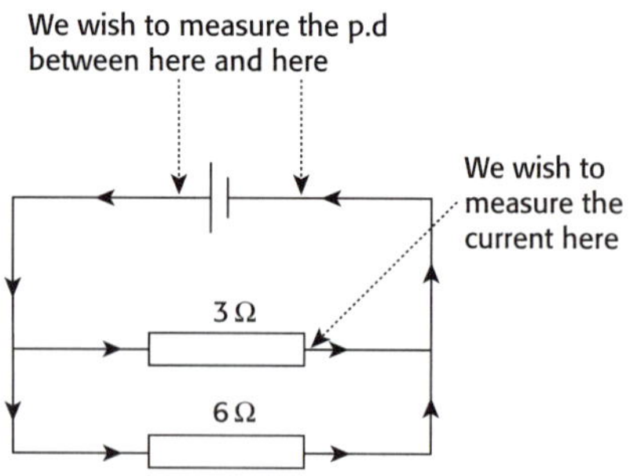

Figure 7

The ammeter

Ammeters measure the current flowing at a particular point in the circuit and thus the addition of an ammeter requires the meter to be added in series with the rest of the circuit. If we want to measure the current flowing out of the 3 Ω resistor, we would break the circuit at this point and add an ammeter as shown in Figure 8.

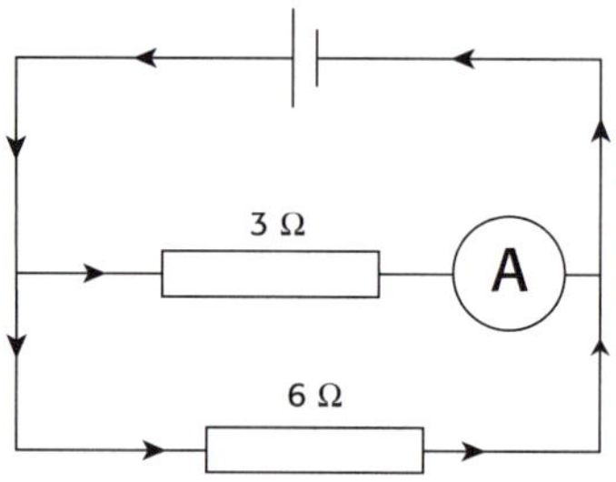

Figure 8

If the perfect ammeter is not going to affect the circuit with its addition, it should have zero resistance. If the ammeter has a non-zero resistance this means the total resistance of the upper branch would be increased slightly from 3 Ω and the current would also change.

The voltmeter

Voltmeters measure the potential difference between two particular points in the circuit and thus the meter must be added in parallel with the device (or devices) in question.

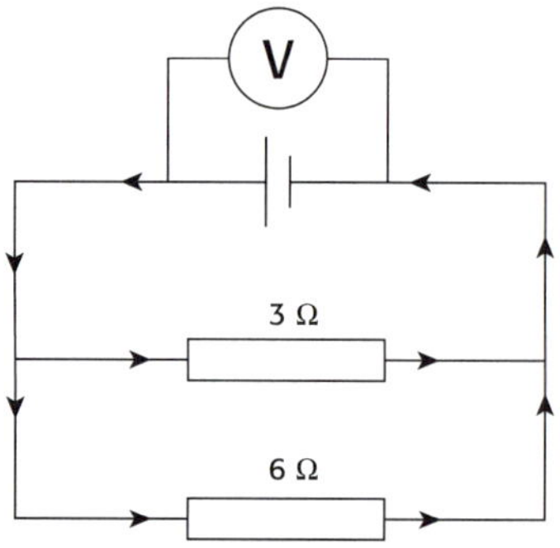

Figure 9

The perfect voltmeter should have infinite resistance – in other words, no current should flow through it. If the voltmeter does allow a current to flow, this means the total current in the circuit must have changed.

1 The circuit in Figure 10 connects two bulbs, a switch and a 6 V battery together in series:
   (a) When the switch is closed, which bulb lights up first? Is it A, B, or do they come on together?
   (b) With the switch open (i.e. no current flowing) what would a voltmeter read if it was connected across
      (i) bulb A
      (ii) the battery
      (iii) the switch?

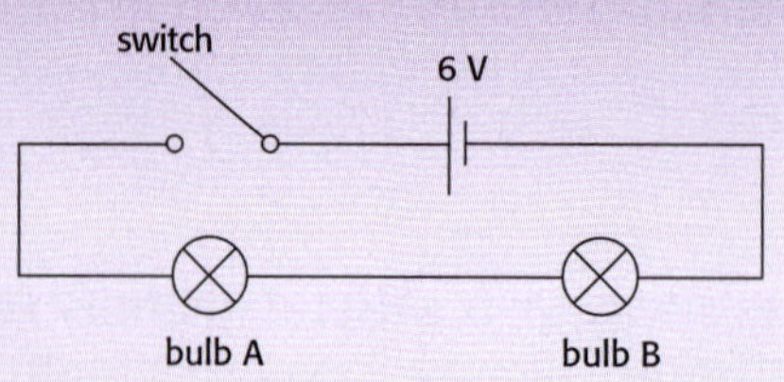

Figure 10

Science tools: factors that affect the resistance of a material

The aim of this section is to provide a framework within which you can attempt to plan and carry out an investigation involving electric currents. It has been designed so that as long as you are able to attempt this without further help or assistance, this task could be formally assessed and count towards your final IB Diploma Programme score.

When planning an experiment, a good approach is for you to imagine that you have suddenly turned into a teacher for the day. Unfortunately, nobody left you any instructions and you have to work out the content of the next physics practical lesson. Not only does this mean that you have to invent the experimental procedure that your students will use, but it needs to be reliable enough to produce data that can be usefully analysed. On top of this, you also have to choose and specify the necessary equipment so that the technician can get it ready for you. Finally, you also have to write out the method so that your class has good instructions to follow. Your instructions need to be sufficiently clear to allow everybody to understand exactly what they have to do. Make sure you leave nothing out.

There are three different aspects to the skill of experimental design. These could be thought of as three stages in the experiment design process:

- Defining the problem and selecting the variables – this involves formulating a focused research question from the general aim stated above and identifying the relevant independent and dependent variables.
- Controlling the variables – this involves designing a method for the effective control of all other possible variables.
- Developing a method for collection of data – the method must allow for the collection of sufficient relevant data.

Power dissipation in circuits

When designing, analysing, or inspecting any electrical circuit, safety considerations should always be the top priority. Two very common sources of electrical fires could be easily prevented if homeowners knew the relationship between power, current, and potential difference.

We want to calculate the power that is **dissipated** in an electrical component. We will start by considering a simple portable heating circuit (see Figure 11). In this circuit there are two resistors (3 Ω and 6 Ω) connected in series to an 18 V battery.

In order to work out the power dissipated in any device, we need to multiply together the p.d. across it and the current going through it.

power (in watts) = potential difference (in volts) × current (in amps)

$P = VI$

In the circuit,

the power dissipated in the 3 Ω resistor = 2 A × 6 V = 12 W

the power dissipated in the 6 Ω resistor = 2 A × 12 V = 24 W

∴ the total power dissipated by the resistors = 12 + 24 = 36 W

Of course we could have calculated the power developed by the battery and we would have found the same answer

the power developed by the battery = 2 A × 18 V = 36 W

This equation comes directly from the definitions of power, p.d., and current.

? Show that the current flowing in the circuit is 2 A, the p.d. across the 3 Ω resistor is 6 V, and the p.d. across the 6 Ω resistor is 12 V.

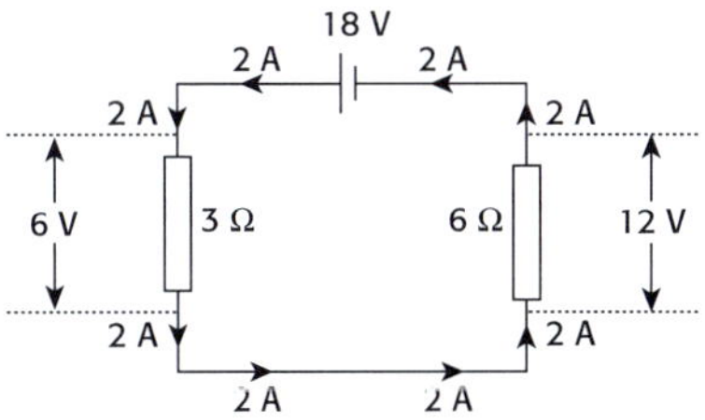

Figure 11

$$\text{power} = \text{rate of energy consumption} = \frac{\text{energy used}}{\text{time taken}}$$

$$\text{current} = \text{rate of flow of charge} = \frac{\text{charge flowed}}{\text{time taken}}$$

$$\text{p.d.} = \frac{\text{energy used}}{\text{charge flowed}}$$

$$\text{so p.d.} \times \text{current} = \text{energy used} \times \frac{\text{charge flowed}}{\text{time taken}}$$

$$= \frac{\text{energy used}}{\text{time taken}} = \text{power}$$

Since p.d., current, and resistance are all interrelated, this equation can be stated in several equivalent ways.

We obtain:

$$P = VI = I^2R = \frac{V^2}{R}$$

We have worked out that the two resistors receive a total of 36 joules every second from the battery. Energy dissipated in a resistor is in the form of thermal energy so the resistors will warm up. When you buy a resistor, it comes with a certain power rating. If you are intending to build the circuit in Figure 18 then you must make sure that the battery is capable of delivering 36 W safely, and the resistors are each capable of safely dissipating the calculated amount of power.

Even if the battery and components were all perfectly safe, other parts of the circuit can go wrong. For example, the wires must be capable of carrying the current that flows in the circuit.

The current and the power calculations were based on the assumption that the wire joining these devices had zero resistance. Even though it probably has a low resistance, it will certainly not be zero. A thick wire has less resistance than a thin wire and this means that when carrying a current, a thick wire will get less hot than a thin wire. A low-power device can be connected using thin wires but a component that requires a lot of power must be connected using thicker cables.

Not only does this principle apply to the connection wires inside an electrical device, it also applies to any power cables or extension leads that are used to connect devices to a source of power. The wiring in your home must be capable of allowing the current to flow without getting too hot. If the wires get too hot, they can cause a fire (see page 141).

Real devices: power supplies

Now we will look at the ways in which real batteries and other power supplies depart from their assumed ideal behaviour.

We can use equation 1 on page 135 to predict the current, I, that would be drawn from a 6 V battery by a 2 kΩ resistor.

?

By using equation 1 on page 128, show that the equation $P = VI$ leads directly to $P = I^2R$ and $P = V^2/R$.

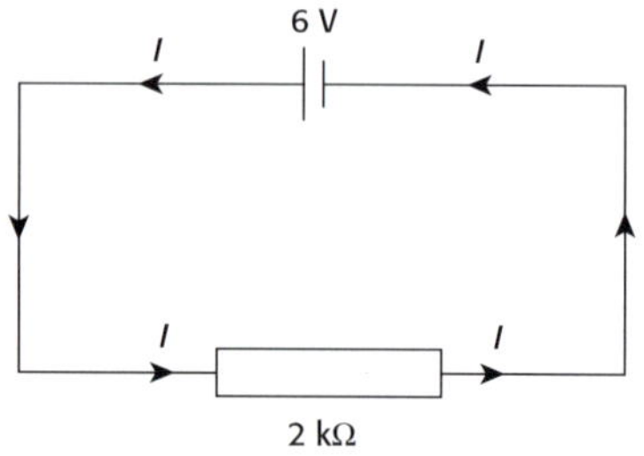

Figure 12

6 volts across the resistor means that the current is

$$I = V / R = 6\ \text{V} / 2000\ \Omega = 0.003\ \text{A} = 3\ \text{mA}$$

This prediction can be checked experimentally. Table 2 shows the results of such an experiment.

Table 2

Resistance / Ω	Predicted current / A	Actual current / A
2000	0.003	0.003
200	0.03	0.03
20	0.3	0.29
2	3	2.4
0.2	30	8.8
0.02	300	11.5
0.002	3000	12.0
0.0002	30000	12.0

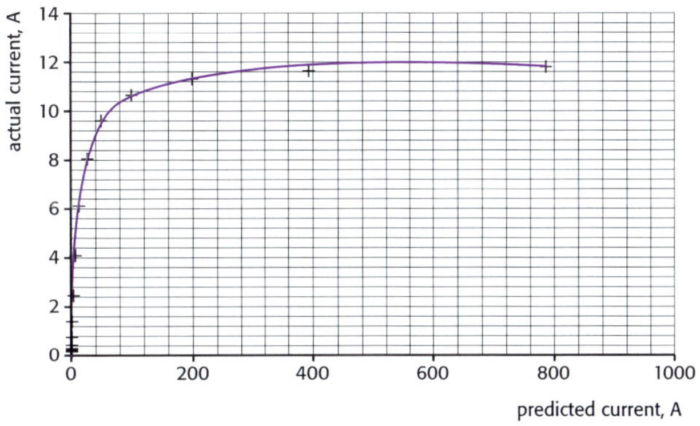

Figure 13 Actual current vs. predicted current

The actual current provided by the battery seemed to be limited to 12.0 A. In order to check that the battery was producing 6.0 V, the experimenter checked to see if the potential difference across the terminals (the **terminal p.d.**) was remaining stable. Table 3 shows the results.

Table 3

Resistance / Ω	Actual current / A	Voltage across battery / V	"Lost" voltage/ V
2000	0.003	6.00	0.00
200	0.03	5.99	0.01
20	0.29	5.85	0.15
2	2.4	4.80	1.20
0.2	8.8	1.71	4.29
0.02	11.5	0.23	5.77
0.002	12.0	0.02	5.98
0.0002	12.0	0.00	6.00

Some of the 6 V seem to have been "lost". Remember that voltage is a measure of the energy per charge so missing voltage means that some energy has apparently gone missing. We know from the law of conservation of energy that this is impossible, so we must find an explanation.

The solution to this problem is that the battery actually has some **internal resistance**. The *total* energy difference per unit charge around the circuit does still remain fixed at 6 V, but some of this energy is used up inside the battery. This is because the energy difference per unit charge from one terminal of the battery to the other is less than the total energy made available by the chemical reaction inside the battery. To distinguish between the terminal p.d. and the total energy difference per unit charge, we introduce a new term—the **electromotive force (emf)**. The emf is equal to the terminal p.d. when no current flows.

Note that the term emf is actually an old phrase, still in use for historical reasons. It is a misleading name because the emf is not actually a force you can measure in newtons. The emf of a battery can be defined as the total energy per unit charge given by the battery. This will be the same as the total energy per unit charge dissipated by a charge during one complete circuit.

An easier way to imagine this situation (and one that works to a very good approximation) is to say that there is an additional resistance in the circuit, an **internal resistance**, which is located between the terminals of the battery. In the example above, the maximum current was limited to 12.0 A, thus the effective minimum resistance of the circuit was given by:

$$\text{internal resistance } r_i = \frac{6.0}{12.0} = 0.5\ \Omega$$

This value is the internal resistance of the battery. All power sources have internal resistance. When a battery becomes "flat" and is unable to provide as high a voltage as it used to provide, its internal resistance must have increased.

The two batteries in Figure 14 are both sold as 12 V batteries – this means they both have an emf of 12 V. The car battery (Figure 14a) has a much lower internal resistance and thus can provide a much larger current.

(a)

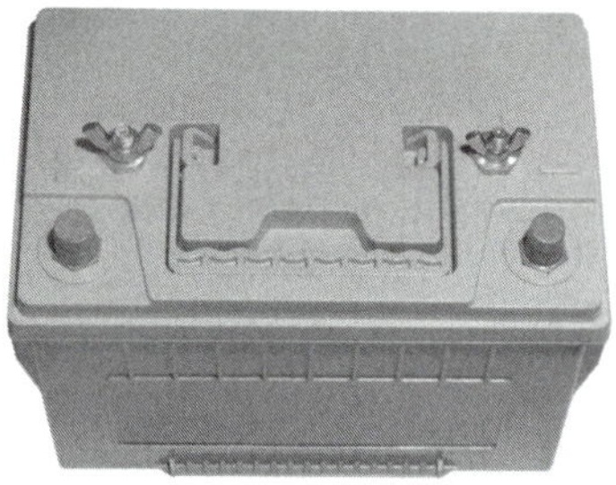

(b)

Figure 14 Two 12 V batteries

Figure 15 is similar to Figure 12 on page 134 but shows a way of visualizing the battery as a combined power source and resistor.

Applying $V = IR$ and equation 4 on page 131 to the whole circuit, we get the following equation:

$$\varepsilon = I(R + r_i) = IR + Ir_i$$

IR: p.d. available to the external circuit V

Ir_i: "lost volts" across internal resistance

or $V = \varepsilon - Ir_i$

Science tools: data collection and processing

You can use the circuit in Figure 15 to take experimental data that can be processed to calculate the emf and internal resistance of an everyday battery (a dry cell). A simple variable resistor is connected to the dry cell and readings can be taken of terminal p.d., V, and current I. The different values are obtained by varying the external resistance. You should not leave the battery running too long at the higher currents, and you should also take care that nothing overheats. The range of currents that you are able to measure will depend on the dry cell that you are using but, as is always the case, you should aim to measure 10 or more different values of the independent variable, current. Ideally, each value should be repeated two or three times. Note that if this experiment is to be used as an assessed practical you should be working on your own.

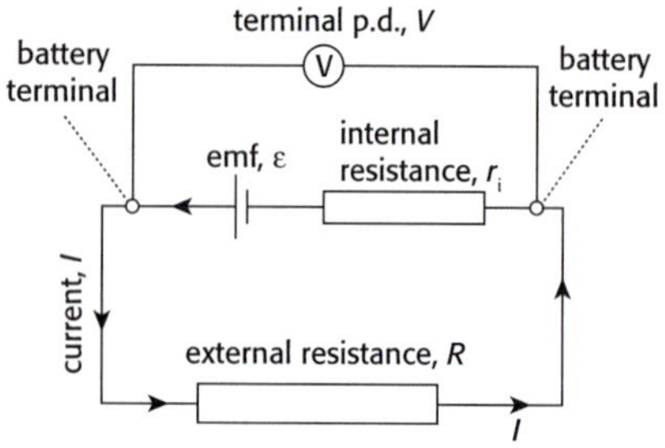

Figure 15

In general, the three aspects for the internally assessed skill of data collection and processing are

- recording raw data including units and uncertainties where relevant
- processing raw data correctly
- presenting processed data including errors and uncertainties

In this investigation, the raw data consists of an appropriate set of readings for terminal p.d., V, and current, I. You must choose the best method of measuring and recording this data, and you should also use an appropriate number of significant figures for the data that you record. This data then needs to be processed, so you will need to decide on the best way to analyse the numbers to allow you to answer your initial question. Finally, when processing data, you need to take into account the uncertainties that you assessed in the raw data. Often, this is most clearly presented as a separate section in your write-up.

You can also take this experiment on to the conclusion and evaluation stage.

Real meters

As we saw on page 132, perfect voltmeters have infinite resistance and perfect ammeters have zero resistance. What happens if this is not the case?

The effect of adding the meter needs to be taken into consideration in the circuit calculations. It is usually necessary to start again with all of your calculations whenever a circuit is altered – it is not always possible to assume that an alteration (for example adding a voltmeter) in one part of the circuit will not effect another part of the circuit. It might help to image a real meter as a combination of a perfect meter and an "additional" resistor. For a real ammeter, the "additional" resistor is in series with the perfect meter, for the voltmeter it is in parallel.

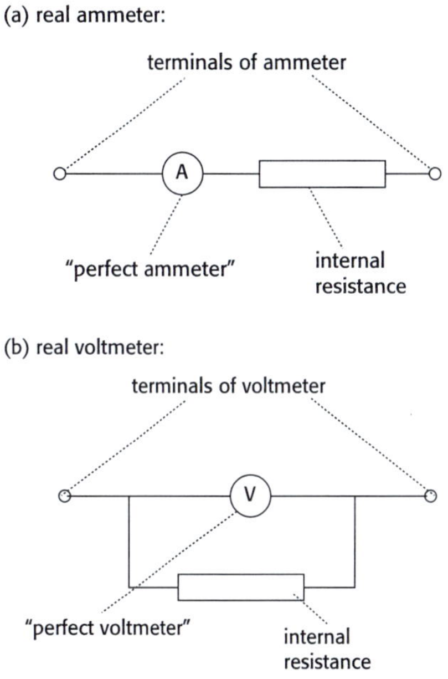

Figure 16

Potential dividers and sensors

For a circuit to be useful, it must interface with the real world; it must have at least one input and one output.

An input device could be as simple as a button or a switch. In a computer circuit, the input devices are things such as the keyboard, the mouse, or a DVD reader. Possible output devices include lights, buzzers, or, in the case of the computer, the screen, the printer, or the loudspeakers. The general name given to any electrical interface that converts energy between different forms is a **transducer**.

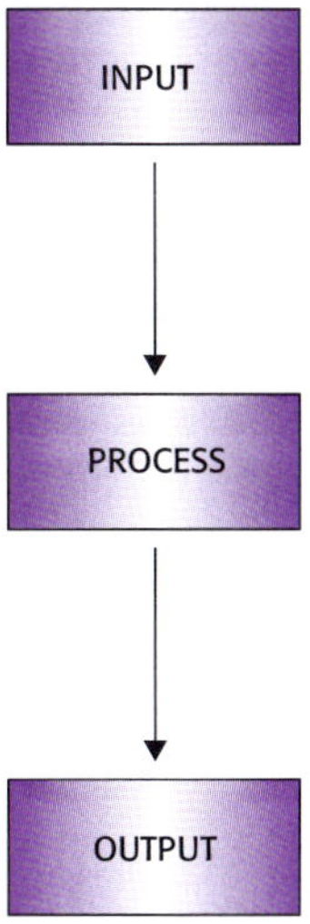

An electrical circuit involves electrical energy and thus input transducers, or **sensors**, are devices that convert some real everyday aspect of the physical world into electrical signals. Examples include microphones (sound into electrical), light sensors (light into electrical), and temperature sensors (thermal energy into electrical).

Sensors can facilitate a whole range of measurements that would be impossible for a lone physicist to observe with a single measuring device. Electronic circuits can be designed in which the measurements are:

- **Automated**: multiple readings of a variable can be recorded over very short response (e.g. air pressure variations during a pulse of sound) or very long time intervals (e.g. the automatic recording of a patient's heart rate over a week).
- **Detailed**: they allow for a large number of precision multiple recordings to be taken at exactly the same time (e.g. the minute variations in the stresses and strains that take place in the body of an aircraft during takeoff).
- **Remote**: they allow for readings to be taken in difficult or even hostile locations (e.g. temperature measurements inside a reactor core or observations on another planet).

So what is happening when, for example, the breaking of a beam of light causes an alarm system to issue a warning?

Many sensors are just a resistor whose resistance varies in a predictable way in response to external factors. Two very important examples of this are *light-dependent resistors* (LDRs) and a type of temperature-dependent resistor called a *negative temperature coefficient* (NTC) *thermistor.*

In both of these devices, an increase in energy input to the device causes a large decrease in resistance. Thus when the amount of light shining on the LDR increases or the temperature of the thermistor goes up, the resistance goes down. In both cases this is because the devices are acting as semiconductors and the number of charge carriers has been affected (see page 129).

Figure 17

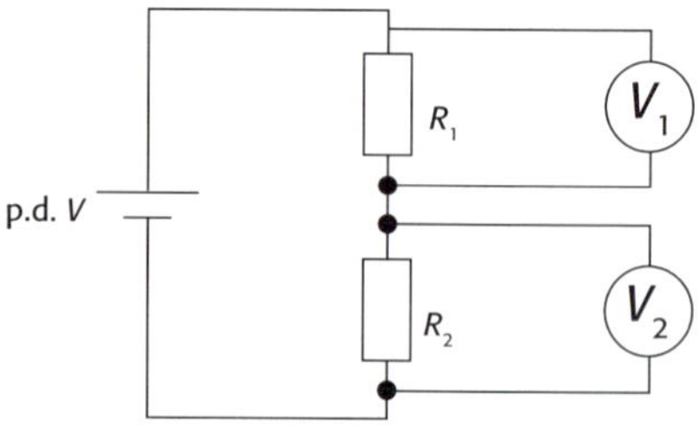

Figure 18

A simple way of utilizing this change in resistance is by employing a **potential divider** circuit using two resistors, R_1 and R_2 (see Figure 18).

An analysis of the current flowing around the circuit, shows that the total potential difference that is available, V, divides up (hence the name of the circuit) with each resistor taking its "share" in proportion to its resistance.

Since $V_1 = IR_1$, $V_2 = IR_2$ and $I = \dfrac{V}{R_1 + R_2}$

$$V_1 = \left(\frac{R_1}{R_1 + R_2}\right)V \text{ and } V_2 = \left(\frac{R_2}{R_1 + R_2}\right)V \qquad (5)$$

By analysing equation 5 we can see that the larger resistance takes the larger share of the total available p.d.

If either of these resistors was swapped for one of a different value, then the way the potential difference was shared would be different. The share of the p.d. would also vary if one of the resistors varied as a result of external conditions. For example, consider the circuit in Figure 19.

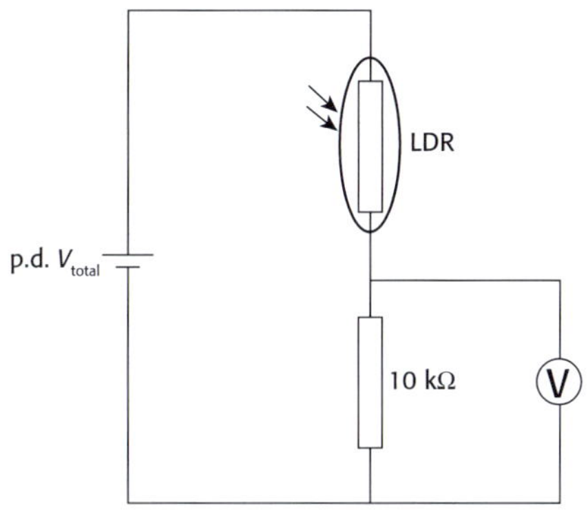

Figure 19

When light shines on the LDR, its resistance is comparatively small, say 200 Ω, but if no light shines then its resistance is greatly increased – perhaps to 20 kΩ.

In the light,

resistance of LDR ≈ 200 Ω
resistance of fixed resistor = 10 kΩ

∴ the resistance of the fixed resistor >> resistance of LDR
so $R_1/(R_1 + R_2) \approx 1$
and p.d. across the fixed resistor ≈ V_{total}

In the dark,

resistance of LDR ≈ 20 kΩ
resistance of fixed resistor = 10 kΩ

∴ the resistance of the fixed resistor ≈ resistance of LDR

$$\text{p.d. across the fixed resistor} \approx \frac{V_{total}}{3}$$

So when light stops shining on the LDR there will be a decrease in p.d. across the fixed resistor. An electrical circuit can be designed to compare this p.d. with a known fixed potential difference. We can create a system that will sound an alarm whenever the measured p.d. goes down. The system we are considering will sound an alarm whenever a beam of light that is arranged to shine on an LDR is broken.

To be useful, the circuit we have described still needs further modifications – the simple system would stop sounding the alarm as soon as the light shines on the sensor again. As long as each addition does not fundamentally affect this basic section, we can modify parts of the circuit to make it useful for our purpose.

Remember that it is not possible to assume that the current or potential difference remains constant after a change to the circuit. After a change, the only way to ensure a correct answer is to start the calculations again.

Sometimes the sensing device can be designed around a **potentiometer** circuit. This is just a potential divider created from a fixed resistor with a movable connection.

As the slider (represented by the arrow in Figure 20) moves from one end to the other, the p.d. measured by the voltmeter varies from the maximum, V_{total}, (when the movable connection is at the top end of the resistor) down to zero (when the movable connection is at the bottom end of the resistor). A simple circuit that uses this approach is the petrol gauge in a car. A device is arranged to float on top of the petrol in the tank. As it moves up and down, it varies the position of the sliding contact on a potentiometer circuit and thus the p.d. recorded is dependent on the amount of petrol in the tank. This reading is calibrated so that it accurately represents the amount of petrol left.

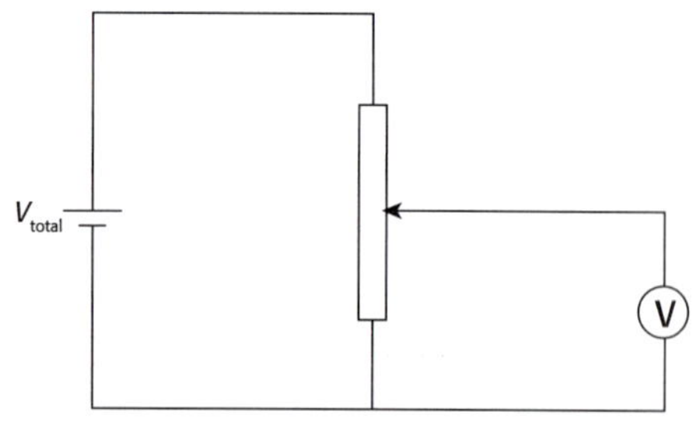

Figure 20

A common sensor in everyday engineering use is the strain gauge. Linear strain is defined in the following way:

$$\text{strain} = \frac{\text{increase in length}}{\text{original length}}$$

A strain gauge measures any small extension or compression that occurs by measuring changes in resistance that take place when a resistor is increased or decreased in length. Often these changes in resistance are very small and so clever techniques need to be incorporated to allow these changes to be measurable.

Real voltmeters have high but not infinite resistance so that when we connect a voltmeter in parallel with a component we change the circuit. The p.d. that we end up measuring is different to the p.d. across the component before we connected the meter. If the resistance of the voltmeter is 10 times greater than the resistance of the device in question, then the maximum error in the reading of p.d. will be 10%. A voltmeter with a higher resistance would further reduce this error.

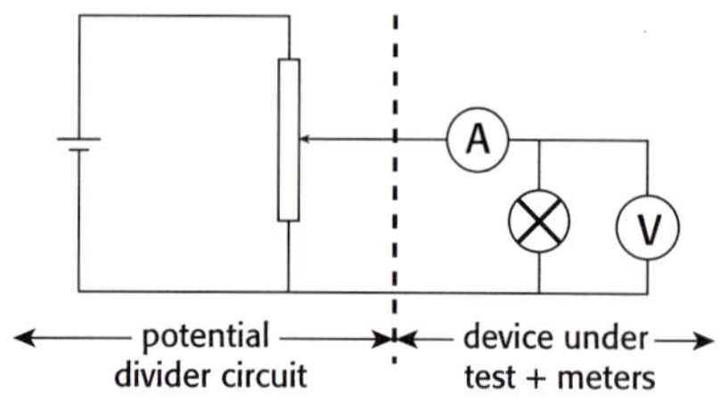

Figure 21

The same consideration needs to be borne in mind when choosing a potentiometer to use in a potential divider circuit. For example, the standard circuit in Figure 21 could be used for measuring the electronic characteristics for a bulb.

You can see that the circuit can be split into two halves. The left-hand side of the circuit allows you to fix a p.d. of your choice. When the right-hand side is added, this p.d. is connected to the bulb and meters. We have thus chosen and placed a p.d. across the bulb that will cause a current to flow through the bulb. Even if these meters were perfect, this way of visualizing the circuit is only useful if the resistances of the potentiometer and the bulb were appropriately selected in the first place.

We assume that the p.d. "fixed" by the left-hand side remained constant when the right-hand side of the circuit was added on. This will only be the case if the resistance of the right-hand circuit is significantly greater than the resistance of the potentiometer (like example (e) above). Normal laboratory potentiometers have a resistance across each end of about 1 Ω or 2 Ω and they are constructed to be able to carry a reasonably large current (5 A or so). This means that the resistance of the object under test (the bulb in the above example) should certainly be more than 10 Ω and ideally much larger.

This principle is also used when designing more complicated electric (and electronic) circuits. It is possible, within reason, to add different system "blocks" onto already existing circuits. If the resistance of the extra block is at least 10 times the resistance of the existing block, this addition can be done knowing that the first circuit will be largely unaltered by the addition.

Physics issues: electrical risks

What are the most common causes of cables overheating?

Using inappropriate power cables

If you want to extend the length of a power cable, it is very dangerous to use wire that is too thin. The circuit can appear to work, but the wires can heat up very quickly.

Commercial extension leads should always have their maximum power rating clearly written on them and the total power rating of all the devices connected using the extension lead must not exceed this limit. Very long cable extensions (the sort that wind up on a drum) have different maximum ratings depending on whether the cable is wound up or unwound. An unwound cable loses thermal energy more readily that one that is wound up.

Using "splitters" wrongly

Each electrical socket in your home is connected to other electrical sockets. The arrangement of these sockets must be in parallel as each socket is arranged to provide exactly the same p.d. – 230 V in Europe and most of Asia, and 110 V in North America. Electronic equipment such as TVs and stereos are designed to operate on this voltage and the more devices you plug into the same socket, the greater the total load and so the socket has to provide a greater total current.

If, for example, you plug in a kettle and this takes 4 A when operating, then two similar kettles must take a total of 8 A. Most domestic wiring is protected by fuses and designed to be able to cope with 15 A safely. If you connected 4 kettles to one socket then the cable connecting the sockets could get too hot and start a fire.

The results of an electrical fire can, of course, be devastating.

A full risk assessment of a domestic electrical circuit might include some of following possibilities:

1 The power leads to most electrical devices are insulated to prevent anybody from accidentally receiving an electric shock. This insulation can degrade with time and with constant movement. It is common for power leads in a plug to be connected to the electrical contacts by trapping the bared copper cables under a screw connection. If the power cable is regularly moved, it is possible for these wires to eventually become frayed or loose.

2 A **short circuit** takes place when power leads are accidentally connected together. If this happens, the current that flows is only limited by the internal resistance of the power supply. Short circuits can be a result of a wiring mistake or poor connections working loose.

3 If you were to touch one of the conductors in an electrical circuit, it would be possible for a current to flow through you to earth. With a low-voltage circuit, the current would be small and you would probably not notice anything. At high voltages, a larger current might flow and this could cause damage, injury, or even death. For this reason, we often carry out risk assessments before working with electrical devices.

1 Explain why it is perfectly safe for a mechanic to touch both the terminals of a 12 V car battery, but it would be dangerous if she touched a metal tool that was dropped so it touched both terminals.

2 Why is it important to warn the public of high-voltage electrical installations (e.g. 1000 V) when even higher voltages can be easily generated by friction and these do not cause any harm, e.g. brushing hair or removing nylon clothing.

3 Choose one domestic piece of electrical equipment and make a comprehensive list of all potential risks associated with its use. Do not just include risks that are associated with electric currents as a large number of domestic injuries result from problems such as tripping over the power cables.

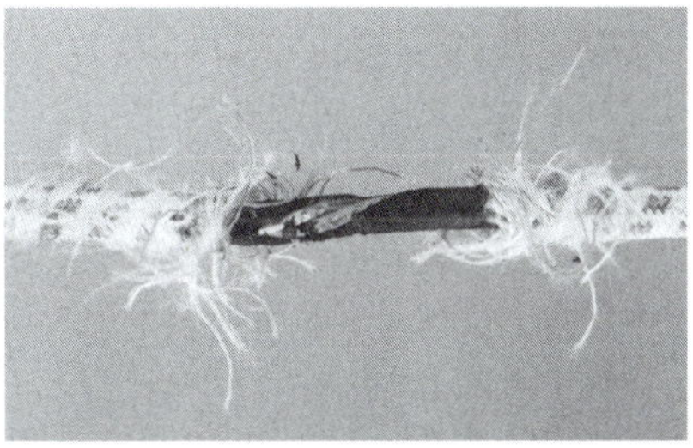

Figure 22

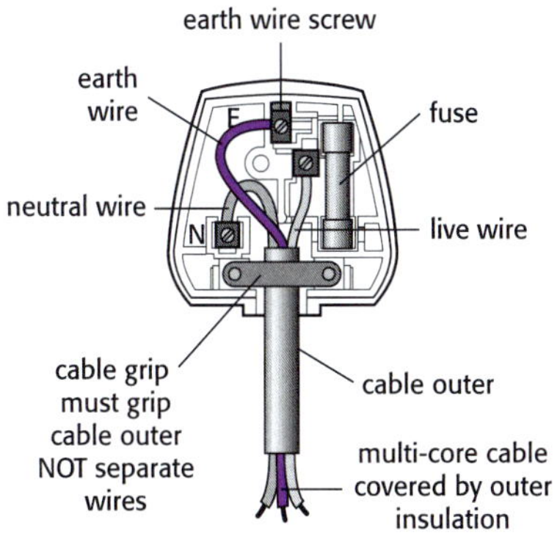

Figure 23

13 Three different non-contact forces

In this chapter we will discuss the similarities and differences between three fundamental forces: the force of gravity, the electrostatic force (between charged objects), and the magnetic force (between magnets and/or currents).

All of these forces have an infinite range (the forces all reduce in magnitude as the objects are moved further apart, but the force never quite gets to zero). It is also worth noting that the objects involved in each of these forces do not have to come in contact for the forces to act between them.

Surprisingly, every force that you observe on a day-to-day basis is an example of one (or more) of these three forces. This is true whether you are considering pushes and pulls (e.g. the push on the tennis ball from your racket when playing a game of tennis, or the friction of your shoes on the ground) or other forces (e.g. the forces that result in you seeing the Sun rising everyday). In fact, the only category of forces that cannot be explained using these three forces are those that exist inside the nucleus of an atom.

Figure 1

Although electric forces and magnetic forces seem quite separate from each other, it turns out that they are related to each other and can both be thought of as different aspects of one fundamental force, called the **electromagnetic force**. It can be helpful, initially, to think of the two effects as entirely separate from one another; after all, the forces each involve different types of object (charges and magnets). Most of the time there is no observable link between the two – a stationary positive charge does not feel a force when next to a magnet.

Before reading any further in this chapter it would be useful to:

- list some examples of everyday forces and categorize them as gravitational, electric, or magnetic (or a combination).
- review the concept of gravitational field strength and recall that $g = 10\ \mathrm{N\ kg^{-1}}$.
- review the basic force laws between charges (opposite charges attract, like charges repel) and between magnets (opposite poles attract, like poles repel).
- review the shape of the magnetic field around a simple bar magnet.
- review any electromagnetic effects that you have studied.

Force fields

A powerful tool for helping us to understand gravitational, electric, and magnetic forces is the concept of a **force field**. When we use these words, you might think about the sort of invisible force field that you might encounter in cartoons or in science fiction.

Conveniently these fictional force fields can be switched on or off and anybody trying to enter the protected region feels a strong resisting force. As far as we know, unfortunately, no such localized strong resisting force field can exist.

Figure 2

In general terms, a particular force field is defined as the region where a force acts. Gravitational fields, electric fields, and magnetic fields are all infinite but the strength of the field varies depending on distance from the mass, charge, or magnet. We can broaden this idea into a mathematical definition.

Figure 3 A character from the Fantastic Four protects herself with a force field

The actual value of the force depends on the object that is placed into the field. The general idea is to imagine that you could place a small test "object" into the field and measure the value of the resulting force. If you put two test "objects" in the same place, you would expect to measure an equal force acting on each of them so the total force would be doubled. Using ratios, you can calculate the forces acting on any object that you place in the field.

The field strength at a given point in the field is defined as the force per unit test "object" that feels the force at that point:

$$\text{field strength} = \frac{\text{force (N)}}{\text{object feeling the force (kg, C, or T)}}$$

An important point to note is that the test "object" must be small in magnitude, as we have assumed that it does not change or affect the field that we are testing. It must also be small in physical size, as we are trying to define the field at a particular point in space.

Gravitational field strength, *g*

Gravitational fields act on masses so the unit test "object" in this situation is a point mass. If we place a small point mass m in a gravitational field of magnitude g, it feels a force F acting on it.

$$\text{gravitational field strength } g = \frac{F}{m}$$

As F is a force (in N) and m is a value of mass (in kg), the units of g must be N kg^{-1}. At the surface of the Earth the gravitational field strength is approximately 9.8 N kg^{-1}, which means that a 1 kg mass will feel a force of about 9.8 N and 74 kg will feel a force of about 730 N.

On the surface of the Moon, the gravitational field strength is: g_{moon} = 1.6 N kg^{-1}. This means that a mass of 74 kg will feel a force of about 120 N.

?

Show, by substitution, that the units N kg^{-1} are equivalent to the units m s^{-2}.

Figure 4

Figure 5

Electric field strength *E*

Electric fields act on charges so the unit test object is a point charge. If we place a small point charge q in an electric field of magnitude E, it feels a force F. Notice that if q represents a negative charge, this will mean that q is negative and so is E.

$$\text{electric field strength } E = \frac{F}{q}$$

As F is a force (in N) and q is a charge (in C) the units of E must be N C^{-1}. We can also define the electric field as a potential gradient – i.e. the change in the voltage per unit distance moved. The unit of E defined in this way is volts per metre (V m^{-1}), which can be shown to be equivalent to N C^{-1}.

For example, there is an electric field between the base of a thundercloud and the ground. If the strength of this electric field is approximately 3.5×10^5 N C^{-1} this means that an electron on the ground whose charge is 1.6×10^{-19} C, will feel a force of about 5.6×10^{-14} N.

Note that even though the unit of electric field strength is N C^{-1}, we cannot define the electric field strength as the force acting on a charge of 1 C. The reason for this is that 1 C is a huge amount of charge. In fact 1 C is so large that it would be impossible to collect this amount of positive or negative charge in one place as the repulsive forces between the individual charges would be too large.

What effect does reversing the sign of *E* have on the physical situation represented?

Magnetic field strength *B*

The magnetic field definition is slightly more complex than the other two, as magnetic fields act on several different types of object. There are three basic things that can feel a magnetic force:

- magnets
- currents
- moving charges

A current is, of course, made up of moving charges, but as the situations are very different it is easy to forget that they are, in fact, equivalent. There are some additional complications:

- the direction of the resultant magnetic force is not usually the same as the direction of the magnetic field.
- there can be additional factors that affect the resultant magnetic force for a given field – magnitude of the current, direction of velocity of moving charge, etc.
- there is no such thing as a "unit magnet".

The unit test object in this situation can involve either currents or charges. When considering currents, the length of the piece of wire under consideration will also affect the size of the force.

For a given current, if a certain length of wire feels a force F then double the length of wire will feel double the force, $2F$. The total magnetic force is proportional to the current flowing down the wire, and to the length of the wire.

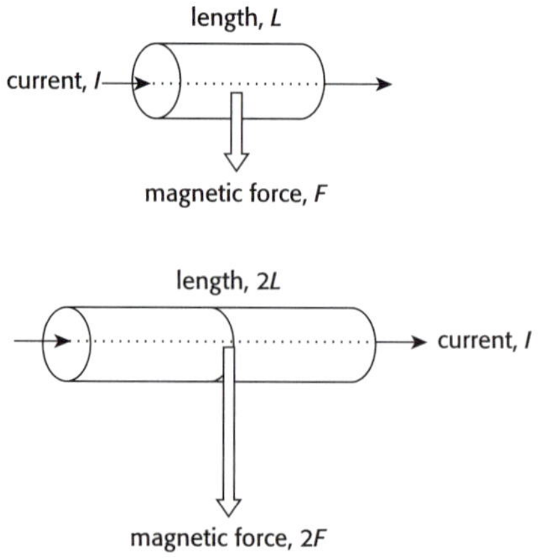

Figure 6 In the example above, the magnetic field must be out of the plane of the paper

In this situation, the unit test object is a small **current-length** which is the value of the current multiplied by the length of wire ($= I \times L$). The units of current-length must be A m (amps × metres). A small current-length *IL* placed in a magnetic field of magnitude *B* can experience a range of forces depending on the orientation of the current-length. The maximum force that can act on it, *F*, is used to define the magnetic field strength, *B*. Do not assume that the dependency on orientation means that this is a vector equation.

$$\text{magnetic field strength } B = \frac{F}{IL}$$

F is a force (in N) and *IL* is a value of current-length (in A m) so the units of *B* must be N A^{-1} m^{-1}. For example, the magnetic field strength close to a particular bar magnet is 7.9×10^{-3} N A^{-1} m^{-1}. A 5.2 cm wire that has a current of 6.1 A flowing along it placed next to this magnet will feel a maximum force of 2.5 mN.

An alternative, and equally valid, definition of magnetic field strength involves a positive point charge *q* moving at a velocity *v*. In this case, the test object is the charge-velocity (in C m s^{-1}) and the maximum force, *F*, is defined using:

$$\text{magnetic field strength } B = \frac{F}{qv}$$

In this version, the units of magnetic field strength must be N C^{-1} m^{-1} s. These two definitions (and their units) are exactly equivalent, so you can always use SI units. If, for example, an electron (1.6×10^{-19} C) is moving at a speed of 2.3×10^{7} m s^{-1} in a magnetic field of 7.9×10^{-3} N A^{-1} m^{-1}, then it can feel a *maximum* force of up to 2.9×10^{-14} N.

There are two other possible SI units for the magnetic field strength, *B* – the tesla (T) and the weber per m^2 (Wb m^{-2}). These different possibilities all turn out to be useful in different situations and are absolutely identical – it does not matter which one you use.

$$1\ \text{N A}^{-1}\ \text{m}^{-1} = 1\ \text{N C}^{-1}\ \text{m}^{-1}\ \text{s} = 1\ \text{Wb m}^{-2} = 1\ \text{T}$$

Direction of the force

Force is a vector quantity, which means that we need to specify both its magnitude and its direction. The definitions of the three field strengths discussed so far can be used to calculate the magnitude of a force, but we have not yet considered the direction of the force. All three fields (gravitational, electric, and magnetic) are vectors but the direction of the force field and the direction of the force are not necessarily the same thing.

In gravitational fields, the situation is straightforward. When we put a mass in a gravitational field, it feels a force. The magnitude of the force that is felt is calculated using $g = \frac{F}{m}$. The direction of this force is just the same as the direction of the gravitational field. Objects experiencing a force due to gravity are, as far as we know, always attracted to each other.

Electric fields are also straightforward. The only slight complication is that there are two types of charge that we could put in the field – positive and negative – and the forces that they feel will be in opposite directions. By convention, the direction of the electric field is the same as the direction of the force on a positive test charge. The force on a negative charge must be in the opposite direction to the field.

The magnetic field due to a magnet or charge is the most complicated of the three. We will begin by discussing magnets. In the same way that there are two types of electric charge, there are two types of magnetic pole (north and south). By convention, the direction of magnetic field represents the force that acts on the north pole of a magnet. The direction of force on a south pole must then be in the opposite direction to the field.

We now need to generalize this result to the two other possible situations that involve magnetic forces – a current in a magnetic field and a moving charge in a magnetic field. In both of these situations the maximum force arises when the components are perpendicular to each other. When a conventional current, I, flows at right angles to a magnetic field, B, the force, F, on this current is at right angles to both of these directions. The directions of the three variables B, F, and I define the x, y, and z-axes of a three-dimensional coordinate system.

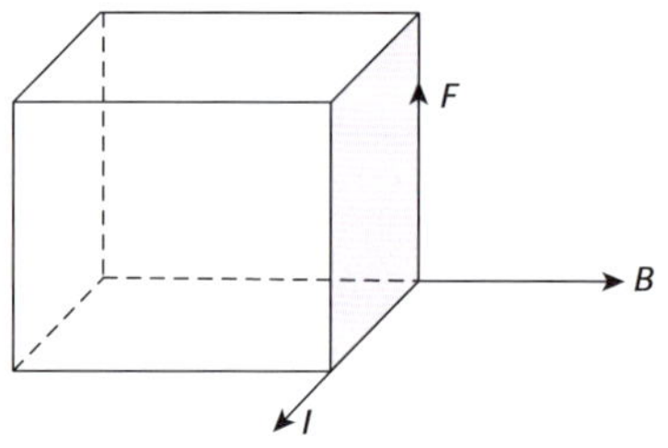

Figure 7

This rule could not have been predicted – it has only been discovered by experimental measurement. There are several different ways that you can remember this relationship – one possible way is Fleming's left-hand rule, another is the right-hand palm rule (see Figure 8).

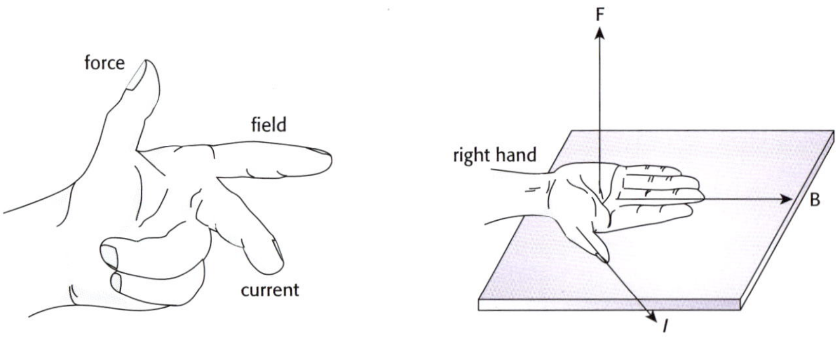

Figure 8

If the current isn't exactly perpendicular to the magnetic field then we must resolve either the current or the magnetic field into two components. In the situation shown in Figure 9, the current has been resolved into a component that is perpendicular to B ($I \sin \theta$) and a component that is parallel to B ($I \cos \theta$).

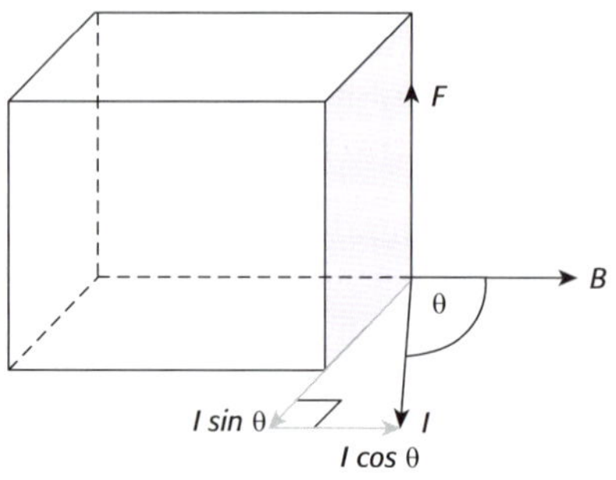

Figure 9

Instead of using $F = BIL$ to calculate the force, we now need to use the appropriate component. Since the component of I parallel to B will not affect the resultant force, F can be calculated using:

$$F = BIL \sin \theta$$

An analogous situation exists when we consider the force on a charge that is moving in a magnetic field. If the charge is moving in the same direction as the magnetic field then it will not feel a force. If the motion of the charge has a component that is perpendicular to the field then it will feel a force given by

$$F = qvB \sin \theta$$

Representation of a field: a general approach

In order to pictorially represent the variation in field strength around the source of a field, we need to find a way of representing both the magnitude and the direction of the force on the test object. The general approach is to introduce the idea of **field lines**. These are called **flux lines** in some situations.

At some time in the past, you have probably already used field lines to draw the variation in magnetic field strength around a bar magnet but you might not have thought about what the lines physically represent. The field lines can be drawn in such a way as to represent the field in a mathematically accurate way.

Typically, the direction of the force on the test object is shown by the direction of the field lines and the variation in the magnitude of the force on the test object is shown by how close together the lines are drawn. Consider the familiar example of the magnetic field around a bar magnet.

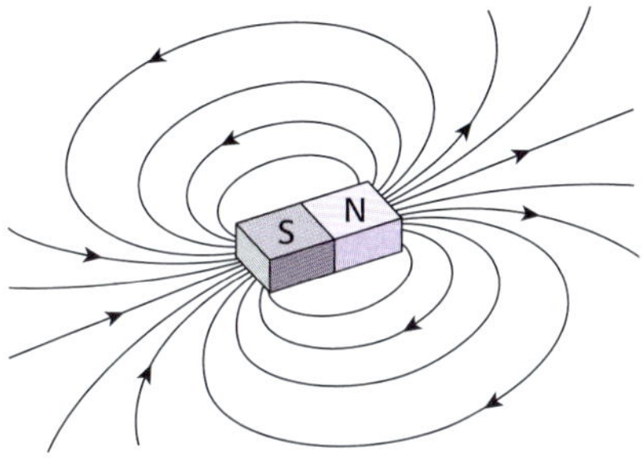

Figure 10 The magnetic force is stronger near the poles of the magnet as shown by the field lines being closer together

The diagram in Figure 10 is a two-dimensional representation of a three-dimensional field. In general, the number of field lines that pass though a given area that is perpendicular to the field lines represents the strength of the force on the test object.

Figures 11a and 11b both represent a uniform field because the field lines are evenly spaced and parallel to one another. In a field that increases with distance, the lines get closer together and a weakening field is shown by the lines getting further apart.

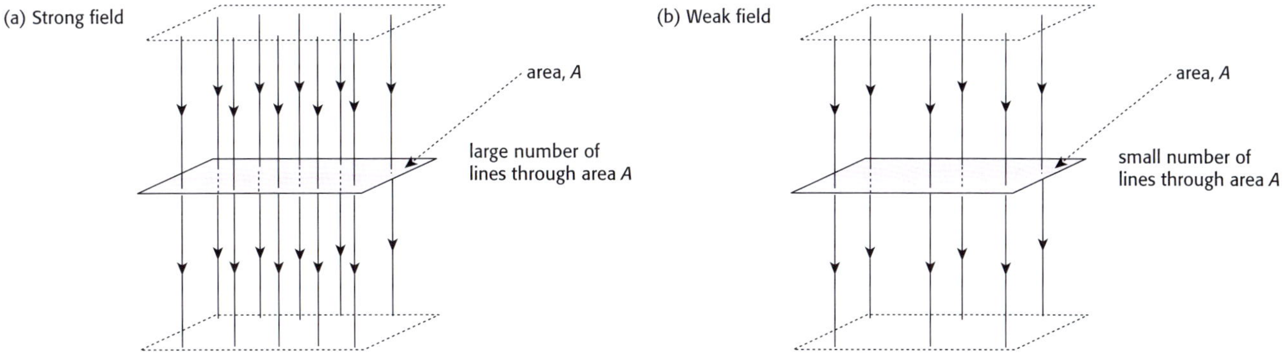

Figure 11

Gravitational field lines

A mass attracts all other masses towards it with a force that diminishes with distance of separation. The arrows on the field lines give the direction of the force acting on a small test point mass.

The variation in magnitude of the gravitational field for a point mass is shown in Figure 12.

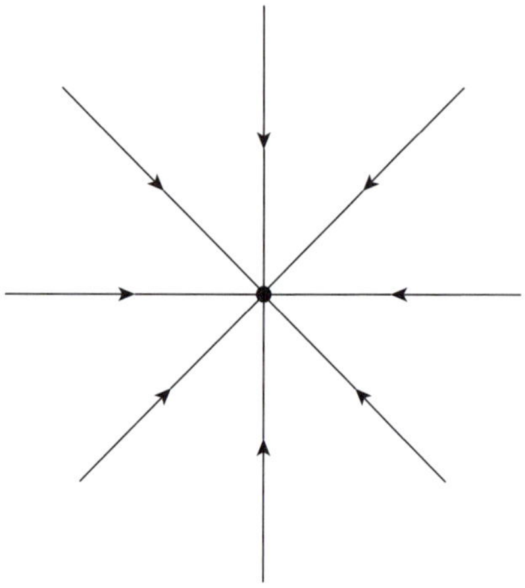

Figure 12

This is called a **radial** field. The closer you get to the source of the gravitational field, the stronger the field. The field lines show this by getting closer together towards the source of the field. The field varies according to an inverse square law (see page 167) and Figure 13 accurately represents this mathematical relationship because the number of field lines that pass through a given area also decreases according to an inverse square law.

In practical situations, the masses under consideration are not point objects. However, the overall gravitational field outside any uniform spherical object turns out to be the same as that for a point object with the same mass placed at the centre of the sphere. Near the surface of the Earth, we often assume that the gravitational field strength is constant with $g = 9.81\ \mathrm{N\ kg^{-1}}$. A constant gravitational field would be represented by evenly spaced, parallel field lines (Figure 13).

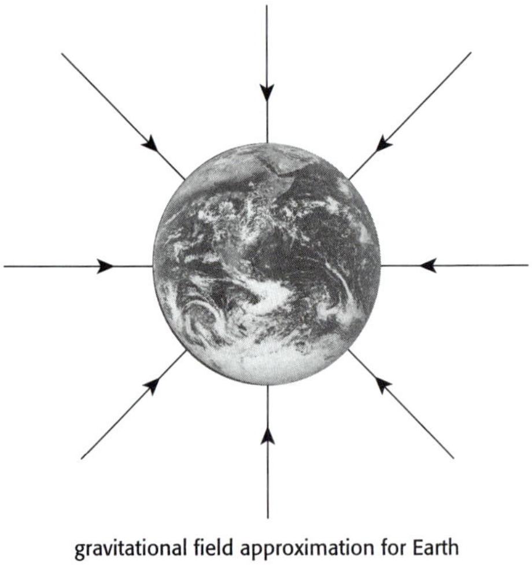

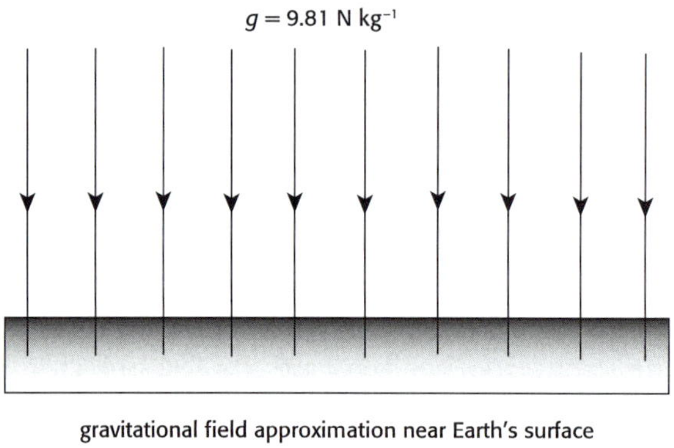

Figure 13

Electric field lines

The electric field pattern around a point charge has the same shape as the gravitational field pattern around a point mass, but the direction of the field lines depends on the sign of the charge. By convention, the direction shown indicates which way a positive test charge placed in the field would move, so field lines point towards a negative charge and away from a positive charge.

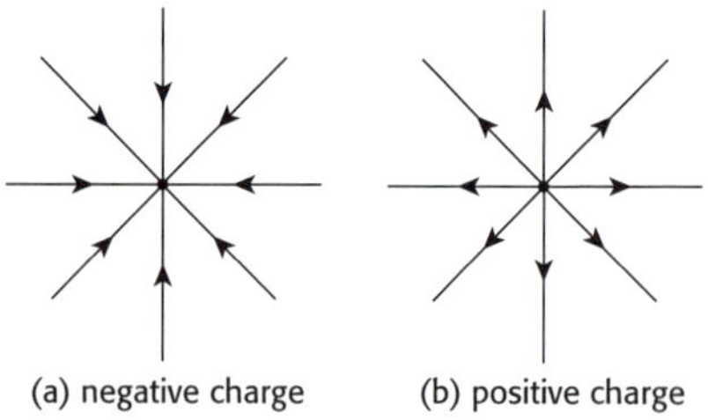

Figure 14

Magnetic field lines

The direction of a magnetic field line represents the direction of the force felt by the north pole of a small test magnet. There is, however, no such thing as an isolated magnetic pole so magnetic field strength is defined in terms of the force on a moving charge, or on a small current-length. Not only does the field depend on the distance to the moving charge, but the geometry of the situation is also important.

Once again, the field that results from a small current-length obeys an inverse square relationship in terms of distance, but the dependence on the geometry of the situation means that the overall magnetic field does not necessarily vary in this way.

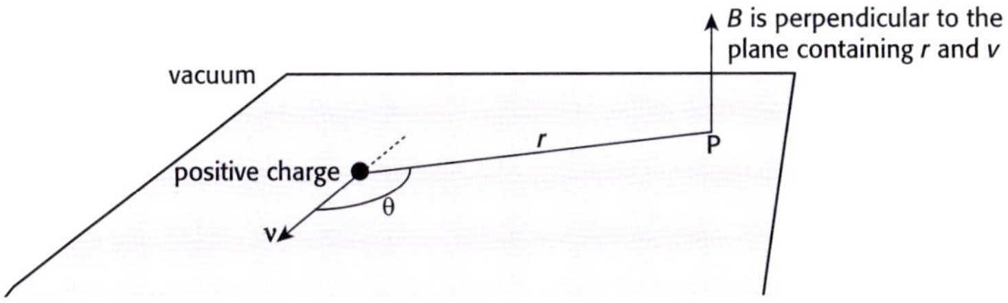

Figure 15 Field at point P due to single moving charge is proportional to the inverse square of the distance away, *r*.

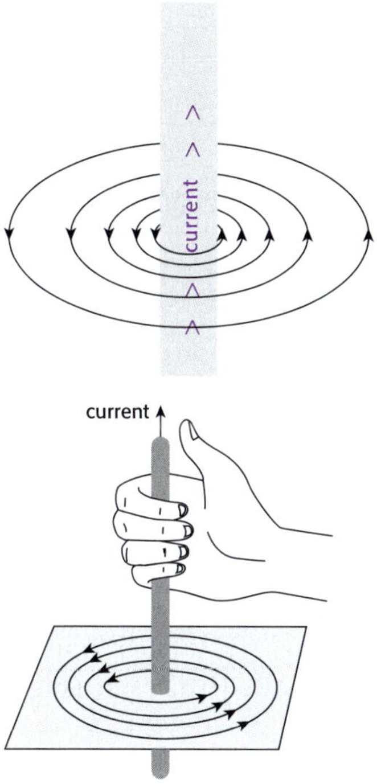

Figure 16

The overall magnetic field at P is the result of the additions of the magnetic fields due to each small section of the wire. The general method for adding fields together is shown on page 151, but in the case of a long straight wire this results in the overall magnetic field pattern shown in Figure 16.

The direction of the overall magnetic field is remembered using the right-hand grip rule.

Calculating gravitational field strength, *g*

At a point P some distance away from a single point object, experiments show that the gravitational field g is:

- proportional to the mass of the point object causing the field, m_1
- inversely proportional to the square of the distance r from the point object to the point P.

The relationship is described as an **inverse square** relationship. Mathematically this is written as:

$$g \propto \frac{m_1}{r^2} \qquad \text{or} \qquad g = -\text{constant} \times \frac{m_1}{r^2}$$

The negative sign is there because the gravitational force is always attractive, so the direction of the force is towards the source of the field. The constant is given the symbol G and is called the *universal gravitational constant*.

$$g = -G\frac{m_1}{r^2} \qquad G = 6.67 \times 10^{-11}\ \text{N m}^2\ \text{kg}^{-2}$$

So, the force acting on another mass, m_2 that is placed at point P is given by:

$$F = g \times m_2 \qquad \text{or} \qquad F = -G\frac{m_1 m_2}{r^2}$$

This relationship is a mathematical statement of **Newton's law of universal gravitation**. In words, this law states that the gravitational force of attraction between two point masses is proportional to the product of the two masses and inversely proportional to the square of the distance of separation between the masses.

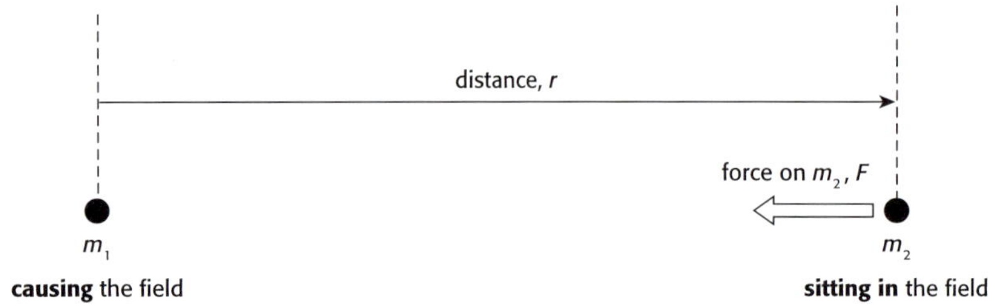

Figure 17

Calculating electric field strength, E

At a point P some distance away from a single point object, experiments show that the electric field E is:

- proportional to the charge causing the field, q_1
- inversely proportional to the square of the distance r from the point object to the point P.
- dependent on the material between the object and point P.

Mathematically this is written as:

$$E \propto \frac{q_1}{r^2} \qquad \text{or} \qquad E = \text{constant} \times \frac{q_1}{r^2}$$

By convention, we define the field in relation to a positive charge, and the field is directed away from a positive charge, so this time there is no negative sign. If we once again assume that the charges are in a vacuum, the constant is given the symbol k and is called the Coulomb constant.

$$E = k\frac{q_1}{r^2} \qquad k = 8.99 \times 10^9 \text{ N m}^2 \text{ C}^{-2}$$

This equation is very straightforward, but it sometimes turns out to be useful to write the constant in a different way:

$$k = \frac{1}{4\pi\varepsilon_0} \qquad \varepsilon_0 = 8.85 \times 10^{-12} \text{ C}^2 \text{ N}^{-1} \text{ m}^{-2}$$

The constant ε_0 is called the permittivity of a vacuum (often referred to as the permittivity of free space). If there is some substance between the charges, this will affect the value of k because different materials have different values of permittivity, ε.

The force acting on another charge, q_2, that is placed at point P is thus given by:

$$F = E \times q_2 \qquad \text{or} \qquad F = k\frac{q_1 q_2}{r^2} = \frac{q_1 q_2}{4\pi\varepsilon_0 r^2}$$

This relationship is a mathematical statement of **Coulomb's law**. In words this law states that the electrostatic force between two point charges is proportional to the product of the two charges and inversely proportional to the square of the distance of separation between the charges.

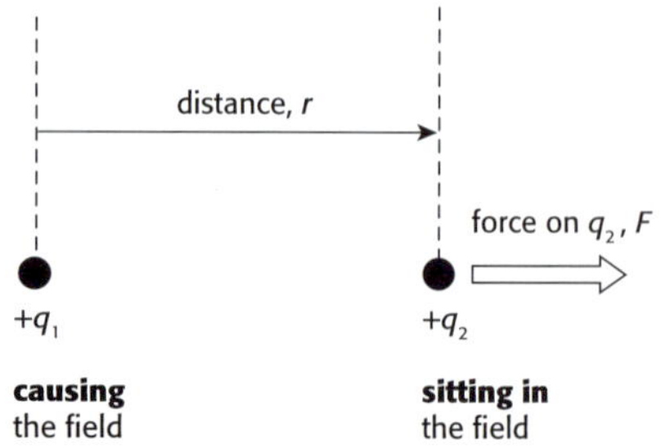

Figure 18

Mathematical physics: how to add fields

All fields are vector quantities (see page 32) so to add two (or more) fields we must correctly add together two (or more) quantities that have both a magnitude and a direction. One key idea is that field lines can never cross one another. It may also be helpful when imagining a combined pattern to remember that field lines repel each other if they are parallel and attract each other if they are anti-parallel.

Another useful property comes from the inverse square nature of all these fields. Although they all have an infinite range, the strength of a field drops rapidly as the separation between source and test object increases. The field due to a nearby object often dominates the resultant field. Figure 19 shows how the gravitational field strength decreases with distance from a point object, starting at a distance of r_0 (the radius of the object).

Whatever the combination of masses, charges, or current-lengths, it should be possible to estimate the overall field felt a large distance from the individual objects. At a large distance we can consider them all together; for example, an object that is a long way from three charges, +3 mC, −5 mC, and +1 mC, will feel an overall field that is equivalent to one charge: +3 − 5 + 1 = 1 mC, no matter how the three individual charges are locally arranged.

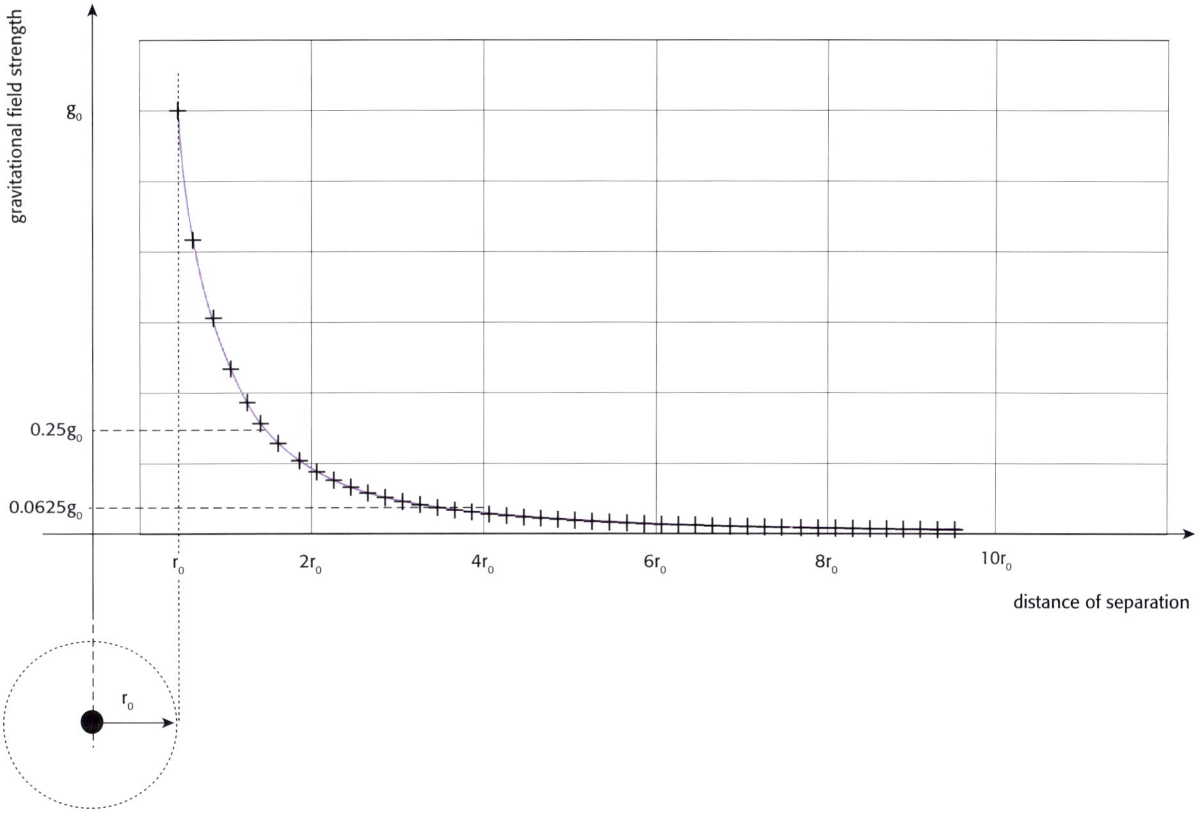

Figure 19 Inverse square relationship

Gravitational field patterns

Field due to a planet

Newton's law of universal gravitation gives us a method for calculating the field strength due to a single point mass, but a planet is clearly not a point mass. Theoretically, the overall field of the planet at any point is the addition of the fields for all the individual point masses that make up the planet.

In principle it is possible to work out the distance to every single point and use Newton's law of universal gravitation to calculate the contribution that each point makes to the overall gravitational field. Once we have calculated all the contributions we could then perform a vector sum. As you can imagine this is a complicated mathematical addition.

Fortunately, you do not need to perform this calculation. In fact, the gravitational field anywhere outside a uniform sphere of constant density is exactly the same as if all the mass had been concentrated at its centre. The same idea can be used for an object placed in the field. The force between two spherical masses (whose separation is large compared to their radii) is the same as if the two spheres had their masses concentrated at a point at the centres of the spheres. This simplification means that planets are often treated as point masses (even though they do not have constant density).

For example, the Earth attracts every point on the Moon and vice versa. The situation in Figure 20a can be simplified to the situation illustrated in Figure 20b.

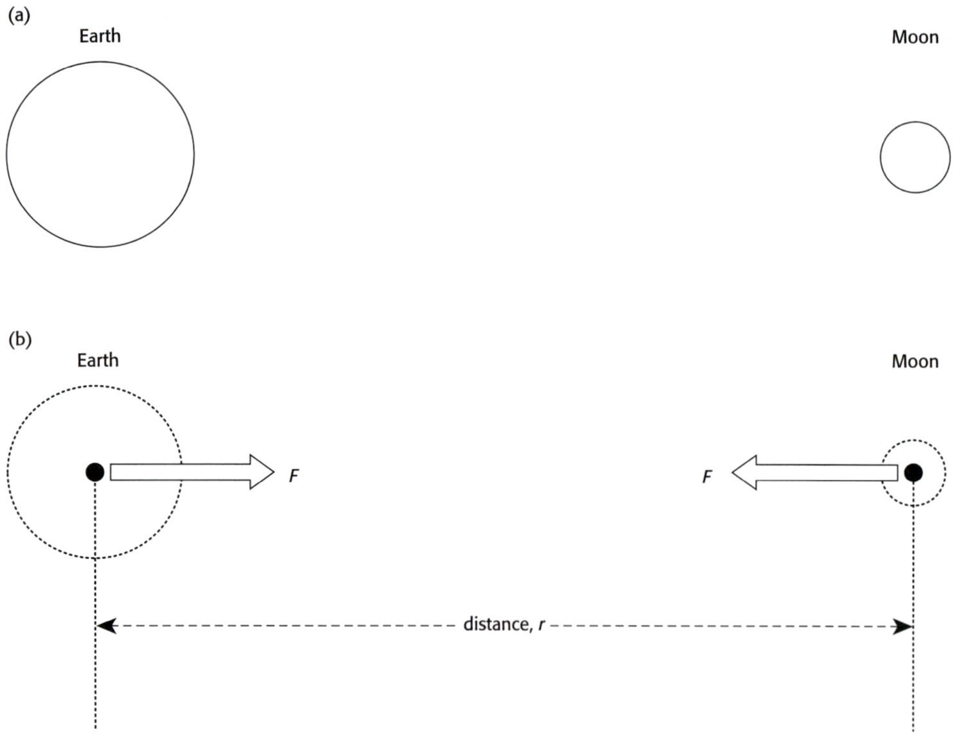

Figure 20

Combined fields of two masses

Figure 21 shows the process for calculating the resultant gravitational field between two masses (e.g. the Earth and the Moon) at one particular point P in space.

To find the shape of the overall resultant field, we must repeat this process at every point in space.

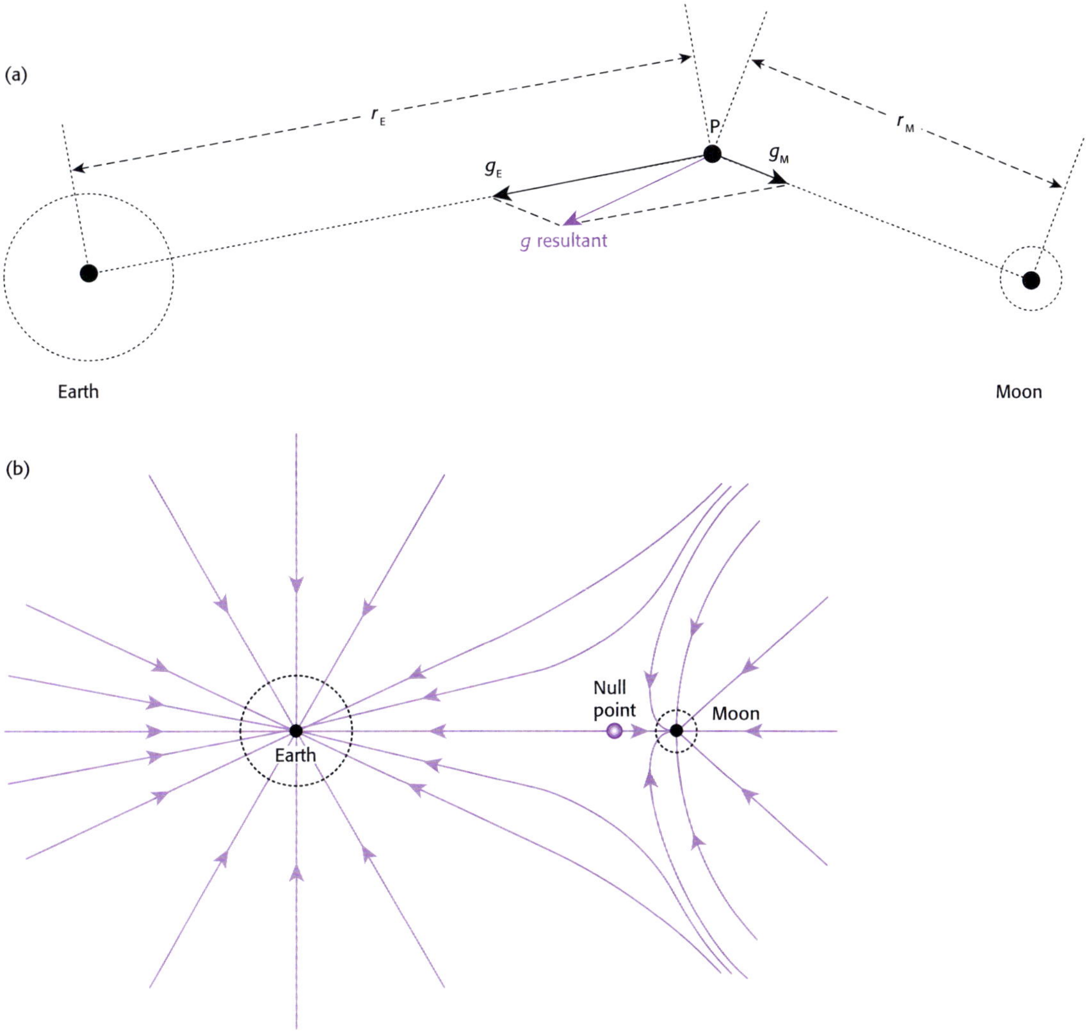

Figure 21

There will be one point somewhere between the Earth and the Moon where the two fields cancel and there will be no resultant gravitational field – the **null point**.

Electrostatic field patterns

A charged sphere

Outside a charged conducting sphere, the resultant electric field is identical to the field of an equally charged point charge concentrated at its centre. This statement is analogous to the gravitational field for a spherical mass, but the situations are slightly different. The mass is distributed uniformly throughout the sphere, whereas charge can be shown to always reside on the outside of the sphere. This fact does not affect the pattern of the fields outside the sphere but it does have an effect on the inside. Inside the mass, there is still a gravitational field but inside a hollow conducting object, there is no field at all.

Outside the sphere, the fields are the same as a point charge concentrated at the centre. Both fields are radial.

Electrostatic potential, V_e

Electrostatic forces are felt by charges, so the definition of electrostatic potential at any point P in the field is

$$V_e = \frac{\Delta E_p}{q}$$

where

ΔE_p is the potential energy change needed to bring a small point positive test charge from infinity to point P, measured in J

q is the value of the test charge, measured in C

V_e is the electrostatic potential at point P, measured in J C^{-1} (or V)

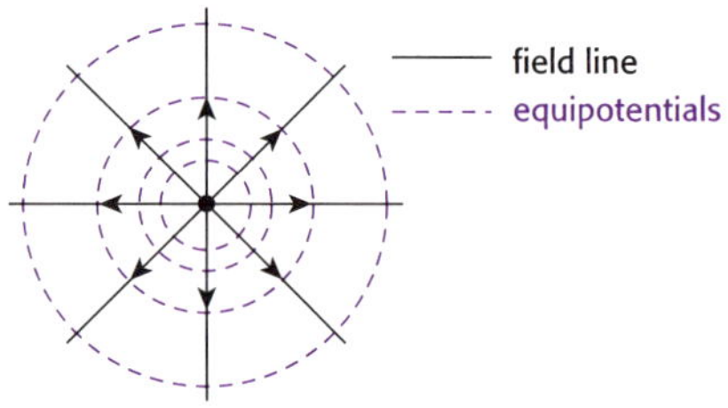

Figure 4 Electric field lines and electric equipotentials around a positive point charge

As charges can be positive or negative, the potential at any point also can be positive or negative. By convention a positive test charge is chosen, so the potentials resulting from the field associated with a positive charge will be positive, and those for a negative charge will be negative.

The concept of difference in potential – potential difference (pd) or voltage – has already been introduced in Chapter 12. An alternative unit for J C^{-1} is V. One possible definition of a conductor is an object for which the electric potential must be constant.

?

The diagram below shows lines of electric equipotential. The change in potential on moving from one line to the next is always the same. At which point does the electric field strength have its greatest magnitude?

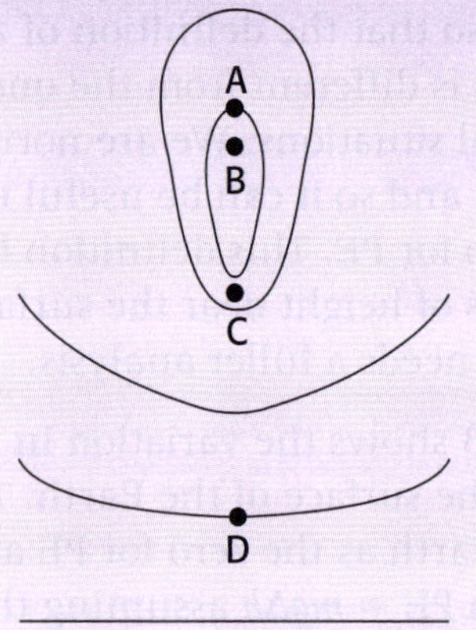

Representation of potential

Potential is a scalar quantity. The variation of potential is most often represented in two dimensions by drawing lines of equipotential and in three dimensions by equipotential surfaces. In either case, the potential is the same at all points on the equipotential.

Contour lines on a map are an example of gravitational equipotential lines, as they join together places that are at the same vertical height.

When a test object is moved along an equipotential, the potential energy must remain the same, and so no work is done. There is a force on the test object, but it is always at right angles to motion along the equipotential. In other words, equipotential surfaces and field lines must be at right angles to one another. For example, Figure 4 represents the electric field lines and electric equipotentials around a positive point charge.

The uniform electric field pattern between two oppositely charged parallel plates was introduced on page 155. In order to calculate, from first principles, the potential at any point between the plates, the individual contributions of each of the point charges on the plates would need to be taken into consideration. This is a complicated mathematical process.

The relationship between field and potential gradient provides a quick way of analysing the situation. See Figure 5. The equipotential lines are always at right angles to the field lines. The constant field in the centre of the plate means that the potential gradient is constant – the equipotential lines are uniformly spaced apart.

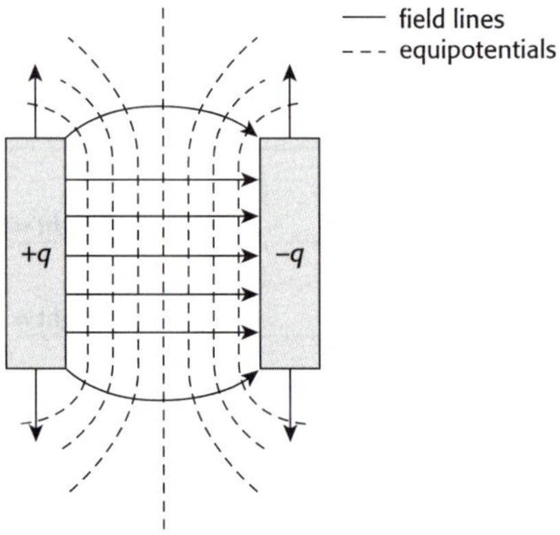

Figure 5 Field lines and equipotentials for parallel plates

Potential gradient

There is a direct mathematical relationship that links the field and the way in which the potential varies. Figure 6 shows a positive test object δq being moved a distance δx between two points A and B in the field due to a positive charge q. In this situation the external force ($= -E\,\delta q$) does work in moving from A to B and so the potential of the system is increased.

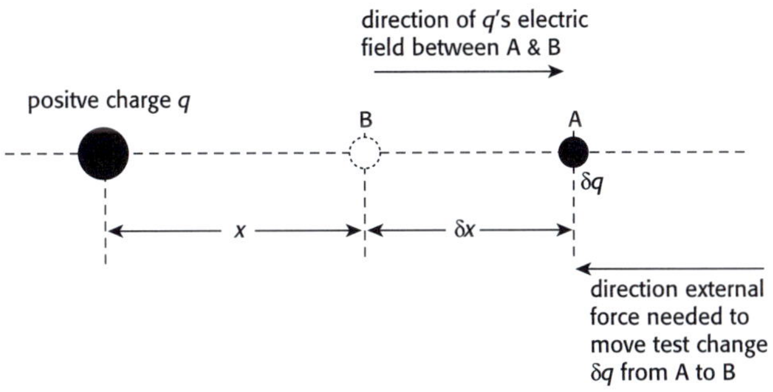

Figure 6

The field between A and B is assumed to be constant and equal to E.

The external work done *on* the system, $\delta W = -$work done *by* the system $= -E\,\delta q \cdot \delta x$

$$E = -\frac{1}{\delta q}\frac{\delta W}{\delta x} \qquad \text{but} \qquad \delta V = \frac{\delta W}{\delta q}$$

Therefore

$$E = -\frac{\delta V}{\delta x}$$

If δx is very small ($\delta x \to 0$), then this becomes

$$E = -\frac{\mathrm{d}V}{\mathrm{d}x}$$

$\mathrm{d}V/\mathrm{d}x$ represents the **potential gradient**, measured in V m^{-1}
E represents the electric field strength, measured in N C^{-1}

A similar relationship can also be derived for gravitational fields.

In both cases:

field strength = −potential gradient

The negative sign indicates that if potential increases with a displacement, then the direction of the field is in the opposite direction to the displacement.

Calculation of potential

To calculate the potential at a given distance from a source of field (either gravitational or electric), we need to calculate the total work done in bringing a test object from infinity. As the value of the field is constantly changing, this calculation involves a technique called integration. This is not discussed in this course companion, but the result is shown below.

$$V_{\text{gravitational}} = -\frac{Gm}{r}$$

?

A conducting rod is introduced between oppositely charged parallel plates and is held parallel to the plates. Deduce the changes that will take place in the field and potential pattern as a result of the conductor being placed inside.

?

1. Show that the units N C^{-1} are identical to V m^{-1}.
2. Using the above derivation as a model, show that a similar relationship applies for gravitational fields.

In both cases the electric and gravitational potential is inversely proportional to the distance away from the point object causing the field.

Mathematical physics: energy and potential

The resultant potential at any point in space is just the addition of all the potentials resulting from all the point objects involved. As shown on page 153–4, a spherical object can often be treated as a point object with all the mass or charge concentrated at the centre of the object. In general this approach will work only when considering significant distances away from the spherical object.

Fields and potentials provide equivalent ways of describing the same situation. In order to describe a gravitational or electrostatic situation fully, we can either

- state values of the electric field strength *E*, calculate the force on a charge at a point, and draw electric field lines, or
- state values of the potential *V*, calculate the potential energy of a charge at a point, and draw equipotential lines

?

On a sheet of graph paper, use both of the above approaches to sketch representations of the field and the potential in one of the following situations. Equipotential lines should be labelled with their values. You need to choose the appropriate number of equipotential lines to show in your diagram, and this will depend on the complexity of the situation you choose.

(a) An isolated +6 nC point charge (up to 10 cm away).
(b) An isolated −6 nC point charge (up to 10 cm away).
(c) An isolated 6 kg point mass (up to 10 cm away).
(d) Two point charges, both +5 nC, separated by 20 cm.
(e) Two point charges, one +5 nC the other −5 nC, separated by 20 cm.
(f) Two point charges, one +5 nC the other +10 nC, separated by 15 cm.
(g) Two point charges, one +5 nC the other −10 nC, separated by 15 cm.
(h) An isolated conducting sphere of radius 5 cm carrying a charge of −8 nC (up to 5 radii away).
(i) The Earth, mass 5.98×10^{24} kg and radius 6370 km (up to 10 radii away).
(j) The Moon, mass 7.35×10^{22} kg and radius 1740 km (up to 10 radii away).
(k) The Earth–Moon system (they are separated by 3.84×10^{8} m).

A good approach would be for each student to work out a different situation, and then present and explain their drawings to their fellow students.

The electronvolt

A potential difference (p.d.) is the difference in potential between two points.

$$\text{potential difference p.d.} = \frac{\text{energy dissipated}}{\text{charge flowed}}$$

This equation was used to define the unit of p.d. to be the volt: hence voltage is an alternative term that means the same as potential difference (or p.d.), and potentials are measured in volts. The same equation is used to create a new unit for measuring energies – the electronvolt. This unit is particularly useful at the atomic level. At this scale, 1 joule is an impossibly large energy for a particle to have, so this new unit is used instead. The definition is as follows:

$$\text{energy difference} = \text{potential difference} \times \text{charge flowed}$$

so 1 electronvolt = 1 volt × charge on one electron

$$1\ \text{eV} = 1.6 \times 10^{-19}\ \text{J}$$

Note that this unit is not an "electron per volt" but an "electron × volt". For example, when an electron is accelerated from rest through a potential difference of 1 kV it must gain a kinetic energy of 1.6×10^{-16} J, or 1 keV.

Escape speed

The gravitational potential at a planet's surface provides an easy way of calculating the energy needed by any mass to escape away from the planet's gravitational influence – for example, to escape the Earth's gravitational pull and send a space probe to Mars.

Consider a small mass *m* at the surface of a planet of mass *M* and radius *r*. The small mass has a gravitational potential of $V_{\text{gravitational}}$ and a total energy that is equal to the product of the potential and mass:

$$\text{gravitational potential energy gPE} = V_{\text{gravitational}} \times m$$

All gravitational potentials are negative, because all gravitational forces are attractive, and so the above quantity is negative. If we assume the planet behaves like a point mass, then

$$\text{gPE} = -\frac{GMm}{r}$$

In order to escape the gravitational influence of this planet, the small mass would need enough energy to be able to get an infinite distance away. The minimum amount of positive kinetic energy that is needed is $+\frac{GMm}{r}$. The mass will be able to escape if it has a value of velocity v_{escape} given by:

$$\frac{1}{2}mv_{escape}^2 = \frac{GMm}{r}$$

$$v_{escape} = \sqrt{\frac{2GM}{r}}$$

A mass with less velocity than the above escape velocity must have a negative total energy and will eventually fall back to the planet. A mass with a greater velocity than the escape velocity would theoretically be able to get an infinite distance away from the planet and still be moving.

?

1 Use the information given in the question on the previous page to compare the escape velocity for the Earth and for the Moon. *Note that the above derivation of escape velocity, and so the answer to question 1, has assumed that there is only one planet exerting a gravitational influence on the mass. In reality, of course, all planets in a system contribute to the total potential energy of a mass on the surface of one of the planets.*
2 Use the information given in the question on the previous page to calculate the escape velocity for the Earth–Moon system from (a) the surface of the Earth and (b) the surface of the Moon.
3 Explain why: (a) your answers to the two questions above are different; (b) why a mass with either of these velocities would still fail to escape from the gravitational influence of the Earth–Moon system.

Thinking about science: Kepler's law

In 1619 a mathematician called Johannes Kepler published *De Harmonices Mundi* (*Harmony of the Worlds*). This book contained a statement of his third law: "The square of any planet's orbital period, *T*, is proportional to cube of its mean distance *r* from the Sun".

$$T^2 \propto r^3$$

?

The table below gives information regarding the masses and distance from the Sun for different planets. The information is given relative to the Earth's mass ($M_\oplus$) and to the Sun–Earth distance (defined as one astronomical unit, 1 AU).

Analyse this information and outline the extent to which the data verify Kepler's third law.

Planet	Orbital time	Distance to Sun / AU
Mercury	87.97 days	0.387
Venus	224.70 days	0.723
Earth	365.26 days	1.000
Mars	687 days	1.523
Jupiter	11.86 years	5.203
Saturn	29.46 years	9.539
Uranus	84.02 years	19.19
Neptune	164.8 years	30.07

Kepler had discovered this mathematical link between the two measured variables, but he could not explain why the relationship was true. Although Kepler's third law made progress towards a deeper understanding of planetary motion, there is no reason to believe that this relationship may not be applicable outside our solar system.

In 1687 Isaac Newton published *Philosophae Naturalis Principia Mathematica* (*Mathematical Principles of Natural Philosophy*). This book (often just called the *Principia*) showed how Kepler's third law could be derived from Newton's laws of motion and universal gravitation. The derivation for circular orbits is shown below, but the mathematics works for elliptical orbits as well.

Consider a satellite of mass *m* in orbit around a planet of mass *M* (Figure 7). The radius of the orbit is *r* and the speed of the satellite in its orbit is *v*.

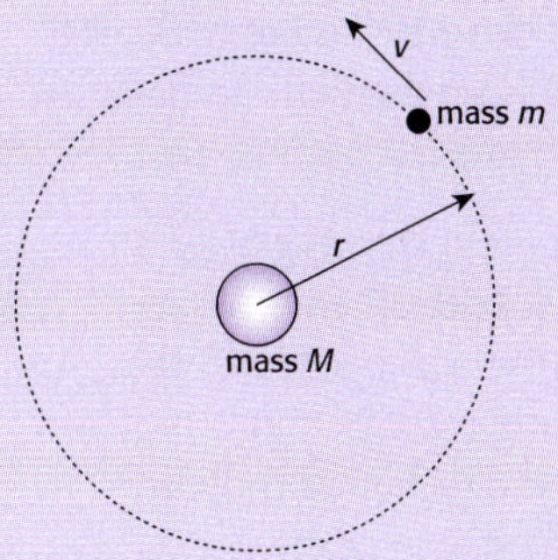

Figure 7 Satellite of mass m orbiting a planet of mass M

The gravitational force of attraction on the satellite provides the centripetal force needed to stay in orbit:

gravitational force = CPF

$$\therefore \frac{GMm}{r^2} = \frac{mv^2}{r} \qquad \therefore \frac{GM}{r} = v^2$$

The time period for the orbit is T, where

$$v = \frac{2\pi r}{T} \qquad v^2 = \frac{4\pi^2 r^2}{T^2} \qquad \therefore \frac{GM}{r} = \frac{4\pi^2 r^2}{T^2} \qquad \therefore \frac{T^2}{r^3} = \frac{4\pi^2}{GM}$$

Since 4, π, G, and M are all constants (M is the mass of the Sun, which is orbited by all the planets in the solar system),

$$\therefore \frac{T^2}{r^3} = \text{constant} \qquad \text{or} \qquad T^2 \propto r^3$$

Newton's laws of motion and of universal gravitation are elegant and beautiful mathematical statements. They are extremely powerful and truly universal statements that can still be used to predict accurately the motion of any planets, stars, and galaxies throughout the universe. They do not, however, explain *why* masses gravitationally interact in this mathematical way – see page 156.

Total energy in orbit

An object that is in orbit has both kinetic energy and potential energy, and the total energy is the addition of the two. Using the same symbols as above:

$$\text{KE} = \frac{1}{2}mv^2 = \frac{1}{2}m\frac{GM}{r} = \frac{1}{2}\frac{GMm}{r}$$

$$\text{PE} = V_{\text{gravitational}} \times m = -\frac{GMm}{r}$$

The KE is numerically half the value of the PE, but the KE is positive and the PE is negative. The addition gives the total energy of an object in orbit.

$$\text{Total energy} = \text{KE} + \text{PE} = \frac{1}{2}\frac{GMm}{r} - \frac{GMm}{r} = -\frac{1}{2}\frac{GMm}{r}$$

1 On the same axes, use the above equations to sketch graphs showing the variation with orbital radius of (a) the kinetic energy, (b) the gravitational potential energy and total energy of a satellite.
2 Use the above equations to explain why a satellite in low Earth orbit will *speed up* as a result of the energy it loses doing work against friction with the atmosphere.

In your view, is the ISS worth the money that is being spent on it? Research some facts and information about the project in order to back up your view.

Thinking about science: apparent weightlessness

Astronauts appear to be weightless. If the spacecraft were a long way away from any massive object, then it would be possible for the force of gravity to be zero, but the spacecraft's orbit is close to the Earth. So she cannot be far enough away from the Earth to have escaped its gravity.

The explanation for the apparent lack of gravity is that the astronaut and the spacecraft are both in **free fall**. For both objects (the spacecraft and the astronaut), gravity is providing the CPF that is required to constantly accelerate (the direction of motion is constantly changing) and so stay in orbit.

On the surface of the Earth, the astronaut feels the effect of her weight pulling her towards the ground. The spacecraft also has weight, directed towards the ground. As both the astronaut and the spacecraft are accelerating at 9.8 m s^{-2} in the same direction, there is no normal contact force between the astronaut and the "floor" of the spacecraft, and so she feels weightless.

A similar sensation would be felt by a passenger in a lift accelerated downwards at 9.8 m s^{-2}. The two forces acting on the passenger are the force up from the floor, F, and their weight, W.

- When the lift is at rest or moving at constant velocity the passenger is doing the same, and the resultant force on the person must be zero, so $F = W$.
- When the lift is accelerating upwards then the passenger is doing the same and the resultant force on the person must be upwards, so $F > W$.
- When the lift is accelerating downwards, then the passenger is doing the same and the resultant force on the person must be downwards, so $F < W$.

If the lift cable broke, the lift and the passenger would free-fall together towards the Earth, both with an acceleration of 9.8 m s^{-2}. In each case the force of gravity would be doing the accelerating and there would be no additional contact force between the lift and the passenger. The passenger would appear weightless during the fall.

This effect is used to train astronauts to get used to the effects of apparent weightlessness. With careful piloting, an aircraft can be made to execute a series of dives towards the Earth, each with a vertical acceleration of 9.8 m s^{-2}. If the acceleration is too small, the passengers will feel an upward force from the floor. If it is too great, the ceiling will seem to push the passengers down. But during a properly executed dive the passengers will appear weightless and float. When the aircraft comes out of the dive then they will feel the contact force between them and the floor.

It is interesting to note the difference between the perception that trainee astronauts have (that gravity has been temporarily switched off) and the explanation of the effect (the acceleration of the aircraft). Einstein realized that it would be impossible for any observer to tell the difference between a gravitational force of attraction in one direction and an acceleration in the opposite direction. This **principle of equivalence** allowed him to develop a new way of visualizing gravitational attractions – the curvature of space-time. These ideas are studied further in the relativity option.

15 Electromagnetic induction

The generation of electrical energy and the transformation of alternating currents between different voltages both rely on electromagnetic induction.

Chapter 13 introduced the motor effect, in which a current placed in a magnetic field feels a force. This force can produce motion if the current is free to move (e.g. in an electric motor), and to get the maximum force the directions of the current, field, and force all have to be at right angles to one another.

Electromagnetic induction is the process by which an e.m.f. is generated. If a complete circuit is available then the e.m.f. can cause a current to flow.

HL

Generating e.m.f. using motion

When a conductor (e.g. a length of wire) is moved through a magnetic field, a small e.m.f. (i.e. a p.d.) is induced across the ends of the conductor, and can be measured with a sensitive voltmeter (Figure 1). The magnitude of the e.m.f. is maximized when the direction of motion, the magnetic field, and the conductor are all at right angles to one another.

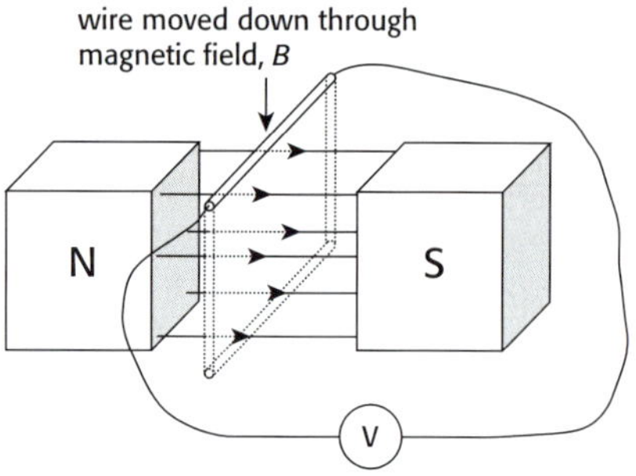

Figure 1 Motion of a conductor through a magnetic field generates a current

In the situation in Figure 2, a voltage will be recorded on the voltmeter when the wire is moving through the magnetic field. Effectively, the wire behaves like a small cell as it moves through the magnetic field. The value of the voltage depends on three different factors:

- the strength of the magnetic field, B
- the speed of the wire, v
- the length of the wire, l.

The generation of e.m.f. as a result of the motion of a conductor through a magnetic field needs to be explained from both the microscopic and macroscopic points of view.

Microscopic point of view

The moving wire must contain equal numbers of positive and negative charges. Although the resulting movement of charge is not along the wire it may be considered as a virtual current, as shown in Figure 2.

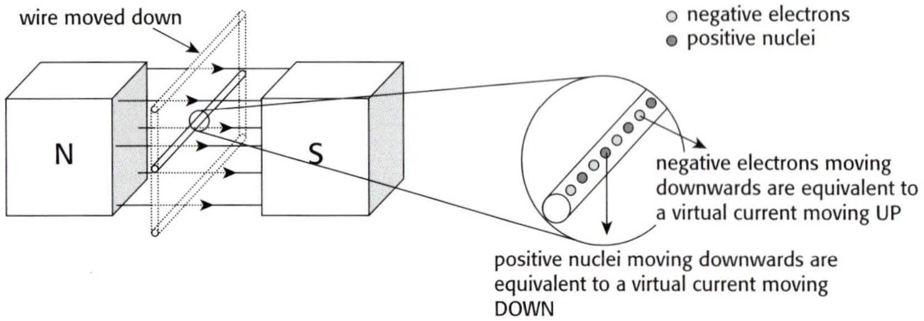

Figure 2 Motion of the wire can be considered to be equivalent to virtual currents

Note that these virtual currents are not *along* the wire, but are in the same direction that the wire is moving. These virtual currents are in the magnetic field, so they feel a force, and the direction of the force can be determined. In this example you should confirm that, as the wire moves down, the electrons feel a force *into* the page (the plane of the paper), whereas the nuclei feel a force out of the page (Figure 3).

The nuclei cannot move, so the magnetic force on a nucleus has no effect. The force on the electrons, however, will cause them to move: in this example they move into the page, making the far end negative and leaving the near end positive.

As the electrons move, a p.d. is generated along the wire. Equilibrium is reached when the value of this p.d., V, produces an electric force on electrons in the wire that exactly balances the magnetic force (Figure 4).

At equilibrium, the forces on an electron in the wire are balanced.

$$F_E = F_M$$

The length of the wire is l, so (using the relationship introduced in Chapter 14) the electric force F_E on an electron can be calculated from the electric field, E, and the charge on the electron, e:

$$F_E = Ee = \frac{V}{l}e$$

The magnetic force can be calculated from the equation given on page 145:

$$F_M = Bev$$

$$\therefore V = Blv$$

where
V is the p.d. (e.m.f.) generated, measured in V
B is the magnetic field strength, measured in teslas (T)
l is the length of the wire, measured in m
v is the velocity of the wire, measured in m s^{-1}

Note that this p.d. has been induced with a specific orientation. If the two ends of the wire were joined to complete a circuit outside the magnetic field then a conventional current would flow, as shown in Figure 5.

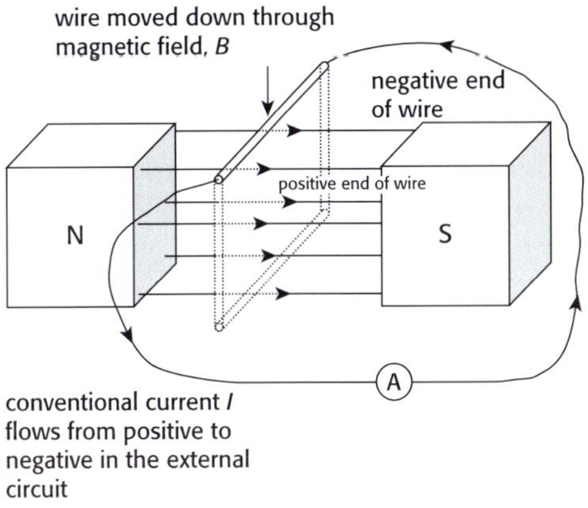

Figure 5 Direction of induced current

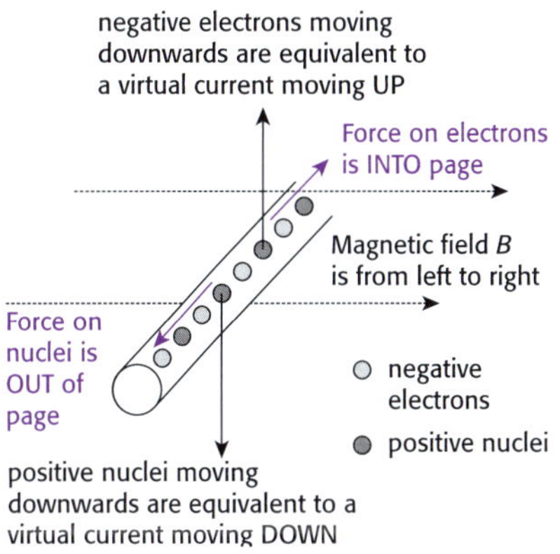

Figure 3 Force on electrons and protons in a moving wire

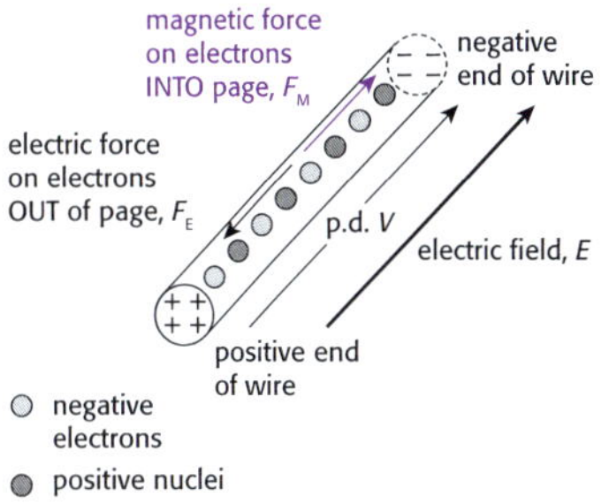

Figure 4 Forces on an electron in the wire are balanced

The north pole of a permanent bar magnet is pushed along the axis of a coil as shown below.

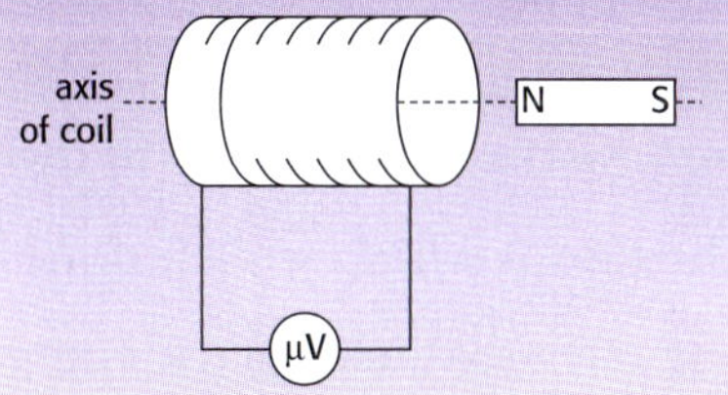

The pointer of the sensitive voltmeter connected to the coil moves to the right and gives a maximum reading of 8 units. The experiment is repeated, but on this occasion the south pole of the magnet enters the coil at twice the previous speed.

Which of the following gives the maximum deflection of the pointer of the voltmeter?

A 8 units to the right
B 8 units to the left
C 16 units to the right
D 16 units to the left

The current that flows is in a magnetic field, so it will feel a force. The direction of this force can be predicted using the motor rule; it always opposes the motion that set it up. This is known as **Lenz's law**:

The direction of the induced e.m.f. is such that, if a complete circuit were available, the current direction would oppose the change that caused the induced e.m.f.

In the above example, the motion is down, which induces a current that flows out of the page. This current feels an opposing force up. If the motion had been up, then the current would have been out of the page and the force would be down – still opposing the change that caused the induced e.m.f.

Macroscopic point of view

Energy considerations can be used to deduce the value of the induced e.m.f. of a moving wire. Provided there is a completed circuit, the induced e.m.f. will cause a current to flow and energy to be dissipated. This energy must come from the work done by the external force that is needed to move the wire. The diagram in Figure 6 represents a rectangular coil that is being pulled at a steady velocity v through a region of magnetic field B.

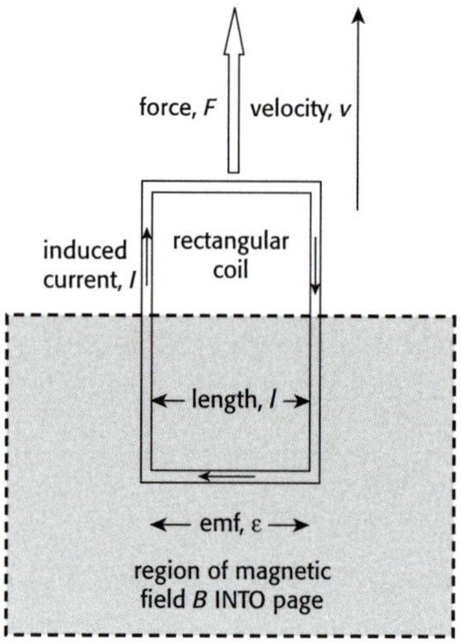

Figure 6 A coil pulled at a steady velocity

The force pulling the coil, F, is opposed by a force of equal magnitude in the opposite direction. This force arises because the induced current in the left-hand section of the coil is in a magnetic field and must therefore feel an opposing force, as predicted by Lenz's law. On this macroscopic scale, Lenz's law is just an application of the law of conservation of energy.

Using the motor rule, the magnitude of the force opposing the motion must be

$$F = BIl$$

Since the wire is moving at constant velocity, the resultant force is zero, so the force pulling to the right must have magnitude $F = BIl$.

The rate at which work is done by the external force $= F \times v$

The rate at which electric energy is dissipated in the wire $= \varepsilon \times I$

From the conservation of energy,

$$\varepsilon \times I = BIl \times v$$

$$\therefore \varepsilon = B\,l\,v$$

where
ε is the induced e.m.f., measured in V
B is the magnetic field strength, measured in T
l is the length of the vertical wire, measured in m
v is the velocity of the wire, measured in m s^{-1}

Note that, in the above derivation, current flows around the coil because the right-hand side of the coil is outside the field. If the magnetic field covered the whole circuit, then a similar e.m.f. would be generated in the right-hand side coil. Since the two e.m.f.s are

?

Use the law of conservation of energy to deduce that it is impossible for the induced current direction to be in the opposite direction in Figure 8.

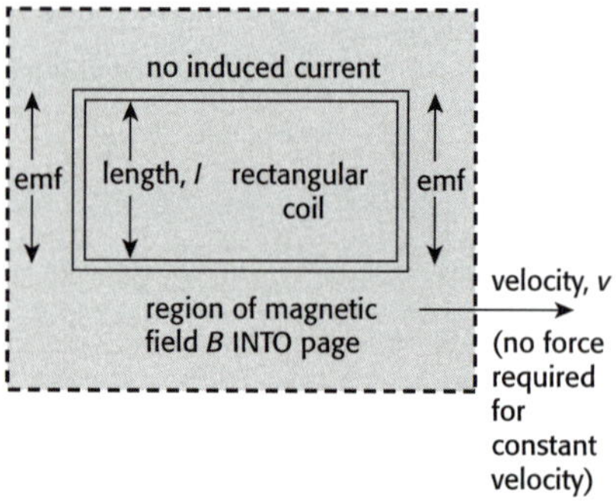

Figure 7 A moving coil completely inside a magnetic field does not induce a current

exactly the same, they would cancel each other, and no current would be induced in the coil.

Mathematical physics: concept of flux

When a conductor moves at right angles to a magnetic field, an e.m.f. is induced. A useful method for calculating the value of the e.m.f. involves the definition of a new mathematical concept associated with magnetic fields –the **flux** ϕ. Chapter 13 introduced the idea of the strength of field being represented using flux lines. Flux lines that are close together represent a strong field (Figure 8).

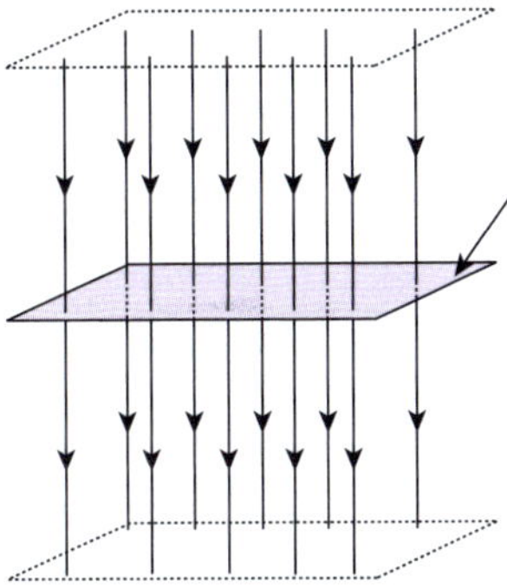

Figure 8 Flux lines represent field strength

We define flux ϕ to be the product of the magnetic field strength B and the area A:

$$\phi = B \times A \times \cos\theta$$

where
B is the magnetic field strength, measured in T
A is the cross-sectional area, measured in m^2
θ is the angle between the field lines and the normal to area, measured in degrees or radians (Figure 9)
ϕ is the flux, measured in T m^2

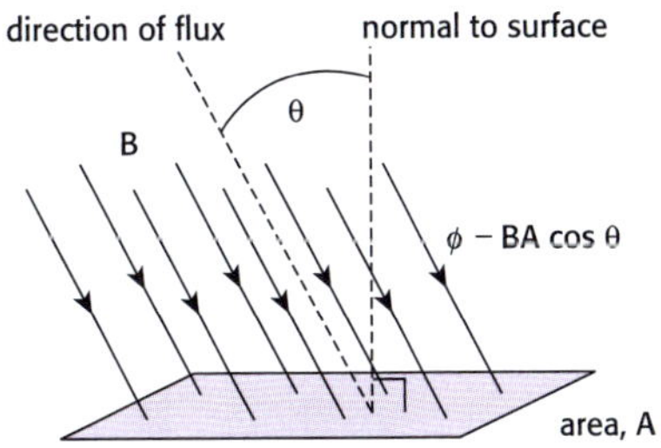

Figure 9 Definition of flux

If the area is normal to the field, then $\cos\theta = 1$ and so the $\cos\theta$ term disappears.

A new unit is defined for flux – the **weber** (Wb).

$$1 \text{ Wb} = 1 \text{ T} \times \text{m}^2$$

This means that an alternative unit for magnetic field strength is the Wb m^{-2}, which corresponds closely to the concept of representing field strength by the number of flux lines going through an area. The number of lines drawn in a field pattern represents the flux. An alternative name for magnetic field strength B is the **magnetic flux density** – that is, the flux per unit area.

This definition can be used to express the equations that we have already derived for the induced e.m.f. in a useful new way. In a time Δt, the wire generating the e.m.f.'s velocity v moves a distance Δx, which sweeps out an area of flux (Figure 10).

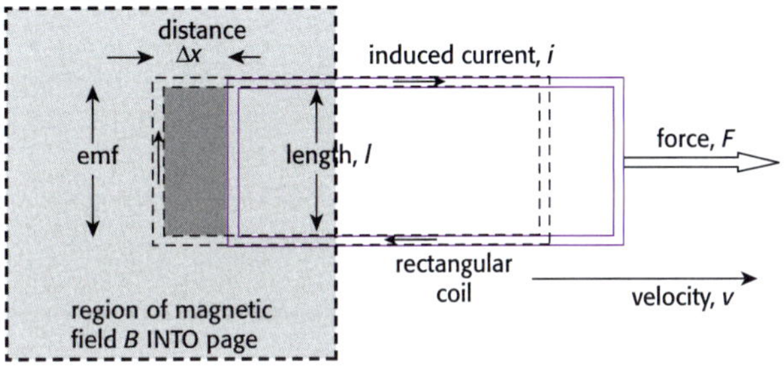

Figure 10 Change of flux through a coil

In a time Δt,

area swept out, $\Delta A = l \times \Delta x = l\,v\,\Delta t$

flux swept out, $\Delta\phi = B\,\Delta A = B\,l\,v\,\Delta t$

rate of change of flux through wire $= \dfrac{\Delta\phi}{\Delta t} = B\,l\,v$
$=$ magnitude of induced e.m.f. ε

Flux can be imagined as the number of field lines. An e.m.f. is generated whenever flux lines are "cut", and the above relationship shows that the induced e.m.f. in a wire is equal to the rate of cutting of flux.

Flux linkage

It is possible to increase the induced e.m.f. in a given situation by using a moving coil of wire (with several loops) instead of one single length of wire. The concept of **flux linkage** is introduced to take this into account. When the plane of a coil is right angles to the field then

flux linked with a coil = flux going through the coil × number of turns

The e.m.f. induced in a coil is equal to the rate of change of flux linkage with the coil. This is known as

Faraday's law of electromagnetic induction. Faraday's law and Lenz's law can be combined into one statement (sometimes known as **Neumann's equation**):

$$\varepsilon = -N\frac{\Delta\phi}{\Delta t}$$

where
ε is the induced e.m.f., measured in V
N is the number of turns on the coil
$\Delta\phi$ is the change in flux through the coil, measured in Wb
Δt is the time taken for the change of flux, measured in s

The full calculus version also applies:

$$\varepsilon = -N\frac{d\phi}{dt}$$

where $d\phi/dt$ is the rate of change of flux through the coil, measured in Wb s^{-1}.

As explained above, adding extra turns to the loop increases the e.m.f. induced only if part of the coil remains outside the field. The identification of flux with the number of field lines make this easy to visualize.

Consider a coil of wire moving out of a region where a magnetic field acts, as shown in Figure 11.

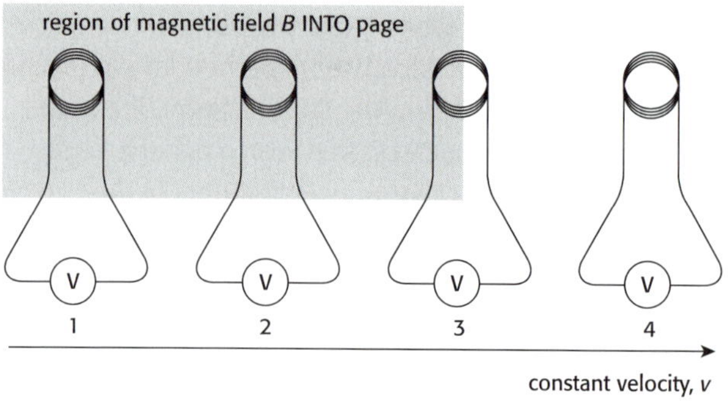

Figure 11 A coil moves out of a magnetic field

As the coil moves between positions 1 and 2, no p.d. will be registered on the voltmeter. The flux linked with the coil does not change, so overall no e.m.f. is generated. An alternative explanation is that identical e.m.f.s are induced in the left-hand and right-hand sides, making the overall induced e.m.f. equal to zero. The only time when e.m.f. is generated is when there is a change in flux in the coil at position 3.

HL

Generators

The rotation of a coil in a magnetic field must also generate an e.m.f., as the total amount of flux linked with the coil changes over time. Figure 12 represents a flat coil rotating at a constant angular velocity ω in a uniform magnetic field B.

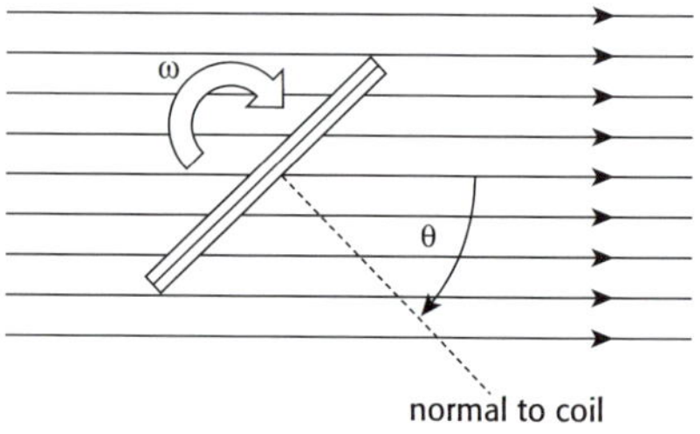

Figure 12 A coil rotating in a magnetic field changes the flux linked

The flux linking with each turn in the coil in the position shown is

$$\phi = BA\cos\theta \qquad \text{where } \theta = \omega t$$

The e.m.f. ε is worked out using Faraday's law:

$$\varepsilon = -N\frac{d\phi}{dt}$$

$$\therefore \varepsilon = N\omega AB\sin(\omega t) = \varepsilon_0 \sin(\omega t) \qquad (\text{where } \varepsilon_0 = N\omega AB) \qquad (1)$$

A coil rotating at a uniform angular velocity in a constant field generates an e.m.f. that is sinusoidal. The maximum e.m.f. is proportional to four factors:

- the number of turns
- the angular frequency of rotation (which is related to the frequency of rotation)
- the cross-sectional area of the coil
- the magnetic field strength.

Transformer e.m.f.

So far, we have considered induced only e.m.f.s caused by the physical motion of a conductor through a magnetic field. E.m.f.s are generated whenever there is a change of flux linkage, even if there is no physical motion. The following examples all give rise to an induced e.m.f. In the final two examples, the objects are both stationary and yet an e.m.f. is still induced. This is knows as a **transformer-induced e.m.f.**

(a) A coil of wire is moved towards the north pole of a bar magnet.

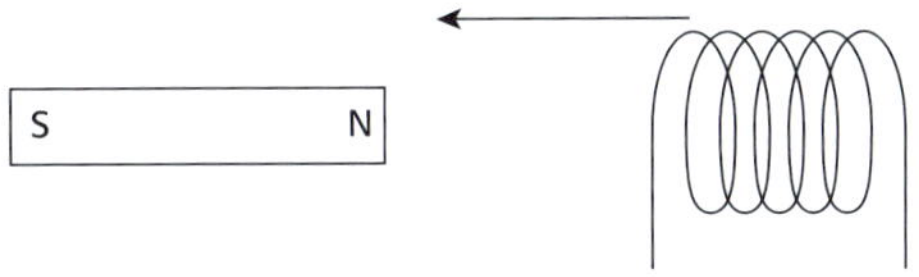

(b) A coil of wire is moved away from the north pole of a bar magnet.

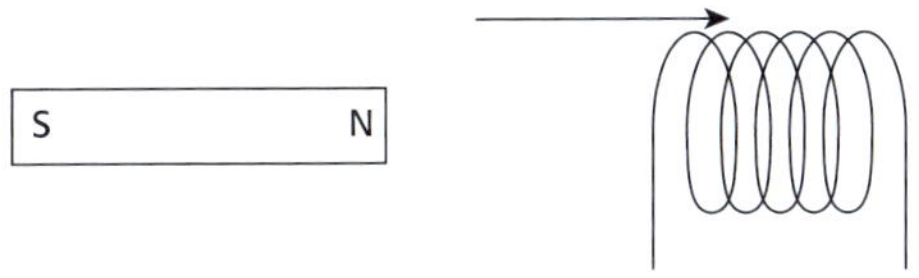

(c) The north pole of a bar magnet is moved towards a coil of wire.

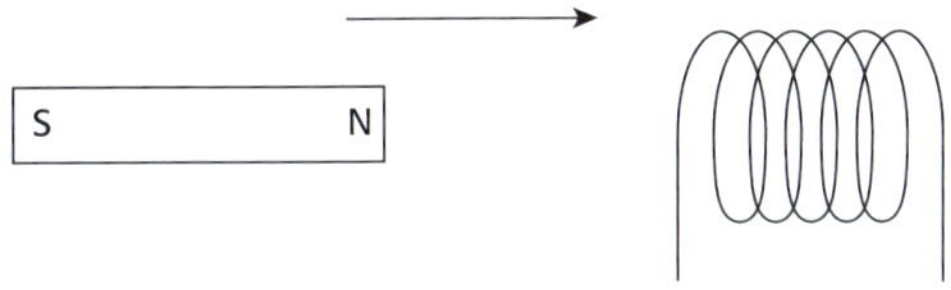

(d) The north pole of a bar magnet is moved away from a coil of wire.

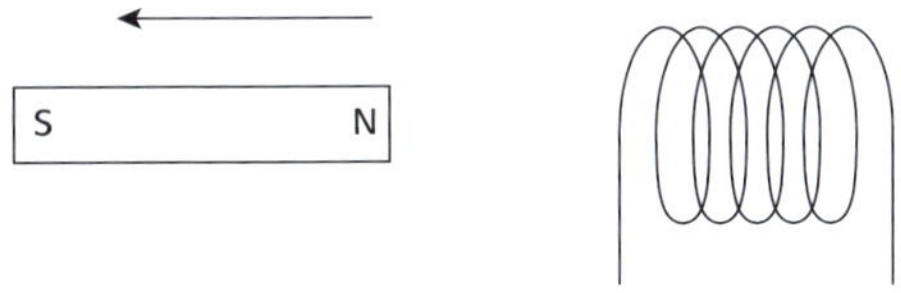

(e) The current in an electromagnet is increased. There is a coil of wire nearby.

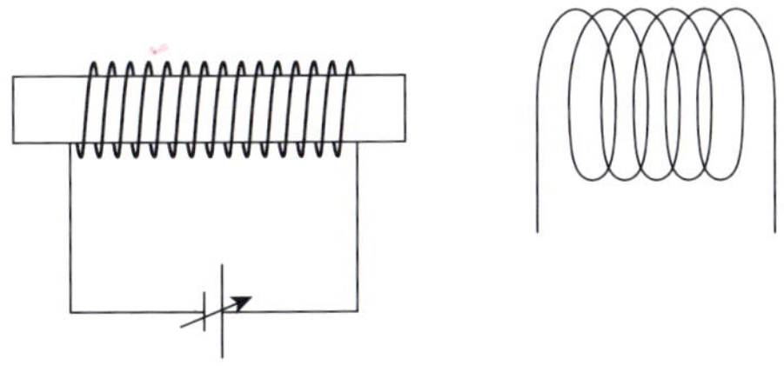

(f) The current in the above electromagnet is decreased.

Although it seems very reasonable to assume that any change of flux must induce an e.m.f., the mechanism by which we explained the generation of the e.m.f. relied upon the actual physical motion of a conductor through a magnetic field. The final two situations appear directly related, but we have not identified a mechanism by which a changing current in one coil can induce a current in a nearby coil.

The analysis of the interaction between magnetic and electric effects is highly complex, and is beyond the scope of IB Diploma Programme physics. The full link can be explained using Maxwell's equations and relativity.

?

1. State the units for each of the variables on the left-hand side and the right-hand side of equation on the previous page. Show that the units for the left-hand side are the same as for the right-hand side.
2. The axes in Figure 13 represent the output of a simple a.c. generator with a coil rotating 50 times a second. On the same axes, sketch the output you would expect if:
   (a) the magnetic field is doubled
   (b) the angular speed is doubled
   (c) the angular speed is halved

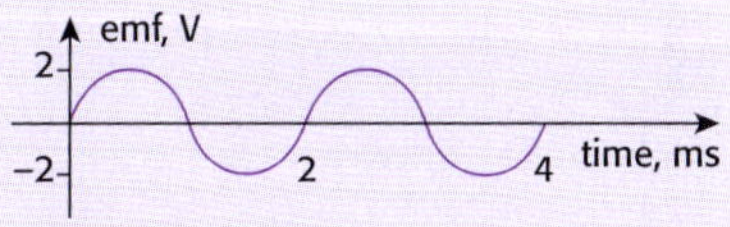

Figure 13 Output of an a.c. generator

?

For each of the above situations (a) to (f) use Lenz's law to predict the direction of the induced current.

Mathematical physics: r.m.s. values

Mains electricity provides alternating current (a.c.), whereas a battery supplies direct current (d.c.). These two types of sources are treated the same, however, when analysing the energy transfers that take place in circuits. A statement such as "The mains electricity provides 110 V a.c." may cause confusion, as the voltage value must be changing all the time, so it is not clear what value is being quoted.

A cell's voltage does not vary greatly with time (Figure 14).

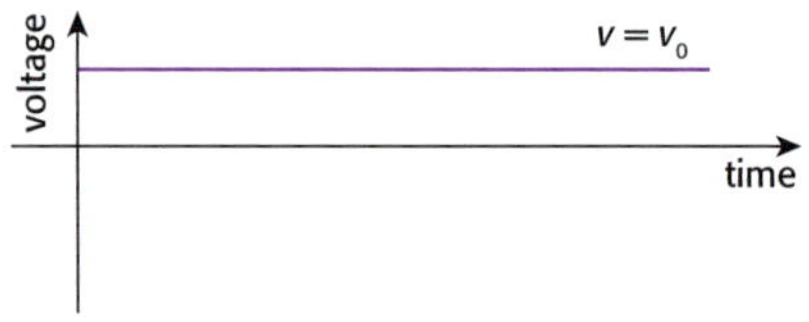

Figure 14 Output of a cell

A power pack output will typically vary quite a lot, but the polarity is always the same (Figure 15).

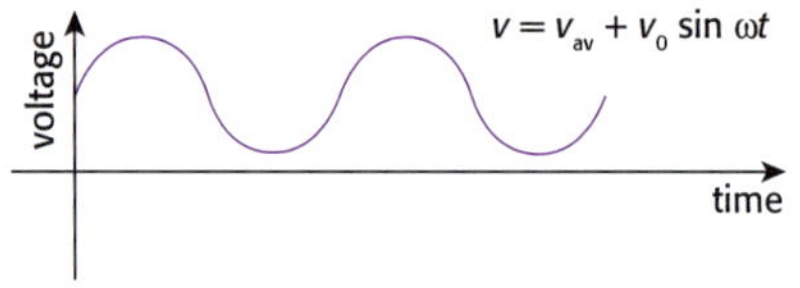

Figure 15 Output of a power pack

In mains electricity the variation with time of the voltage is sinusoidal, and it alternates in polarity (Figure 16).

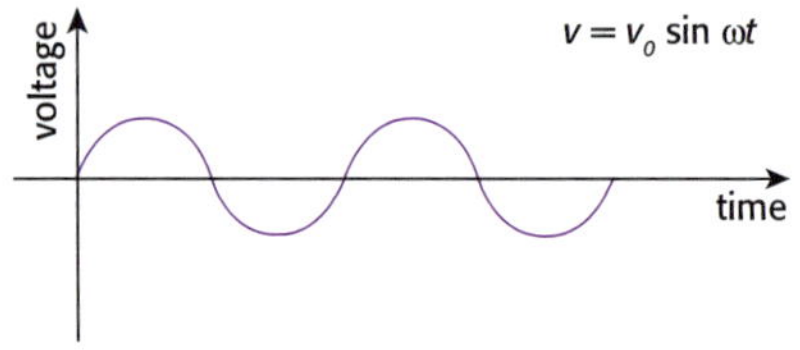

Figure 16 Mains electricity is sinusoidal

The quoted value for mains electricity is not the peak value, nor the mean (this is zero, as the voltage is equally positive and negative), nor the average over half a cycle. It is a value called the **root mean square** or **r.m.s.** value.

The definition of r.m.s. involves three different stages in the calculation:

1 Determine the variation with time of the square of the voltage. This is the square value.
2 Determine the average (mean) value of the squared voltage. This is the mean square value.
3 Determine the square root of this average value. This is the root mean square value.

This process may seem over-complicated, but there is a reason for it: the power P dissipated by a resistor R is proportional not to the voltage V put across it, nor to the current through it, I, but to the square of the voltage.

The r.m.s. value of any alternating electrical quantity is equal to the constant d.c. value that would give the same average power dissipation. In other words, when we are calculating the energy or power for an a.c. circuit, we can continue to use the d.c. equations so long as we substitute the r.m.s. values.

At any given instant, the instantaneous power can be calculated using

$$P = IV = I^2R = \frac{V^2}{R}$$

The resistance R is constant, but all other quantities vary with time. Since $V = IR$, the instantaneous current is given by $I = I_0 \sin\omega t$, where $I_0 = V_0/R$.

The variation with time of power dissipated is

$$P = \frac{(V_0 \sin\omega t)^2}{R} = \frac{V_0^{\,2}}{R}\sin^2 \omega t$$

The power varies all the time, because the $\sin^2(\omega t)$ term introduces a factor that varies between 0 and 1.

Over one cycle,

the average value of power dissipation $= \frac{V_0^{\,2}}{R} \times$ average value of $\sin^2(\omega t)$.

A full calculus analysis shows that the average value of $\sin^2(\omega t) = ½$. This means that

$$P_{\text{average}} = \frac{1}{2} \times \frac{V_0^{\,2}}{R}$$

The r.m.s. value of an a.c. voltage V_{rms} is the constant d.c. voltage that would give the same average power dissipation as the sinusoidal a.c. voltage.

$$\therefore \frac{V_{\text{rms}}^2}{R} = \frac{1}{2} \times \frac{V_0^{\,2}}{R}$$

$$\therefore V_{\text{rms}} = \frac{V_0}{\sqrt{2}}$$

A similar relationship applies for sinusoidal currents:

$$I_{\text{rms}} = \frac{I_0}{2}$$

1 An a.c. voltage of frequency 50 Hz and peak value 325 V is connected across a 500 Ω resistor. Use a computer spreadsheet, a graphics display calculator, or graph paper to:
   (a) plot a graph (over at least one complete cycle) of the variation with time of
      (i) the voltage across the resistor
      (ii) the current flowing through the resistor
      (iii) the power dissipated in the resistor.
   (b) Use your answer to (a)(iii) to estimate
      (i) the energy dissipated over a whole cycle
      (ii) the average power dissipated over a whole cycle.
   (c) Use your answer to (b)(ii) to calculate the equivalent constant d.c. voltage that would give the same power dissipation as (b)(ii).
   (d) Compare your answer to (c) with the r.m.s. value calculated using the relationship between the peak value and r.m.s. value. Comment on your answer.

2 Using a GDC, or otherwise, estimate the r.m.s. value for the following a.c. voltages:
   (a) a square wave of amplitude 2 V:

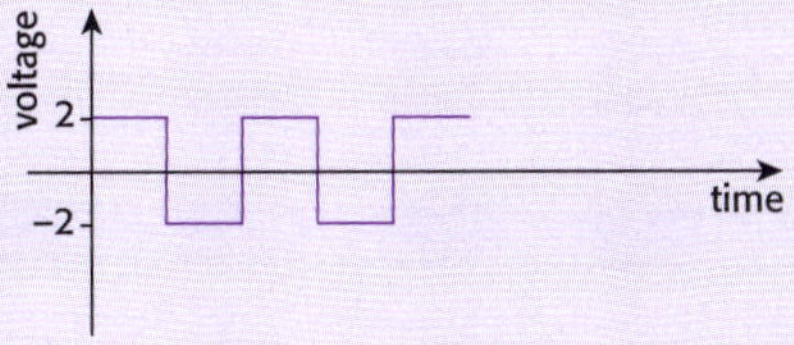

   (b) a triangular wave of amplitude 2 V:

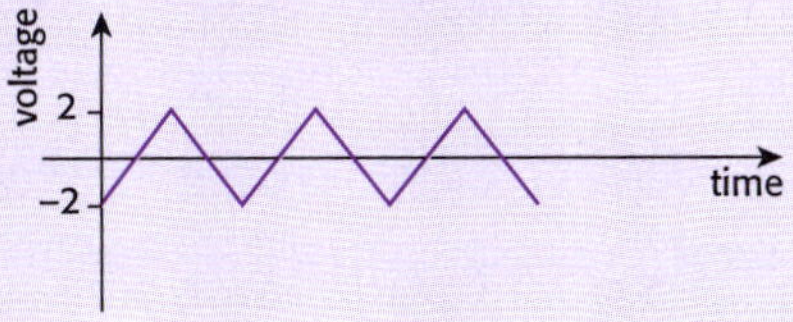

   (c) a "sawtooth" wave of amplitude 2 V:

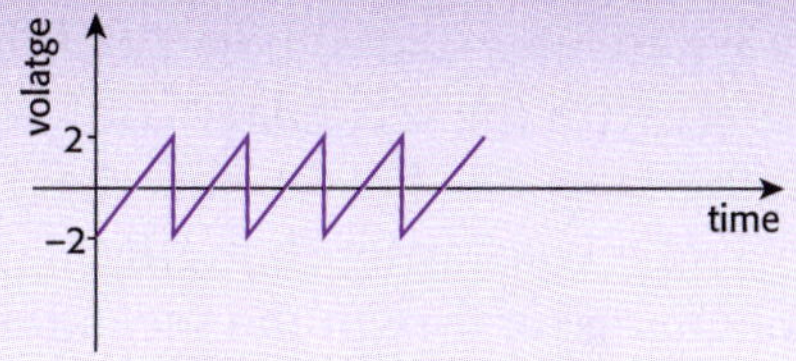

Phase of V and I

At any given time, the potential difference across a resistor, V, and the current through it, I, must be in phase, because they are linked by $V = I R$.

There are, however, devices in which the current and the p.d. are not in phase – **capacitors** and **inductors**.

The full study of these devices (and thus the response of a general electrical circuit to a.c.) is beyond the IB Diploma Programme syllabus. In general, however, there is a phase difference between the alternating potential difference across a group of components and the current flowing through the components. The concept of resistance was introduced for d.c. The a.c. equivalent definition is called the **impedance** Z of a circuit, and is defined as the ratio of the r.m.s. voltage to the r.m.s. current.

Transformers

We saw on page 166 how the magnetic variation that results from a change of current in one circuit can be used to induce a variation of e.m.f. in another unconnected circuit. The transformer is a device that uses this effect to alter voltages.

Figure 17 shows the principal components in a transformer. Two coils of wire – the **primary** and the **secondary** – are wrapped around the same iron core. An alternating voltage is placed across the primary coil, and this induces an alternating voltage across the secondary coil. Adjusting the number of turns in either coil can change the r.m.s. value of the induced voltage that is generated on the secondary.

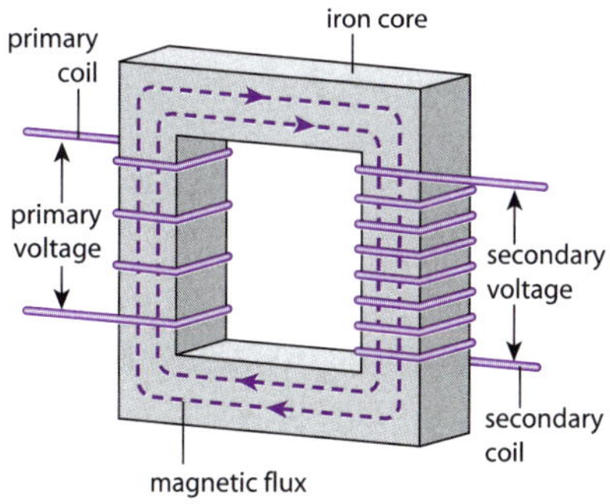

Figure 17 A transformer

The transformer circuit symbol is shown on the right:

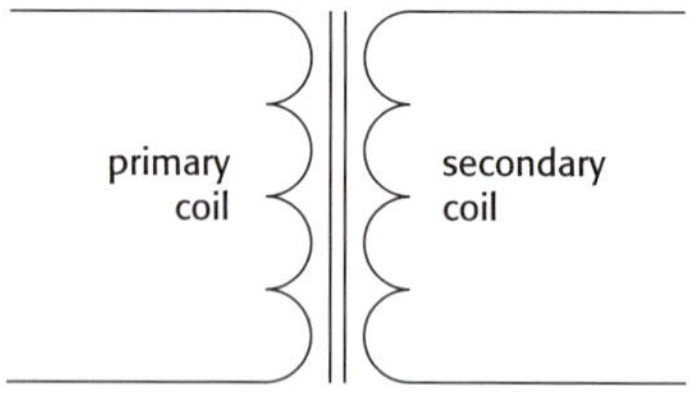

Figure 18 The electric circuit symbol for a transformer

The process by which an induced voltage is created across the secondary is as follows.

1 The alternating voltage input on the primary creates an alternating magnetic field.
2 The iron core ensures that almost all of the flux created by the primary links with the secondary coil.
3 The changing magnetic field means that the flux linkage with the secondary coil is constantly changing.
4 The rate of change of flux in the secondary induces an alternating potential difference across the secondary.
5 Increasing the number of turns on the secondary increases the flux linked with the secondary, and so the value of the alternating p.d. across the secondary.

The net effect is that an alternating voltage placed across the primary coil can be transformed into any chosen value of alternating p.d. out at the secondary, depending on the turns ratio chosen.

In a **step-up transformer**, the output voltage is greater than the input voltage – for example, an adapter that allows devices designed for operation in Europe, and thus requiring 230 V, to operate normally using the mains in the USA, which is at 110 V. In a **step-down transformer**, the output voltage is less than the input voltage – for example, a mobile phone (cellphone) charger that plugs into the mains.

The following relationship applies to an ideal transformer:

$$\frac{V_s}{V_p} = \frac{N_s}{N_p}$$

where
V_s is the r.m.s. value of the p.d. induced across the secondary, measured in V
V_p is the r.m.s. value of the p.d. across the primary, measured in V
N_s is the number of turns in the secondary
N_p is the number of turns in the primary

At first sight it can seem as though the ability to alter the p.d. contradicts the law of conservation of energy, but the currents that flow in the primary and the secondary are also fixed by the turns ratio. Assuming the transformer is 100% efficient:

$$\text{power out} = \text{power in}$$

$$V_s I_s = V_p I_p$$

$$\therefore \frac{V_s}{V_p} = \frac{I_p}{I_s}$$

where
V_s is the r.m.s. value of the p.d. induced across the secondary, measured in V
V_p is the r.m.s. value of the p.d. across the primary, measured in V
I_s is the current flowing in the secondary, measured in A
I_p is the current flowing in the primary, measured in A

?

1 (a) Estimate the diameter of an overhead power cable.
(b) Assuming the wire is made of aluminium (resistivity $\rho = 5 \times 10^{-8}\ \Omega$ m), calculate the resistance of a 100 km section of the power line.

2 A small factory is directly connected to a power station by power cables of total resistance 50 Ω. The output voltage of a power station is 25 kV, and the current drawn by the factory is 100 A. Calculate:
(a) the power output of the station
(b) the power lost in the power cables
(c) the efficiency of the power transmission.

In order to calculate the voltages and currents that are involved with the use of ideal transformers, the following procedure should be used:

- The output voltage is fixed by the input voltage and the turns ratio.
- The load that is connected to the secondary device fixes the output current.
- The turns ratio can be used to calculate the input current that is drawn from its supply.

Physics issues: power considerations in real situations

When power is being transmitted, for example from a power station to a customer, the aim is to avoid as much energy loss as possible. Transformers play a significant role in reducing the losses to an acceptable level.

Loss in power lines

An ideal power line would have zero resistance. Lines with non-zero resistance would dissipate energy by warming up, and real power lines will inevitably have some resistance. For a wire of given material and length, a decrease in resistance can be achieved by increasing the thickness of the wire. In practice there has to be a trade-off between the extra material costs associated with lower-resistance power lines and the cost of the energy lost to the environment.

?

The efficiency of the energy transfer system described in question 2 above is improved by adding 2 transformers of turns ratios 5 and 0.2 appropriately into the system.

(a) Explain how the circuit should be modified to include the transformers.

(b) Calculate

(i) the new power loss in the power cable

(ii) the improved efficiency of the system.

Reasons for using transformers

The main reason for using a.c. and transformers in the transmission of electrical energy is the reduction of power loss. The power lost, P, in transmission lines is calculated from the current flowing, I, and the resistance of the lines, R, using $P = I^2R$. When designing the transmission system used, the operating voltage has to be chosen.

A higher operating voltage would mean that the current necessary for a given overall power would be less, and so the power loss would be reduced. A halving of current would result in the power loss dropping to a quarter of its original value.

Unfortunately, high voltages are potentially dangerous, and so are unsafe to use in the home. Transformers are used to step up the output just before transmission to a high voltage in order to reduce power lost in the cables. The voltage can then be stepped down for safe consumer use.

A practical transmission energy grid around a country will involve many different users being supplied at different voltages (Figure 19).

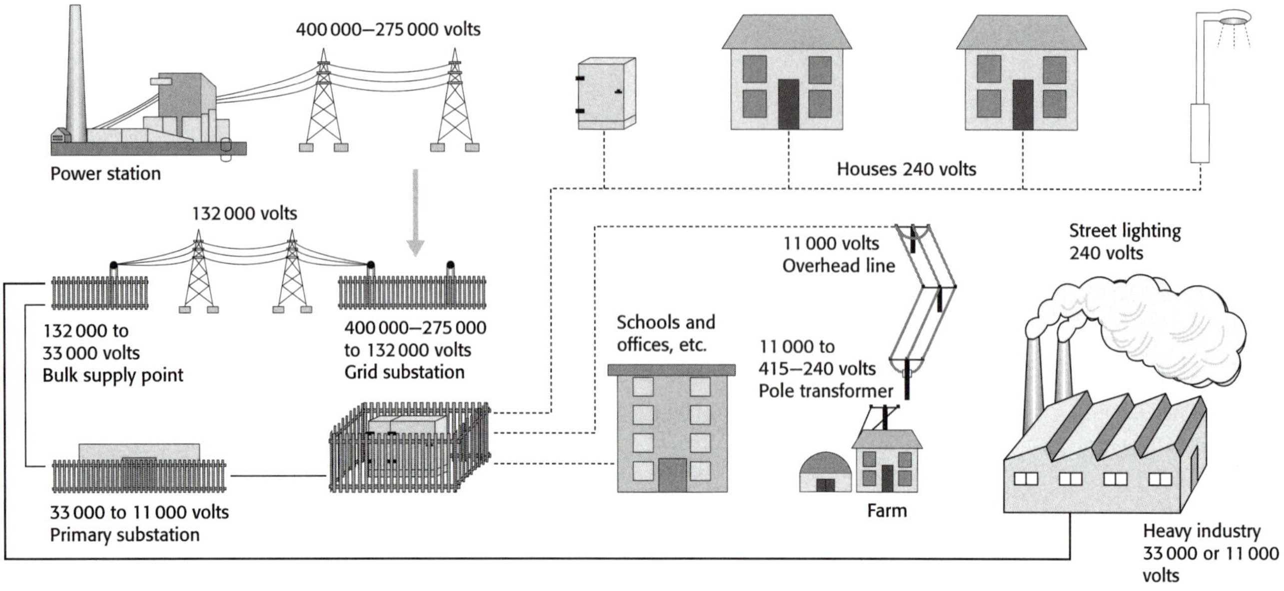

Figure 19 Typical energy distribution grid

Transformers are also used to change the mains voltage into lower, even safer voltages for individual devices such as battery chargers for mobile phones, radios, and cameras. Occasionally the need exists to step the voltage back to higher voltages (e.g. within the older models of television).

Losses also happen in transformers

Transformers prevent a great deal of energy loss in power cables, but unfortunately they are not ideal, and some energy is lost in the transformer itself. The possible sources of loss include:

- heating in the wires of the primary and secondary coils resulting from the resistance of the wires
- heating in the iron core resulting from eddy currents (see below)
- **hysteresis** losses in the core. The core warms up as a result of the cycle of magnetic changes taking place. An iron core will result in less energy lost compared with, say, a steel core.
- losses associated with magnetic "leakage". Any magnetic flux that is produced by the primary coil but does not link with the secondary coil represents energy being lost. The iron core is there to ensure that most of the magnetic field is concentrated within the core.

An e.m.f. will be induced in any conductor that links with the magnetic field produced by the primary. The core itself is a conductor, and thus an e.m.f. will be induced in the iron. Unless the design of the transformer can prevent a current from flowing in the core, this voltage would cause large currents to flow around the core, as shown in the cross-sectional diagram in Figure 20.

These unwanted induced currents are known as **eddy currents.** The complete iron core has a low resistance, so the induced eddy currents would be large, and a great deal of the input energy would be wasted in heating the core. The solution is to prevent the current from flowing by splitting the core into very thin independent sections that are electrically insulated from one another. Each thin section is known as a **lamina**, and the core is said to be **laminated** (Figure 21).

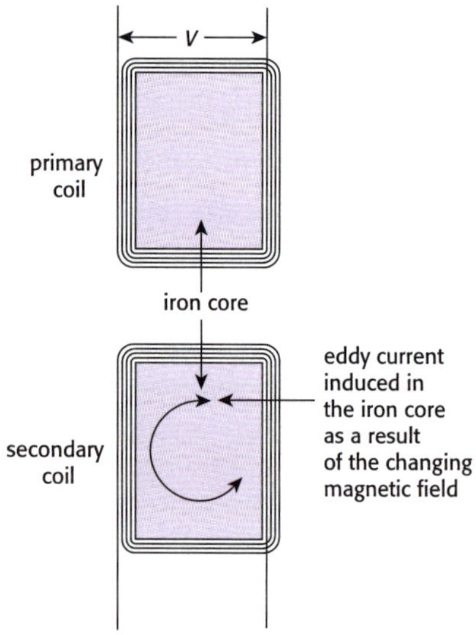

Figure 20 Transformer in cross section

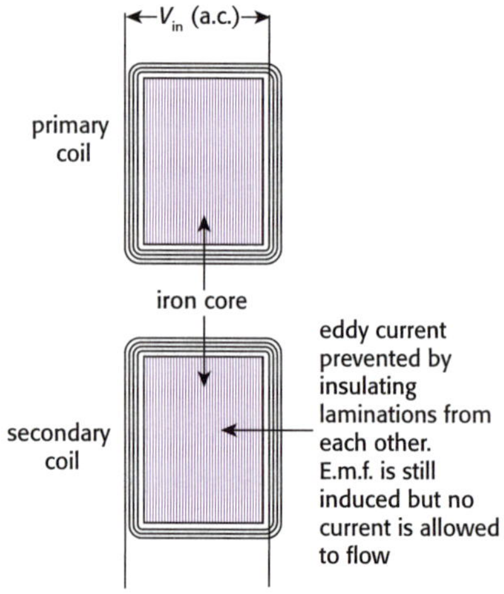

Figure 21 Laminated core of transformer

Inquiry: dangers of power lines

Many people have strong feelings about whether (or not) it is safe to live very close to an overhead power line. The large alternating currents that flow in power lines must have EM radiation associated with them, and this invisible radiation is a worry to many.

As discussed on page 85 (Chapter 8), any individual photon associated with the field will not have enough energy to cause ionization. The frequency of the a.c. is extremely low (50 or 60 Hz), so the EM frequencies are also low. The effects of ionizing radiations are discussed further in the next chapter (page 197), but the extremely low frequency means that these fields are non-ionizing.

The extra-low-frequency EM radiation is still capable of inducing small alternating currents in the body, and some argue that these could cause a detrimental health effect. If this effect does exist, it could be hypothesized that the risks would be dependent on the frequency, the length of any exposure-induced current, and the degree to which the induced current is localized within the body, as measured by the **current density**, defined by the equation below:

$$\text{current density} = \frac{\text{induced current}}{\text{cross-sectional area through which the current flows}}$$

If this effect does exist it should be identifiable with research, but so far no evidence of a clear link has been established between low-frequency fields and any harm caused to genetic material in the laboratory nor detrimental health effects in a population.

Some statistical studies, however, have located regions where more children are diagnosed with leukaemia – a cancer of the blood – than would be expected if the causes were random. The reason for these **leukaemia clusters** has not yet been discovered, but some have suggested a link with overhead power cables.

In 1994 the Royal Society of Edinburgh held a Symposium on Leukaemia Clusters. The report of this meeting is available on the Leukaemia Research Foundation's web page (www.lrf.org.uk). It includes the following statement:

Ionizing radiation is responsible for less than 10% of childhood leukaemias. It is generally agreed that neither waste from nuclear installations, nor the preconceived exposure of fathers to radiation, is a likely cause of leukaemia in children. The latter theory is biologically implausible on the basis of what is known about the frequency of human gene mutations. At present, the evidence for the implication of other causes is very weak. The possibility that non-ionizing radiation might be responsible is unlikely, but worthy of further exploration. There is increasing evidence that extremely low-energy electromagnetic fields can induce metabolic changes in cells, notably a change in calcium permeability across cell membranes. To suggest that this increases the risk of cancer is, as yet, without evidence, but it is possible that such electromagnetic fields may have biological effects.

More than 10 years later, the debate is continuing. The following article appeared in a UK paper in April 2007:

Power lines link to cancer in new alert

By Nicholas Cecil, *Evening Standard* 20 April 2007

Homes and schools could be banned from being built near power lines.

A secret report has raised fresh fears of a link between power lines and cancer. The confidential study, obtained by the Evening Standard, *urges ministers to consider banning the building of homes and schools close to overhead high-voltage power cables because of possible health risks.*

It says a ban is the best way to reduce significantly exposure to electromagnetic fields from the electricity grid system.

The report was drawn up by scientists, electricity company bosses, the National Grid, government officials and campaigners over two years after the Health Protection Agency accepted there was a weak statistical "association" between prolonged exposure to power fields and childhood leukaemia.

But the 40 members of the panel have clashed over the final details and conclusions. It stops short of specifically recommending a ban on new homes and schools within 60 metres of power lines, or vice versa, which could wipe a total of £2 billion off property prices across Britain and limit land for housing developments.

But the report concludes that the Government should consider such a move, stating: "We urge government to make a clear decision on whether to implement this option or not."

The report, to be signed off by panel members next week, has sparked conflict at a series of hearings, according to a Whitehall source.

Two members of the panel, regulator Ofgem and Scottish & Southern Energy, are understood to have quit.

Some members of the panel took the view – adopted by the Government's health advisers and the World Health Organization – that childhood leukaemia is the only adverse health effect where evidence is strong enough for precautionary measures to be considered.

According to this view, if there is a link, the building ban would cut just one case of childhood leukaemia every year or two, and the costs would outweigh the benefits by a factor of at least 20.

The second group generally backed views highlighted by the California Department of Health Services which suggested electromagnetic fields are "possibly carcinogenic" in terms of childhood leukaemia, and placed four other health effects in this risk category. They were adult leukaemia, adult brain tumours, miscarriages and a form of motor neurone disease, although some scientists believe there are links with more diseases.

"The advice to government from following this 'California' view would therefore be to tend to favour implementing the 'corridors for new build' option," SAGE added, stressing that in this scenario the costs and benefits would be at least comparable.

The panel is set to recommend that the Health Protection Agency should issue more information about how to reduce the impact of exposure to electromagnetic fields. It will also call for a change to the working of overhead lines to reduce the radius of intense electromagnetic fields.

In a response to this article, the Leukaemia Research Foundation released a press statement:

Statement concerning the connection between power lines and childhood leukaemia

"As the UK's leading funder of research into childhood leukaemia we would welcome any new research that may help save the lives of the 500 children a year diagnosed with leukaemia.

"However, it is important to note that the report referred to in the Evening Standard *is not new research; nor does it offer any new evidence as to the cause of childhood leukaemia.*

"We have been funding research for almost 50 years, including one of the largest studies ever carried out into the causes of childhood cancers (UKCCS study). Scientists strongly believe the cause of childhood leukaemia to be an abnormal immune response to an infection.

"There is no biological evidence anywhere in the world that proves a clear link between power lines and childhood leukaemia; nor is there any compelling evidence to suggest a link between any form of cancer and power lines.

"50 years ago, very few children with leukaemia survived. Thanks to decades of research that have resulted in improved treatments, childhood leukaemia survival rates are now at an all time high. Leukaemia Research remains committed to investing in research that will benefit the 110,000 people living in the UK with a blood cancer."

Cathy Gilman, Chief Executive of Leukaemia Research.

?

1. To what extent do the two reports agree, and to what extent do they disagree?
2. Undertake your own research into the dangers of power lines. Your findings should be presented in a format agreed with your teacher in advance (e.g. poster/article for school magazine/web page/podcast/essay/presentation). There are many web pages available on both sides of the argument. You should ensure that any evidence you present is properly cited.

Thinking about science: correlation and cause

Powerful statistical techniques have been devised that allow data to be analysed. They can provide a quantitative measure of the degree to which two (or more) variables are correlated together, but great care needs to be taken when interpreting statistical evidence. It is important to keep in mind the difference between a **cause** and a **correlation**.

Even if two variables are highly linked together, there is no guarantee at all that there is a causal relationship between them. The two factors could be linked by a third factor, or it is possible for the factors to move together by chance. It is important to understand the difference.

For example, an analysis of children's ability to read and their foot size shows a very strong correlation. On average, the bigger the size of a child's feet, the better the child is at reading. This does not mean that large feet cause a child to be better at reading. In this case the children's ages cause the correlation – the older the child, the more likely he or she is to be a good reader (they have learned more), and as the child gets older their feet will also grow.

Although the lack of cause should be obvious in the above example, it is all too easy to fall into the trap of assuming that a link is there even if this is not the case.

There is a correlation between owning a washing machine and the risk of dying in a car accident. It would be poor logic to assume that buying a washing machine increases the chances of an accident. Suggest a reason for this correlation.

Sometimes, however, the lack of a mechanism does not mean that two apparently unrelated events do not have a causal link. In 2001 a Danish study of 7000 women found that those who worked predominantly at night were 50% more likely to develop breast cancer. The figures were adjusted to consider other possible factors, including alcohol consumption and age at birth of first and last children (see http://news.bbc.co.uk/1/hi/health/1108590.stm). One possible theory is that the exposure to light during the night could trigger hormonal changes, which then cause breast cancer.

16 Atomic and nuclear physics

This chapter considers different models for the basic structure of all matter. The discovery of radioactive decay led to a major revision in the models used, but the story does not end there. The most commonly used simple atomic model (involving protons, neutrons, and electrons) can be shown to have major shortcomings, and thus a better, more sophisticated, model is required.

Before starting any of the sections in this chapter it would be useful to answer the following questions.

1. Estimate, in metres, typical sizes of the following objects:
   (a) the width of a human hair
   (b) a strand of DNA
   (c) a molecule of water
   (d) an atom of carbon
   (e) a nucleus of iron.
   Now research the accuracy of your estimates.
2. Explain the difference between an atom and a molecule.
3. There are approximately 100 different elements; explain how they differ from one another.
4. Summarize your knowledge of the atomic model. Where are the protons, neutrons, and electrons located in this simple model? Do you already know any experimental evidence that would back up this model?

Thinking about science: models and experimental evidence

Question 4 asks you to recall a simple model of the atom. In order to explain where the electrons, neutrons, and protons are found, many people draw diagrams like the ones shown in Figure 1.

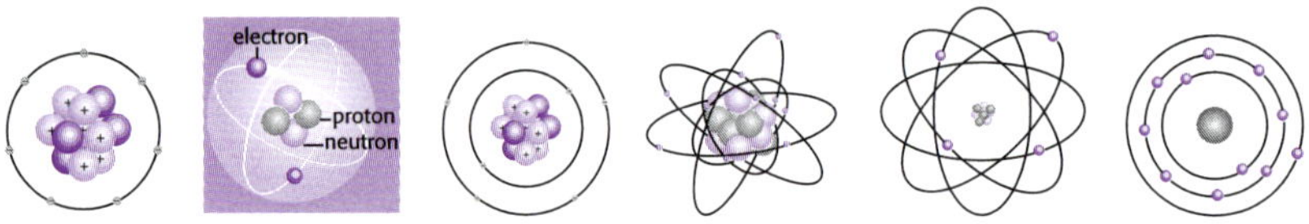

Figure 1 Simple models of the atom

These diagrams are attempts to represent two atoms of nitrogen, one of beryllium, one of helium, one of neon, and one of sodium. Although they have some ideas in common, they are very different: so which, if any, is right? It is not helpful to just ask your teacher to try and identify the one that is closest to reality and learn this for the examination; committing a model to memory is just an exercise in factual recall.

From a theory of knowledge perspective, one individual's opinion is not necessarily a sound basis on which to form a belief about the structure of the universe. Understanding why particular atomic models have been developed allows you to critically question the model in question and develop a real understanding of the nature of the physical universe. As an IB learner, you should already be familiar with the IB learner profile (see page 3), which summarizes the skills you need to develop. All IB learners should strive to become "thinkers" – especially critical thinkers:

"They exercise initiative in applying thinking skills critically and creatively to recognize and approach complex problems, and make reasoned, ethical decisions."

In the context of this chapter, thinking critically means investigating the experimental evidence that we have relating to atomic structure and questioning how well any models fit with that data.

Historical models of matter

Ancient models of the fundamental structure of all matter can seem strange today, but more up-to-date ideas are not always easy to believe. Many people accept that all matter is made out of a combination of just three things – electrons, neutrons, and protons. The common structure that is used to put these three things together – the atom – is mostly empty space: the central nucleus and the tiny electrons occupy a small percentage of the overall size of the atom. This is true whatever the object, be it living, dead, or never alive (e.g. you, the earth on which you walk, the air that you breath, the food that you eat, this book, the computer on which it is being written, a star). Are you comfortable with the idea that most of the volume you occupy is empty space?

Ancient philosophers, whose ideas were not formally based on experimental evidence, were free to propose and justify any model that seemed appropriate. The modern scientific method (which you study as part of your theory of knowledge course) is based on the objective use of measurable experimental data to critically test theories and hypotheses.

The Greek philosopher Democritus (*c.* 460–370 BC) was one of the earliest philosophers to introduce the concept of an **atom** – the smallest particle into which anything can be broken. In this model, each of the different types of atom (water atoms, fire atoms, earth atoms, and so on) had different shapes, and these could be used to explain matter's different properties. For example, fire atoms might be covered in thorns, which is why you can get burned; water atoms would be smooth, which is why water flows easily. This model was not, however, widely accepted at the time.

Empedocles (*c.* 490–430 BC) proposed that all matter was made up of a combination of four **elements** or basic principles: earth, fire, air, and water. For example, when something living dies and decays, it seems to change into earth, so this makes sense.

A fifth element, ether, was later added by Aristotle (384–322 BC) as the basic building block used in "the heavens". This philosophical model was generally accepted up until the late 17th century.

An interesting aspect of this theory was that it might be possible to change one type of object into another. **Alchemy** was the attempt to find a way of changing base metal into gold; often this involved the search for something called the philosopher's stone. This transmutation was envisaged happening at high temperatures.

Figure 2 *The Alchemist*, by David Rychaert (1612–1661)

In the 17th century, the playwright Ben Jonson (1572–1637) wrote the play *The Alchemist*, in which a character called Subtle pretends to be an alchemist and cheats gullible people out of their money on the pretext of being able to achieve this transmutation. In the following extract Subtle is trying to persuade a suspicious character, Pertinax Surly, that his ability to change lead into gold is less surprising than nature's ability to change an egg into a chicken:

SUBTLE Why, what have you observ'd, sir, in our art,
Seems so impossible?

SURLY But your whole work, no more.
That you should hatch gold in a furnace, sir,
As they do eggs in Egypt!

SUBTLE Sir, do you
Believe that eggs are hatch'd so?

SURLY If I should?

SUBTLE Why, I think that the greater miracle.
No egg but differs from a chicken more
Than metals in themselves.

SURLY That cannot be.
The egg's ordain'd by nature to that end,
And is a chicken in potentia.

SUBTLE The same we say of lead and other metals,
Which would be gold, if they had time.

Isaac Newton (1643–1727) is remembered these days as one of the truly great scientists of all time, and as a founder of the scientific method, but he was also interested in alchemy. He conducted many secret experiments in the hope that this theory might be valid. To quote John Maynard Keynes, Newton was "not the first of the Age of Reason; he was the last of the magicians..."

It was the study of gas pressures that led the English chemist John Dalton (1766–1844) to revive Democritus's idea and propose that all matter consisted of small particles called **atoms**. The French chemist Joseph Proust (1754–1826) experimentally verified that some substances were compounds of different elements, and that the proportions of these elements were fixed in precise ratios. This supported Dalton's idea of individual atoms. However, at that time it was not clear how many different types of atom existed. The Russian chemist Dmitri Ivanovich Mendeleyev (1834–1907) arranged the known elements into a table (the **Periodic Table**) based on their masses and on how the elements combined with other elements – that is, their chemical properties.

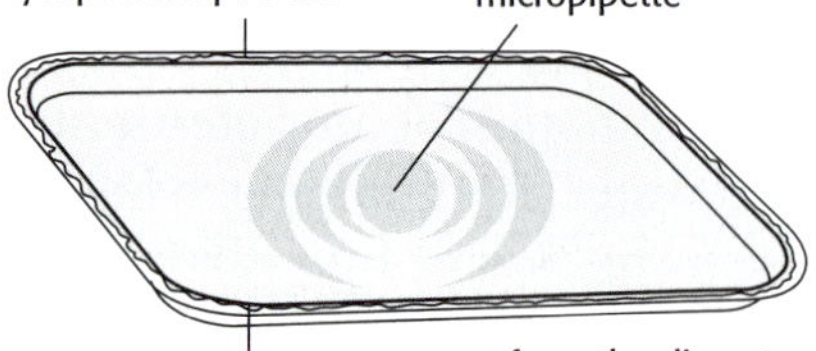

Figure 3 Oil drop on water

We can use very simple experiments to try to set upper and lower limits to the volume of space that atoms and molecules occupy. One famous experiment involves carefully measuring how far an oil drop spreads out when it floats on water (Figure 3).

If we know the volume of the oil drop before we place it on the water, and then measure the area of the oil patch after it spreads out, it is a simple calculation to work out the thickness of the patch. This thickness is an upper limit for the size of atoms and/or molecules.

Another simple experiment is to consider a stone step that has worn down over time. If we assume that each time a shoe slides on a stone step, it can rub off one atom, we can estimate the size of an atom. The calculation starts by finding a way of estimating the number of times that the steps in Wells Cathedral have been walked upon since the steps were built (1319). If we then measure the volume that has been worn away, we can estimate the size of an atom.

Towards the end of the 19th century, the atomic model was based around the number of elements that were known to exist, and all matter was considered to be made of different combinations of these elements. It was only during the 1890s that some progress was made with modelling the *internal structure* of the atom.

The electron

Professor J.J. Thomson (1856–1940) was awarded the Nobel Prize in Physics in 1906 for the discovery of the **electron**. He was investigating the recently discovered phenomena of "cathode rays" – invisible rays that were emitted from the negative electrode when an electrical potential difference was placed across two metal plates in a vacuum. The rays caused the glass or a zinc sulfide screen to glow. By making the anode out of specific shapes (Figure 4), it was possible to show that the cathode rays must travel in straight lines.

Figure 4 Maltese cross tube

Thomson demonstrated that the cathode rays were small negative charges because they were deflected in an electric field. By adding a suitably arranged magnetic field at right angles to the electric field he could bring the cathode rays back to zero deflection. He used this result to calculate the velocity of the negative particles and the deflections measured in the electric field experiment to calculate the ratio of charge to mass for the particles. This value was always the same, whatever substance he used for the cathode. Deflections had been achieved with the charged ions of different elements; the cathode ray particle had a much smaller mass for a similar charge.

When doing these experiments Thomson also discovered that different forms of the same element existed. These had exactly the same chemical properties, but differed slightly in terms of their overall mass. He had discovered **isotopes**.

Thomson obtained the same value of charge-to-mass ratio for cathode rays as for the negative particles that were emitted by some metal surfaces when ultraviolet light was shone on them (the photoelectric effect – see page 207). He suggested that every atom contained these tiny particles, now called electrons.

?

1. Draw the arrangement of electric field and magnetic field that would allow for the magnetic deflection in Thomson's experiment to balance the electric deflection.
2. Derive an expression for the speed of electrons if they show no overall deflection when subjected to an electric field E crossed with a magnetic field B in Thomson's experiment.
3. When cathode rays were first discovered, many physicists thought that they must be a form of wave motion similar to light. Outline the observations this model does explain and those that it does not.
4. Thomson's experiments allowed him to calculate the charge-to-mass ratio for the electron, but not the electron's actual charge or its mass. The first accurate measurement of the charge on an electron was achieved by the American physicist R.A. Millikan. Research this experiment, and summarize both the experiment and Millikan's calculations.
5. All atoms were known to be neutral. The discovery that atoms contained negative electrons meant that some positive matter must also exist. Thomson's atomic model of the time envisaged the electrons held together in an atom, like plums in a pudding – the "plum pudding" model. The "pudding" in the atoms would be some sort of positive "glue" that normally kept the electrons trapped inside the atom. Research this model, and create a presentation showing how this model could explain the experiments of the time.

The nucleus

On Sunday 1 March 1896 the discovery of radioactivity was made by Henri Becquerel. This phenomenon was subsequently investigated by Pierre and Marie Curie, and three types of radioactive emission were identified – alpha, beta, and gamma.

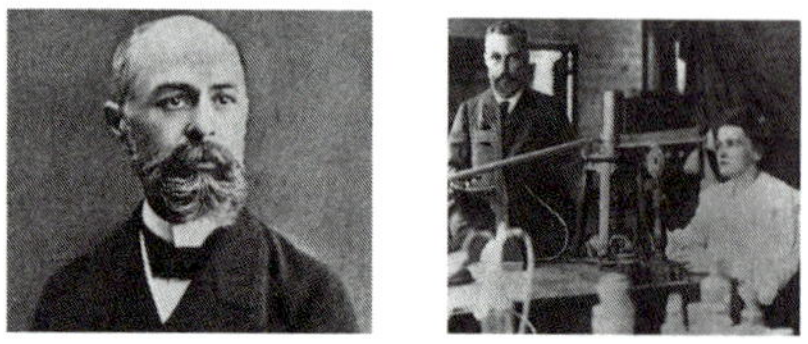

Figure 5 Henri Becquerel (left); Marie and Pierre Curie (right)

Ernest Rutherford, who was a student of J.J. Thomson, worked in collaboration with Thomas Royds and showed that alpha particles were helium nuclei. He also calculated the charge-to-mass ratios for alpha and beta particles using a similar technique to J.J. Thomson. When doing this experiment, he noticed that a small piece of mica (a type of mineral) placed in the path of the alpha particles caused the particles to be scattered, and he designed an experiment to measure the degree of scattering that took place when alpha particles passed through a thin piece of gold foil (Figure 6).

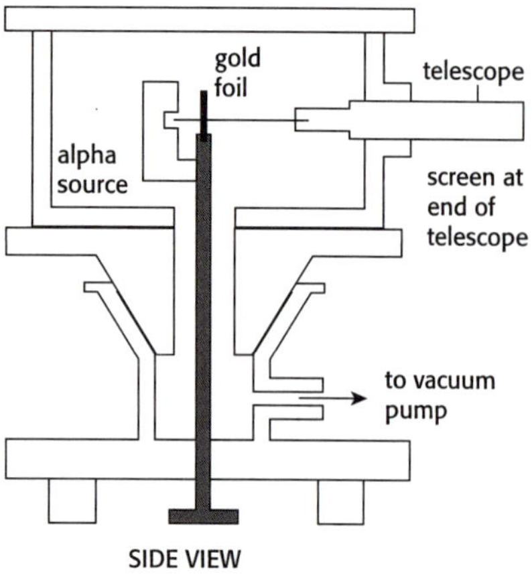

Figure 6 Rutherford's experiment on scattering of alpha particles

The vast majority of the alpha particles went through the foil virtually undeflected. A member of Rutherford's research group, Hans Geiger (who later developed the Geiger counter), was looking for a project for one of his PhD students called Ernest Marsden. Marsden was given the task of using the apparatus to check whether there were any deflections through large angles.

The amazing discovery was that a small number of the alpha particles were deflected through very large angles indeed: in effect, some of them "bounced back". Rutherford described this discovery as astonishing: "It was almost as incredible as if you fired a 15-inch shell at a piece of tissue paper and it came back and hit you." The mathematics of the situation was analysed in detail, and the evidence could only be explained if all the positive charge in the atoms of the gold foil was concentrated at the centre, along with all the effective mass: i.e. a nucleus. Rutherford calculated that the diameter of the nucleus was 100 000 times smaller than the diameter of the atom, and the positive nucleus would repel the incoming alpha particle according to Coulomb's inverse square law. He was able to calculate the theoretical percentage of alpha particles that would be deflected through any particular angle range according to this model, and the predictions agreed exactly with the experimental measurements.

Figure 7 Left to right: Ernest Rutherford (1871–1937); Hans Geiger (1882–1945); Ernest Marsden (1889–1970)

The proton

Rutherford's experiments continued to advance understanding of nuclear physics. In 1919 he was able to show that a fast-moving alpha particle could interact with a nitrogen nucleus. The nitrogen nucleus was transmuted into oxygen, and the result was the emission of a hydrogen nucleus.

In Rutherford's own words: "We must conclude that the nitrogen atom is disintegrated under the intense forces developed in a close collision with a swift alpha particle, and that the hydrogen atom which is liberated formed a constituent part of the nitrogen nucleus."

$$^{14}_{7}\mathrm{N} + ^{4}_{2}\mathrm{He} \rightarrow ^{17}_{8}\mathrm{O} + ^{1}_{1}\mathrm{H}$$

This transmutation meant that Rutherford had succeeded in doing something that the alchemists had unsuccessfully attempted to do

?

1. Do some research into Rutherford and Royds' experiment. Explain how alpha particles were shown to be helium nuclei.
2. Outline how you could experimentally prove that beta particles are, in fact, electrons.
3. Draw a careful and accurate diagram to show the paths taken by alpha particles as they are deflected by gold nuclei.

– change one element into another. He went on to show that other elements could also be transmuted, but the hydrogen nucleus was always emitted. He therefore proposed that the hydrogen nucleus must be contained inside all nuclei, and it was renamed the **proton**. However, it was clear from the beginning that something else must be inside the nucleus for the calculations of total mass to make sense. Although a proton + electron combination would be neutral, and was originally proposed as the missing mass, detailed calculations showed that a nucleus formed of just protons and electrons could not be possible. An additional neutral particle inside the nucleus was proposed – the **neutron**.

Figure 8 Sir James Chadwick (1891–1974)

The neutron

James Chadwick was awarded the 1935 Nobel Prize in Physics for his discovery of the neutron in 1932. His research built on an observation by two German physicists and a subsequent French experiment.

In 1930 the German physicists Walther Bothe and Herbert Becker found that a very penetrating radiation was emitted when they bombarded beryllium with alpha particles. This radiation was non-ionizing. In 1932 Irène Joliot-Curie (the daughter of Pierre and Marie Curie) and her husband Frédéric let the radiation hit a block of paraffin wax, and found it caused the wax to emit protons.

The debate at the time was to decide whether this neutral radiation was a sort of high-energy radiation (like gamma rays) or whether it could be a stream of neutrons. Chadwick and Rutherford were able to show that gamma rays would not have sufficient energy to cause protons to be emitted, but that neutrons would be able to do this. Chadwick's elegant analysis involves applying the conservation of momentum to the collisions. The equation representing the creation of the neutrons is

$$^{4}_{2}\text{He} + ^{9}_{2}\text{Be} \rightarrow ^{1}_{0}\text{n} + ^{11}_{6}\text{C}$$

?

Do some research into Chadwick's experiment and his analysis. Explain how he used the law of conservation of momentum in his calculations.

The Bohr model of the hydrogen atom

Before the neutron had been discovered, a Dutch physicist, Niels Bohr, had proposed an important mathematical model of the hydrogen atom. This brought together Rutherford's proposed nucleus with the ideas of electric fields and forces into a mechanism by which the phenomena of atomic spectra could be understood.

Figure 9 Niels Bohr (1885–1962)

Successes of the model

It had been known for some time that, when different elements became hot enough, they gave off light. If this light was analysed in detail it was seen to contain specific frequencies of light, which corresponded to particular colours – the **emission spectrum** for each element (Figure 10). Different elements had different frequencies.

Light from some other sources (e.g. the Sun) showed most frequencies but with a few missing – an **absorption spectrum** (Figure 11).

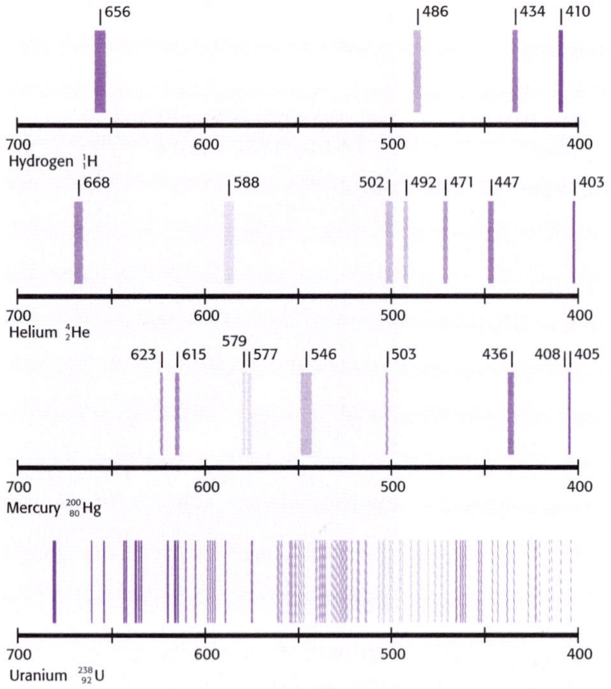

Figure 10 Absorption spectra. Top to bottom: hydrogen, $^{1}_{1}H$; helium, $^{4}_{2}He$; mercury, $^{200}_{80}Hg$; uranium, $^{238}_{92}U$.

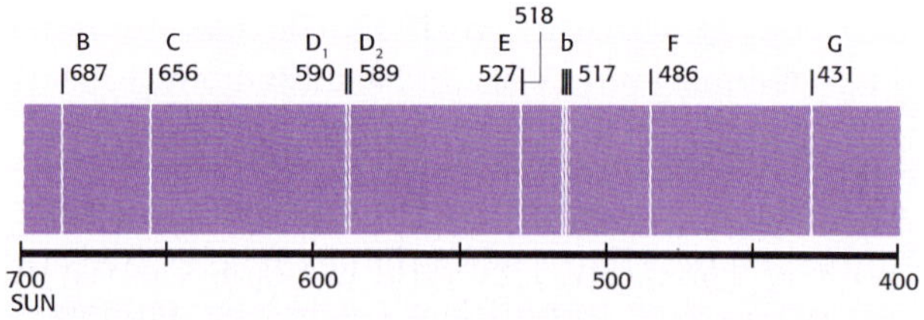

Figure 11 Absorption spectrum from the Sun

Emission and absorption spectra are discussed further on page 213. The different colours of light can be explained using an aspect of quantum theory that was just beginning to be developed at the same time. The relevant idea that needs to be understood here is that light is not a continuous wave but is emitted as "packets" of energy. These "packets" are called **photons**. The energy of a photon, E, is related to its frequency f by the equation $E = hf$, where h is a constant called **Planck's constant** ($h = 6.63 \times 10^{-34}$ J s). The different colours of light correspond to different photon energies.

The fact that specific colours exist in an emission spectrum implies that the electrons in atoms can only have specific energies. Not all energies are possible; there are only a limited number of energy levels available.

Bohr proposed that it was the electron that moved between different energy levels. In order to move up an energy level, an electron needs to absorb a photon. When an electron moves down energy levels it emits a photon (Figure 12).

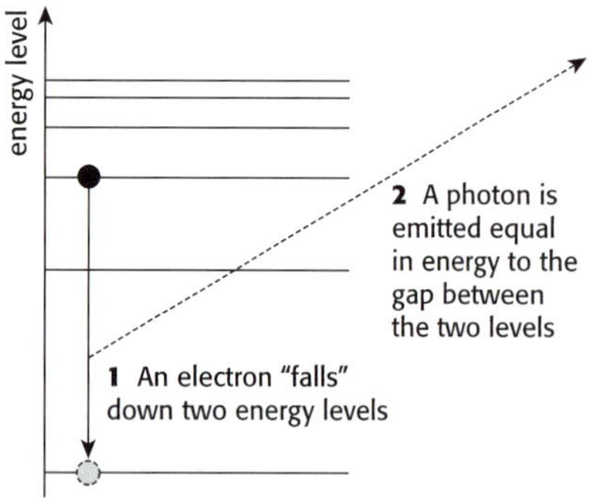

Figure 12 The emission of a photon when an electron drops to a lower energy state

Using Coulomb's law and the mathematics of circular motion, Bohr was able to show that different radii for the electron's orbit in the hydrogen atom would correspond to different energies. He discovered that, provided he made a few simple assumptions about which orbits were available and which were denied, he could calculate all the energy levels that existed in the hydrogen atom. His calculated available orbit energies were in close agreement with many of the energies calculated from analysis of the hydrogen emission spectrum.

With the discovery of the neutron, the final piece of a jigsaw seemed to have fallen into place. The model for an atom involved tiny electrons (containing all the negative charge and hardly any mass) continuously in orbit around a tiny nucleus (containing all the positive charge and all the mass). The force that provided the centripetal force necessary to retain these electrons was the electrostatic attraction between the electrons and the nucleus.

To summarize, over a period of less than 40 years at the beginning of the 20th century the whole of atomic theory had been proposed, experimentally researched, and refined. Teams of physicists from around the world, working together and sharing results with one another, were able to significantly improve our understanding of the basic building blocks of the universe. This incredibly rich period of scientific research needs to be put in perspective against the backdrop of the First World War, which took place between 1914 and 1918.

At the same time as new atomic models were being created, two fundamentally important branches of physics were also being developed: **quantum theory** (which was associated with the atomic models being created, including the Bohr model), and **relativity**. Before this, physics had not undergone such a radical new approach to the problem of understanding the universe since the time of Isaac Newton in the 17th century.

Known problems of the model

Unfortunately, things are not as simple as they seem. Although the above simple model can explain many experimental observations, it is unable to explain others, and there are some serious problems. The following issues need to be addressed if a proper understanding of atomic structure is going to be developed based upon this model.

- There is no available mechanism by which the protons and neutrons would stay inside a small nucleus. The only known force that might help balance the force of repulsion between the protons is the force of gravitational attraction. At the atomic and subatomic level the gravitational force is so small that it is often ignored. There must be at least one other type of force that isn't electromagnetic or gravitational.
- Even if another force or interaction is used to explain why nuclei are kept together, we still don't have a mechanism for radioactive decay.
- Even if we ignore the stability of the nucleus for a short time, there is another problem with the model. Accelerating charges radiate energy. An electron that is moving in constant circular orbit around a proton must be accelerating (it is changing direction all the time), and so it should radiate energy all the time. As it loses energy it should spiral into the nucleus. This predicts that atoms can't exist! Clearly they do exist, so something is very wrong with the model.
- The mathematical ability of the Bohr model to be able to predict the hydrogen emission spectrum frequencies is impressive, but if we try to use the same approach with other elements, the predictions do not agree with experiment. Moreover, the Bohr model fails to predict some of detail of the lines: the so-called "fine structure" and the relative amplitudes of different frequencies.

?

1 Do some research into how emission spectra and absorption spectra are observed. You do not need to memorize the details of any particular experimental setup, but if you understand how the spectra are observed then you can better understand the physics behind their creation.
2 Use the mathematics of circular motion and of electrostatic fields to calculate the speed v of an electron moving in a circular orbit around a proton at a radius r.
3 The proton, the neutron, and the electron are needed to explain the structure of an atom. This chapter started with some diagrams representing how they are arranged in different atoms. For each diagram try to list ways in which the picture helps understanding, and ways in which the picture might cause misunderstandings.

taking place, radioactive half-lives are used to compare different nuclear stabilities. It is true that a nucleus can be bombarded with other particles in order to make a decay take place, but these **stimulated emissions** involve the capture of the incoming particle, which creates a new but unstable nuclide, and then its own decay.

> Define the following terms as precisely as possible. If there are any terms that you do not know, do some quick research to find out what they mean:
> (a) nucleus
> (b) nuclide
> (c) atomic number Z
> (d) mass number A
> (e) nucleon number
> (f) neutron number
> (g) alpha decay
> (h) beta decay
> (i) gamma decay
> (j) half-life.

Of the 1600 or so different nuclides that either exist or have been created, only about 370 can be called stable, and of these there are thought to be about 100 that can still decay but have half-lives greater than 10^{10} years. No element with an atomic number greater than 82 (lead) has any naturally occurring stable isotopes.

An important consideration is the natural relative abundance of different nuclides. For example, there are eight possible nuclides of carbon ranging from carbon-9 to carbon-16. The vast majority of naturally occurring carbon is one of the two isotopes carbon-12 (98.9%) or carbon-13 (1.1%): both of these are stable. A tiny fraction (approximately 10^{-10}%) also exists as carbon-14, which undergoes beta-decay with a half-life of 5700 years. Analysis of this tiny percentage that is radioactive can be used in the process of **carbon dating**, which can be used to estimate the date of some historical items.

A neutron is only relatively stable inside a nucleus; outside this situation, it decays into a proton with a half-life of 889 seconds. Electrons, protons, and photons seem to be stable, though some theories predict mechanisms by which the proton might decay. Recent experiments in Japan seem to have put a lower limit of 10^{35} years on any half-life for the proton.

Predicting the relative stability of a particular nuclide depends on the interactions that are taking place within the nucleus: these are complex, and involve interactions between the protons and neutrons. However, we can generalize the types of decay that are likely to take place by comparing the numbers of protons and neutrons in stable nuclides.

Small nuclei seem to be stable with equal numbers of protons and neutrons. For example, oxygen-16 is stable, and has 8 protons and 8 neutrons. As the proton number increases, stable nuclei tend to have a proportionately larger number of neutrons. For example, lead-208, which is stable, contains 82 protons and 126 neutrons. Graphically this can be seen by plotting the number of neutrons against the number of protons for all the stable nuclides that exist (Figure 15).

Nuclides that have a higher number of neutrons than those in the "band of stability" will tend to decay emitting a negative beta particle. Only large nuclei tend to undergo alpha decay.

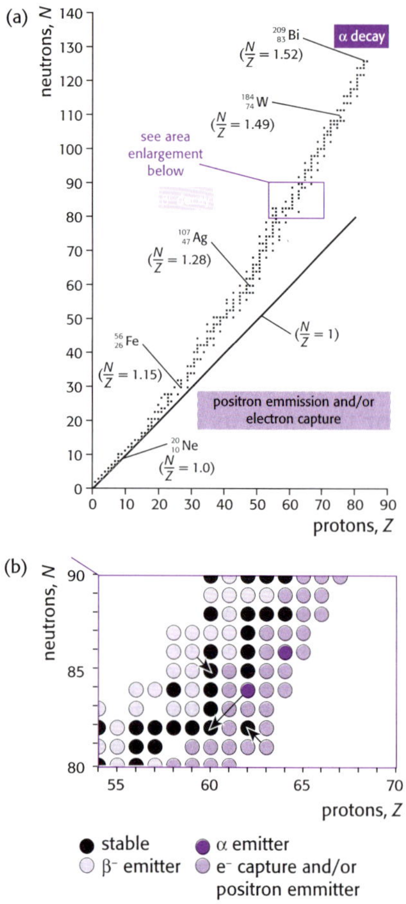

Figure 15 (a) Graph of neutrons against protons for non-radioactive isotopes, showing the band of stability. (b) Close-up of a section of the band of stability, showing the types of decay a radioactive isotope might undergo to become stable.

Is energy conserved in radioactive decay? $E = mc^2$

Perhaps the most famous equation in all of physics is Einstein's mass–energy equivalence.

This most famous of formulae was originally proposed using different symbols: "If a body releases the energy L in the form of radiation, its mass is decreased by L/V^2" (Einstein 1905). In its current form, so long as you know the speed of light c (3×10^8 m s^{-1}), this equation allows you to calculate the amount of energy E in joules needed to create a given amount of mass m in kilograms, or the amount of energy released if matter is converted into energy.

Whatever form the equation is quoted in, it is simple enough to apply, but the concepts it introduces should be extremely worrying. The laws of conversation of mass and conservation of energy seem no longer to apply.

Conservation of mass–energy

Rather than think in terms of the conservation of mass, or the conservation of energy, we can now think in terms of the conservation of mass–energy. There is a **binding energy** associated with every nucleus. Whenever bonds are formed, be they chemical or nuclear, energy is released, and this energy must come from somewhere. The mass of any nucleus is less than the sum of the masses of the protons and neutrons that are contained within the nucleus. The difference between these two masses is called the **mass defect**. Whenever a nuclear reaction takes place (fission, fusion, or artificial transmutation), any energy change that takes place must be understandable in terms of the conservation of mass–energy.

The following list highlights the orders of magnitude of some of the possible mass changes that take place:

- The mass of a single carbon nucleus is 1.992×10^{-26} kg, whereas the total mass of six individual protons and six individual neutrons is 2.009×10^{-26} kg.
- Uranium-238 radioactively decays into thorium-234 with the release of an alpha particle. The mass of the uranium nucleus is 7.614×10^{-30} kg greater than the combined mass of the thorium nucleus and the alpha particle.
- In order to create an electron–positron pair, a minimum of 1.635×10^{-13} J are required.
- The Sun radiates energy from its surface at a rate of 3.8×10^{26} W. Its temperature is constant, so this is the rate of energy release as a result of hydrogen fusion, which equates to a mass loss of 4.2×10^{9} kg s^{-1}.

?

1. For each of the first two bullet points, calculate the energy associated with the quoted mass difference.
2. For each of the last two bullet points, deduce the mass associated with the quoted energy change.

Mass units

The SI unit of mass is the kilogram, but this is a very large unit when considering particle interactions. In order to simplify comparisons on the atomic scale, the unified atomic mass unit (u) is often used. One unified atomic mass unit is defined as one-twelfth of the mass of a carbon-12 atom. A proton and a neutron at rest both have an approximate mass of 1 u. An object's overall mass increases whenever it gains energy and decreases whenever it loses energy. In order to break a carbon-12 nucleus up into its constituent parts energy would need to be added, and so the proton and the neutron have masses slightly larger than 1 u even after the masses of the electrons have been taken into consideration:

Proton rest mass = 1.007 276 u

Neutron rest mass = 1.008 665 u

Electron rest mass = 0.000 549 u

It is very common to find atomic mass units used in tables comparing different isotopic masses. Remember that most tables quote the overall *atomic* mass (i.e. the mass of the atom including the electrons). In order to work out individual nuclear masses we need to subtract the appropriate number of electron masses from the total given in the table.

All physics equations must balance in terms of units as well as numbers. Einstein's mass–energy equivalence equation is no exception, and it can be used to define new units for mass that are often more suitable to use when considering particle interactions.

$$\text{units of } m = \text{units of } \frac{E}{c^2}$$

A common unit of energy at the atomic scale is the electronvolt (eV) or its SI multiples keV or MeV. 1 MeV = 1.6×10^{-16} J. This unit, the mega-electronvolt, is used to define a unit for mass: MeVc^{-2}, which includes the speed of light squared.

$$\text{units of } m = \frac{\text{units of } E}{c^2} = \text{MeV}\, c^{-2}$$

Using these units, an electron has a mass of 0.511 MeVc^{-2}, and thus an electron–positron pair has a total mass of 1.022 MeVc^{-2}, which means that an energy of 1.022 MeV is needed to create an electron–positron pair. In general, an energy of x MeV is needed to create a particle with rest mass = x MeVc^{-2}.

Thinking about science: what is energy?

Energy is a fundamental concept in physics. It is not something physical that can be studied, but rather an abstract idea that arises from the mathematical principles that seem to govern all physical changes. There are many situations in which we can use a formula for calculating the energy before and after a physical change, and to date no exception has ever been found in which the total energy before a change has not equalled the total energy after the change. Energy considerations allow us to analyse and predict the outcome of physical changes, but they do not necessarily help us understand the mechanisms by which these changes take place.

Over time, energy formulae have been developed to acknowledge that energy can take different forms (gravitational potential energy, kinetic energy, thermal energy, elastic energy, electrical energy, chemical energy, radiant energy, nuclear energy, etc.) but the principle of energy conservation has always held.

For example, when an electrical heater is placed in liquid water, connected to an appropriate battery and the current is switched on, the temperature of the water increases. The processes involved at the microscopic level are complex. From the energy perspective, chemical energy in the battery has been converted (via electrical energy) into thermal energy in the water. Equations have been developed that relate the increase in thermal energy in the water to its temperature change, and which relate the electrical energy to the current, potential difference, and time taken.

In a simple experiment to verify the law of conservation of energy, any "missing" energy can be explained in terms of energy being used to heat the wires in the electrical circuit, or energy that has been dissipated to the surroundings.

If the heater remains connected to its power source, the water's temperature will continue to increase until it gets to the boiling point. While the water is boiling, electrical energy is still being provided by the power source, but no temperature change is taking place. This does not mean that the law of conservation of energy has been broken, but rather that we need to modify the equations to include another form that is not immediately obvious. This form of energy is called latent heat ("latent" means hidden). The breaking of the intermolecular bonds in water uses energy, and the making of bonds will release the same amount of energy if the gaseous water turns back into liquid water later on. The law of conservation of energy applies to this situation so long as we include all the possible forms.

One of the very best descriptions of energy was given by the Nobel Prize winning physicist Richard Feynman (1918–1988), in his famous published series of undergraduate lectures given at the California Institute of Technology in the 1960s. (Feynman *et al.* 1963).

Figure 16 Richard Feynman (1918–1988)

In his analogy he imagines a child playing with toy building blocks. The mother of this naughty child develops an experimental law that helps her work out where the child has placed the blocks – whether they are locked inside his toy box (she can measure the change in the weight of the box) or thrown into the dirty water in the bath (she can measure the change in the height of the water). Her abstract mathematical formula will accurately predict where the blocks (representing the energy) have gone, but says nothing about the nature of the blocks themselves.

Einstein's formula does not contradict the law of conservation of energy; it just provides us with another factor that we need to include when considering possible energy changes – an object's mass. The formula represents the idea that mass and energy are equivalent: if the energy of an object changes, then so does its mass. In everyday chemical changes the energy changes involved are equivalent to only tiny changes in mass, so no such change is noticed. For example, 42 kJ of energy is required to increase the temperature of 1 kg of water by 10 °C. The mass change associated with this amount of energy is calculated as follows:

$$m = \frac{E}{c^2} = \frac{42000}{\left(3 \times 10^8\right)^2} = 4.7 \times 10^{-13}\ \text{kg}$$

This is a tiny mass increase compared with the 1 kg, but it is very large compared with the mass of an individual water molecule (approximately 3×10^{-26} kg). In this context water molecules are not being created or destroyed, but the mass of each water molecule has increased slightly. Mass can no longer be imagined as just the amount of matter present in an object, but a mathematical quantity that is related to the total energy concentrated in an object (see page 190).

In all nuclear changes, the energies involved can be understood in terms of the mass changes that are taking place. In extreme situations (e.g. high-energy particle collisions) it is possible for new particles to be created. Other conservation laws (e.g. conservation of linear momentum, conservation of charge) can help explain the creation or annihilation that is observed.

Thinking about science: the neutrino

When a particular nuclide decays via α-decay, the kinetic energy of the α-particle that is emitted is always the same. Similarly, γ-decays always have fixed frequencies for any particular decay under consideration. In each case, the measured energy release associated with the decay is in agreement with the measured mass changes. β-decay is problematic, however, because for any particular decay, experimental measurements demonstrate that the emitted β-particle comes out with a range of possible energies. The technical language describing this is that the β-spectrum is continuous.

For example, chlorine-36 decays into argon-36, emitting a β-particle. The β-particles have a range of kinetic energies from almost zero up to a maximum value of about 0.71 MeV.

Analysis of the masses of the particles involved in this decay predicts that all the β-particles should have the maximum energy of about 0.71 MeV. What is happening to the "missing" energy?

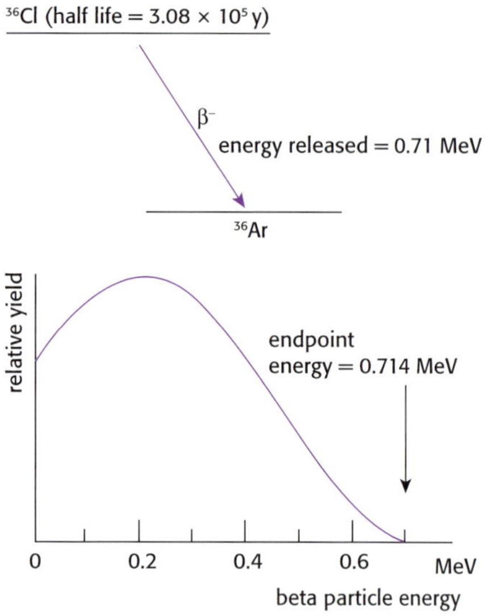

Figure 17 Decay of chlorine-36 and the resulting beta particle energy distribution

It is tempting to answer this question in the same way as most everyday energy calculations and state that the missing energy must have gone into thermal energy ("heat") and/or sound, but this answer is not appropriate here.

Remember that we are talking about a nuclear decay. On this scale, thermal heat is just the random kinetic energy of the atoms involved, and the only particles that concern us are a chorine nucleus, an argon nucleus, and an electron. We are considering the kinetic energy of each of these particles; there is no such thing as extra "heat". Sound is also the movement of atoms or molecules and thus is not an appropriate answer either. No gamma rays are emitted (nor is there any other electromagnetic radiation), so there does not appear to be any mechanism by which this energy could have become "lost".

Data books rarely give values for atomic nuclei as the original data come from measurements with ions. Atomic masses can be used to calculate nuclear masses by subtracting the mass of the appropriate number of electrons. The following data apply to the β-decay of $^{36}_{17}Cl$.

Atomic mass of $^{36}_{17}Cl$ = 35.968 307 u

Atomic mass of $^{36}_{18}Ar$ = 35.967 545 u

Mass of $^{0}_{-1}\beta$ = 0.000 549 u

Mass defect

= mass of $^{36}_{17}Cl$ nucleus – mass of ($^{36}_{18}Ar$ nucleus and β-particle)

= (35.968 307 – 17 × 0.000 549) – (35.967 545 – 18 × 0.000 549) – 0.000 549 u

= 0.000 762 u

Energy released = 0.000 762 × 931.5 MeV

= 0.7098 MeV

This energy will be shared between the argon atom and the β-particle. The share can be calculated by applying the law of conservation to the decay. Given the much larger mass of the daughter product, the argon will essentially be at rest if the chlorine particle was at rest when the decay took place. The energy of the decay should effectively all go into the kinetic energy of the β-particle.

Thus the mass defect predicts the kinetic energy of the emitted β-particle, but most have less energy than this. Some of the energy available cannot be accounted for. Does this mean that the conservation of energy – a firm tenet of all science for more than 400 years – is false? If so, a lot of physics will have to be rewritten. To quote from a lecture that Niels Bohr delivered in 1930:

At the present stage of atomic theory we have no argument, either empirical or theoretical, for upholding the energy principle in β-ray disintegrations, and are even led to complications and difficulties in trying to do so.

In a now famous letter to a meeting of physicists at Tübingen University the Austrian-Swiss physicist Wolfgang Pauli wrote (in German) suggesting a desperate remedy to allow physics to hang on to the law of conservation of energy:

Dear Radioactive Ladies and Gentlemen,

I beg you to listen most favourably to the carrier of this letter… I have hit upon a desperate remedy to save … the law of conservation of energy. Namely, the possibility that there could exist in the nuclei electrically neutral particles, that I wish to call neutrons, which … do not travel with the velocity of light. The mass of the neutrons should be of the same order of magnitude as the electron mass and in any event not larger than 0.01 proton masses. The continuous beta spectrum would then become understandable by the assumption that in beta decay a neutron is emitted in addition to the electron such that the sum of the energies of the neutron and the electron is constant ...

I agree that my remedy could seem incredible because one should have seen those neutrons long ago if they really exist. But only he who dares wins and the serious situation, due to the continuous structure of the beta spectrum, is lightened by a remark of my honoured predecessor, Mr Debye, who recently said to me in Bruxelles: "Oh, it's much better not to think about this at all, just like the new taxes."

... Thus, dear radioactive people, look and judge. Unfortunately, I cannot appear in Tübingen personally since I am indispensable here in Zurich because of a ball on the night of 6/7 December.

With my best regards to you ...

Your humble servant

W. Pauli

Pauli is suggesting is that there is another particle that is emitted in this decay. If a neutral particle of very low mass were to be emitted at the same time as the β-particle, then the two particles would share the energy that is available. The only problem is that this particle had not been detected by anybody at the time. The choice was between the particle not existing and the law of conservation of energy being wrong, or the particle existing but being virtually undetectable. At the time Pauli proposed calling it a neutron (the particle we now know as a neutron had not yet been discovered), but this name was changed a few years later to the **neutrino**. To be absolutely precise, the particle that is emitted at the same time as a β-particle is called an **anti-neutrino** (see page 196 for a discussion of antimatter in general).

$$^{36}_{17}\mathrm{Cl} \rightarrow {}^{36}_{18}\mathrm{Cl} + {}^{0}_{-1}\beta + \bar{\nu}$$

anti-neutrino

In order for this hypothesis to be plausible, the properties of the anti-neutrino read more like science fiction than agreed science fact. To be undetected it must not feel the strong nuclear force at all (electrons also do not feel the strong force), and the fact that it is electrically neutral means that it is also unaffected by an electromagnetic field. Zero, or close to zero, mass means that it is also unaffected by a gravitational field. With these three properties it can therefore pass through the whole Earth, and the chance of any interaction taking place is very small indeed.

The nuclear reactions happening in the Sun mean that millions upon millions of these particles arrive at every square metre of the surface of the Earth every second. They pass straight through all the buildings, everything else (including you), go all the way through our planet, come out on the other side and head off into space. In 1963 John Updike (an American novelist and poet) published his poem "Cosmic Gall" in his collection *Telephone Poles and Other Poems*:

Neutrinos, they are very small.
 They have no charge and have no mass
And do not interact at all.
The earth is just a silly ball
 To them, through which they simply pass,
Like dustmaids down a drafty hall
 Or photons through a sheet of glass.
 They snub the most exquisite gas,
Ignore the most substantial wall,
 Cold shoulder steel and sounding brass,
Insult the stallion in his stall,
 And, scorning barriers of class,
Infiltrate you and me! Like tall
and painless guillotines, they fall
 Down through our heads into the grass.
At night, they enter at Nepal
 and pierce the lover and his lass
From underneath the bed—you call
 It wonderful; I call it crass.

Without any supporting experimental evidence, the neutrino hypothesis would have just remained an interesting or crass proposal (depending on your point of view). Indeed if neutrinos "do not interact at all", then their existence could be only an unverifiable hypothesis. Luckily, Updike's third line is not scientifically accurate, and neutrinos *do* interact, but only rarely. Nowadays it is possible for these chargeless and virtually massless particles to be detected occasionally.

In 1987 an extraordinary piece of luck along with careful measurements convinced many scientists of the existence of neutrinos. On 23 February that year a supernova (SN1987a) was observed in the Large Magellanic Cloud.

One hundred and seventy thousand years ago a blue giant star (50 times larger than our Sun), Sanduleak-69 202, exploded. The light had been travelling through space since the explosion and started to arrive on Earth in February 1987.

Supernovae that are visible to the naked eye are extremely rare events in our galaxy, and this was the first such event to take place since 1604. Galileo built and made observations with his new instrument, the telescope, in 1609, so this supernova was the first "local" one to be observed since stars had been observed with telescopes.

The explosion also released a huge number of neutrinos (10^{56}) in a couple of seconds, and these had also been travelling through space for thousands of years. Two neutrino detectors (neither had originally been designed to look for cosmic neutrinos) had both recently undergone refurbishments that allowed them to be able to detect the extra burst of activity.

One detector was the Kamioka Underground Observatory, located in Kamioka-cho, Gifu, Japan. In 1985 the detector (Kamiokande) was upgraded so as to be able to observe neutrinos of cosmic origin. The other neutrino detector, the IMB (Irvine–Michigan–Brookhaven) experiment is located in Mentor, Ohio, USA. Its refurbishment was completed in September 1986.

At 07:35:41 GMT on 23 February 1987 each detector received a burst of approximately 10^{16} neutrons passing through it. As a result, a grand total of 19 neutrino interactions were recorded over 13 seconds. Eight were recorded in the IMB experiment and 11 in Kamiokande. The linking of these observations to the detailed theories of the nuclear reactions taking place inside a supernova provides strong evidence that these ephemeral particles really do exist.

According to Richard Feynman (see page 193):

The principle of science, the definition, almost, is the following: The test of all knowledge is experiment. *Experiment is the* sole judge *of scientific "truth".*

Feynman et al. 1963, p. 1-1.

The process of how scientific ideas are modified and developed was analysed in an important work, *The Structure of Scientific Revolutions*, which was published by Thomas Kuhn in 1962. The book aims to analyse the history of science. Its publication was a landmark event in the sociology of knowledge.

1 Have you been persuaded that neutrinos exist? Explain your beliefs as succinctly as possible to somebody who does not study physics.
2 How would you answer John Updike?
3 Why is a total of 19 events out of 10^{16} seen as statistically significant?
4 Research what is meant by the term "paradigm shift" as used by Thomas Kuhn. Is the discovery of neutrinos such a shift?
5 The fictional character Sherlock Holmes remarked to his assistant Dr Watson: "How often have I said to you that when you have eliminated the impossible, whatever remains, *however improbable*, must be the truth" (Conan Doyle 1890). How far does this sentence apply to the discovery of the neutrino? Is it possible that the law of conservation of energy will be shown to be incorrect at some time in the future?

HL

Antimatter

In August 1932 a photograph was published that resulted in an important realization (Figure 18). The photograph showed a **particle track** (the path taken by an atomic particle) that demonstrated that, in addition to the matter that we observe on a day-to-day basis around us all the time, there also exists a "parallel" type of particle called antimatter.

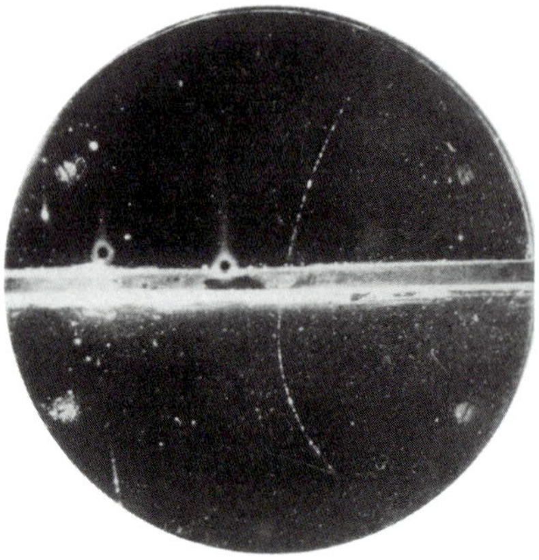

Figure 18 The first piece of evidence for the existence of a positron

A magnetic field was arranged at right angles to the plane of the paper, going into the page. The track in Figure 18 shows a particle passing through a 6 mm lead plate. We know that the particle must lose energy as a result of going through the lead plate, and thus it slows down. The upper portion of the track is more curved, so the particle must have entered from the bottom and be moving up the page. It curves to the left, so we can deduce that the particle must be positive. It can be shown (by more detailed analysis of the track) that it has exactly the same charge-to-mass ratio as an electron, but it is positive. The track shows the path of a previously unknown particle – the positive electron or **positron.**

Every form of matter that exists has an equivalent "opposite" form of antimatter. Any property that the particle has (e.g. charge) will have the same magnitude but the opposite sign in the antiparticle. Although the properties of antimatter are not easy to verify, most theories predict that an antiparticle would still gravitationally attract a normal matter particle. The gravitational force is, however, negligibly small on the atomic and nuclear scale.

Antimatter is not a regular everyday observation, because, if matter and antimatter come into contact with one another, they will annihilate one another. All their combined mass–energy will be converted into another form – often electromagnetic radiation. But antimatter is definitely a real part of the composition of the universe around us. Many naturally occurring nuclides radioactively decay with the emission of a positron, or β^+ decay. In this case a neutrino is emitted at the same time (as opposed to the antineutrino that is emitted during β^- decay). For example, chlorine-36 undergoes β^- decay, whereas the isotope chlorine-34 undergoes β^+ decay:

$$^{36}_{17}\mathrm{Cl} \rightarrow {}^{36}_{18}\mathrm{Ar} + {}^{0}_{-1}\beta + \overline{\nu}$$

$$^{34}_{17}\mathrm{Cl} \rightarrow {}^{34}_{16}\mathrm{S} + {}^{0}_{+1}\beta + \nu$$

All the antimatter that has been observed to date exists only in minute quantities, but there is no reason to believe that antimatter atoms or antimatter molecules would not be stable. Antihydrogen (consisting of a positron in orbit around an antiproton) has already been produced. It was first produced at CERN (the European Centre for Nuclear Research) towards the end of 1995. Experimenters are hoping to be able to design experiments to measure its emission/absorption spectrum, which is predicted to be identical to that of normal hydrogen.

If the spectral lines of matter and antimatter were the same, there would be no way in which we could tell whether some of the stars and galaxies that we observe are made of matter or antimatter! Current theories of the evolution of the universe propose that all the matter we currently observe is just the small amount "left over" after an initial early period in which much larger (and very nearly equal) amounts of matter and antimatter were annihilated.

Physics issues: tracking particles

Subatomic particles are too small to be seen with an optical microscope, but there are many techniques that can be used to work out where they used to be. The most common technique for visualizing a particle track is to use the fact that many of these particles cause ionizations. As a moving particle passes through a medium, it can cause a neutral atom to lose an electron and thus become a positive ion. All the ionizations caused by a subatomic particle can be used to find out where the particle has been. For example, one technique arranges for a volatile liquid to boil. The bubbles that form do so around the ions that the particle has created, and so the ionization tracks become visible as a trail of bubbles (Figure 19).

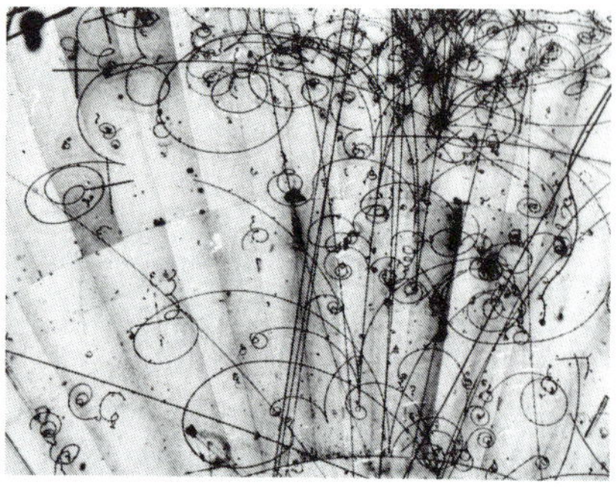

Figure 19 Fermilab bubble chamber: 4.6 m in diameter in a 3 T magnetic field

Unfortunately, neutral particles rarely cause ionizations, so their paths cannot be seen in many detectors, but their existence can often be inferred from the other tracks that *do* exist.

A common technique used to identify the various particles is to ensure that a constant magnetic field is acting all the time while the particles are being tracked. As a result of the magnetic force that acts on moving charged particles, a positive particle will curve in one direction and a negative particle will curve in the opposite direction. The angle of curvature depends on the particle's charge, mass, and velocity. Detailed measurements of the tracks can be used to calculate the ratio of the particle's charge to its mass, which often allows the particle to be identified.

Physics issues: background radiation, biological effects, and risk

Radioactive decay is a natural phenomenon, and is going on around you all the time. The activity of any given source can be measured in terms of the number of individual nuclear decays that take place in a unit of time. This information is quoted in **becquerels** (Bq).

1 Bq = 1 nuclear decay per second

Experimentally, this would be measured using a **Geiger counter**, which detects and counts the number of ionizations taking place inside the **GM tube** (Figure 20). Some Geiger counters display only the total count in a given time, whereas others can display the measured count rate directly.

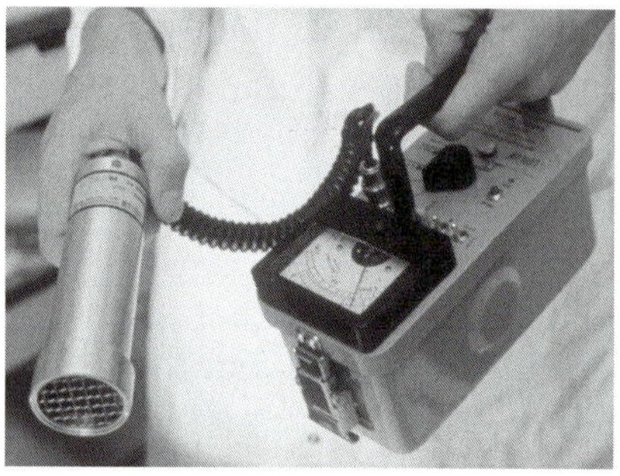

Figure 20 Geiger counter and GM tube

A working Geiger counter will always detect some radioactive ionizations taking place even when there is no identified radioactive source: there is a **background count** as a result of the **background radiation**. A reading of 30 counts per minute, which corresponds to the detector registering 30 ionizing events, would not be unusual. Some cosmic gamma rays will be responsible, but there will also be alpha, beta, and gamma radiation received by the detector as a result of radioactive decays that are taking place in the surrounding materials. The pie chart in Figure 21 identifies typical sources of background radiation, but remember that the value of background radiation will vary from country to country and from place to place.

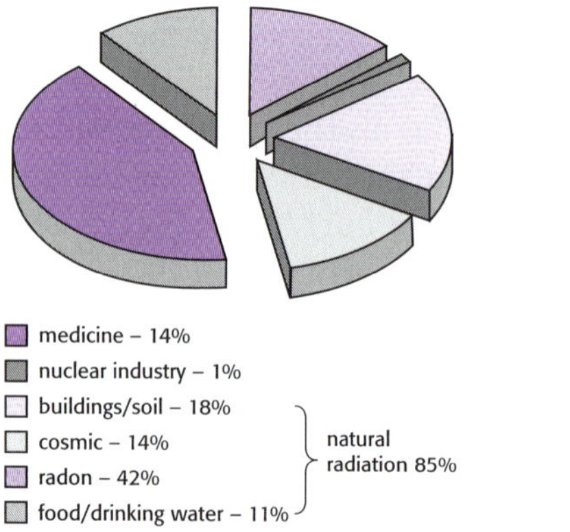

Figure 21 Typical sources of background radiation

Most people are concerned about any exposure and want to know whether the background radiation where they live is safe. Unfortunately, the process of estimating the potential risk that results from exposure to ionizing radiation is complex, and full of uncertainties.

One aspect that contributes to the difficulty in making predictions is that many of the effects of radiation are **stochastic**. This means that a greater dose will, in general, be more likely to have a detrimental effect, but there is an underlying randomness in the process. Given this randomness, it is impossible to assume that there is a minimum dose that is safe. Typically effects (such as cancer) can develop many years after the exposure that caused them. There is an analogy here with smoking. Some of the effects of smoking are stochastic, and although some smokers manage to avoid developing lung cancer, there is a direct correlation between the number of cigarettes that have been smoked and the likelihood that cancer will develop.

One aspect that potentially adds to the fear of ionizing radiation is its invisible nature, and thus the possibility of individuals being harmed without realizing anything has happened. The total energy associated with a harmful radioactive dose is very small: for example, a lethal dose might transfer only about 750 J to an individual. This same energy, if given to a mug of water, would cause a temperature rise of much less than one degree Celsius. How can this tiny amount of energy be so harmful?

Much of the biological damage caused by radiation can be related to DNA damage as a result of the ionizations. DNA is a long and complex molecule, and the biochemistry of its interactions with other chemicals will be affected by the breaking or making of bonds, which can result from the ionizations caused. The viability of any cell in our bodies depends on the DNA that it contains. A typical cell has the structure shown in Figure 22.

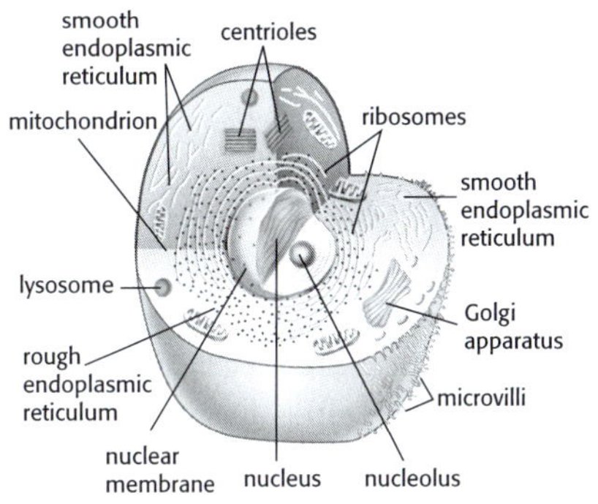

Figure 22 A typical human cell

When incoming radiation causes an ionization in the cell nucleus it may cause damage to the DNA, called DNA mutation (Figure 23).

To put this into perspective, there are approximately 10^{14} cells in a human. Naturally occurring background radiation means that ionizations are occurring all the time. On average there will be one ionization in the DNA molecule in every cell in your body every year. The human body has impressive processes for dealing with this damage. In general, if a mutation occurs, there are three possible outcomes, depending on the amount of damage: repair, cell death, or cell mutation.

A simple break, or **lesion**, is not usually a problem; the vast majority of these lesions are repaired by the body's own biochemical processes, and the cell can survive. Many cells cannot be repaired, and die, but this is not a problem, because millions of cells die every day in every human. The third possibility is that the cell continues to survive, but the damage is not repaired, or is incorrectly repaired. Some cells continue to survive in their damaged state. If damaged cells continue to multiply in an uncontrolled way then it is called cancer.

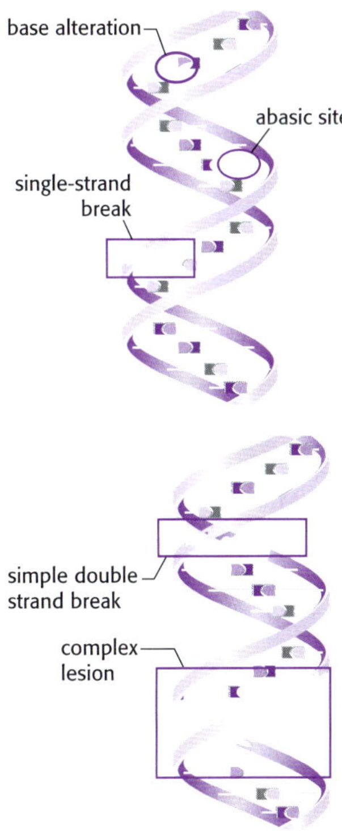

Figure 23 Possible DNA damage

At very high doses a large number of cells would be damaged and die, and this would affect the ability of an organ or of the whole body to be able to survive. Some types of cell are more susceptible to damage than others. At lower doses it may take some time for effects to become noticeable. The list below indicates the likely effects of a range of whole-body radiation doses and dose rates on individuals.

The unit used to compare radioactive doses is the **sievert** (Sv). Technically a quantity called the **dose equivalent** is being measured. This takes into account the fact that different amounts of damage will result from different types of radiation, even if the total energy received per unit mass is the same.

- **10 000 mSv** (10 sieverts) as a short-term and whole-body dose would cause immediate illness, such as nausea and decreased white blood cell count, and subsequent death within a few weeks. Between 2 and 10 sieverts in a short-term dose would cause severe radiation sickness, with increasing likelihood that this would be fatal.
- **1000 mSv** (1 sievert) in a short-term dose is about the threshold for causing immediate radiation sickness in a person of average physical attributes, but would be unlikely to cause death. Above 1000 mSv, severity of illness increases with dose.
  Doses greater than 1000 mSv occurring over a long period are less likely to have early health effects, but they create a definite risk that cancer will develop many years later.
- Above about **100 mSv** the probability of cancer (rather than the severity of illness) increases with dose. The estimated risk of fatal cancer is 5 of every 100 persons exposed to a dose of 1000 mSv (i.e. if the normal incidence of fatal cancer were 25%, this dose would increase it to 30%).
- **50 mSv** is the lowest dose at which there is any evidence of cancer being caused in adults. It is also the highest dose that is allowed by regulation in any one year of occupational exposure. Dose rates greater than 50 mSv/yr arise from natural background levels in several parts of the world but do not cause any discernible harm to local populations.
- **20 mSv/yr** averaged over 5 years is the limit for radiological personnel such as employees in the nuclear industry, uranium or mineral sands miners, and hospital workers (who are all closely monitored).
- **10 mSv/yr** is the maximum actual dose rate received by any Australian uranium miner.
- **3–5 mSv/yr** is the typical dose rate (above background) received by uranium miners in Australia and Canada.
- **3 mSv/yr** (approx.) is the typical background radiation from natural sources in North America, including an average of almost 2 mSv/yr from radon in air.
- **2 mSv/yr** (approx.) is the typical background radiation from natural sources, including an average of 0.7 mSv/yr from radon in air. This is close to the minimum dose received by all humans anywhere on Earth.
- **0.3–0.6 mSv/yr** is a typical range of dose rates from artificial sources of radiation, mostly medical.
- **0.05 mSv/yr**, a very small fraction of natural background radiation, is the design target for maximum radiation at the perimeter fence of a nuclear electricity generating station. In practice the actual dose is less.

The chance of developing a fatal cancer from a dose of 1 Sv is rated at about 5% or 0.05 (5 people in 100 would die as a result of this dose). The risk associated with smaller doses is assumed to be proportional. A typical dose from a lung investigation involving technetium-99 might be about 1 mSv: thus the associated risk is 0.005% or 5×10^{-5}. This means that 5 fatalities would be expected in 100 000 people exposed to this dose.

Many treatments involve much lower doses and the calculated risk may well be lower, e.g. 1×10^{-6} (one in a million). It is extremely hard to put this level of risk into context and assess how concerned one should be. Mathematically, the following activities all have a fatality risk associated with them (not, of course, as a result of radioactive decay) of approximately one in a million:

- smoking 1.4 cigarettes
- living 2 days in a polluted city
- travelling 6 min in a canoe
- 1.5 min mountaineering
- travelling 480 km in a car
- travelling 1600 km in an airplane
- living 2 months together with a smoker.

The technetium-99 lung investigation mentioned above is thus 50 times more dangerous than each of the above.

There is no such thing as a safe dose. Radioisotopes have many medical applications, both in terms of diagnosis and in terms of treatment, and the risk of not treating a particular condition needs to be weighed against any extra risk that involves ionizing radiation. In general, all additional exposure needs to be as low as can be reasonably achieved, and needs to show a positive overall benefit to the patient.

In order to make informed decisions, it is important to understand the risks and the benefits of any proposed procedure, and to compare these with everyday risks. Sometimes our own perceptions of the risks are widely different from the statistical analysis of the data. For example, Professor Gerd Gigerenzer of the Max Planck Institute in Berlin has analysed the road accident statistics from the US Department of Transportation following the appalling events of 11 September 2001. In the months that followed, many people chose to drive rather than fly, and thus exposed themselves to a greater risk of being involved in a car accident. Professor Gigerenzer's analysis (Gigerenzer 2004) suggests that the number of Americans who lost their lives on the roads by avoiding the risk of flying was higher than the total number of passengers killed on the four fatal flights.

Working with data: how do you tell if it's exponential?

It is all too common for students to start thinking that all graphs that curve downwards towards the *x*-axis must be further examples of exponential decay, and all graphs that curve upwards must be exponential growth. This is not true, so if you are given some data that relate two quantities, how can you tell whether the relationship is an exponential one? There are essentially three different techniques:

1 *Find/verify the constant ratio property.*

All exponential functions must show the constant ratio property. A plot can be analysed to see whether this is the case as follows:

- Chose a ratio: halving, doubling, or anything appropriate.
- Use the graph to calculate at least three values (preferably more) of increase in the *x*-value that equates to this change in the *y*-value.

- Given the uncertainties involved, it is important to use all sections of the graph for this calculation.
- The values should be equal within an appropriate uncertainty range.

2 *Plot a graph that has one axis logarithmic and the other linear.*

The mathematics of this technique do not need to be understood by SL students, but it can always be used. Suppose we have an exponential decrease governed by the following relationship:

$$N = N_0 e^{-kt}$$

$$\ln N = \ln\left(N_0 e^{-kt}\right)$$

$$= \ln\left(N_0\right) + \ln\left(e^{-kt}\right)$$

$$= \ln\left(N_0\right) - kt$$

Thus a plot of ln N on the y-axis against t on the x-axis will give a straight line going through the origin with y-intercept = $\ln(N_0)$ and gradient = $-k$.

This plot can be achieved in two ways. Using normal graph paper it is simply a matter of calculating the natural logarithm of each value, using the ln function key on a calculator, and then plotting this directly on a graph. Small numbers (less than e) will work out to be negative. Alternatively, special **log-linear graph paper** can be used to plot the numbers directly.

If the line is straight to within an appropriate amount of uncertainty, then the original data must be exponential. Calculating the gradient is straightforward on normal graph paper, but log-linear graph paper can prove slightly more complex.

Note that logs to base e (also called natural logs or $\ln(x)$) and logs to base 10 ($\log(x)$) will both produce straight-line graphs. The only difference between the two is a scaling factor:

$$\ln(N) = 2.3026 \times \log(N)$$

Table 1

Number N	log (N)	ln (N)
1	0.0	0.0000
10	1.0	2.3026
100	2.0	4.6052
1000	3.0	6.9078

3 *Show proportionality between the rate of change and the quantity itself.*

In some situations it may be possible to demonstrate an exponential relationship by, for example, verifying the basic random nature of a change by demonstrating that the rate of change is proportional to the quantity that is changing.

The radioactive decay equation

$$\frac{dN}{dt} = -\lambda N$$

shows that the *rate of decay* is proportional to the number of atoms available to decay (Figure 24).

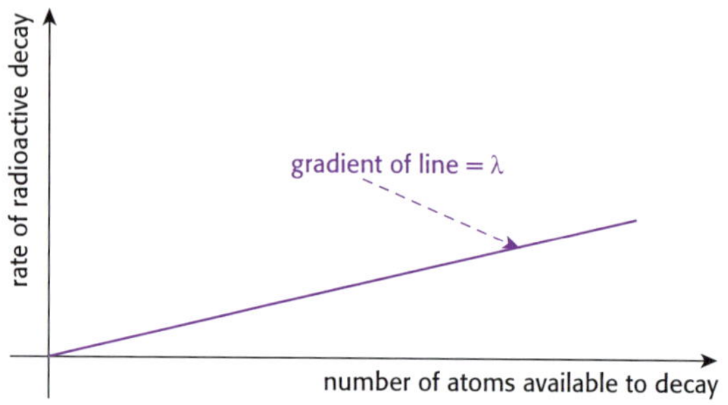

Figure 24 Graph of the radioactive decay equation

Sometimes merely proving the exponential nature of a change is not enough; you may need to analyse the data further. For radioactive decay the aim of data collection may well be to find the half-life of a particular radioactive substance. If the rate of decay varies significantly over a reasonable period of time, then methods 1 or 2 can be used to show its exponential nature. With method 1 the half-life can be estimated directly from the graph, but method 2 would provide a better way of utilizing all the data, and thus a more precise method for estimating the half-life. The gradient of the log-linear graph would equal the decay constant λ, and there is a direct mathematical relationship between the decay constant and the half-life that allows the half-life to be determined.

In many situations, however, the rate of decay is effectively constant, because the overall decay rate is low (the decay constant is a very small number). For example, the half-life of uranium-238 is 4.5 billion years. Because this is such a long time period, no laboratory experiment will be able to collect data to allow the half-life to be determined using methods 1 or 2. In this situation the half-life can be calculated using method 3. The number of atoms that are available to decay comes from a measurement of the mass of the sample, and a measured activity can be used to calculate the rate of decay.

There are many good simulations that show how random processes can result in an exponential relationship. For example, the planetqhe site (www.planetqhe.com) provides a simulation of radioactive decay (see link under "compound events I"). Analyse the data provided by the spreadsheet to see how well it fits exponential decay.

1 Radioactive decay is a *random* process. This means that
  A a radioactive sample will decay continuously.
  B some nuclei will decay faster than others.
  C it cannot be predicted how much energy will be released.
  D it cannot be predicted when a particular nucleus will decay.

2 An isotope of radium has a half-life of 4 days. A freshly prepared sample of this isotope contains N atoms. The time taken for $7N/8$ of the atoms of this isotope to decay is
  A 32 days. C 12 days.
  B 16 days. D 8 days.

3 A nucleus of the isotope xenon, Xe-131, is produced when a nucleus of the radioactive isotope iodine I-131 decays.
  (a) Fill in the boxes below in order to complete the nuclear reaction equation for this decay.

$$^{131}_{\square}\text{I} \rightarrow {}^{131}_{54}\text{Xe} + \beta^- + \square$$ [2]

The activity A of a freshly prepared sample of the iodine isotope is 6.4×10^5 Bq and its half-life is 8.0 days.

  (b) Using the axes, draw a graph to illustrate the decay of this sample.

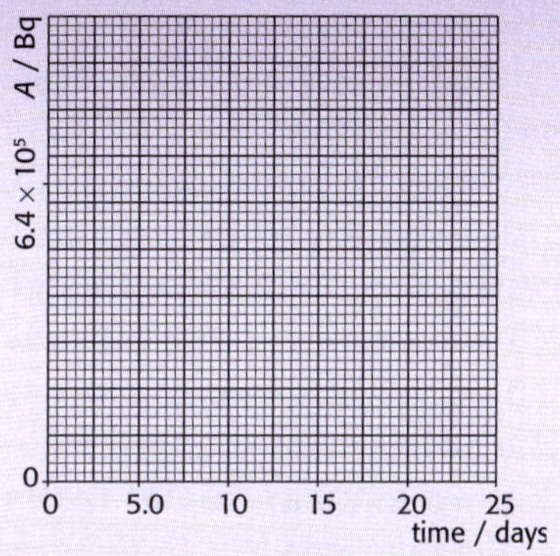

[3]

  (c) Determine the decay constant of the isotope I-131. [2]

The sample is to be used to treat a growth in the thyroid of a patient. The isotope should not be used until its activity is equal to 0.5×10^5 Bq.

  (d) Calculate the time it takes for the activity of a freshly prepared sample to be reduced to an activity of 0.5×10^5 Bq. [2]

[Total 11 marks]

Artificial transmutations

Is it possible to make a particular atom become radioactive? Strawberries are sometimes exposed to gamma radiation so they stay fresh for longer, but this does not make them become radioactive. It is possible to contaminate a substance with radioactive material, but this just means that atoms that are already radioactive have been mixed with those that were originally present.

For example, as a result of the 1986 accident at Chernobyl in the Ukraine, restrictions were placed on the movement and sale of sheep grown in Wales (2700 km away). The details of this accident are discussed in more detail on page 266. The aim of the restrictions was to prevent sheep with high levels of radioactive caesium-137 (caused by eating contaminated grass) to enter the human food chain. The European agreed safe maximum for meat is 1000 Bq kg^{-1}. Individual sheep within defined restricted areas were monitored, and the resulting measured count rate converted to an estimated count rate per mass. As radioactive decay is a random process, a working limit of 645 Bq kg^{-1} was applied. In practice this means that the chance of a sheep giving a "false low" one-off reading is only 1 in 40. Any sheep that exceeded this limit had to remain within the restricted area, and could not be used for human consumption. Immediately after the accident the restrictions and monitoring affected two million sheep in Wales. By 2006, because of the reduction in background (caesium-137 has a half-life of 28 years) the number of sheep at risk had fallen to 180 000.

Note that no atoms in the sheep had been turned radioactive; it is just that the sheep's diet means that they had potentially ingested radioactive material, which then became part of its body mass. Your own body contains some radioactive elements, which contribute to the average background count.

Under certain circumstances it *is* possible for nuclear reactions to take place that alter a previously stable nuclide into an unstable one. Typically the reaction involves the collision between an individual nucleus and something else – e.g. a neutron, an alpha particle, or even (in the case of nuclear fission) another nucleus. In this situation a successful reaction can be thought of as one in which the incoming particle interacts with the nucleus to form an intermediate state, which then decays into two or more fragments. Overall, the atom of one element has been changed into another element, and the reaction is known as an **artificial transmutation**. One model of how this reaction takes place is to imagine the nucleus behaving like a drop of liquid (Figure 25).

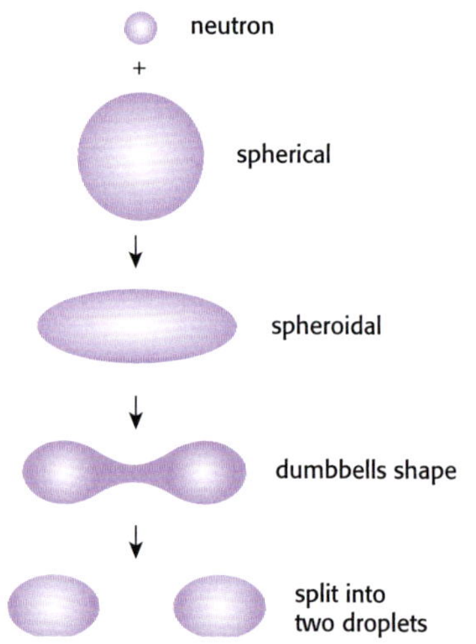

Figure 25 Liquid drop model of the fission mechanism

The first artificial transmutation reaction was achieved by Rutherford in 1919. Nitrogen nuclei that are bombarded by alpha particles form oxygen nuclei:

$$^{14}_{7}\text{N} + ^{4}_{2}\alpha \rightarrow ^{14}_{8}\text{O} + ^{1}_{1}\text{H}$$

The discovery of the neutron is another example, as is the physics of nuclear fission reactors (page 260), and the nuclear fusion that is taking place in our Sun.

Data-based question: graph of binding energy per nucleon

In order to break up a nucleus, a specific amount of energy needs to be provided. There is a binding energy associated with different nuclides. The larger the nucleus, the bigger the total binding energy. The relative stability of different nuclides can be compared by calculating the binding energy per nucleon for each nuclide. The resulting graph is shown in Figure 26. The nucleus of iron-56 turns out to have the most binding energy per nucleon, is radioactively stable, and does not decay. Remember that this graph is comparing different nuclides; it cannot be used to compare the chemical stability of different elements.

For each of the following situations, take readings from the graph in Figure 26 to quantitatively explain the energy changes involved. If a reaction can take place, calculate the energy released per reaction; if no reaction happens, explain why. You should start by writing the appropriate nuclear equation in each case.

(a) In the Sun, the fusion of different isotopes of hydrogen to form helium releases energy. The three-stage process involves
  - the fusion of two protons to form deuterium (hydrogen-2)

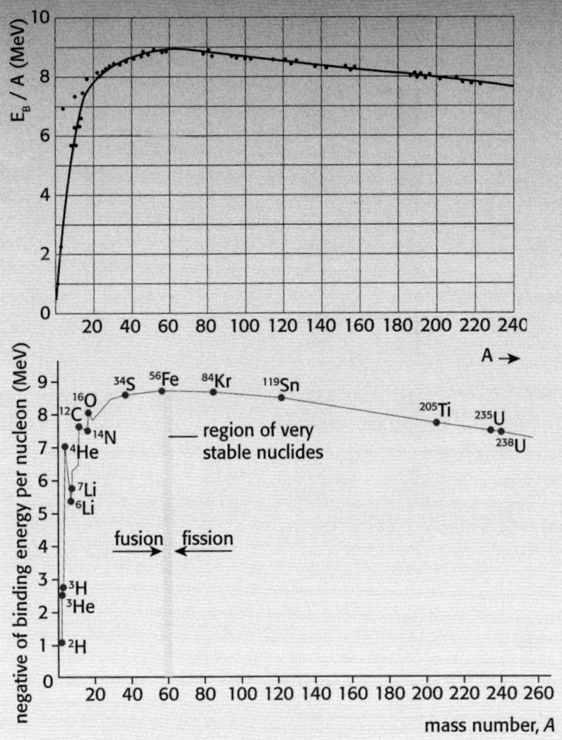

Figure 26 The stability of different nuclides

- the fusion of deuterium and a proton to form helium-3
- the fusion of two helium-3 nuclei to form a helium-4 nucleus with the emission of two protons.

(b) In a red giant star other possible reactions include the fusion of:
- two helium-4 nuclei to form beryllium-8
- beryllium-8 with another helium-4 nucleus to form carbon-12
- carbon-12 with another helium-4 nucleus to form oxygen-16.

(c) When a red giant star stops shinning, its core is iron.

(d) Uranium-238 decays naturally into thorium-235 with the emission of an alpha particle

(e) If bombarded with a neutron, uranium-235 can break up to form barium-141, krypton-92, and three other neutrons. This reaction can produce a chain reaction.

(f) If bombarded with a neutron, uranium-235 does not completely break up to form helium nuclei and/or other neutrons.

This question considers some aspects of the atomic and nuclear physics associated with isotopes of the element helium.

Atomic aspects

(a) The element helium was first identified from the *absorption spectrum* of the Sun.

  (i) Explain what is meant by the term *absorption spectrum*. [2]

  (ii) Outline how this spectrum may be experimentally observed. [2]

(b) One of the wavelengths in the absorption spectrum of helium occurs at 588 nm.

  (i) Show that the energy of a photon of wavelength 588 nm is 3.38×10^{-19} J. [2]

  (ii) The diagram below represents some of the energy levels of the helium atom. Use the information in the diagram to explain how absorption at 588 nm arises.

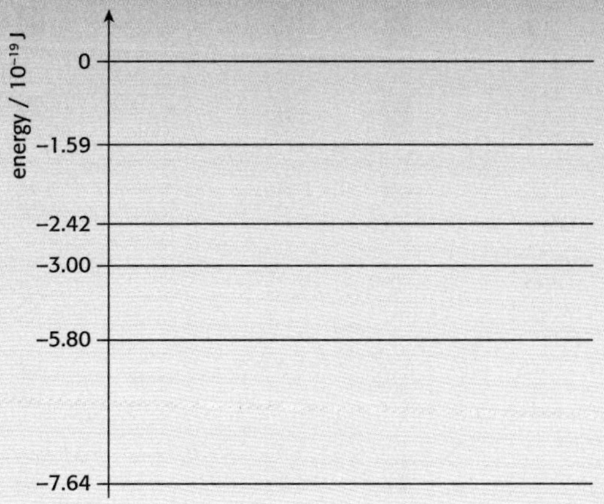

[3]

Two different models have been developed to explain the existence of **atomic** energy levels. The **Bohr model** and the **Schrödinger model** are both able to predict the principal wavelengths present in the spectrum of atomic hydrogen.

(c) Outline
   (i) the Bohr model, and
   (ii) the Schrödinger model. [6]

Nuclear aspects

(d) The helium in the Sun is produced as a result of a nuclear reaction. Explain whether this reaction is burning, fission, or fusion. [2]

At a later stage in the development of the Sun, other nuclear reactions are expected to take place. One such overall reaction is given below.

$${}^{4}_{2}\text{He} + {}^{4}_{2}\text{He} + {}^{4}_{2}\text{He} \rightarrow \text{C} + \gamma + \gamma$$

(e) (i) Identify the atomic number **and** the mass number of the isotope of carbon C that has been formed.

Atomic number: ....

Mass number: .... [2]

(ii) Use the information below to calculate the energy released in the reaction.

Atomic mass of helium = $6.648\,325 \times 10^{-27}$ kg
Atomic mass of carbon = $1.993\,200 \times 10^{-26}$ kg [3]

Another isotope of helium ${}^{6}_{2}\text{He}$ decays by emitting a β^{-}-particle.

(f) (i) State the name of the other particle that is emitted during this decay. [1]
   (ii) Explain why a sample of ${}^{6}_{2}\text{He}$ emits β^{-}-particles with a **range of energies**. [2]
   (iii) The half-life for this decay is 0.82 s. Determine the percentage of a sample of ${}^{6}_{2}\text{He}$ that remains after a time of 10 s. [2]

[Total 27 marks]

References

Doyle, A.C. 1890. *The Sign of Four*. London. Spencer Blackett.

Einstein, A. 1905. "Is the inertia of a body dependent on its energy content?" *Annalen Der Physik*. Vol 18, number 13. pp 639–641.

Feynman, R.P., Leighton, R.B. and Sands, M. 1963. *The Feynman Lectures on Physics*. Vol 1, section 4–1. Reading, MA. Addison-Wesley.

Gigerenzer, G. 2004. "Dread risk, September 11, and fatal traffic accidents". *Psychological Science*. Vol 15, number 4. pp 286–287.

Kuhn, T. *The Structure of Scientific Revolutions*. Chicago. University of Chicago Press.

Updike, J. 1963. *Telephone Poles and Other Poems*. New York. Knopf.

17 Quantum physics

At the beginning of the 20th century, there was a crisis in physics. The latest experimental observations on the properties of light and atoms could not be explained using the "classical" ideas of the time, such as those dealing with motion and waves. A new approach was needed, and it led to the creation of a completely new branch of physics: **quantum theory**, or **quantum mechanics**. The last time that such a complete scientific revolution had taken place was when Newton had put forward his theories on motion and gravity some 300 years earlier. In this chapter, you will discover more about this new revolution and why it became necessary.

Thinking about science: the nature of light

For centuries, the nature of light had been a mystery. By Newton's time, two opposing models had emerged. One treated light as a stream of **particles**, the other as a set of **waves**. To be successful, a model should be able to explain all the observed properties of light including the energy transfer, shadows, reflection, refraction (the change in direction when it enters a different medium), diffraction (the spreading of light around an object), and interference (the addition of light from two sources to form bright and dark "fringes"). This is how the models compare.

Light as a stream of particles

In the early 1700s, in his book *Opticks*, Newton proposed that light is a stream of tiny *corpuscles* or particles. In this model, a typical light beam consists of a large number of particles, each travelling in a straight line and each carrying energy. During reflection, the particles bounce off a surface – all in the same direction if the surface is smooth like a mirror. Newton also proposed a mechanism to explain refraction but, unfortunately, this was based on the assumption that the particles speed up when entering a medium such as glass or water, which we now know to be wrong. Newton was unable to explain the diffraction of light without using some wave concepts.

Light as a wave motion

Huygens first put forward a wave model of light in the late 1600s. In this model, energy is carried, not by a stream of particles, but by waves travelling out from their source. Over the years, the wave model superseded Newton's particle model because it was able to explain effects such as diffraction and interference. In the late 1800s, James Clerk Maxwell, doing theoretical work on electric and magnetic fields, predicted the existence of electromagnetic (EM) waves which would have the same speed as that measured experimentally for light. As a result, by the beginning of the 20th century, scientists felt confident that all the properties of light could be explained by treating it as an electromagnetic wave. However, their confidence was about to be undermined, as you will find out below.

Quantum explanations to replace classical ideas

This section considers four problems: the photoelectric effect, black body radiation, the Compton effect, and electron diffraction. With these, classical models do not work. New ideas are needed to explain them. These are the **particle properties of light**, the **wave properties of matter**, and the concept of **quanta**.

Photoelectric effect

This is the emission of electrons from a surface, such as polished metal, because light or ultraviolet (UV) is shining on it. The effect was discovered by Hertz in 1887. The emitted electrons are sometimes called **photoelectrons** and, in a suitable setup, they can flow as a current in a circuit.

The electrons need energy to leave the surface. In the wave model of classical physics, the energy is supplied as a constant flow by the incoming light waves. However, this fails to explain the following experimental observations:

- In general, there is a minimum frequency, called the **threshold frequency**, below which no photoelectrons are emitted, whatever the intensity of the light.
- Different metals have different values for this threshold frequency.
- When photoemission is taking place, the magnitude of the photoelectric current depends on the intensity of the light source.
- No time delay is observed between the arrival of the light and the emission of the electrons, even at very low intensities.
- At a given frequency of ultraviolet light, the energy of the photoelectrons is independent of the intensity of the light but varies linearly with the frequency of the incident light, as in Figure 1.

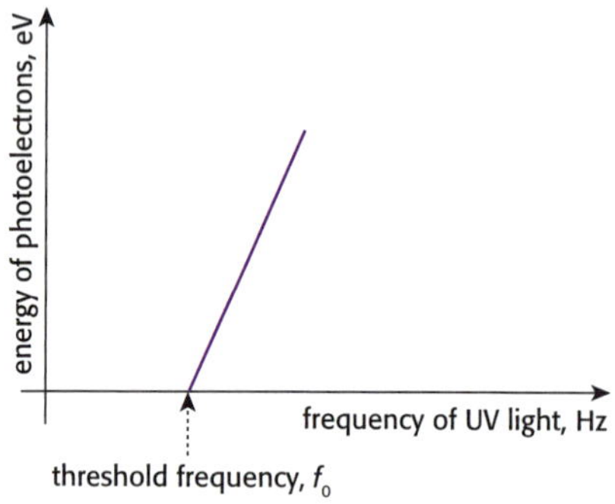

Figure 1

In 1905, Einstein came up with a radical model to explain the features of the photoelectric effect. He considered the incoming light not as a constant flow, but broken up into "packets" of energy. Each packet is called a **quantum**. Its energy E is given by this equation:

$$E = hf$$

h is the **Planck constant**: 6.63×10^{-34} J s
f is the frequency in Hz

In Einstein's model, the light travels in the form of particles called **photons**, each carrying one quantum of energy. For any given frequency, the more photons arriving per second, then the greater the intensity of the beam. A 60 watt light bulb emits about 1.8×10^{20} photons every second.

If an electron absorbs a photon, the energy it gains is hf. But for any particular metal, there is a minimum amount of energy needed to make an electron break free of the surface. This amount is called the **work function**, W. If the photon provides *less* than this minimum, no electron is emitted. If it provides *more*, then the excess is transferred to the electron as kinetic energy, as given by this equation:

$$\frac{1}{2}mv^2 = hf - W$$

In 1921, Einstein was awarded the Nobel Prize for Physics for this equation and his explanation of the photoelectric effect.

Black body radiation

Hot objects radiate electromagnetic (EM) energy at a range of frequencies. More accurately, *all* objects do. But if the temperature is increased, two changes happen. First, the total energy radiated per second increases. Second, there is an increase in the proportion of energy emitted at the higher frequencies – in other words, shorter wavelengths. At room temperature, an object mainly radiates in the infrared part of the EM spectrum. A hotter object may emit some visible radiation, becoming "red hot" or, if the temperature is high enough, "white hot".

Good emitters of radiation are also good absorbers. A **black body** is a theoretical object that will absorb all frequencies falling on it. So, if light is falling on it, it will absorb all frequencies, reflect none, and therefore appear black. As a black body is the best possible absorber, it is also the best possible emitter. Odd though it may sound for something so bright, the Sun is a black body emitter.

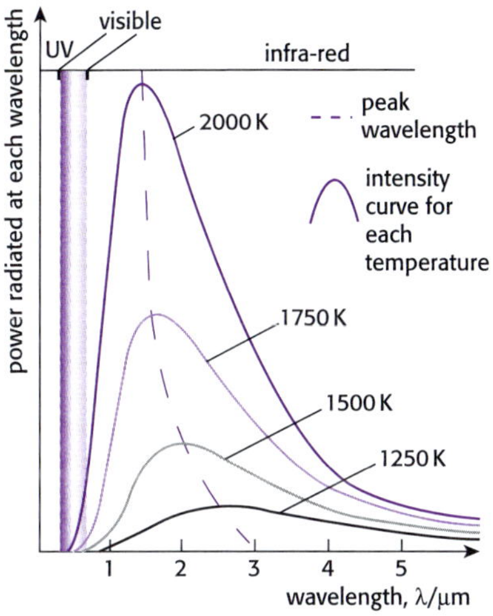

Figure 2 Black body radiation curves for a range of temperatures

Electron diffraction

The effects described above suggest that EM waves have particle-like properties. This raises an important question: can particles have wave-like properties? In 1927, Davisson and Germer scattered a stream of electrons from the surface of a crystal of nickel and found that, at some angles, the amplitude of the scattering was increased. This was equivalent to the diffraction patterns that can occur with light waves. In other words, electrons can behave like waves.

Working with data: Davisson and Germer experiment

Figure 3 shows the principle behind the Davisson and Germer electron diffraction experiment.

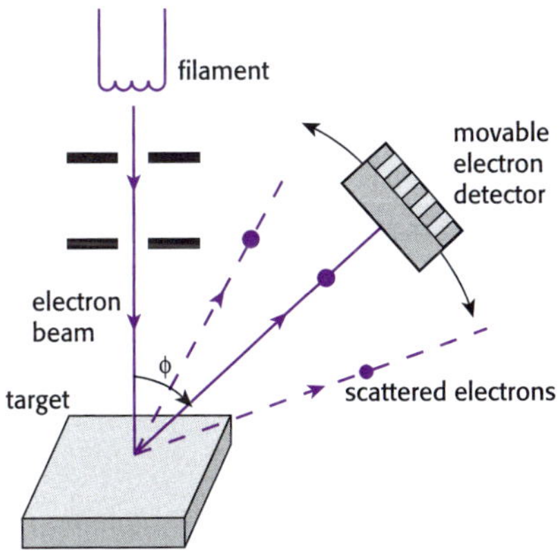

Figure 3

A beam of electrons strikes a target nickel crystal. The electrons are scattered from the surface. The intensity of these scattered electrons depends on the speed of the electrons (as determined by their accelerating potential difference) and the angle.

A maximum scattered intensity was recorded at an angle of 50° when the electron beam had an energy of 54 eV. The experiment data can be represented in one of two ways. Figure 4 is a traditional representation of the data.

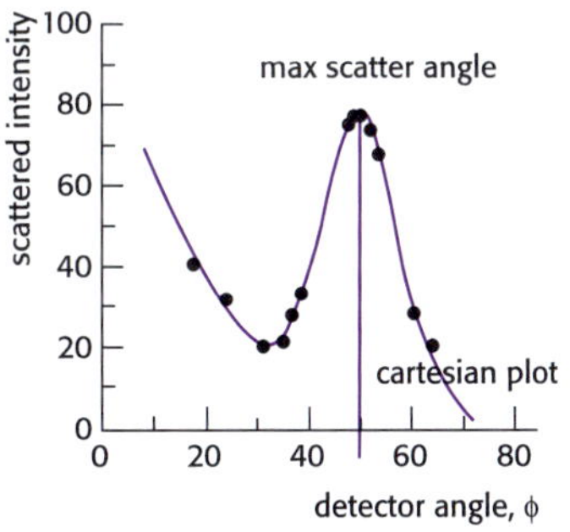

Figure 4

Figure 5 is known as a polar plot. In this graph, the scattered intensity is represented by the distance (the **radial distance**) from the origin to the line (i.e. along the dashed line) at any given angle.

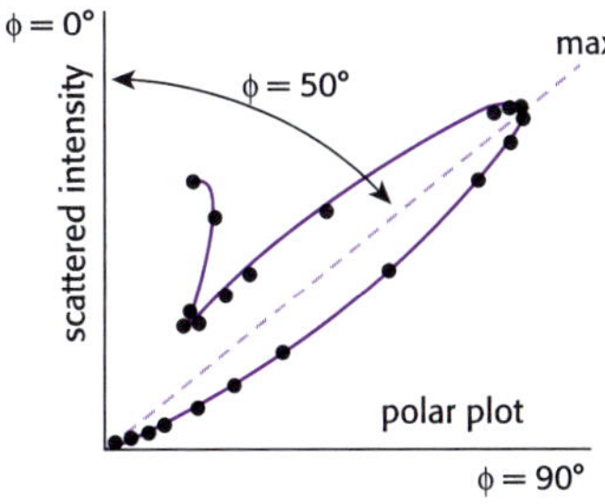

Figure 5

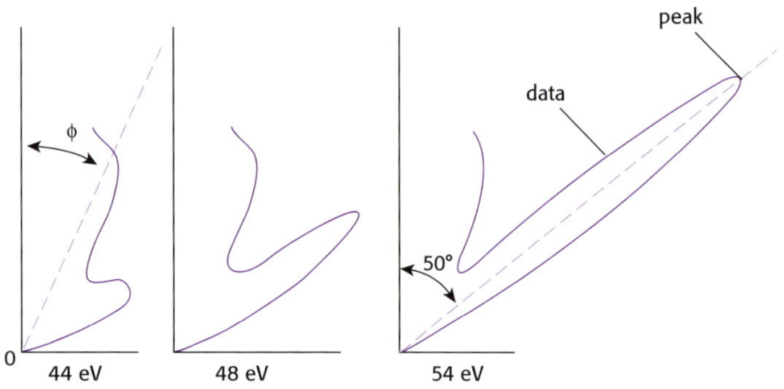

Figure 6 Polar plots for different energies

The peak amplitude of scattering takes place at a given energy of electron beam (see Figure 6). The reason for this increased amplitude is constructive interference of the electrons' matter waves at that particular angle. For an atomic spacing of d, the path difference AB between two adjacent atoms is $d\sin\phi$ (see Figure 7).

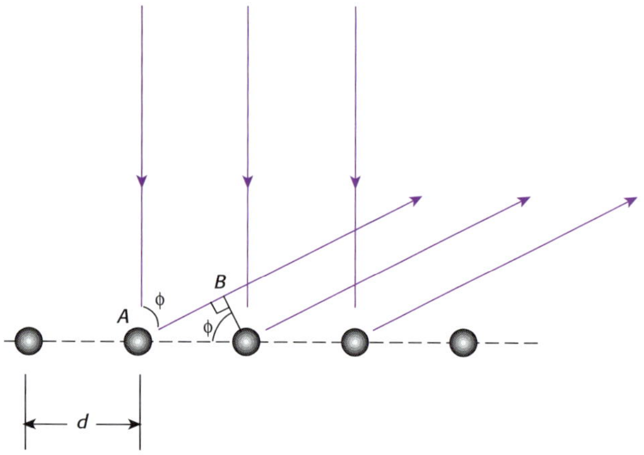

Figure 7

For the electrons that have been accelerated through a p.d. of 54 V:

1 Summarize the information represented by the radial graph plots in Figure 6.
2 Calculate the momentum of an electron.
3 Use the de Bröglie relationship to calculate the wavelength of the electrons.
4 The separation of the nickel atoms is known to be 2.15×10^{-10} m.
   (a) Calculate the path difference between two nickel atoms at the maximum intensity.
   (b) Comment on your answer.

The experiment was repeated for many different accelerating potential differences V. Figure 8 shows how the effective electron wavelength varies with the accelerating PD.

5 Deduce a general expression for the wavelength of the electron in terms of its mass m, its charge e, and the accelerating pd V.
6 Use the graph in Figure 8 to estimate a value for Planck's constant h.

Further investigation work on the Davisson–Germer experiment is readily available on the Internet. One good example is the simulation published by the Physics Education Technology Project, which is available at http://phet-web.colorado.edu/simulations/schrodinger/dg.jnlp

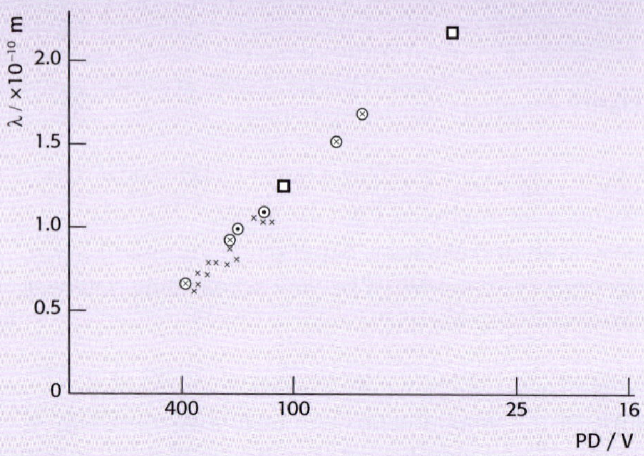

Figure 8

Thinking about science: wave–particle duality

Refraction, interference, and diffraction can be explained by assuming that EM radiation travels as waves. On the other hand, the photoelectric and Compton effects, and nature of black body radiation, require explanations based on quanta and particles called photons. Is this a contradiction? No. It just means that each wave or particle model is incomplete. A complete model must include both aspects. Similarly, some properties of electrons require that they be treated as particles. But electron diffraction shows that they can have wave-like properties. Overall, the evidence suggests that waves can behave like particles, and particles like waves. This is called **wave–particle duality**.

Probabilities

When light is diffracted by a slit, the interference between different waves causes a pattern of bright and dark fringes whose positions be calculated using the traditional wave model. Quantum mechanics should predict the same pattern. However, its mathematics does not give the precise locations and paths of particles or waves. Instead, it calculates probabilities. Each photon can take many possible paths. Each path has a probability associated with it, and it is the combination of all the probabilities that produces the pattern.

Matter waves

According to the **de Bröglie hypothesis**, all matter – for example, an electron or a proton – has a **probability wave** (a **matter wave**) associated with it. The wavelength of this probability wave can be calculated from a particle's momentum, *p*.

$$p = \frac{h}{\lambda}$$

This matter wave provides a way of calculating the probabilities of any future positions or interactions. The mathematics is complicated, but the information is described by an equation called the particle's **wave function**. The probability of a particle being found in any particular region depends on the amplitude of this wave function (to be more precise, it is proportional to the square of the amplitude). With many particles, the wave functions combine and can interfere.

When electrons are scattered by the surface of a crystal, they are more concentrated at certain angles. These correspond to the regions of constructive interference that result from the electrons' wave functions combining. Once again, the path of an individual electron cannot be predicted, but the relative probabilities of its arrival at different angles can be calculated using quantum mechanics. The experimental measurements are in close agreement with the theoretical predictions.

Thinking about science: probabilities and quantum states

The objects around us exist on a macroscopic scale: they are large enough to be seen. On this scale, we can use classical mechanics (Newton's laws) to predict the paths taken by things as they move around. One characteristic of classical mechanics is that its laws are **deterministic.** This means that the future is absolutely determined by the past. For example, if you know the mass, velocity, and position of a ball and the forces on it at any instant – in other words, the starting conditions – you could, in principle, work out exactly where the ball was going to be at all times in the future.

At the atomic level, classical mechanics must be replaced by a model using the quantum theory. This has two key features:

The first feature is that, in quantum theory, certain states – for example, the amount of energy a particle has – are only "allowed" to have specific values. For more about this, see the section on orbitals below.

The second feature has already been described: the model is not deterministic but **probabilistic**. This means that whatever information we have about the starting conditions of a system, the best we can do is to calculate the relative probabilities of various outcomes. We can never predict with certainty what will happen.

Both features are apparent in the quantum mechanics model of the atom, as in the following example:

In a hydrogen atom, the nucleus has just one electron in orbit around it. However, you should not think of this as a point object, fixed in a particular orbit. There are different possible quantum states for the electron, each of which corresponds to a different total energy. In each state, or **orbital**, the electron has a range of different possible locations. One can picture the electron as being "smeared-out", with its wave function being used to calculate the probability distribution of the different locations.

The use of wave functions for electrons may seem complicated, and can only be used to give the probability (rather than the certainty) of finding the electron located in a region of space. But, to date, all experiments have confirmed that this is the best method of predicting what takes place. The measurements agree with the calculated probabilities. Fundamental particles really do seem to behave in this strange and mysterious way, and interactions between individual atoms can only be accurately described by using quantum theory.

The probabilistic rules of quantum physics do not mean that classical physics is wrong. The effects we see in everyday objects are the result of many billions of quantum events. Classical physics is a very good, and useful, approximation for the averaged-out effects, on a larger scale, of many quantum events.

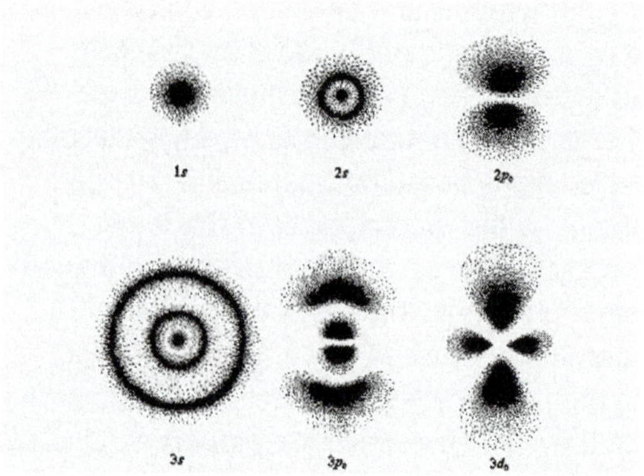

Figure 9 Probability density plots of some hydrogen atomic orbitals. The density of the dots represents the probability of finding the electron in that region.

Thinking about science: the uncertainty principle

Quantum theory suggests that one cannot determine outcomes with any certainty. It also imposes a limit on the accuracy with which any experimental measurement can be achieved, no matter how hard one tries. This limit is given by **Heisenberg's uncertainty principle**.

In quantum mechanics, there is a wave function associated with each electron, proton, photon, or other particle. These wave functions are used to calculate the relative probabilities of various future outcomes. Using this idea, Heisenberg was able to show that certain variables are linked in a way that limits how much one can know about both of them at the same time. For example, for a moving electron, position and momentum are two such variables. To quote Heisenberg: "The more precisely the position is determined, the less precisely the momentum is known in this instant, and vice versa" (*Zeitschrift für Physik, 43*:1927). In symbolic form the relationship is:

$$\Delta x \Delta p \geq \frac{h}{4\pi}$$

Δx and Δp are the uncertainties in position and momentum (respectively), and h is the Planck constant (6.63×10^{-34} J s).

Two other related measurements are energy E and time t. For these, the uncertainty relationship has the same format as the previous one:

$$\Delta E \Delta t \geq \frac{h}{4\pi}$$

There are three additional points to make here:

- For each variable, the uncertainty refers to a statistical quantity called the **standard deviation**.

- The value of the Planck constant is so small that, in everyday events, the effects of quantum uncertainty are not usually observed.
- The uncertainties in the above equations come from the mathematics of quantum mechanics. They are associated with the particles themselves, and *not* due to limits imposed by experimental techniques.

Practical implications

The practical implications of both of the above relationships can be seen in particle physics experiments, and in some modern electronic devices. For example, some radioactive materials emit alpha particles. According to the rules of classical physics, these particles should remain trapped in the nucleus because they do not have enough energy to escape. However, the uncertainty principle "allows" them to borrow energy, provided they can "pay it back" in the time limit imposed by the equation. As a result, the particle can escape its trap. This is an example of a process called **quantum tunnelling**. Quantum tunnelling by electrons is used in **scanning tunnelling microscopy**, a procedure that allows individual atoms to be imaged.

Inquiry: electron diffraction tube

When a stream of electrons hits a sample of graphite powder, the electrons form a pattern of rings:

Figure 10

The rings correspond to regions of constructive interference. The path difference between different layers of atomic spacing corresponds to a whole number of electron wavelengths.

1 Do some research to find out how the X-ray diffraction patterns are used to determine the atomic structure of different crystalline substances.
2 Design a poster that aims to give an introduction into the physics behind the phenomena of electron diffraction. The information should aim to make the topic understandable by non-physicists.
3 Predict what will happen to the size and brightness of the circles if:
   (a) the accelerating pd is decreased
   (b) the accelerating pd is increased.

Emission and absorption spectra

If chemists want to identify the metals present in a compound, they can use a flame test. A sample of the metal salt is placed in a flame, and the colour observed helps identify the metal involved.

One common type of street light gives off an orange colour because it uses low-pressure sodium vapour to produce the light.

Table 1

Metal	Flame colour
barium	light green
calcium	brick red
copper	blue/green
lead	blue/white
potassium	lilac
sodium	bright orange

Emission spectra

In a spectrum, the different colours are associated with different frequencies of light. Each element has its own **emission spectrum**: the particular frequencies produced when the element is hot enough to radiate light.

To analyse a spectrum, the emitted light needs to be split into its component frequencies. The instrument used is called a **spectrometer**. It uses either a prism to refract and disperse the light or a **diffraction grating**.

Different frequencies of light are recorded at different angles. The measurement of each angle is used to calculate the observed wavelengths, from which the frequency can be calculated. Typically, the practical arrangement involves shining light through narrow slits so that the resulting spectrum is displayed as a series of lines at different wavelengths. This is sometimes called a **line spectrum**.

Experimentally, as the temperature of the element is increased, the bands become broader. An extremely hot source of light will tend to give off a **continuous spectrum** of light – white light.

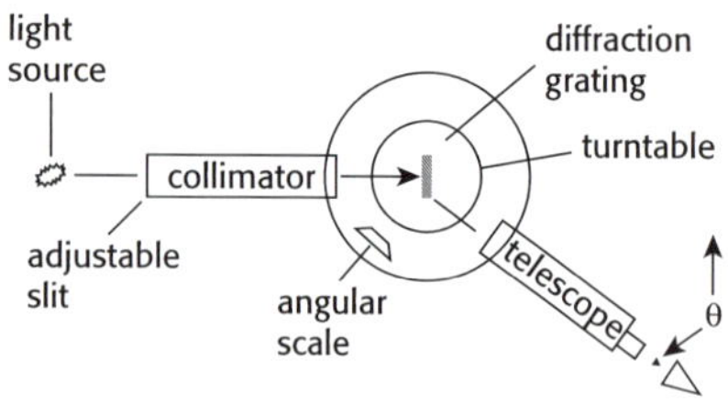

Figure 11 Spectrometer

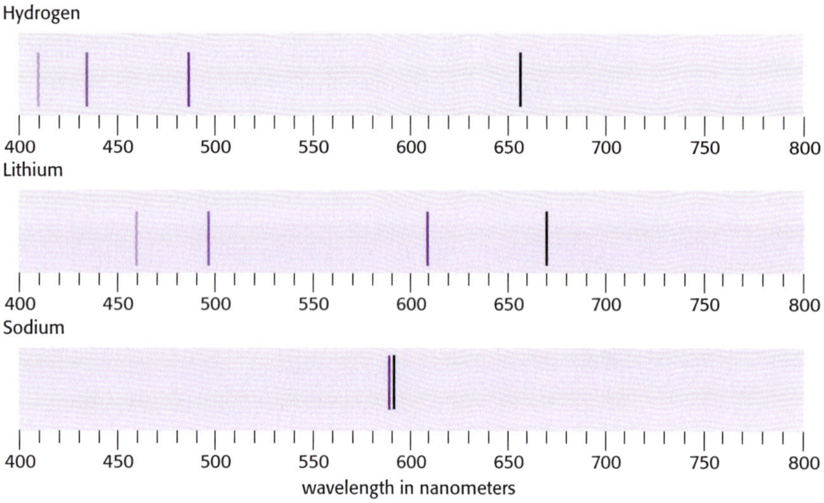

Figure 12 Atomic emission spectra for H, Li, and Na

Absorption spectra

Newton was one of the first scientists to analyse the light from the Sun. He observed that the spectrum is not, strictly speaking, continuous. It is nearly a continuous spectrum of white light but there are some frequencies missing (or of reduced intensity). This is typical of an **absorption spectrum.** The missing frequencies correspond to those found in emission spectra.

There is a link between emission and absorption spectra. An element emitting particular frequencies of light is also able to absorb the same frequencies. If a continuous spectrum (white light) shines through a gas, then it will preferentially absorb certain frequencies. This absorbed energy will be re-radiated, but if that happens, it will take place in all directions. So the intensity of the characteristic frequency in the forward direction will be reduced.

The light from the Sun passes through its outer, colder, layers where the particular frequencies are absorbed. These are characteristic of the elements that exist in the outer layers.

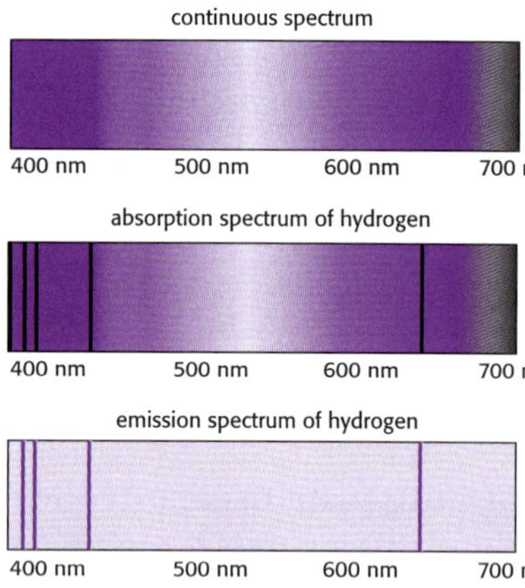

Figure 13 Absorption and emission spectra of hydrogen

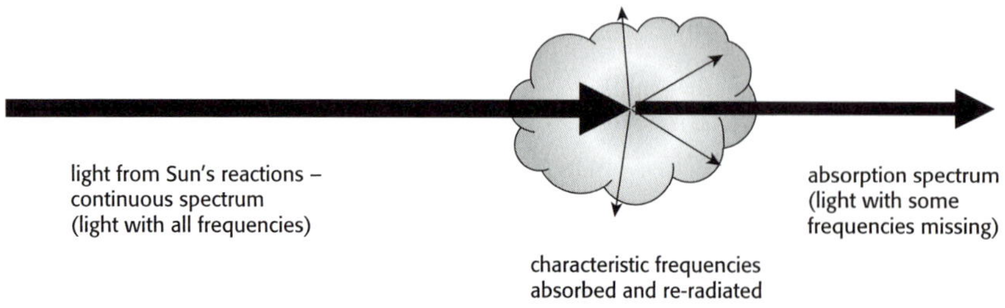

Formation of a line spectrum

One key question is "how is a line spectrum formed?". In other words, why are only certain specific frequencies present? To understand that, one needs to go back to the link between the energy of a photon and its frequency:

$$E = hf$$

If only certain frequencies are present in a spectrum, then it follows from the equation that only certain photon energies are possible. The quantum theory explains this as follows. For the electrons in an atom, there are only a limited number of discrete (separate) energy quantum states. Electrons are "allowed" to occupy these, but not have any energy in between. If an electron loses energy by making a transition to a lower level, then a photon of equivalent energy (and of one particular frequency) is emitted. Similarly, an incoming photon can lift an electron to a higher state, provided the energy change is the correct match.

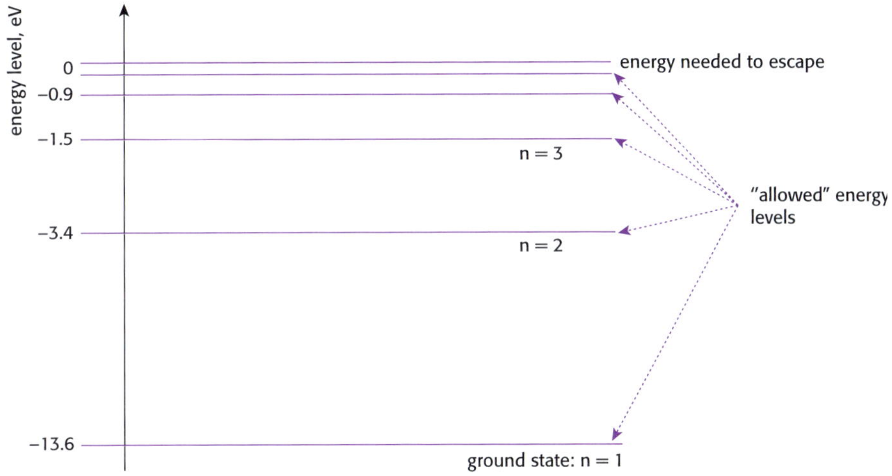

Figure 14 Electron energy levels in hydrogen

The electron energy levels in an atom are often represented by lines on an energy plot. The chart in Figure 15 is for a hydrogen atom. The lowest energy level represents the **ground state** of the electron. The top level (defined as zero) is for an electron that has enough energy to escape from the nucleus. It is the transitions between the different energy levels that correspond to particular frequencies in the hydrogen spectrum. For example, if an electron moves between the n = 4 and the n = 2 energy levels, this must involve an energy loss:

energy change $\approx -0.9 - (-3.4)$ eV $= 2.5$ eV

This "lost" energy must go into a photon of energy 2.5 eV

So: frequency of photon $\approx 2.5 \text{ eV} / h = \dfrac{2.5\times1.6\times10^{-19}}{6.6\times10^{-34}}$

$$= 6.0 \times 10^{14} \text{ Hz}$$

So: wavelength of photon $\approx 5 \times 10^{-7}$ m ≈ 500 nm

The experimentally measured wavelength turns out to be 486 nm.

In the hydrogen spectrum, the lines can be arranged into sets, each representing a different series of possible transitions. The series are named after the scientists who first discovered them.

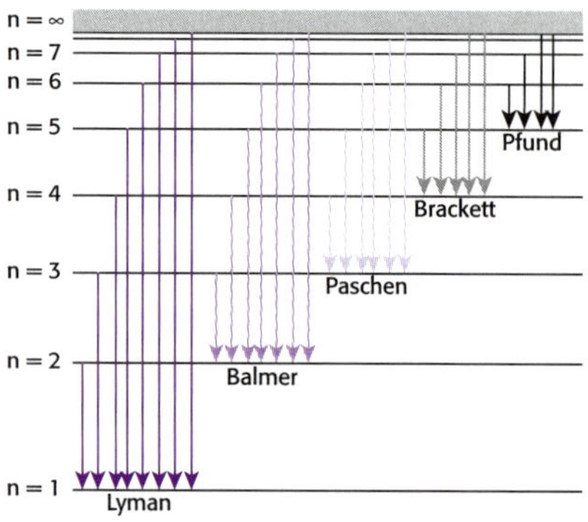

Figure 15 Electron energy levels

The above model is much simplified. In a more detailed analysis using the quantum theory, different energy levels are associated with different orbitals (see earlier section) and the electrons are assumed to behave as standing waves. This method of modelling electrons is known as **wave mechanics**.

Modelling electron orbits

In the most simple model of an atom, electrons orbit a nucleus made up of protons and neutrons. This offers no explanation for the existence of atomic energy levels. However, by making some additional assumptions about quantization, Bohr was able to accurately predict many aspects of the hydrogen emission spectrum.

Bohr started by quantizing a quantity called the angular momentum of an electron, in units of $\frac{h}{2\pi}$. In other words, the allowed values for the angular momentum are $\frac{h}{2\pi}, 2\frac{h}{2\pi}, 3\frac{h}{2\pi}$ ... and so on.

Bohr argued that an electron with one of these amounts of angular momentum must be in a stable orbit. Using Newtonian mechanics, and assuming that the orbits were circular, he calculated their energies. He found that these matched the values needed to produce the emission spectrum of hydrogen. This was an impressive result but, unfortunately, it did not work well for more complicated atoms with more protons and electrons. Also, it failed to explain some of the fine detail of the observed spectrum and the relative intensities of the various lines. It took a wave model of electron orbits to overcome these difficulties.

The mathematics of an accurate wave model is very complicated, but an insight into the principles involved can be gained by starting with a simple "electron in a box" model. We know that the electron is trapped in the atom because it is attracted by the nucleus. This model imagines that the electron exists somewhere within a box whose sides are of length L. For simplicity, we imagine an atom to be cubic (rather than spherical), and start by considering how a probability wave (a matter wave) might fix in the box. From an earlier section, you will remember that the probability of finding an electron at any particular location depends on the amplitude of this wave.

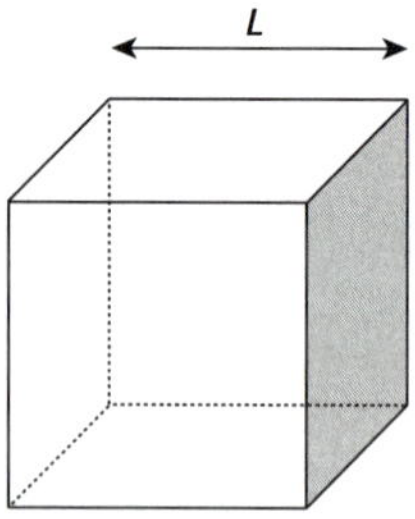

In this model, the electron is trapped inside the box, so the probability of finding it outside the box must be zero. This means that the wave must have zero amplitude outside the box, but can have a value inside. Only a standing wave will be able to fit these conditions, and there must be a node at each wall. Possible standing waves are shown here.

In a "real" atom, an electron is not, of course, trapped inside a box. However, the concept of the wave function can be modified to calculate the chances of finding the electron located at any given distance from the nucleus. **Schrödinger's model** of the hydrogen atom used wave mechanics to develop a more accurate wave function for the electron. It took various other properties into account, including angular momentum, spin, and the variation in electrical potential energy associated with the charges on the electron and the proton. In this model, the wavelength of the electron's wave function is not fixed, but varies with distance.

For the wave function, **boundary conditions** apply. The amplitude must be zero at infinity otherwise the electron will not be bound to the nucleus. A wave function has no physical meaning, but the square of its amplitude is proportional to the probability of finding the electron in the region being considered.

For some problems, you also need to understand the link between an electron's kinetic energy (KE) and electrical potential energy (PE). The total energy associated with its wave function is the sum of these two. When an electron is an infinite distance from the nucleus, its PE is zero. When it is close to the nucleus, its PE is negative, so its KE must be positive. However, for a bound electron, its total energy is negative.

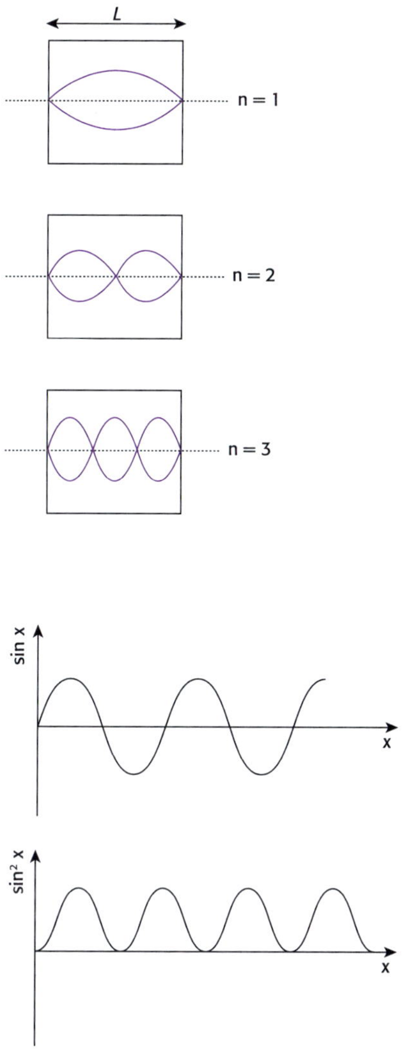

Figure 16

?

1. Use the "wave in a box" diagrams to calculate the first three possible values for the wavelength, λ, of the electron in this model in terms of L.
2. Use the de Bröglie relationship to predict the first three possible values for the momentum p of the electron in this model in terms of λ.
3. The mass of the electron is m_e. Use the relationship between momentum and kinetic energy to calculate the first three possible values for the KE of the electron in this model.
4. Derive a relationship for the possible values for the KE of the electron in this model in terms of n, m_e, h, and L.

18 Global energy sources

This chapter covers a crucial area of physics, which impacts on all of us every day, no matter what part of the world we live in. Our societies are completely dependent on energy resources and power generation in order to function.

Before starting this chapter, make sure you:

- understand the difference between renewable and non-renewable energy sources
- know how electrical energy is generated from fossil fuels
- understand the principle of conservation of energy and how to apply it in a range of examples
- understand how thermal energy is transferred from place to place.

This chapter will compare energy use, electricity generation and the pollution they cause, between the richer and poorer nations of the world. You will need to understand the topics covered from the perspective of different nations, and show some empathy for the varying economic and social conditions around the world.

As a useful starting point we shall take the member countries of the OECD (Organization for Economic Cooperation and Development) as examples of richer, more developed countries, and non-OECD countries as the less developed or developing nations.

?

Use the OECD website, www.oecd.org/, and other sites to answer the following questions.

1 Which countries are currently members of the OECD?
2 What are the requirements for membership of the OECD? List some common features of the countries that are OECD members.
3 Brazil, Argentina, India, and China are not members of the OECD, but Mexico, South Korea, and Japan are. Explain why this could be, giving suitable reasons.

World energy consumption

All aspects of the global economy, including industrial production, financial and other services, agriculture, and transport, require an adequate supply of **primary energy**, predominantly in the form of fossil fuels, and of **secondary energy**, in the form of electricity and consumable fuels. A primary energy source such as coal, crude oil, or gas is fuel in its initial form. Secondary energy sources such as electricity and petrol (gasoline) are produced from primary sources. The thirst for economic growth and technological development, particularly in the more developed countries, has led to an **unsustainable** growth in the demand for energy, with the unwelcome side effect of pollution and its possible impact on the world's climate.

In 1987 the UN World Commission on Environment and Development (the Brundtland Commission) defined the term **sustainable development as**:

"Development which meets the needs of the present without compromising the ability of future generations to meet their own needs".

Although this definition is 20 years old, it is still regarded as applicable today.

World economic growth has been uneven. The rate of growth in the richer areas has far outpaced that in the poorer, less developed countries. In the race for development, partly to alleviate poverty, the less developed countries have often been forced to make choices about their energy generation that have resulted in accelerating pollution with few controls on output through legislation.

Globally, society has to balance the ever-increasing demand for energy with the need to make changes sustainable so that future generations can enjoy the benefits of new technologies. The world map in Figure 1 illustrates some of the differences between the main regions of the world. It shows total annual electrical energy consumption per head of population.

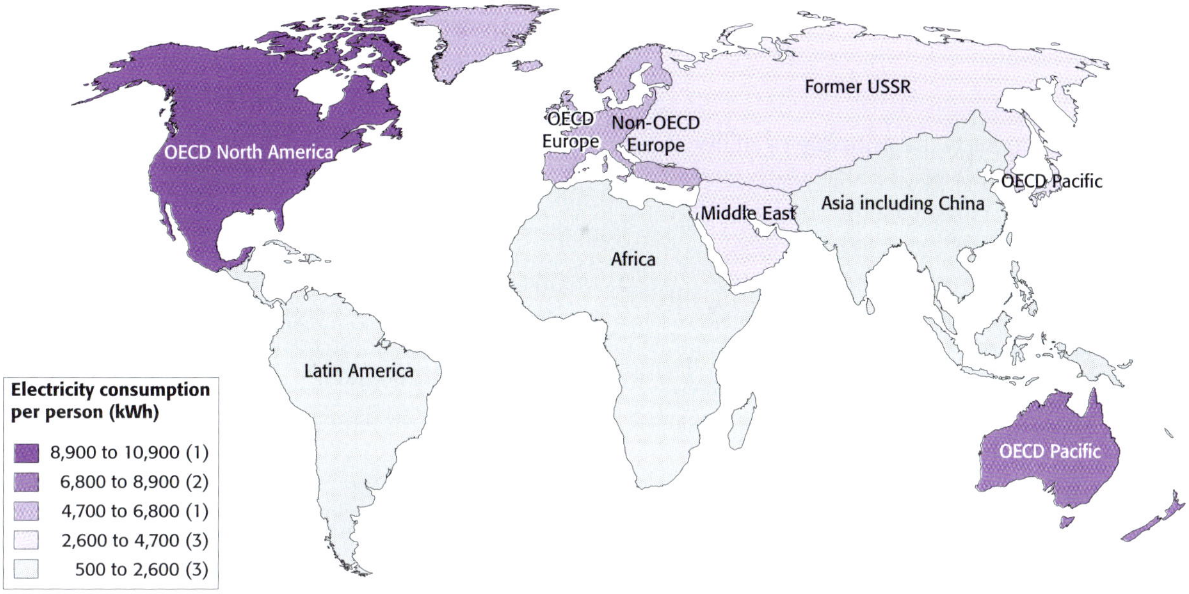

Figure 1 World map illustrating annual electricity consumption per capita

Inquiry

Data such as that contained in Figure 1 can be misleading. For example, it does not show us which areas of the world use the most electricity. To know this you will need information on human populations. A good website to start researching this would be the UN site, www.un.org/

1 Use the data in the map in Figure 1 to outline differences between OECD and non-OECD countries in terms of electricity consumption.
2 How do you think a map showing total electricity consumption would differ from Figure 1?

Thinking about science: ethics and morality

The IB Diploma Programme TOK guide describes ethics as involving,

"a discussion of the way we ought to live our lives, the distinction between right and wrong, the justification of moral judgments, and the implications of moral actions for the individual and the wider group."

This chapter gives you plenty of opportunity to discuss the ethical issues raised by energy generation and its consequential contributions to pollution and climate change. Ask yourself: is there such a thing as absolute right and wrong? Your answer may depend on your personal moral values, which will be influenced by your cultural background, education, and conscience. In a situation when two "moral" people disagree on an ethical judgment, how do we decide who is right or wrong?

Consider the two statements below, concerning a coal-fired power station and a plan to increase its output.

A Increasing the output of power station X will provide cheaper energy because of economies of scale, and will enable more people to have access to an affordable and reliable supply of electricity.

B Increasing the output of power station X will increase carbon dioxide and sulfur dioxide levels locally, which will contribute to global climate change.

Each statement identifies winners and losers, whether the output is increased or whether it stays the same. There are different winners and losers depending on the point of view. If moral judgments are based on values, and values differ from one person to another, there is an ethical dilemma here. How can we resolve this dilemma?

One simple method is to apply the **principle of utility**, which states that the supreme good is that which tends towards the greatest good of the greatest number of people. In this sense "good" refers to happiness, satisfaction, or pleasure. If an action tends to achieve this, it is deemed good; if it does not, it is considered wrong. This can lead to problems, though: for example, cutting power output in order to reduce climate change globally may lead to unemployment, hardship, and eventual poverty for power workers.

1 Apply the *principle of utility* to the power station dilemma mentioned above and suggest whether scenario A or B is preferable.
2 Explain whether you think the *principle of utility* gives the "best" outcome in this case.

The Three Gorges Dam on the Yangtze River in China should produce 22 000 MW of electrical power when at full capacity in 2011. This will replace at least 20 conventional coal-fired power stations. The reservoir formed behind the dam will have an area greater than 1000 km^2, and will displace over 1.4 million people from their cities, towns, and farms (see Chapter 22).

3 Use the principle of utility to decide whether the construction of the Three Gorges Dam can be justified. Can the displacement of so many people be considered an ethically correct decision?

The conservation of energy

A foundation stone of much of physics is the law of conservation of energy:

Energy may not be created or destroyed; but it can be converted from one form into another.

The common phrase "the generation of energy" is misleading, because it implies energy can be created out of nothing. We should instead be focusing on energy changes. Rather than say that we generate electrical energy when a fuel is burned, it would be better to say:

"We release the stored chemical energy from a fuel in the form of thermal energy, which may then be converted into kinetic energy that can turn a generator, which converts this into electrical energy."

Figure 2 shows this process graphically.

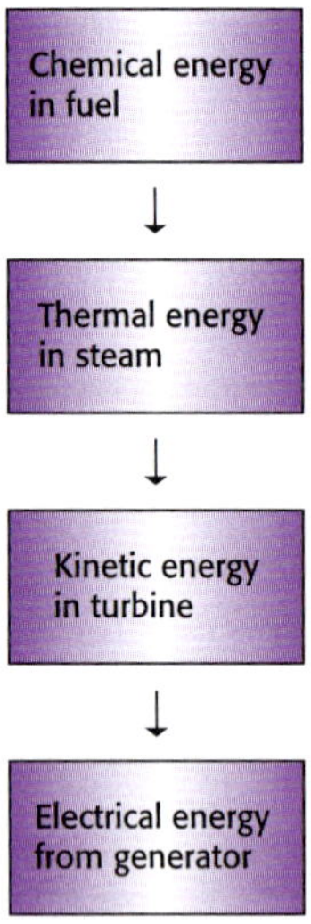

Figure 2 Energy flow from fuel to electrical energy in a thermal power station

We are not creating energy; we are simply converting it from one form into another. Inevitably there are losses of useful energy along the way (often in the form of thermal energy or sound). This leads to **energy degradation**, in which thermal energy is transferred to the surroundings, as increased particle kinetic energies that cannot be recovered. The law of conservation of energy can always be strictly applied in any physical process, but there will be some losses that cannot be prevented: some energy will be degraded into thermal energy that can no longer usefully be used.

From your previous studies you will know that the efficiency of any energy transfer process is defined as follows:

$$\text{efficiency} = \frac{\text{useful energy obtained}}{\text{total energy in}}$$

The efficiencies of most coal-fired power stations can be as low as 30%. Reductions in energy degradation can improve efficiencies significantly (see Chapter 21).

Renewable and non-renewable sources

Renewable sources of energy are those which are produced at a quicker rate than they are consumed (Table 1).

Some biofuels such as wood could be renewable or non-renewable.

Figure 3 illustrates the world's reliance on fossil fuels. Most of the category "other" in the figure is hydroelectricity, but it also includes the combustion of wastes and biofuels, as well as solar, geothermal, wave, and wind energy.

Table 1 Renewable and non-renewable sources of energy

Renewable sources	Non-renewable sources
Hydro • water stored in dams • tidal • pumped storage	Fossil fuels • coal • oil • gas
Wind	Nuclear (uranium)
Solar	Waste
Wave	
Geothermal	
Biofuels, e.g. ethanol	

The primary source

Nearly all the sources of energy mentioned above derive their energy from the Sun. Electromagnetic **radiation** arrives at the Earth's surface. Some of this is then re-emitted back into space, but a small amount is absorbed by plants through the process of **photosynthesis**, in which water and carbon dioxide in the atmosphere combine to form glucose, a chemical energy store. The overall reaction can be summarized as

$$\text{carbon dioxide} + \text{water} \rightarrow \text{glucose} + \text{oxygen}$$

or in chemical notation

$$6CO_2 + 6H_2O \rightarrow C_6H_{12}O_6 + 6O_2$$

The key idea here is that carbon dioxide is removed from the atmosphere by plants, which act as **carbon stores**.

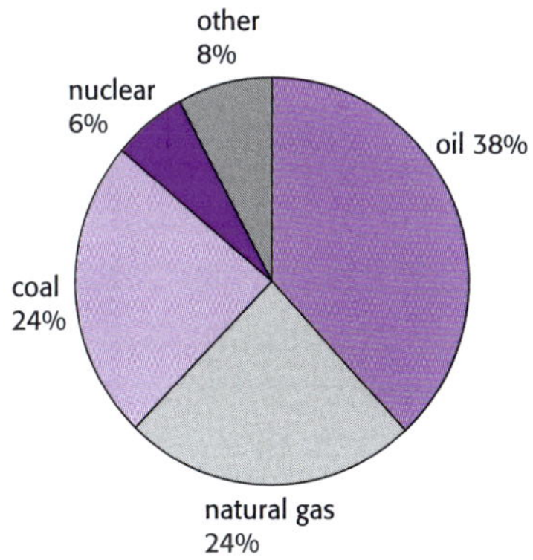

Figure 3 World's proportional energy consumption by fuel in 2003

The fuel crisis

A frequent cause for concern over the last 40 years or so has been the so-called "energy crisis". As we mentioned earlier, energy is conserved, and so does not run out. However, it *does* become degraded, and is therefore no longer conveniently usable.

A fuel is a store of energy in a convenient form. When we talk about fuels we usually think about fossil fuels – combustible sources that may be burnt in the presence of oxygen to produce thermal energy – and nuclear fuels. The **fuel crisis** indicates that we are running out of fossil fuels. Figure 4 illustrates how quickly we have exploited oil resources, for example.

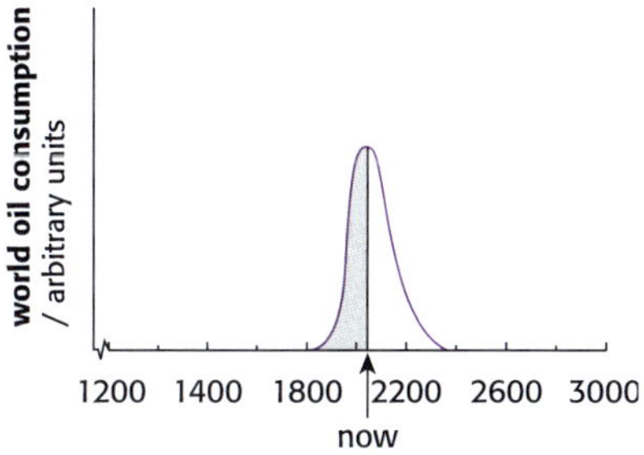

Figure 4 Historical illustration of world oil consumption

It seems that all oil will be exhausted within a few centuries at current estimates – assuming that, as reserves start to run out, so our consumption falls. At current rates of consumption oil reserves will be exhausted significantly sooner.

Oil companies and governments invest significant amounts of capital in predicting the amount of available crude oil. These predictions are complicated, and expensive geophysical surveys are needed to determine how much oil is actually stored under the ground and how much is physically recoverable. Figure 5 illustrates how the estimates of known oil reserves have changed over time.

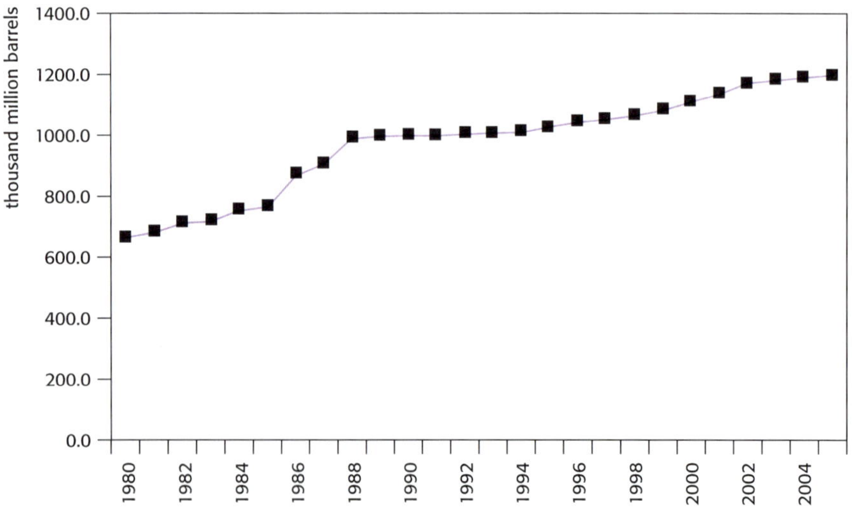

Figure 5 World oil reserves over the last few decades

There are several reasons why the estimate of known reserves has actually increased.

- More oil has been discovered, in new deposits.
- Thanks to advances in technology, known as enhanced oil recovery, more oil can now be extracted from reservoirs that were previously thought to be exhausted.
- The rising price of oil has meant that some previously discovered sources are now economically worth exploiting.

Oil companies use the term "**ultimately recoverable resource**" as an estimate of the total amount of oil that will ever be recovered and produced. This resource can be broken down into three main categories:

- cumulative production
- discovered reserves, and
- undiscovered resources.

Cumulative production is an estimate of all of the oil produced up to now, **discovered reserves** are an estimate of known oil deposits with a good probability of recovery, and **undiscovered resources** are based on geological knowledge together with some estimation.

Despite these predictions, it is widely projected that the world's fossil fuel dependence will continue to increase over the next few decades, as shown in Table 3. All values are in quadrillion British thermal units (10^{15} Btu), or quads. (The quad is an internationally used unit of energy; 1 quad = 1.1×10^{18} J.)

?

1. Briefly outline how fossil fuels are formed, and suggest a timescale for this formation.
2. Describe how the chemical energy stored in fossil fuels originated from the Sun.
3. Outline how the following energy sources derive their energy from the Sun
   (a) hydroelectricity – water stored behind dams
   (b) wind power
   (c) ethanol.
4. Nuclear fuels do not derive their energy from the Sun. Describe the origin of the energy stored in uranium.
5. Outline the origins of geothermal energy.

?

1. Explain the difference between the terms resources and reserves when applied to fossil fuels.
2. Use Table 2 to estimate how long fossil fuels are expected to last at current rates of production.

Table 2 Fossil fuel reserves and production in 2005

Fuel (and unit used)	Proved reserves in 2005	Production in 2005
Coal (10^9 tonnes)	909.1	5.85
Oil (10^9 barrels)	1200.7	29.6
Gas (10^{12} m^3)	179.8	2.76

Source: BP Statistical Review of World Energy June 2006

3. Give reasons why the fuels may last for either (a) a longer time or (b) a shorter time than your estimate in question 2.

Table 3 Projection of fossil fuel dependence

	1990	2002	2003	2010	2015	2020	2025	2030
Total non-OECD								
Oil	52.7	63.3	65.1	82.3	91.6	100.4	109.8	120.0
Natural gas	38.0	44.7	47.2	63.1	75.1	87.0	99.6	113.1
Coal	45.9	51.8	54.8	79.1	92.9	105.7	118.4	132.4
Nuclear	3.1	4.0	4.2	5.5	7.0	8.7	10.0	10.6
Other	10.3	14.6	15.2	23.6	26.8	2 9.6	33.1	36.7
Total	150.0	178.4	186.4	253.6	293.5	331.5	371.0	412.8
Total OECD								
Oil	83.4	95.4	97.0	103.3	107.5	110.3	114.5	119.1
Natural gas	37.2	51.2	51.9	58.0	64.7	69.1	72.9	76.8
Coal	43.5	44.9	45.6	49.7	51.4	54.3	58.3	63.1
Nuclear	17.3	22.7	22.3	23.4	24.0	24.3	24.1	24.1
Other	15.9	17.6	17.4	21.7	22.3	23.5	24.7	25.7
Total	197.4	231.9	234.3	256.1	269.9	281.6	294.5	308.8

Source: History, Energy Information Administration (EIA), International Energy Annual report 2003 (May-July 2005), website www.eia.doe.gov/iea/. Projections: EIA, *Annual Energy Outlook 2006.*

Inquiry

The data in Table 3 would be much easier to analyse in graphical or visual form. Use suitable graphs to prepare presentations on the following topics.

1 Demonstrate how the world total use of fossil fuels is projected to increase over this timescale, showing clearly the change in share of each fossil fuel.
2 Compare the total energy dependence of OECD countries with that of non-OECD countries.
3 Compare the growth in the use of coal with the use of nuclear energy in both OECD and non-OECD countries.
4 The "other" category refers to renewable sources of energy. Compare the projected growth of these renewable sources with that of non-renewable sources.

Describe the trends in detail, and use your knowledge and suitable websites to explain your findings. Think carefully about the best type of graph to use to display your data, and remember that appropriate use of colour can have a strong visual impact on your audience. Some good websites to start with are:

www.eia.doe.gov/iea/
http://www.bp.com/statisticalreview

The suggested topics above are merely ideas, and should not limit your research. The data can be used to suggest many interesting hypotheses and conclusions.

During your research you may have encountered several different energy units.

5 Find out the value of the following energy units in joules:
(a) megatonne of coal equivalent (Mtce)
(b) megatonne of oil equivalent (Mtoe)
(c) cubic metre of natural gas (m^3)
(d) barrel of oil (bbl)
(e) kilowatt-hour (kWh).

Fossil fuel dependence

Most forecasts of global energy consumption over the next few decades predict that:

- Global energy demand is going to increase.
- Fossil fuels will continue to provide most energy.

This may seem surprising: fossil fuels are running out, and there is a long-term need for any future development to be sustainable. Are governments being irresponsible in planning to use up valuable resources? Are they planning with due consideration for the environment?

Sustainable development is built on three key ideas or pillars: economic aspects, social aspects, and environmental aspects. There is a tendency to focus only on environmental issues when considering sustainable development, but the social and economic aspects are just as important. Growth in energy production, through increased use of coal for example, can contribute significantly to a country's economic growth, generating much-needed revenue. Although increased fossil fuel use has a detrimental effect on the environment, electrification can do much to reduce poverty in developing countries, improve healthcare, and provide employment. Improving efficiencies of power production and modern technologies may significantly reduce the impact of fossil fuel use on the environment.

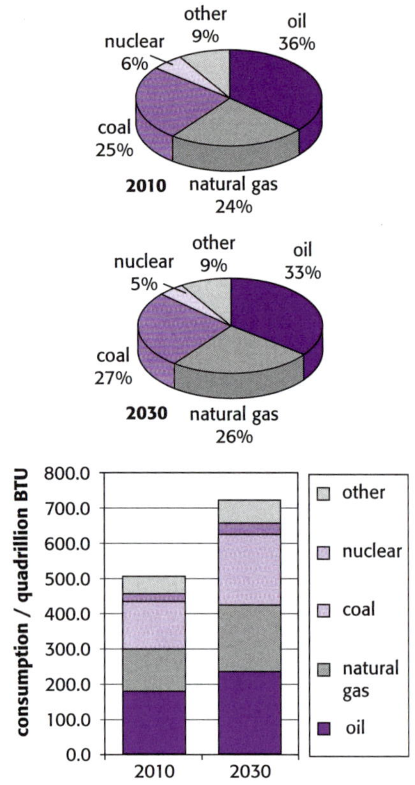

Figure 6 Predicted world energy consumption by fuel, in 2010 and 2030

1. Making predictions such as those above is very challenging. Outline the information a forecaster might need in order to make these predictions.
2. Outline the circumstances that may cause the predictions to be:
   (a) an overestimate
   (b) an underestimate.
3. Calculate the predicted dependence on fossil fuels in 2010 and 2030 as a percentage.
4. Outline some reasons why increased fossil fuel use may be considered:
   (a) sustainable
   (b) unsustainable.
5. Explain, with reference to economic, social, and environmental developments, why governments planning for increased use of fossil fuels can still be considered to be responsible.
6. One way to reduce greenhouse gas emissions and still plan for increased energy demand would be to increase the share of renewable fuels and nuclear power. However, if you compare the projections for 2010 and 2030, the nuclear and renewable share (shown as "other" in the graphs above) remains virtually constant. Outline some reasons why the proportions of
   (a) nuclear power, and
   (b) renewable energy generation

   are not expected to increase despite the pressure for development to be more sustainable.

Inquiry: historical use of fossil fuels

The use of coal has a long history. The first commercial use is thought to have been in China around 1000 BC, where it was used to smelt copper and cast coins. Coal proved to be the main fuel in the Industrial Revolution in the West in the 18th century, when it was used in the iron-making process. The 1960s saw oil overtake coal, with the growth in the transportation sector, and the use of gas is likely to increase rapidly in the future as gas turbines to generate electricity are proving cleaner and more efficient than coal or oil (see page 248), although in 2006 coal is still the major fuel used for electricity production.

Research the historical use of fossil fuels and produce a timeline of your findings – this may help you understand why we rely so much on these sources today.

Thermal generation of electricity

About 20% of primary energy is converted into electrical energy. For many people, electricity provides a relatively cheap and reliable source of energy. However, more than 1.6 billion people in 2006 had no access to electricity – about 25% of the world population – and this is not predicted to fall significantly (1.4 billion by 2030).

Figure 7 illustrates the proportions of different energy sources used to generate electrical energy in 2006.

The "other" category under renewables includes wind, solar, and geothermal together. Although only 0.8% at present, it is likely that the use of wind turbines will increase dramatically in the future (see page 256).

In Chapter 15 you learnt about the physics of electricity generation. The basic principle is to rotate a magnet inside a coil. In a typical thermal power station coal is used to produce steam at high pressure, which then turns a turbine connected to the generator. The process is illustrated in the block diagram shown in Figure 8.

Burning coal in power stations contributes significantly to global carbon dioxide emissions. In addition, coal usually contains impurities, often sulfur. Nitrogen from the air also reacts in the high temperatures used for combustion. Additional waste products include sulfur dioxide (SO_2) and various nitrogen oxides indicated by the formula NO_x, which produce acid rain. Despite this, coal is – in 2007 – used more than any other fuel in the generation of electrical energy, and this is likely to remain true for some time.

Technological improvements in the use of coal and design of power stations mean that coal has the potential to be cleaner and more efficient in the future: see Chapter 21.

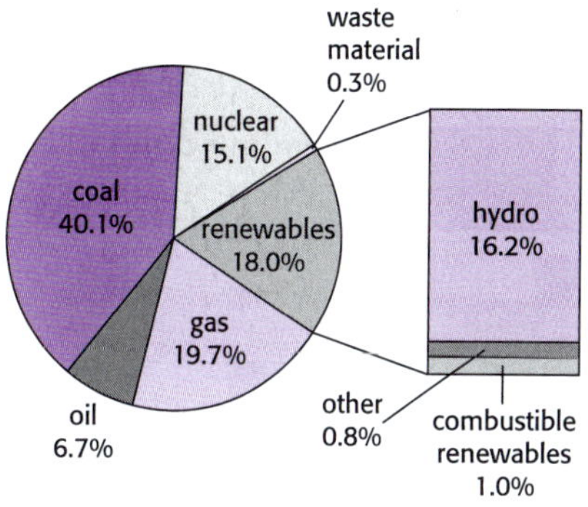

Figure 7 The proportions of different energy sources used to generate electrical energy in 2006

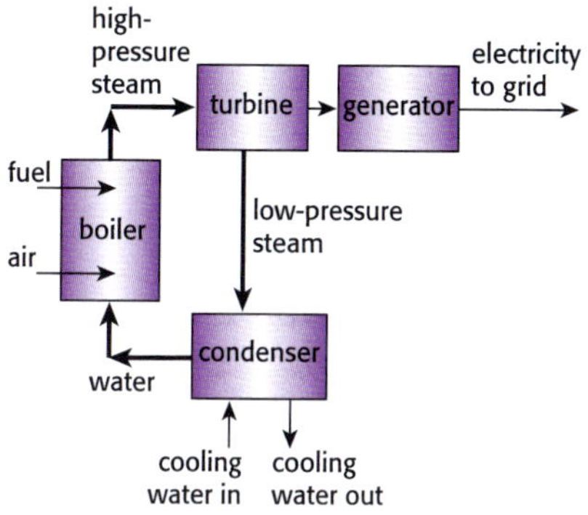

Figure 8 Block diagram of a thermal power station

Estimating the energy density of fuels

Fossil fuels have a relatively high **energy density** compared with other fuels. This is defined as the total energy released per kg burnt, and is measured in J kg^{-1}.

Energy is released from fossil fuels by combustion, as summarized in highly simplified form by the following equation:

fuel + oxygen → carbon dioxide + water

It is quite straightforward to design and carry out an experiment that gives a reasonable estimate of the energy density of various fuels using apparatus available in your school science laboratory. One way to estimate energy density would be to set up an arrangement so that a fuel is used to heat a known mass of water (Figure 9).

Figure 9 Heating water to find the energy density of methylated spirits

The formula $Q = mc\Delta T$ enables you to estimate the thermal energy required to raise the temperature of the water by a measured amount.

If the mass of fuel is measured before and after heating, then you can calculate the mass of fuel used to produce the temperature rise in the water, and hence the amount of thermal energy needed.

Any experiments involving heating will involve the loss of a significant amount of energy to the surroundings – the air, the vessel containing the water, the thermometer, and so on. Great care is needed to ensure that the water is heated in the same way by each of the fuels selected and, to ensure some reliability, tests have to be repeated.

This investigation would be easily assessible for the three skill areas of:

- data processing and presentation (DCP)
- concluding and evaluating (CE)
- manipulative skills (MS).

Any experimental work involving the burning of fuels will require careful attention to safety, and this is assessed in MS. Suitable precautions must be taken to remove the fumes produced, either by ensuring adequate ventilation in the laboratory with extractor fans or, preferably, by doing the whole experiment in a fume cupboard.

Table 4 can be used to gain an insight into the energy content of some fuels, particularly those available to many people living in developing countries.

?

Use the data in Table 4 to answer the following questions.

1 Calculate the mass of (a) anthracite coal, (b) automotive gasoline (petrol), and (c) methane that must be burnt to raise the temperature of 500 g of water from 15 °C to 85 °C. (The specific heat capacity of water is 4200 J kg^{-1} K^{-1}.)

2 Outline any assumptions you have made in answering question 1.

3 Without considering any other factors, explain which fuel would be the fuel of choice for a power-generating company.

4 What other factors would need to be considered when selecting a suitable fuel for the generation of electrical energy?

The Colbert fossil plant in Alabama, USA, uses anthracite coal to generate up to 7.8 billion kWh of electrical energy annually, supplying over half a million homes.

5 Assuming a maximum efficiency of 30%, estimate the mass of coal consumed in: (a) one day; (b) one year.

6 Estimate: (a) the rate of electrical energy production; (b) the mass of coal used per household.

The German power company RWE is planning to build a more efficient coal-fired power station in Germany, using clean coal technologies, possibly by 2014. It is likely to use lignite as the fuel, and generate electrical energy at 450 MW with a planned efficiency as high as 45%. Carbon dioxide produced will be captured and stored deep underground – see page 252.

7 Estimate the minimum mass of lignite used: (a) in one day and (b) over 1 year to operate this power station.

8 Explain why the mass of lignite calculated is a minimum value.

The Pratum Rice Mill, located in Pathumthani Province, Thailand, generates electrical energy in an innovative way. The primary fuel is rice husks, which are used to run a generator rated at 9.2 MW. The electricity is used to power the rice plant itself, and a surplus of up to 5.5 MW will be used in the Thailand national grid.

9 Assuming an overall efficiency of 28%, estimate the mass of rice husks needed to operate the plant annually.

10 Compare the energy density of biofuels and fossil fuels. Discuss whether large-scale or small-scale use of biofuels in electrical energy generation is most appropriate. Give reasons for your answer.

Table 4 Energy densities of a range of fuels

Simple compounds	Energy density / MJ kg^{-1}
carbon (to CO_2)	32.8
ethanol	29.7
hydrogen	142
propane	50.3

Food	Energy density / MJ kg^{-1}
carbohydrates	17.2
fats	38.9
proteins	17.2

Biomass fuels	Energy density / MJ kg^{-1}
charcoal	29.0
coconut husks	9.8
dung, air-dry	12.0
rice husks	15.5
maize cobs	18.9
maize stalks	18.2
peat	14.6
wood, green	10.9
wood, air-dry	15.5
wood, oven-dry	20.0

Fossil fuels	Energy density / MJ kg^{-1}
gas (methane)	55.5
coal, anthracite	31.4
coal, bituminous	>23.9
coal, sub-bituminous	17.4–23.8
coal, lignite	<17.43
diesel	45.3
gasoline, automotive	45.8
gasoline, aviation	43.1
jet fuel	43.4
kerosene	46.3
oil, crude (petroleum)	41.9
oil, heating	42.5

Source: http://hypertextbook.com/physics/matter/energy-chemical/

Inquiry: Energy end use

It is convenient to categorize the end use of energy into four sectors:

- industry
- commercial
- residential
- transportation.

The US-based Energy Information Authority (EIA) website presents a huge amount of world data for public use: www.eia.doe.gov/. It includes several convenient spreadsheets that you can download for your own inquiry work. The world data

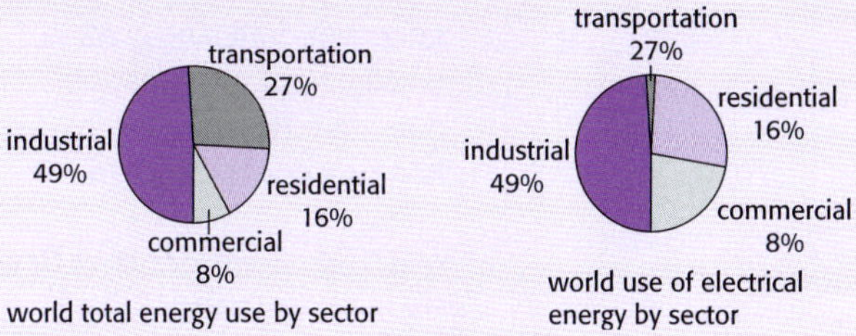

Figure 10 World end use of energy in 2003

for 2003 (Figure 10) is quite useful when considering the end use of primary energy sources.

1 Use the EIA website or other sources to define clearly what each category of end use means.
2 Into which category would educational establishments – schools, universities, and colleges – fit?
3 Find out the proportion of use of each fossil fuel in each sector, and explain why a certain fuel is preferred in each case – it may be for historical, social, or economic reasons.
4 Currently, transportation uses up more than a quarter of our primary energy. Find out: (a) what is predicted to happen in the future to the amount of energy used for transport; (b) what is projected for the proportion of electricity used in transport. Back up your findings with suitable data.
5 A challenging task would be to try and calculate, from the data available, the proportion of electrical energy used in our total end use of energy – one difficulty being the need to convert all energy units into a common unit before making any calculations. (An answer of between 15% and 20% would be reasonable, depending on which data source you use.)

Physics issues: energy and poverty

The alleviation of poverty is a major global challenge. One aim of sustainable development is to provide accessible supplies of affordable energy to the rural poor in developing countries, to help to lift them out of poverty. However, simply providing a supply of electricity will not necessarily reduce poverty. The most successful development projects in rural areas involve local people at all stages, from initial planning to final completion. Such "ownership" is crucial for the long-term viability of such projects.

The Institute of Energy (IEA) states some key observations and predictions:

- Currently, some 1.6 billion people – one-quarter of the world's population – have no access to electricity. Without new policies, 1.4 billion people will still be without any supply of electricity in 2030.
- Four out of five people without electricity currently live in rural areas of the developing world, mainly in South Asia and sub-Saharan Africa, but the pattern of electricity deprivation is set to change: 95% of the increase in population in the next three decades will occur in urban areas.
- Some 2.4 billion people currently rely on traditional biofuels – wood, agricultural residues, and dung – for cooking and heating. That number will *increase* to 2.6 billion by 2030. In developing countries, biomass use will still represent over half of residential energy consumption in 2030.
- Lack of electricity and heavy reliance on traditional biomass are hallmarks of poverty in developing countries. Lack of access to electrical supplies exacerbates poverty and contributes to its perpetuation, as it precludes most industrial activities and the jobs they create.

- Investment will need to focus on various energy sources, including biomass, for thermal and mechanical applications to bring productive, income-generating activities to developing countries. Electrification and access to modern energy services do not necessarily guarantee poverty alleviation.
- Renewable energy technologies such as solar, wind, and biomass may be cost-effective options for specific small-scale, local applications, while conventional fuels and established technologies are likely to be preferred to generate electrical energy for the large-scale national grid.

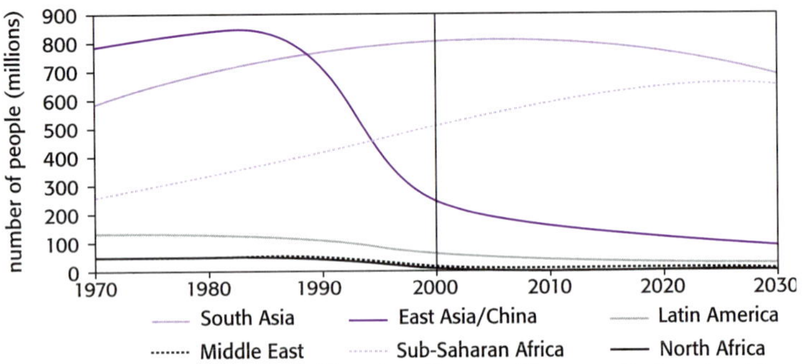

Source: Adapted from http://www.worldenergyoutlook.org/model.asp

Figure 11 Number of people without electricity, 1970–2030

If a convenient supply of electricity is an indicator of development, then many countries still have some way to go to help reduce some of the constraints of the poverty. Poor people have limited access to bank loans or other capital to develop small-scale income-earning schemes. They may be forced to borrow money from friends, family, or unlicensed "loan sharks", often ending up in debt, which impoverishes them further. Lack of access to clean water and sanitation spread disease, thus making poor people less effective at work and so less likely to earn an income. Such poverty traps are often referred to as the poverty cycle.

Two illustrations of how the lack of a reliable supply of electricity can contribute to the poverty cycle are outlined below.

- Dried dung is commonly used in rural South Asia for heating and cooking. It is a cheap and reliable, though inefficient, energy source. The fumes produced are highly polluting, which causes illness, particularly amongst women, who are the major care givers, educators, and providers in the family.
- In many parts of rural Kenya it takes several hours every day to fetch and carry enough water for cooking, drinking, washing, and sanitation, so time for any income-earning activities is limited. Water cannot be pumped to many villages unless there is a reliable supply of electrical energy.

Electrification in China: a case study

China secured electricity access for almost 700 million people in two decades, enabling it to achieve an electrification rate of more than 98% in 2000. From 1985 to 2000, electricity generation in China increased by nearly 1000 TWh, 84% of it was coal-fired, and most of the rest hydroelectric. The electrification goal was part of China's

?

This question is about the sustainable generation of electricity.

(a) State and explain whether the energy sources used in the following power stations are renewable or non-renewable:
 (i) coal fired
 (ii) hydroelectric. [4]

(b) Biogas can be produced from cow dung. This may be used as a fuel for an electricity generator. State and explain whether cow dung may be considered a renewable or non-renewable source of energy. [2]

(c) One cubic metre of biogas produces 6.36 kWh when burnt in oxygen, and is used to generate up to 2 kWh of electrical energy.
 (i) Calculate the number of joules in 1 kWh. [1]
 (ii) Calculate the efficiency of the generator. [1]

(d) Small-scale biogas generators can be used in the rural areas of developing countries to provide a cheap and sustainable source of electrical energy. Explain what the term sustainable means in this context. [2]

(e) Assume that 100 cows produce enough dung to operate a 10 kW generator. Use the data in part (c), and your answers to it, to calculate the volume of gas burnt each second in the generator. [3]

poverty alleviation campaign in the mid-1980s. The plan focused on building basic infrastructure and on creating local enterprises. China's economy grew by an average annual 9.1% from 1985 to 2000.

A key factor in China's successful electrification programme was the central government's determination and its ability to mobilize contributions at the local level. The electrification programme was backed by subsidies and low-interest loans. The programme also benefited from the very cheap domestic production of elements, ranging from hydro generators down to light bulbs. China has also avoided a trap into which many other nations have fallen: most Chinese customers pay their bills on time. If they do not, their connections are cut off.

China's transformation and distribution networks still need very large investment to match standards in more developed countries. Electricity services can be unreliable and wiring and meters in homes and offices are undependable, sometimes even unsafe. Usage is low, especially in rural areas, where consumers tend to restrict their electricity use to lighting their homes because of the cost and inconsistency of supply; power cuts are common in such areas.

Table 5 China's population growth

Year	Population
1950	562 000 000
1960	648 000 000
1970	820 000 000
1980	984 000 000
1990	1 147 000 000
2000	1 264 587 054
2010	1 347 000 000*

**This figure is a projection, made in 2000.*

?

1. Plot a suitable graph of the data in Table 5, and determine the relationship between population and time.
2. Calculate the proportion of the total population, based on the 2000 value, who gained access to a supply of electrical energy during the previous two decades.
3. Estimate the proportion of the population without access to electricity in 1980.
4. The expression "an electrification rate of more than 98% in 2000" used above is ambiguous: it could be interpreted in more than one way. Assuming it means that 98% of the total population have access to a supply of electricity, calculate how many people still do not have such access.
5. If the expression in question 4 refers to the growth in supply of electrical energy from 1980 to 2000, estimate the proportion of the population without access in 1980.
6. Compare your answers to questions 3 and 5, and comment on the consistency of the data within the passage.
7. Such rapid growth in the provision of electrical energy can have both costs and benefits to the country overall. Outline: (a) some of the costs; (b) some of the benefits.
8. Reflect on your answer to question 7 and explain whether, on balance, you think such rapid development has been beneficial to China. Your answer should include references to the three pillars of sustainable development – social, economic, and environmental.

The project planning cycle

Many IB world schools in developed countries have formed relationships with schools in less developed countries with the aim of working together to help contribute, in a small way, to the alleviation of poverty. Creativity, Action, Service (CAS) is an integral part of the IB Diploma Programme, and all students are expected to become involved with the service aspect. Initiating and developing a long-term development project with the aim of alleviating poverty in a sustainable way can be extremely rewarding and worthwhile for everybody involved.

Project planning is remarkably similar to the planning required for a full experimental investigation incorporating all group 4 practical skills, the group 4 project, or even the extended essay.

Identifying the main aims of the project is crucial to the ultimate success. The aims must be sufficiently focused to enable specific goals to be achieved, yet flexible enough to allow for changes in circumstances. The project must have positive impacts on the economic, social, and environmental needs of the communities targeted for intervention. A key aspect of this is that the local community you have chosen to work with must be fully involved in all aspects of the project planning cycle. Partnerships with schools, charities, and NGOs (non-governmental organizations) in the host country could be of great assistance, providing local knowledge and raising awareness of cultural issues on both sides.

Raising funds is one challenge, but the next is to use the funds to complete the project, e.g. building a rural school, drilling boreholes, or building suitable pit latrines to ensure healthy sanitation. It is also vital to ensure that funding for the project can be maintained, so that the capital resources can be kept functioning over several years. Sadly, some projects fail after only a few years. Often this is because funding for maintenance runs out. A visit to the country and getting actively involved in the implementation part of the project can be very rewarding. Many students find the experience of working with people who are significantly disadvantaged compared with themselves very positive, and sometimes even life-changing. This is illustrated in some quotes by students from Sha Tin College, Hong Kong, after working in an orphanage for street children in Cebu in the Philippines.

"I realized that something so little can mean so much to someone, and if you didn't do that small thing it would not be the same."

"After this trip I cannot just ignore the problem of poverty, I cannot possibly see things the way I did before ... I was there with poverty – I hugged it and I changed forever."

"I truly feel that this trip has changed my life and the way I see things."

"I have learned to be truly thankful for everything I have been blessed with."

Perhaps some of these quotes will motivate you to start considering how you can help, in a small way, to alleviate poverty in the world.

References

Thomas, A. et al. 1994. *Third World Atlas*. Buckingham: Open University Press. p 62; quoting from World Commission on Environment and Development (1987), *Our Common Future* (The Brundtland Report). Oxford University Press, Oxford.

Tomkinson, J.L. 2004. *The Enterprise of Knowledge: A Resource Book for Theory of Knowledge*. 2nd edn. Athens. Anagnosis Books. p 348.

World Commission on Environment and Development. 1987. Our Common Future (The Brundtland Report). Oxford. Oxford University Press.

Inquiry

Use a variety of sources, including, the UN Food and Agriculture Organization (FAO) and IEA websites, to complete the research outlined below.

?

1. Briefly outline the predicted trends for electrification in each region illustrated in Figure 11 and try to explain the trend for: (a) China; (b) South Asia (India); (c) Africa.
2. Outline what is meant by a poverty trap, using some specific examples from your own research.
3. Discuss how access to cheap electricity may help rural communities to escape from the poverty trap. You should refer to economic and social development in your answer.
4. Describe any possible negative impacts of electrification, using a range of global examples.

9 Solar radiation and the greenhouse effect

Radiation from the Sun is called **solar radiation**. This chapter considers how its energy is trapped by the Earth's atmosphere, and how certain human activities are adding to this effect. Before starting any of the sections in this chapter it would be useful to:

- know that the combustion of fossil fuels releases carbon dioxide into the atmosphere
- know that our demands for energy are projected to increase, leading to increased carbon dioxide production
- explain how conduction, convection, and radiation transfer thermal energy
- understand that there is evidence of a link between atmospheric carbon dioxide concentrations and global temperatures
- appreciate how scientists use data to support their arguments, but remember that there may be different possible interpretations of the same set of data.

The greenhouse effect

Some of the gases in the atmosphere (notably water vapour, carbon dioxide, and methane) effectively form an insulating blanket around the Earth. Without this, the Earth would be a rather cold and miserable place, with an average temperature of about –20 °C. The extra warming is called the **greenhouse effect**, because greenhouses also trap the Sun's heat. The gases that cause it are called **greenhouse gases**.

The greenhouse effect is necessary for life on Earth. It keeps global temperatures high enough for biodiversity to flourish. However there is evidence to suggest that human activity, mainly burning fossil fuels for energy production, has put significant amounts of extra carbon dioxide into the atmosphere over the last century or so. At present global temperatures are rising and there are predictions that this may cause dramatic climate changes and rising sea levels. Most (but not all) climate scientists believe that this **global warming** is being caused by the human, or **anthropogenic**, addition of greenhouse gases to the atmosphere. They call it the **enhanced greenhouse effect**.

Greenhouses

These buildings, made of glass or clear plastic, are used in many parts of the world to grow plants and food crops where the outside temperature is too low. For example, mangoes, a tropical fruit, can be grown successfully in Japan inside greenhouses.

So how does a greenhouse work? A common explanation suggests that the glass allows shorter wavelength radiation from the Sun to pass through to heat the ground, but reflects back the longer wavelength infrared from the warmed ground. However, if polythene is used for a greenhouse, it performs just as well as glass, even though it is almost as transparent to longer wavelengths as it is to shorter ones. A more likely explanation is that the temperature increase occurs because the warmed air inside a greenhouse is trapped and cannot rise and flow away. In other words, convection is prevented.

The atmospheric greenhouse effect does not work like a greenhouse. So a different model is needed. By the end of this chapter, you should be able to explain the greenhouse effect using the correct physics.

?

1 Briefly outline how thermal energy is transferred by conduction, convection, and radiation.
2 State which of the three mechanisms is normally responsible for transferring thermal energy from soil, plants, etc.
   (i) to the air, and
   (ii) within the air.
3 A small 50 W electrical immersion heater is used to heat some water for 10 minutes (specific heat capacity of water = 4200 J kg^{-1} K^{-1}). Calculate the temperature rise in
   (i) 250 cm^3 of water (a beaker), and in
   (ii) 3.75×10^5 cm^3 (a bath). You may assume that the mass of 1 cm^3 of water = 1 g.
4 Use the ideas in question 3 to suggest why a greenhouse is warmer inside than outside. To do this it might be useful to consider two identical sets of tropical plants and soil, one set inside a greenhouse and the other outside.

Black body radiation

The concept of a black body is covered in detail on pages 208–9. Stars are considered to be black bodies, but many cooler objects also approximate to black body radiators. The Earth is not a perfect black body, and this must be considered when developing a climate model to explain its observed average temperature.

Figure 1 shows some features of the radiation emitted by a black body. All wavelengths are radiated, but at different intensities, depending on the surface temperature of the object. The wavelength of the peak radiation is given by the **Wien displacement law**:

$$\lambda_{max} T = \text{constant}$$

The constant has a value of 2.90×10^{-3} m K.

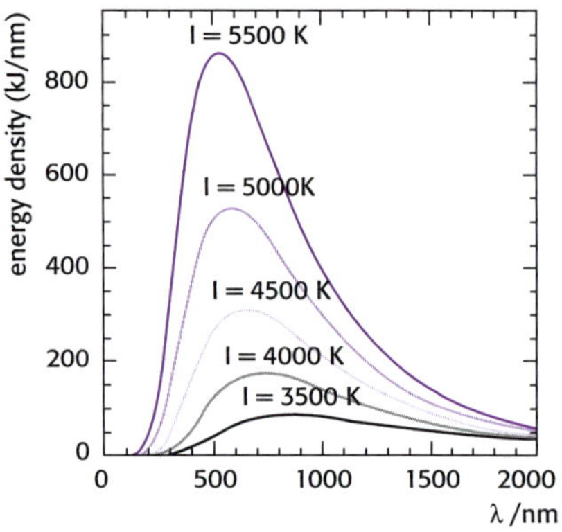

Figure 1 Black body radiation spectrum for a range of temperatures

?

By analysing the radiation curve of a hot, distant object, such as a star, its surface temperature can be estimated using the equation $\lambda_{max} T$ = constant. Doing the calculation for the Sun, for example, gives a surface temperature of just under 6000 K (see question 2 on next page.)

1 Use the graph in Figure 1 to find the wavelength of peak radiation for
   (i) T = 5500 K
   (ii) T = 3500 K.
2 Calculate the peak wavelength radiated by the Sun assuming the surface temperature is 5800 K. State which part of the electromagnetic spectrum this is found in.
3 Compare the curve for 3500 K, shown in Figure 1, with a curve for 280 K, which is similar to the Earth's mean temperature. State how the maximum energy density and peak wavelength positions differ.
4 Suggest a meaning for the physical quantity represented by the area under the black body curve.

It is very difficult to determine the Earth's mean temperature, averaged over one year. Temperatures vary significantly at different latitudes and altitudes, and from season to season. For the next question assume a value of 15 °C.

5 Write down the Earth's average temperature in kelvin (K) and calculate the peak wavelength radiated, assuming black body behaviour. In which part of the electromagnetic spectrum is this?

Solar radiation

To investigate this more closely, one needs to start with some definitions:

Power is the rate at which energy is transferred. A power of one watt (1 W) means a rate of energy transfer of one joule per second (1 J s^{-1}).

The **solar constant**, *S*, is the average solar power per m^2 arriving at the top of the Earth's atmosphere. Its value is 1.37 kW m^{-2}. In other words, 1.37 kilojoules of solar energy reach each square metre of the top of the Earth's atmosphere every second.

The Sun's **luminosity** is the *total* power radiated by the Sun. Although it fluctuates very slightly, the solar constant is normally assumed to be reasonably stable at 1.37 kW m^{-2}. A knowledge of the solar constant enables one to estimate the luminosity of the Sun and hence its surface temperature.

The Sun is approximately 149×10^6 km from the Earth. Astrophysicists call this distance one **astronomical unit** (1 AU). At the distance of the Earth, all the power emitted by the Sun is spread out over a sphere of radius 1 AU.

The solar power flux at the Earth's surface is significantly less than the solar constant.

There is a significant amount of absorption and reflection by the atmosphere. Also, the actual value of the solar power flux in any one region varies depending on the latitude, the season, and the time of day (at night it is, of course, zero). Typically, the average solar power flux at the Earth's surface at mid-European latitudes ≈ 150 W m^{-2}.

Compare your value with this estimate and list some reasons why your value is different. Consider how you might improve the experiment to get a more accurate result. If time allows, redesign your experiment and try again.

1 Use the value of the solar constant *S*, given above, to calculate the Sun's luminosity.

To estimate the temperature at the Sun's surface, you first need to estimate the Sun's surface area.

2 The angular diameter of the Sun measured from the Earth is 0.52°. Calculate
   (i) the angular diameter in radians,
   (ii) the Sun's radius in metres,
   (iii) the surface area of the Sun, assuming that it is a sphere.

The **Stefan-Boltzmann** law gives the relationship between the luminosity *L* of a star, the surface area *A*, and the surface temperature *T*:

$$L = \sigma A T^4$$

where σ is the **Stefan-Boltzmann constant**: 5.67×10^{-8} W m^{-2} K^{-4}

3 Calculate the total power and the power per m^2 radiated by the Earth at a temperature of 228 K. You can assume the Earth is a sphere of radius 6400 km.

Mathematical physics: a simple energy balance model

If the Earth is at thermal equilibrium, this means that it is emitting energy at the same rate as it is absorbing it. So:

power radiated by the Earth = power received from the Sun

Assuming that the Earth is equivalent to a disc of radius *r*, normal to the incoming solar radiation:

total incoming power = $S \times \pi r^2$
(*S* is the solar constant)

Some of this radiation is reflected straight back into space. This fraction is called the **albedo**, α.

So: total power absorbed by the Earth = $S(1 - \alpha)\pi r^2$ (1)

The next stage is to calculate the power radiated by the Earth, assuming that this is emitted over a sphere of area $4\pi r^2$, and that this behaves as a black body radiator. For this, we apply the Stefan-Boltzmann law for an object at temperature *T*:

So: power radiated by the Earth = $\sigma \times 4\pi r^2 \times T^4$ (2)

So: $4\sigma\pi r^2 T^4 = S(1 - \alpha)\pi r^2$ (3)

So:

$$T = \sqrt[4]{\frac{S(1-\alpha)}{4\sigma}}$$

Using $S = 1370$ W m^{-2}, $\alpha = 0.3$, and $\sigma = 5.67 \times 10^{-8}$ Wb m^{-2} K^{-4} gives

$$T = \sqrt[4]{\frac{1370 \times (1-03)}{4 \times 5.67 \times 10^{-8}}}$$

So: $T = 255$ K

This simple analysis predicts the temperature of the Earth to be lower than it actually is. So the model needs to be modified. The atmosphere (and the greenhouse gases in it) have an important role in keeping the planet at the much more acceptable global annual mean temperature of ≈288 K.

The Earth with its atmosphere is not a perfect black body. **Emissivity**, ε, is defined as the ratio of power radiated by an object to the power radiated by a black body at the same temperature. Allowing for this modifies equation 2 to this:

$$\text{power radiated by the Earth} = 4\varepsilon\sigma\pi r^2 T^4$$

You should be able to show that the Earth's temperature will be given by:

$$T = \sqrt[4]{\frac{S(1-\alpha)}{4\varepsilon\sigma}}$$

1 Calculate the emissivity of the Earth assuming the mean temperature is 288 K.
2 Assuming that the Earth is a perfect black body, radiating at 288 K, calculate the power flux radiated.
3 Compare your answer to the previous question with the lower flux value radiated for a black body at 255 K.

Modelling energy absorption

In this section, the aim is to develop a model to explain how the atmosphere helps to keep the Earth warm. Before studying this section, make sure that you have read Chapter 7, dealing with the mathematical models used for simple harmonic oscillations.

Certain molecules in the atmosphere absorb particular wavelengths radiated by the Earth and the Sun. However, the Earth is significantly cooler than the Sun, so its black body radiation spectrum is shifted to regions of longer wavelength compared to that of the incoming solar radiation. Because of the shift in peak wavelength, molecules in the atmosphere absorb more of the outgoing (terrestrial) radiation than they do incoming (solar) radiation.

Several gases in the atmosphere absorb energy in these wavelengths but the most important ones are water vapour, carbon dioxide, and methane. All occur naturally, although human activities have significantly increased the concentrations of the first two.

To see how energy is absorbed, carbon dioxide will be used as an example (Figure 2).

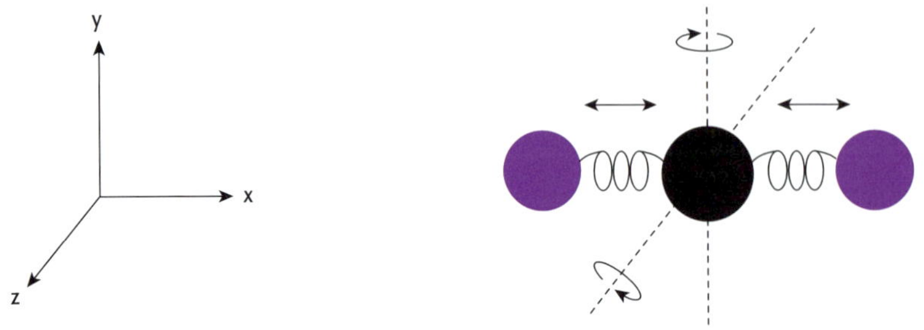

Figure 2 Carbon dioxide molecule and possible modes of translation, vibration, and rotation

Inquiry: temperatures throughout the Solar System

The surface temperature of a planet depend on several factors, including:

- the distance from the Sun
- the composition of the atmosphere
- the albedo (the proportion of incoming solar energy reflected back into space)
- the emissivity.

Choose a planet or moon in the Solar System and carry out some research to find out about the temperature range you would expect to find there. Present your findings to your group in a suitable way.

You should include the following:

- An estimate of the average solar power flux for your chosen planet or moon. Show how you calculated this.
- Details of the atmosphere, including its composition

A good approach would be to research the key data for your object then use a spreadsheet or graphical calculator to develop your model.

Carbon dioxide has a linear molecule with a shape as shown in Figure 2. Each oxygen atom (purple) is joined to the carbon atom (black) by a double bond. Essentially, this bond is formed by clouds of electrons: see the chemistry course companion for further details. In the model in Figure 2, it is convenient to think of the bonds as a spring. When this spring is stretched or compressed, forces try to return the atoms to their equilibrium position.

The above molecule has kinetic energy which may be considered in separate modes:

(i) three linear modes due to movements in the *x*, *y*, and *z* directions;

(ii) two rotational modes due to rotation around the two dotted axes illustrated;

(iii) four vibrational modes (Figure 3), each one excited by a different wavelength.

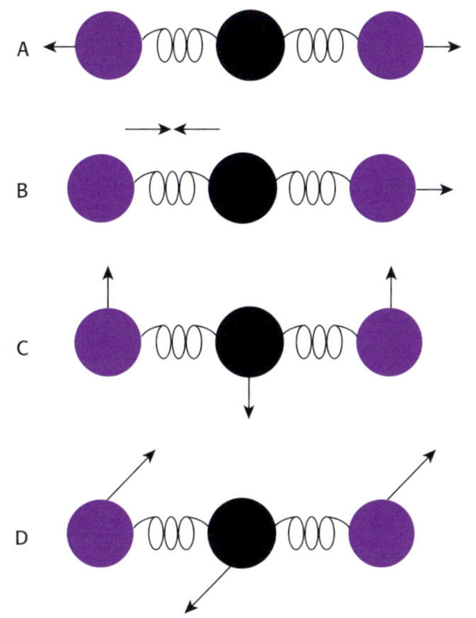

Figure 3 The four vibrational modes of carbon dioxide

1 Describe carefully the movement of each atom in the four modes of vibration shown in Figure 3.

In a linear molecule such as carbon dioxide, the number of vibrational modes is given by $3N - 5$, where N is the number of atoms in the molecule. In a non-linear molecule such as water or methane, the number of modes is given by $3N - 6$.

2 Calculate the number of vibrational modes in
 (i) a water molecule (non-linear)
 (ii) a methane molecule (also non-linear).

3 Draw a simple diagram to illustrate these modes in a water molecule.

From Chapter 7, you will know that the frequency of vibration of a spring is given by this equation:

$$f = \frac{1}{2\pi}\sqrt{\frac{k}{m}}$$ where k is the spring constant.

4 Compare modes A and C in Figure 3. Explain whether it is the stretching mode or the bending mode that
 (i) requires more energy
 (ii) has a higher frequency. (Hint: compare the value of k in each case.)

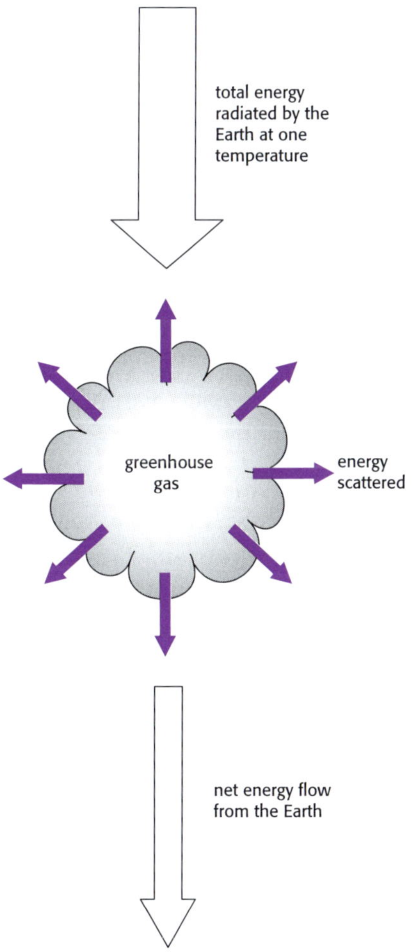

Figure 4 Absorption of energy by greenhouse gases leads to scattering

The rotational and vibrational modes have a **natural frequency** which corresponds to some of the wavelengths radiated by the Earth, particularly those in the infrared part of the spectrum. The carbon dioxide molecules absorb energy from precise wavelengths, and the oscillatory modes are excited. This is an example of **resonance** (see page 76). The internal energy of the molecules increases and they re-radiate this energy in random directions: this is called **scattering**. The net flux of outward energy is therefore reduced, as some of this scattered energy is re-absorbed by the Earth. In this way, the planet warms up until a new equilibrium temperature is reached.

Data-based question: greenhouse gas absorption spectrum

The graph below shows the transmission of different wavelengths from a particular waveband through carbon dioxide gas.

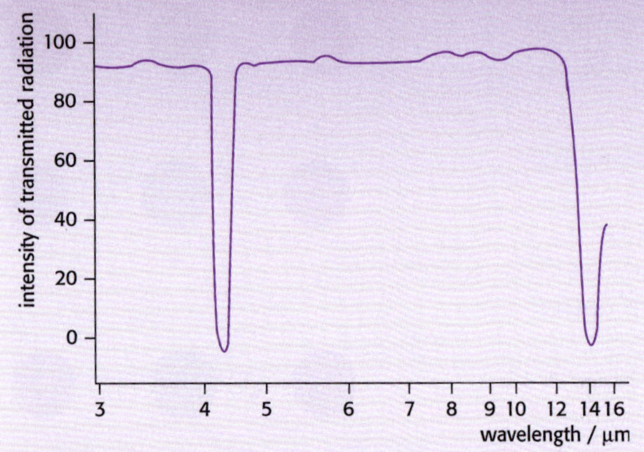

Figure 5 Carbon dioxide absorption spectrum

1. Write down how many μm there are in 1 m.
2. State the range of wavelengths illustrated in the graph and to which waveband in the electromagnetic spectrum they belong.
3. Write down the wavelengths of radiation that are absorbed most efficiently by carbon dioxide gas.
4. Outline the different molecular mechanisms possibly responsible for absorption in each wavelength.
5. Before going any further, refer back to the rather simplistic explanation of the greenhouse effect at the start of this chapter. You should now be able to write a much more detailed and physically accurate description of the process.

Evidence for an enhanced greenhouse effect

The enhanced greenhouse effect is another name for the global warming caused by the activities of humans.

The Inter-Governmental Panel on Climate Change (IPCC), a body within the UN, published its fourth assessment report in February 2007. It stated the following:

"Warming of the climate system is unequivocal, as is now evident from observations of increases in global average air and ocean temperatures, widespread melting of snow and ice, and rising global mean sea level".

The report backed up this statement with evidence from temperature records going back over nearly 200 years. Some of this data is included below. The range of uncertainty in any value is estimated and included in square brackets – [ ].

1. *Eleven of the last twelve years (1995–2006) rank among the 12 warmest years in the instrumental record of global surface temperature (since 1850).*
2. *The updated 100-year linear trend (1906–2005) of 0.74 [0.56 to 0.92] °C is therefore larger than the corresponding trend for 1901–2000 given in the TAR (Third Assessment Report) of 0.6 [0.4 to 0.8] °C.*
3. *The linear warming trend over the last 50 years (0.13 [0.10 to 0.16] °C per decade) is nearly twice that for the last 100 years.*
4. *The total temperature increase from 1850–1899 to 2001–2005 is 0.76 [0.57 to 0.95] °C.*
5. *Urban heat island effects are real but local, and have a negligible influence (less than 0.006 °C per decade over land and zero over the oceans) on these values." (IPCC, 2007, p4)*

?

1. State three of the observed consequences of the enhanced greenhouse effect reported in the IPCC assessment on this page.
2. Use the data in point 2 to calculate the mean temperature increase per decade from 1906 to 2005.
3. Use the data in points 2 and 3 to plot
   (i) the global mean temperature deviation from the 1956 value on a graph over the period 1906–2005
   (ii) on the same axes, plot the global mean temperature deviation from the 1956 value averaged from 1956 to 2005.
4. Use your graph from question 3 to outline any changes in the rate of temperature rise in recent decades.
5. Explain
   (i) the meaning of the term "urban heat island effect" used in point 5, and
   (ii) why this point was included with the other data.

Modelling an enhanced greenhouse effect

Developing an accurate model to predict the consequences of the enhanced greenhouse effect is very difficult. Many of the models produced so far have led to different conclusions. A sensible approach is to look at the predictions of as many of the models as possible, then try to reach a consensus. However, this is not quite the same as reaching an average, because the scientific community may judge some models to be more accurate than others.

In the simple model discussed earlier in the chapter, this formula was used to predict the average global temperature:

$$T = \sqrt[4]{\frac{S(1-\alpha)}{4\varepsilon\sigma}}$$

S, α, and ε are all variables which can change over time.

The value predicted by this equation is too low by about 35 K. We now know that the atmosphere and its composition have a significant effect on the temperature. Also, climate is influenced by other mechanisms, such as the fluid dynamics of the atmosphere and oceans, combined with the effects of the Earth's spin. There is another process that is in action: positive feedback.

Positive feedback

To understand this, consider the following examples:

- Snow and ice reflect significant amounts of solar radiation straight back into space. Global warming has led to melting of the polar ice caps, which reduces the Earth's albedo. This in turn increases the absorption of solar radiation, which increases global warming.
- The oceans store huge quantities of dissolved carbon dioxide. Oceanic warming reduces the solubility of carbon dioxide, so gas is released into the atmosphere. This in turn causes more global warming.

In both cases, a change in one factor causes a change in another which, in turn, "feeds back" to magnify the original change.

?

Global warming may well lead to desertification (the formation of deserts) in certain parts of the planet. Outline an argument of how this may lead to:
(a) positive feedback, increasing the effects of global warming, or
(b) negative feedback, reducing the effects.

Assumptions for a simple model

This model is based on an idea from the climateprediction.net website. A spreadsheet or graphical calculator is used to model changes in the Earth's temperature over time.

The model assumes the following:

1. The initial power flux of incoming solar radiation at the Earth's surface is 390 W m^{-2}.
2. The outgoing power flux depends only on the temperature of the Earth, initially 288 K (the 2007 value), and is given by the Stefan-Boltzmann law:

    outgoing power flux = σT^4

3 The change in temperature of the Earth over time depends on:
- the energy difference between the incoming and outgoing radiation
- the surface heat capacity, C, of the Earth. (C is the thermal energy needed to raise the temperature of 1 m^2 of the Earth's surface by 1 K. In the model, $C = 4.0 \times 10^8$ J m^{-2} K^{-1}.)

Over a period of time, Δt, the Earth absorbs energy which causes a temperature rise ΔT.

The energy absorbed = incoming energy – outgoing energy

But: energy = power × time

So: temperature rise = (incoming power – outgoing power) × time/heat capacity

The new temperature is calculated like this:

new temperature = old temperature + temperature change

The idea is to model what would happen if the effective incoming solar radiation changes. This could happen if:
- the solar luminosity changes due to solar flares or sunspot activity
- the Earth's albedo changes
- the Earth's emissivity changes due to increased greenhouse gas concentrations in the atmosphere.

?

This question is about the enhanced greenhouse effect and climate change.

Here is some data:
solar constant = 1.37 kW m^{-2}
Earth's radius = 6.4×10^3 km
Stefan-Boltzmann constant = 5.67×10^{-8} W m^{-2} K^{-4}
Surface heat capacity of Earth = 4.0×10^8 J m^{-2} K^{-1}

(a) (i) Show that, using a simple energy balance model, the Earth re-radiates about 340 W m^{-2} of power per unit area. [3]
(ii) State two assumptions you have made. [2]

(b) (i) Use the Stefan-Boltzmann law to estimate the mean temperature of the Earth using this data. [2]
(ii) Briefly explain why the Earth is warmer than your answer above suggests. [2]

A change in the atmosphere causes the effective power flux absorbed by the Earth to increase by 10%, causing the planet to become hotter.

(c) Calculate:
(i) the new power flux absorbed by the Earth. [1]
(ii) the total energy gained per m^2 in 1 year. [1]
(iii) the total energy radiated per m^2 in 1 year, assuming that the Earth continues to radiate at 340 W m^{-2} during this time. [1]

(d) (i) Determine the temperature rise of the planet over 1 year. [2]
(ii) State one additional assumption of this model. [1]

[Total 15 marks]

Setting up the spreadsheet

If designed well, a spreadsheet will quickly re-calculate all your dependent variables when you change just one of your independent variables. It is worth taking a little time to plan your spreadsheet design before inputting your variables. An Excel spreadsheet for this simple climate model is illustrated:

	A	B	C	D	E	F
1	Modelingclimate change.					
2						
3	σ	c	Δt (years)	Δt (s)	T_0	% change in energy
4	5.67E-08	400000000	0.5	15768000	288	5
5						
6						
7	time (years)	Incoming radiation	Outgoing radiation	ΔT	T	
8	0	390.07939	390.07939	0	288	
9	0.5	409.58336	390.07939	0.7688465	288.769	
10	1	409.58336	394.26154	0.6039864	289.373	
11	1.5	409.58336	397.57044	0.4735496	289.846	
12	2	409.58336	400.17928	0.3707091	290.217	
13	2.5	409.58336	402.2305	0.2898498	290.507	
14	3	409.58336	403.8398	0.2264114	290.733	
15	3.5	409.58336	405.10023	0.1767253	290.91	
16	4	409.58336	406.0861	0.1378621	291.048	
17	4.5	409.58336	406.85643	0.1074959	291.155	
18	5	409.58336	407.45783	0.0837884	291.239	
19	5.5	409.58336	407.92707	0.0652912	291.305	
20	6	409.58336	408.29299	0.0508664	291.355	
21	6.5	409.58336	408.57825	0.0396218	291.395	
22	7	409.58336	408.80054	0.0308588	291.426	
23	7.5	409.58336	408.97374	0.0240314	291.45	
24	8	409.58336	409.10865	0.0187131	291.469	
25	8.5	409.58336	409.21373	0.0145708	291.483	
26	9	409.58336	409.29557	0.0113449	291.495	
27	9.5	409.58336	409.35929	0.0088329	291.503	
28	10	409.58336	409.40891	0.0068768	291.51	
29						

Figure 6 Spreadsheet to model climate change

1 The constants, σ and *C*, are entered in cells A4 and B4. Note the format for the value of σ.
2 The time increment, Δ*t*, is entered in cell C4 in years for ease of use. As calculations use seconds, a formula is entered in cell D4 to covert years to seconds = $C$4*365*3600. The term $C$4 refers to the number in cell C4. The "$" s are used in Excel to "fix" the reference – see below.
3 The initial temperature is entered into cell E4.
4 The number in cell F4 is the % increase or decrease in the incoming solar radiation.
5 Row 8 is used to set up the initial conditions.
6 The formula = $E$4 is used in cell E8.
7 Cell C8 is calculated using the Stefan-Boltzmann law; = $A$4* E8^4.
8 As the incoming radiation must initially be equal to the outgoing radiation, the formula = C4 is entered into cell B4.
9 Row 9 makes the first set of calculations following the increase (or decrease) in the incoming solar radiation.

?

1 State the formula used in cells D9 and E9.
2 Use the graph to determine
   (i) the new steady state temperature
   (ii) the time to reach this new steady state.

Change the time interval to (i) 0.25 years and (ii) 1 year.

3 Describe any similarities and differences between your new graphs and the one illustrated in Figure 7.
4 State, for each new time interval
   (i) the final steady state temperature
   (ii) the time to reach this steady state.

The formula in cell C9 is = $A$4*E8^4

5 Explain what would happen when filling down if the formula = A4*E8^4 was used. (You can test this out by trying it).
6 Explain the use of the "$" notation throughout the spreadsheet – it is very important to understand this when designing spreadsheets.
7 Explain whether you think this kind of model is useful for scientists to attempt predictions of future climate change.

10 The time interval chosen in cell C4 is generated in cell A9 using the formula = A8 + $C$4.
11 The increased solar power in cell B9 is generated using the formula = $B$8*(1 + $F$4/100)
12 The outgoing radiation in cell C9 is assumed to be constant in the model over the selected time interval. It is calculated using the formula = $A$4*E8^4. The temperature in cell E8 is the previous temperature.
13 The temperature change and new temperature are generated in cells D9 and E9.
14 Highlight the cells A9 to E9, grab the small black square in the bottom right-hand corner of cell E9 and fill down for 20 rows.
15 Plot a graph of temperature (y-axis) against time (x-axis). To do this highlight the cells A7 to A28, press and hold the CTRL key and highlight cells E7 to E28. Click the chart wizard button,  and select the various options to complete the graph.

The graph for this model is shown below.

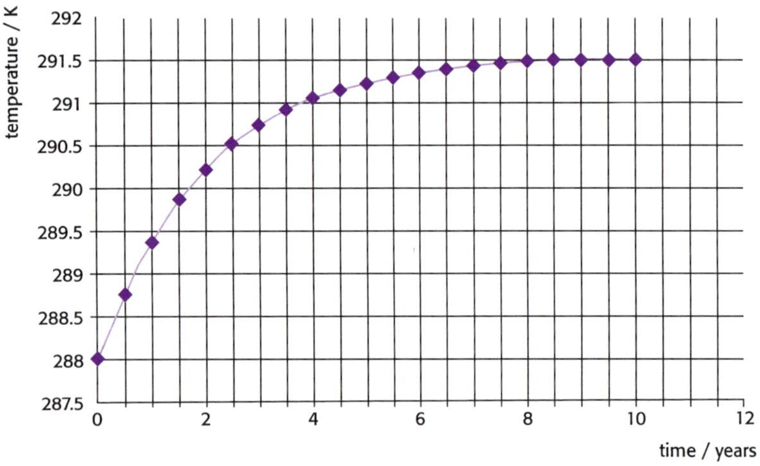

Figure 7 Graph of temperature against time for a simple climate model

This model is very flexible. Try different scenarios and find out what the model predicts. For example:

- Change the % change in solar energy to +8% and −5%.
- Change the initial temperature.
- Change the time interval.
- Try to improve the sophistication of the model by including albedo and emissivity in the calculations.
- Try to find out if there is a relationship between atmospheric carbon dioxide concentration and emissivity. If so, this could be included in the model.

An interesting feature of this model can be explored if you use a large time interval. For example, a time interval of 5 years gives the graph below.

This suggests an unstable model, with an oscillating temperature. Think about why the model predicts this result. Does temperature change in discrete time steps, or is it a continuous change? You might want to investigate this feature of the model further.

?

This question is about black body radiation and the greenhouse effect.

The power absorbed from the Sun at the surface of the Earth is 240 W m^{-2}.

(a) Calculate the mean temperature of the Earth. [4]

(b) State any assumptions you made in your calculation. [2]

The solar constant is measured to be 1.37 kW m^{-2}.

(c) Estimate the mean albedo of the Earth and state any assumptions you have made. [6]

(d) Describe the *greenhouse effect* and compare it with the *enhanced greenhouse effect.* [6]

[Total 18 marks]

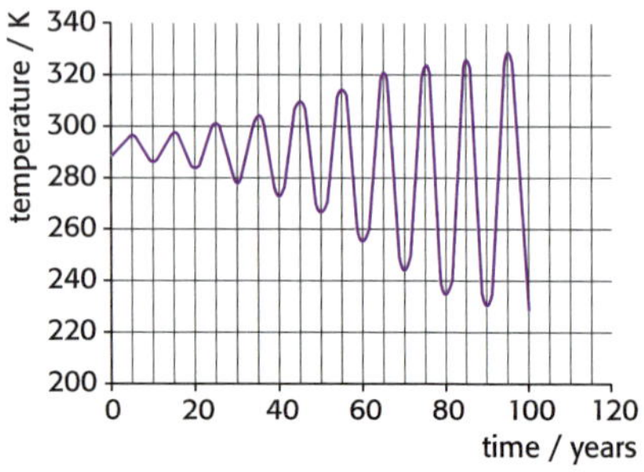

Figure 8 An unstable model of climate change

20 Climate change

This chapter will explore the evidence for the causes of global warming, and then consider some of the possible effects of climate change. The consequences of climate change are so serious on a global scale that tackling global warming is one of the most important challenges facing us all. Failure to do so could drastically affect the entire planetary ecosystem.

Some scientists think that the possible effects of the enhanced greenhouse effect have been underestimated and that climate change may cause even more drastic effects than predicted.

The ice age is coming, the sun is zooming in
Engines stop running and the wheat is growing thin
A nuclear error, but I have no fear
London is drowning—and I live by the river"
(The Clash, Joe Strummer / Mick Jones, "London's Calling", 1979)

Thinking about science: interpretation of climate data

Two things are happening:

- global temperatures are rising, and
- global concentrations of carbon dioxide are increasing.

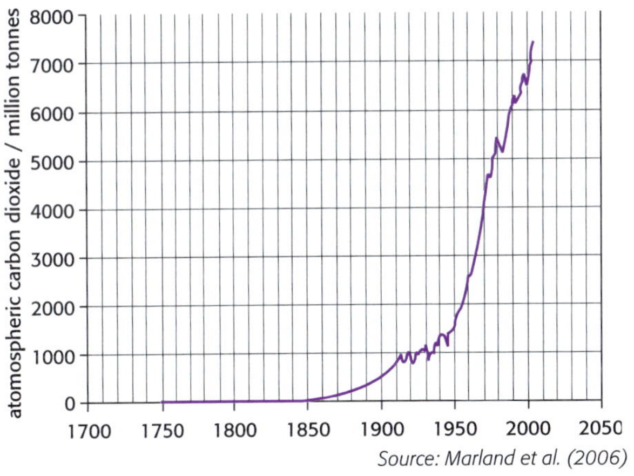

Figure 1 Atmospheric carbon dioxide levels since 1750

One of the important questions asked in the TOK course is "How do we know what we know?" The answer – from the scientific point of view – is found in data.

Figure 1 shows the estimated mass of atmospheric carbon dioxide since 1750, and Figure 2 shows the variations in global annual mean temperatures for the period 1881–2005 from the mean value averaged over 1950–1975.

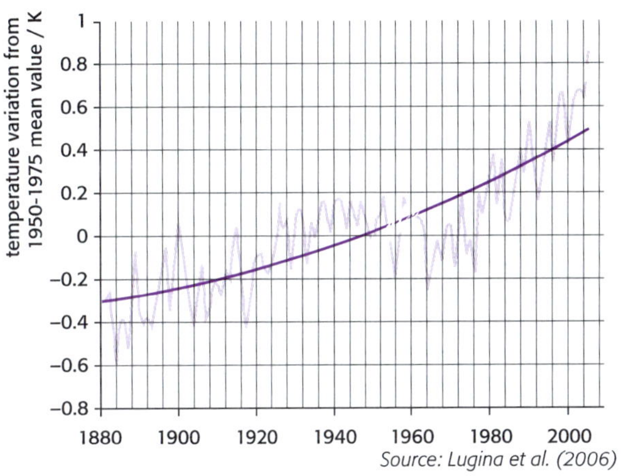

Figure 2 Monthly surface air temperature time series, area-averaged over the 30° latitudinal belts of the globe, 1881–2005

1 Describe the trends shown in Figures 1 and 2.
2 If taken at face value, outline any evidence from the graphs that suggests there is a correlation between the two sets of data.

Question 2 is quite difficult to answer without knowing the specific data values. One way to see if there is indeed a correlation between the two data sets would be to plot the temperature variation against the carbon dioxide levels and see whether there is any meaningful relationship (Figure 3).

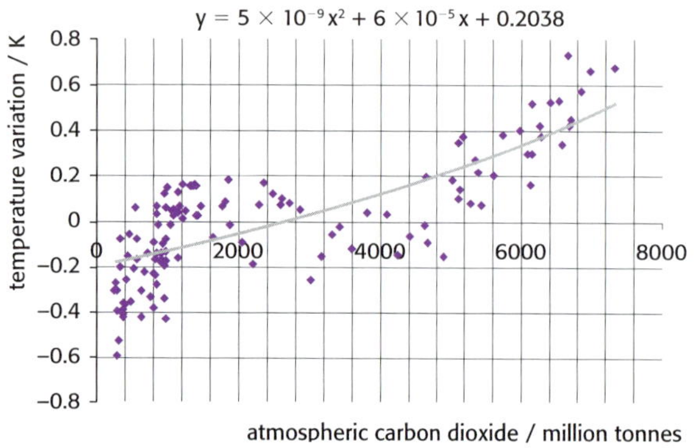

Figure 3 Plot of mean temperature variation against atmospheric carbon dioxide levels from 1881 to 2003

A problem is that the mean temperature may well be chaotic. We can predict values using the deterministic, mechanical rules applied to the fluid dynamics of the atmosphere, but when we apply these rules to such a complex system involving so many variables (atmospheric pressure, wind speed, and so on), the outcome appears chaotic. **Chaos theory** is a relatively new branch of science, and it has produced much fascinating research.

To see whether there is any correlation between temperature and carbon dioxide levels, and so decide whether recent changes may be **anthropogenic** (that is, originating in human activity) we need to go much further back in time.

Long-term trends

The data illustrated in Figures 1–5 was determined from ice core records drilled at the Russian Antarctic base Vostok (Figure 4). In an example of international cooperation between Russia, France, and the USA, several ice cores were drilled between 1987 and 1998, the deepest going more than 3600 m below the surface.

Ice core data are unique: every year the ice thaws and then freezes again, forming a new layer. Each layer traps a small quantity of the ambient air, and sophisticated analysis of this trapped air can determine mean temperature variations from the current mean value and carbon dioxide concentrations. The depth of the cores obtained at Vostok means that a data record going back more than 420 000 years has been built up through much painstaking and careful analysis. In Figure 5 the time axis goes back in time as you move to the right, and is measured in kiloyears (kyr) BP. The term BP means *before present* and refers to the arbitrary date of 1950 when radioisotope records were first established.

The two graphs in Figure 5 show many similarities, which suggests that there is a correlation between the two sets of data. The data have been analysed precisely, and the trends are summed up on the Carbon Dioxide Information Analysis Centre website, as quoted below.

There is a close correlation between Antarctic temperature and atmospheric concentrations of CO_2.... The extension of the Vostok CO_2 record shows that the main trends of CO_2 are similar for each glacial cycle. Major transitions from the lowest to the highest values are associated with glacial–interglacial transitions. During these transitions, the atmospheric concentrations of CO_2 rise from 180 to 280–300 ppmv.... The extension of the Vostok CO_2 record shows the present-day levels of CO_2 are unprecedented during the past 420 kyr. Pre-industrial Holocene levels (~280 ppmv) are found during all interglacials, with the highest values (~300 ppmv) found approximately 323 kyr BP. When the Vostok ice core data were compared with other ice core data ... for the past 30 000–40 000 years, good agreement was found between the records: all show low CO_2 values [~200 parts per million by volume (ppmv)] during the Last Glacial

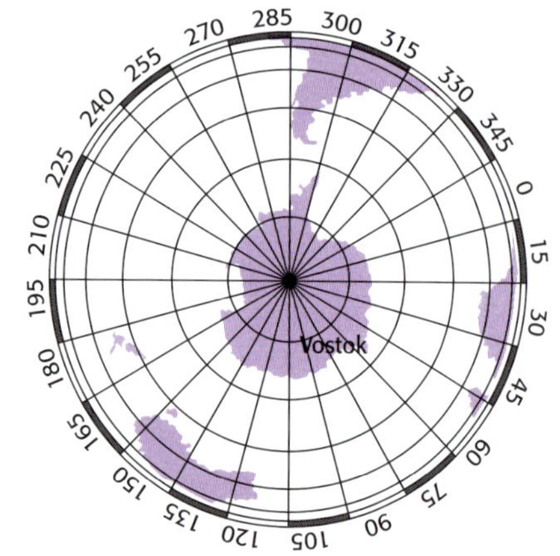

Figure 4 Map showing the location of the Vostok Antarctic base

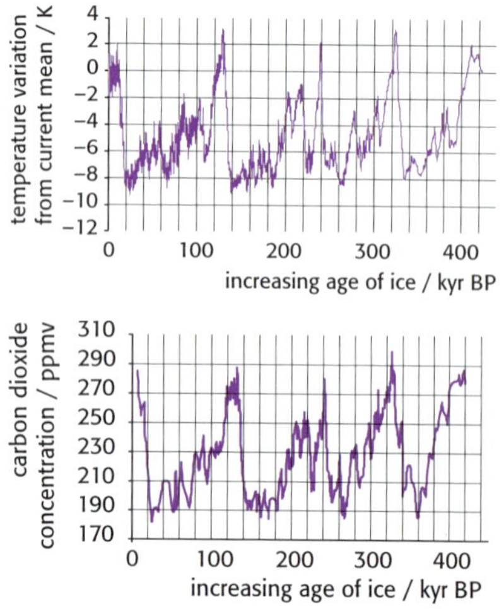

Figure 5 Temperature and carbon dioxide data from the Vostock ice cores. Source: (top) Petit *et al.* (2000); (bottom) Barnola *et al.* (2003)

Maximum and increased atmospheric CO_2 concentrations associated with the glacial–Holocene transition.... these measurements indicate that, at the beginning of the deglaciations, the CO_2 increase either was in phase or lagged by less than ~1000 years with respect to the Antarctic temperature, whereas it clearly lagged behind the temperature at the onset of the glaciations.

Source: Barnola et al. (2003)

Use the temperature and carbon dioxide graphs determined from the Vostock Ice cores, and the passage above, to answer the questions in the margin.

The balanced view

The IPCC report (mentioned on page 264) suggests not only that global warming is leading to climate change, but that human activities are directly to blame. It concludes that:

Most of the observed increase in globally averaged temperatures since the mid-20th century is very likely due to the observed increase in anthropogenic greenhouse gas concentrations. Discernible human influences now extend to other aspects of climate, including ocean warming, continental-average temperatures, temperature extremes and wind patterns. (IPCC, 2007, p.8)

Some scientists argue against this, but in comparison with the work of the IPCC there is very little supporting evidence for their point of view. Professor Richard Lindzen of MIT acknowledges that:

- global mean temperatures are 0.5 K higher than a century ago,
- carbon dioxide levels have risen over the past two centuries, and
- carbon dioxide is a greenhouse gas.

However, he goes on to say that there is much uncertainty in making predictions, because:

...the climate is always changing; change is the norm. Two centuries ago, much of the Northern Hemisphere was emerging from a little ice age. A millennium ago, during the Middle Ages, the same region was in a warm period. Thirty years ago, we were concerned with global cooling.

Distinguishing the small recent changes in global mean temperature from the natural variability, which is unknown, is not a trivial task. All attempts so far make the assumption that existing computer climate models simulate natural variability, but I doubt that anyone really believes this assumption.

We simply do not know what relation, if any, exists between global climate changes and water vapour, clouds, storms, hurricanes, and other factors, including regional climate changes, which are generally much larger than global changes and not correlated with them. Nor do we know how to predict changes in greenhouse gases. This is because we cannot forecast economic and technological change over the next century, and also because there are many man-made substances whose properties and levels are not well known, but which could be comparable in importance to carbon dioxide.

What we do know is that a doubling of carbon dioxide by itself would produce only a modest temperature increase of one degree Celsius. Larger projected increases depend on "amplification" of the carbon dioxide by more important, but poorly modelled, greenhouse gases, clouds and water vapour. (Lindzen, The press gets it wrong, our report doesn't support the Kyoto treaty, 2001)

?

1 Explain whether the graphs as shown can be used to determine which comes first: changes in carbon dioxide levels or changes in temperature.

2 From the passage:
   (a) Determine which comes first: the change in temperature or the change in carbon dioxide levels.
   (b) Explain how a change in this variable causes the other variable to change.

3 What evidence from the passage or in the graphs suggests that the current interglacial warmer period is different from the previous natural cycles?

Inquiry

Fossil and geological records over many thousand of years suggest there has been a natural cycle of warming and cooling, leading to ice ages and corresponding warmer periods in between. Some possible causes of this oscillation in global temperatures are:

- periodic wobbling of the Earth's axis of rotation
- increased or decreased solar luminosity due to solar flare or sunspot activity
- changing volcanic activity.

Choose one of these ideas and research it thoroughly. Describe the physical mechanisms involved, and explain how these explain the historical cooling–warming cycles deduced.

Physics issues: chaos and the effects of climate change

Our climate is driven by solar radiation and the Earth's spin. The changes in the oceans and the atmosphere are complex, and difficult to predict. In principle, all these changes are deterministic: that is, they obey well-known mechanical and thermodynamic rules, but the interactions between all possible climate "drivers" can only be modelled to some extent on very large computers.

Inquiry

The following are some predictions of climate change:

- temperature change
- erratic precipitation
- sea level rise
- increased number and strength of tropical cyclones.

There is a tendency to blame all unusual climate phenomena on global warming, when there is no evidence to do so. Extreme weather has occurred throughout the Earth's history, but can any recent events be blamed on climate change?

Consider some of the predictions of climate change made above, and carry out some research on a particular region to see whether there is any evidence of climate change. Some examples of specific events are given below, but you may want to choose something relevant to where you live.

- Hurricane Katrina caused great destruction in New Orleans in the Southern USA in August 2005.
- The summer of 2003 was the hottest ever on record in Europe – possibly the hottest for 500 years. In addition, the ten hottest years have occurred since 1990.
- Much of Queensland and New South Wales in Australia have received significantly lower rainfall over the last ten years than would be expected from long-term climate records, leading to severe water shortages and an increased risk of bush fires.
- Early 2007 saw excessive flooding in Ethiopia and Somalia in North-East Africa, displacing over a quarter of a million people. This followed a period of drought affecting around 11 million people in Kenya, Ethiopia, Eritrea, Somalia, and Djibouti.
- The Chinese government estimates that floods in Fujian province in June 2006 were the worst for 50 years.
- Many residents in Mumbai, India, reported in July 2005 that the flooding was the worst they could remember.
- In July 2001, more than a million peasant farmers in Honduras, Nicaragua, El Salvador, and Guatemala were severely affected by crop failure due to a lack of rain. At the same time around 9000 Miskito Indians living on the Atlantic coast of Nicaragua had their rice crops destroyed by flooding.

Make sure your research is thorough, and that you avoid looking at just individual events, though the list above provides some possible starting points. You should consider climatic patterns and trends over the longer term. The IPCC website is an excellent source of climate data – see http://www.ipcc.ch/

Data-based question: rising sea levels

Sea level rises provide one of the most immediate and direct consequences of global warming. Two islands of the Pacific nation of Kiribati have already been covered by rising seas, and several other nations are under threat. Most of the Maldives, an island nation in the Indian Ocean, are less than 2 m above sea level, and a sea level rise of 1 m would cause severe difficulties. However, it is not only small island nations that are under threat. More than 70% of the world's population live on low-lying coastal plains, which include 11 of the 15 biggest cities.

You can find out about the threat posed by rising sea levels where you live with the website http://flood.firetree.net/

Source www.greenpeace.org.uk

Figure 6 Pita Meanke, of Betio village, stands beside a tree as he watches the "king tides" crash through the sea wall into his family's property, on the South Pacific island of Kiribati

The data in Table 1 are from the IPCC fourth assessment report.

Table 1 Estimated rate of sea level rise due to different sources

Source of sea level rise (Observed rate of sea level rise and estimated contributions from different sources)	Rate of sea level rise (mm per year)	
	1961–2003	1993–2003
1. Thermal expansion	0.42 ± 0.12	1.6 ± 0.5
2. Glaciers and ice caps	0.50 ± 0.18	0.77 ± 0.22
3. Greenland ice sheets	0.50 ± 1.2	2.1 ± 0.7
4. Antarctic ice sheets	1.4 ± 4.1	2.1 ± 3.5
Sum of individual climate contributions (above) to sea level rise	1.1 ± 0.5	2.8 ± 0.7
Observed total sea level rise	1.8 ± 0.5	3.1 ± 0.7
Difference (observed minus sum of estimated climate contributions)	0.7 ± 0.7	0.3 ± 1.0

Source: IPCC (2007, p5)

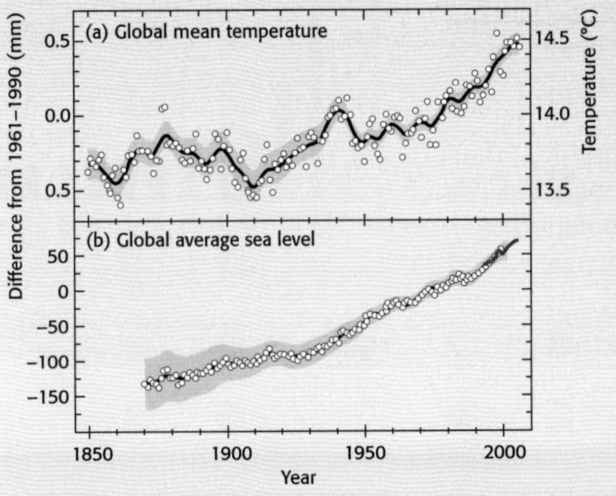

Figure 7 Sea level data from the IPCC report

1 Compare the observed mean sea level rise during 1961–2003 with that for 1993–2003. Suggest a reason (or reasons) for the difference.
2 Describe the trends shown in the graphs in Figure 7 and suggest whether there is a correlation between the global mean temperature and sea level rises.

Some of the sea level rise is due to melting ice sheets in Greenland and Antarctica.

3 Calculate the percentage uncertainties from both time periods in the sea level rise due to (a) the Greenland Ice sheet and (b) the Antarctic ice sheet.
4 Comment on the value of these uncertainties, compared with data you might normally measure in the lab.
5 Suggest a reason for the reduction in these uncertainties over the two time periods.
6 Explain the apparent discrepancies between the sum of the individual values in rows 1–4 in Table 1 with the sum of the individual climatic contributions and the observed sea level rise.

A major contribution to sea level rise is thermal expansion. Volume expansivity, γ, is defined in the equation $\Delta V = \gamma V \Delta T$, where ΔV is the change in original volume V caused by a temperature change ΔT.

7 Deduce the units of volume expansivity.

Water has a rather unusual expansivity in that it contracts between 0 °C and 4 °C before expanding. Use $+210 \times 10^{-6}$ (in SI units) as the mean value of γ for water in the following questions.

8 A swimming pool of length 25 m and width 10 m is filled to an average depth of 1.5 m. Calculate the increase in depth due to volume expansion when the temperature rises by 10 K.

The Indian Ocean is estimated to have a total area of 73 556 000 km^2 and a volume of 292 131 000 km^3.

9 (a) Calculate the average depth of the Indian Ocean.
(b) Estimate the mean sea level rise in the next 50 years, assuming a temperature rise of 0.16 K per decade.
10 List some of the assumptions you made in your answer to question 9.

You can decide for yourself whether the people of the Maldives should be concerned. It is quite clear that decisive, international action is needed in the short term in order to reduce the effects of global warming.

Physics issues: global efforts to reduce climate change

Reaching the conclusion that the current phase of global warming is most likely anthropogenic is a big step forward. The unanswered question is: Can anything be done in time to slow down and reverse any harmful climate change? If changing practices are to have any effect, the nations of the world need to unite and work together in planning and initiating concrete action.

The graph in Figure 8 illustrates the top ten carbon dioxide emitters in the world. It is no surprise that these countries have some of the largest economies and energy needs. Many are OECD countries, but China and India are developing countries with rapidly growing economies. Any reductions in carbon dioxide output from all these countries will clearly have the most significant effect on the global total.

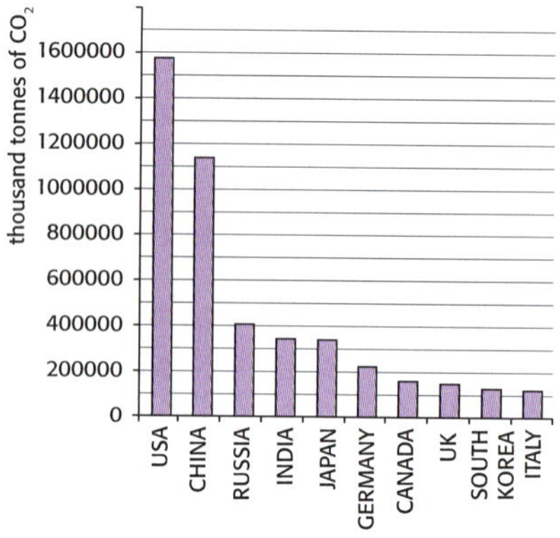

Figure 8 The world's top ten carbon dioxide emitters in 2003

A similar graph of carbon dioxide emissions per person (Figure 9) shows quite a different story. The second set of countries forms a different group. Only the USA is found in both. These countries are arguably the most inefficient in their energy production.

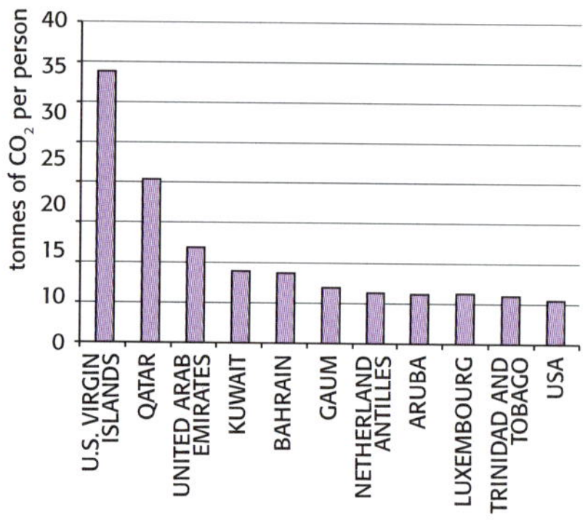

Figure 9 Per capita carbon dioxide production

The Kyoto Protocol was first approved in December 1997 by more than 160 of the world's countries, in Kyoto, Japan. The basic principle is for developed countries to reduce their greenhouse gas emissions back to 1990 levels during the period 2008–2012. The protocol allows developing countries still to increase output, as they are not responsible for the current situation, but it is hoped that China and India will voluntarily cut their emissions, though they have no legal obligation to do so under the current treaty.

It is still to be seen how effective this agreement will be, but it is widely believed to be the only way forward to reduce overall greenhouse gas emissions.

Inquiry: global efforts to reduce climate change

Carry out some research into the aims of one of the following global agreements or international organizations and present your findings back to your class. In each case consider the guiding questions included.

1 The Kyoto Protocol
  (a) Which countries have not signed up, and why?
  (b) What is the clean development mechanism (CDM)?
  (c) What are certified emission reductions (CERs), and how are they used?
  (d) Has any progress been made in reducing greenhouse gas emissions to date?

2 Asia-Pacific Partnership on Clean Development and Climate (AP6)
  (a) Which countries have signed this agreement?
  (b) What are its aims?
  (c) How will its objectives reduce climate change?

3 The Inter-Governmental Panel on Climate Change (IPCC)
  (a) How is this body organized, and who funds it?
  (b) What are its aims?
  (c) What has it achieved so far?

4 The United Nations Framework Convention on Climate Change (UNFCC)
  (a) What is the role of this body within the United Nations?
  (b) What has it achieved?
  (c) How does it support efforts to reduce global warming?

The following websites should be useful, but try to use several sources in your research.

- The United Nations: www.un.org/
- UNFCC: www.unfccc.int/2860.php
- IPCC: www.ipcc.ch/
- AP6: www.asiapacificpartnership.org/
- Kyoto Protocol: www.unfccc.int/kyoto_protocol/items/2830.php

1 This question is about the evidence that links increased levels of greenhouse gases to global warming.

(a) The Vostok and similar ices cores provide evidence of the climate in the past. Two useful measurements have been atmospheric carbon dioxide levels and temperature variations. With reference to how the carbon dioxide is trapped, briefly outline how historical carbon dioxide levels can be determined. [3]

(b) Describe any relationship found between carbon dioxide levels and temperature variation. [2]

(c) The records indicate several previous ice ages and inter ice age warm periods. We are currently in a warmer period. Explain why many climate scientists think that this inter ice age warm period is anthropogenic, and not a natural occurrence as has happened in the past. [2]

(d) Some scientists think that global warming is caused by natural phenomena. State two such natural phenomena, and explain how they may cause global warming. [4]

[Total 11 marks]

2 This question is about the effects of climate change, and about international efforts to reduce global warming.

(a) One prediction of global warming is that sea levels will rise. State three different causes of sea level rise, and indicate which one is predicted to cause the biggest effect. [4]

(b) Two important international efforts to reduce global warming are the Kyoto Protocol and the Asia-Pacific Partnership on Clean Development and Climate. Compare and contrast the approaches of these two international efforts in the attempt to reduce global warming. [6]

[Total 10 marks]

References

Barnola, J.-M., Raynaud, D., Lorius, C. and Barkov, N.I. 2003. Historical CO_2 record from the Vostok ice core. In *Trends: A Compendium of Data on Global Change*. Oak Ridge, TN, USA. Carbon Dioxide Information Analysis Center, Oak Ridge National Laboratory, US Department of Energy.

IPCC. 2007. *Fourth Assessment Report*, February. Geneva. Intergovernmental Panel on Climate Change.

Lugina, K.M., Groisman, P.Ya., Vinnikov, K.Ya., Koknaeva, V.V. and Speranskaya, N.A. 2006. In *Trends Online: A Compendium of Data on Global Change*. Oak Ridge, TN, USA. Carbon Dioxide Information Analysis Center, Oak Ridge National Laboratory, US Department of Energy.

Marland, G., Boden, T.A. and Andres, R.J. 2006. Global, regional, and national CO_2 emissions. In *Trends: A Compendium of Data on Global Change*. Oak Ridge, TN, USA. Carbon Dioxide Information Analysis Center, Oak Ridge National Laboratory, US Department of Energy.

Petit, J.R., Raynaud, D. Lorius, C., Jouzel, J., Delaygue, G., Barkov, N.I. and Kotlyakov, V.M. 2000. Historical isotopic temperature record from the Vostok ice core. *Trends: A Compendium of Data on Global Change*. Oak Ridge, TN, USA. Carbon Dioxide Information Analysis Center, Oak Ridge National Laboratory, US Department of Energy.

21 Technologies to reduce climate change

Improved efficiencies in the use of fossil fuels can have a significant effect in reducing greenhouse gas emissions.

We can improve energy efficiency in two ways:

- by reducing the amount of energy that we need to support our needs through technological improvements, such as generating more electricity per tonne of coal burnt, or achieving more kilometres of distance travelled by a car per litre of fuel so that less primary energy is required, and
- by changing our behaviour so that we use and waste less energy.

In addition, we need to reduce our dependence on fossil fuels. If any savings are achieved then these can be further invested in research into new technologies to generate further savings.

Energy-use projections (see Figure 3 on page 221 and Figure 6 on page 224) would, however, suggest we are still likely to be heavily dependent on fossil fuels for some time to come. All improvements in efficiency will go some way to reduce the effects on climate change, but other techniques are also needed to reduce carbon dioxide emissions. One promising technique has been to capture and store carbon dioxide at the point of production and prevent its release into the environment.

Efficiency is the key

The idea behind trying to improve the efficiencies of thermal power stations and vehicles is to get more useful energy out for less energy put in. Not only will our fossil fuel resources last longer, but less carbon dioxide will be produced in the combustion process to generate each MJ of energy.

Improvements in end-use efficiencies are likely to have an even bigger impact on overall fuel reduction. For example, energy-efficient buildings could reduce energy needs and hence contribute significantly to the reduction of carbon dioxide emissions.

It can be shown (see page 125 in Chapter 11) that the maximum obtainable efficiency η of a generator is given by

$$\eta = 1 - \frac{T_2}{T_1}$$

where T_1 is the absolute temperature of combustion and T_2 is the ambient or air temperature (measured in kelvin). This maximum value of efficiency is reduced significantly by further energy losses in overcoming resistive forces, heat losses, and noise.

Gas-fuelled generators

The use of gas turbines to generate electricity is becoming increasingly popular. They can operate at higher temperatures than coal-fired and oil-fired power stations, and so potentially

?

1 Steam is heated, under pressure, to 540 °C in a thermal power station. Write down 540 °C in kelvin.

2 Calculate the maximum possible efficiency when the ambient temperature is 30 °C.

One way of increasing the efficiency of a heat engine is to increase the furnace temperature, T_1, or reduce the ambient temperature, T_2. As the air temperature cannot be easily controlled, a sensible solution would be to design generators that operate at different temperatures. (This also works for vehicle engines.)

3 (a) Try the calculation in question 2 again, with an increased value of T_1 and a lower value of T_2.

(b) Summarize the relationship between efficiency and operating temperature.

can be more efficient. Also, the thermal energy produced in the combustion of an equivalent mass of gas is higher than that for coal and oil. Carbon dioxide emissions produced from the different fossil fuels are compared in Table 1 using the unit, the **toe** – tonne of oil equivalent. The toe is the amount of fuel needed to produce the same energy as 1 tonne of crude oil. As well as being more efficient, the combustion of 1 toe of gas produces less carbon dioxide than 1 toe of oil or coal.

Table 1 Comparing production of carbon dioxide from fossil fuels

Fuel	CO_2 produced when burning 1 toe of fuel/t
coal	3.8
fuel oil	2.9
gas	2.1

The technology involved in the design of gas turbines is more complex than for conventional oil and coal generators, mainly because of the higher temperatures and pressures involved. This has involved designing turbine blades of low-density, high-strength alloys into very precise aerofoil shapes. The gas is drawn in through one set of turbines (fan blades), which rapidly increase the gas pressure (Figure 1).

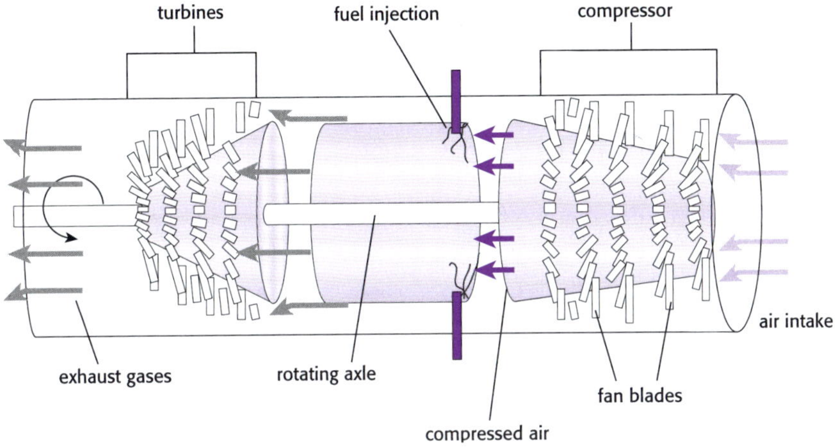

Figure 1 A gas turbine

The gas is then ignited in the combustion chamber and expands rapidly, turning the second set of turbine blades on its way to the exhaust. These blades turn the generator.

Another advantage of gas as a fuel is its relative abundance. New gas fields have recently been discovered, and the largest in the world – just off the coast of Northern Qatar in the Arabian Gulf – only started commercial production in the early years of this century. By 2011 Qatar could very likely supply one-third of the world's liquefied natural gas (LNG) needs (http://www.statc.gov/r/pa/ei/bgn/5437.htm).

With gas now recognized as a cleaner alternative to coal and oil, its use is projected to increase over the coming years, and it is expected to generate an increasing proportion of our electrical energy in the future.

Physics issues: improving our energy use

The following section discusses three examples of different ways in which energy use can be improved. This is achieved either by improving the efficiency of the process involved, or by improving the environmental impact of a process.

?

Explain why the increased use of natural gas for electrical energy generation and transport has the potential to reduce global warming.

Inquiry: making coal more efficient

Use the World Coal Institute website (www.worldcoal.org), the Institute of Energy clean coal centre (www.iea-coal.org.uk), and other sources to research the following questions:

1 Coal is pulverized and ground into a powder before being burnt. Explain why this increases the combustion efficiency.
2 Early coal-fired power stations had efficiencies of only 25%, but it seems likely that this can be increased to greater than 45% in the future. Outline some of the techniques used to achieve these greater efficiencies.

Energy-efficient vehicles

Vehicles, run on fossil fuels, are another major contributor to the enhanced greenhouse effect through the emission of carbon dioxide. Motor manufacturers are designing cars and trucks that are increasingly efficient, based on improvements in aerodynamic design. Similarly, combustion at higher temperatures can allow for increased efficiencies. However, conventional petrol (gasoline) or diesel engines still operate at very low efficiencies.

The best petrol engines rarely transfer more than 13% of the energy input to the wheels; 87% of the energy produced from burning the fuel is dissipated as thermal energy or noise in the engine and other moving parts, as well as additional losses due to running air conditioners, etc. Of the remaining 13%, over half is used to heat up the tyres, leaving only 6% available to produce useful kinetic energy, which is lost as thermal energy in the brakes every time the car stops. The vast majority of this remaining 6% is the kinetic energy of the car itself, and only a very small proportion, less than 1%, is the kinetic energy of the driver – the original intention of the car in the first place.

New and innovative ways are being devised to overcome these losses, including:

- the use of lightweight alloys and advanced polymer composite materials
- the use of electric vehicles
- the use of hybrid vehicles (see below), particularly the plug-in hybrid electric vehicle (PHEV).

Purely electric vehicles do not provide many energy savings. The electric motor is more efficient than a thermal engine, but is not able to transfer as much power to the wheels. Huge battery packs are needed to provide the electrical energy required, significantly increasing the mass of the vehicle. The speeds and maximum range available are currently much less than those obtainable from a conventional petrol-powered car. Electric vehicles also require regular charging, and the energy for this, of course, comes mainly from thermal power stations. The data in Figure 2 show that total carbon dioxide emissions over the lifetime of a purely electric vehicle can be only slightly less than those produced directly by petrol-powered cars, if the power station is coal burning.

?

Outline, with reference to Newton's second law, how the development of lightweight alloys can be used to reduce the energy consumption of a car.

Although carbon dioxide emissions in the manufacture of the different types of vehicle are fairly similar, the production of electric vehicles does involve more greenhouse gas production, because the scale of production is smaller than for petrol-powered cars. The use of hybrid vehicles leads to the emission of far less carbon dioxide than gasoline (petrol) vehicles during their lifetime, but the emissions from electric vehicles depend entirely on the source of electricity. If coal is used to generate the electrical energy then carbon dioxide emissions will be almost as high as using petrol directly. The use of liquefied natural gas (LNG) to generate the electricity leads to significantly lower emissions, and hydroelectric power generation leads to extremely low emissions.

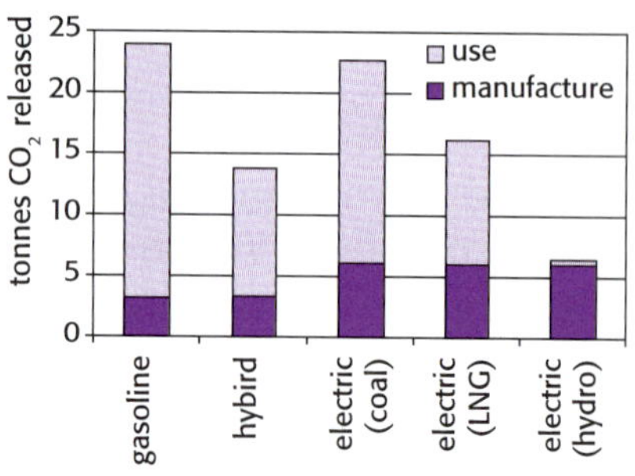

Figure 2 Total carbon dioxide emissions over the lifetimes of gasoline, hybrid, and electric cars. The electric car is shown three times, depending on the method of generating electricity: coal, liquefied natural gas, or hydroelectric.

Hybrid vehicles aim to utilize the best of both worlds. These cars use electric motors, with a petrol engine as back-up to provide additional power when necessary, such as for rapid acceleration, going uphill, or when the batteries have run down. Sophisticated computerized systems have been devised that can switch from the electric motor to the petrol engine and back as required to achieve the best possible efficiency.

The next possible step is the development of **plug-in hybrid electric vehicles** or PHEVs (Figure 3). Manufacturers claim that these may well operate in cities with fuel consumptions over 100 miles per gallon, about 32 km L^{-1}. This compares with about 21 km L^{-1} for the most efficient hybrid vehicle on the market in 2006 and 10 km L^{-1} for the average petrol-powered car. They can be recharged by plugging in to the a.c. mains.

?

1. Explain why the maximum efficiency obtainable from electric motors will be higher than that for a comparable petrol (gasoline) engine.
2. List what factors should be the same to ensure a fair comparison of the two motors in the previous question.
3. State some situations when a hybrid engine will need to switch from the electric motor to the petrol (gasoline) motor.

Figure 3 This plug-in Toyota Prius routinely gets more than 130 mpg. The black circle on the left of the back bumper is a plug socket, allowing an easy connection to the a.c. mains electrical supply.

Hydrogen also has potential as a clean fuel; its combustion produces water as the only waste product. The German car manufacturer BMW has a prototype hydrogen-fuelled car available on the market, although only the very wealthy will be currently able to afford this technology. There are some interesting limitations: with a range of only 200 km you would have to plan your journeys very carefully. An additional slight inconvenience is that at the time of writing there are only five hydrogen filling stations in the world. One more concern is that there is a risk of explosion if the car is parked in a garage!

?

1. State the waste product of burning hydrogen.
2. Briefly outline some safety concerns when using hydrogen as a fuel.

Figure 4 The Elcho combined heat and power plant in Poland, which is capable of supplying 226 MW of electricity and up to 500 MW of district heat.

Combined heat and power (CHP)

One major area of focus in the attempt to reduce our energy use has been to try and find a use for the waste heat from electrical energy generation. One technique is to use this thermal energy to heat buildings and homes in colder climates such as Europe or North America (Figure 4), or for the desalination of sea water in countries with limited fresh water supplies, such as the Arabian Gulf and North East Africa (Figure 5). Such strategies increase the overall efficiency of the system up to between 70% and 90%. In cooler countries, cars can also use the waste heat from the engine to heat the vehicle interior.

Figure 5 The Ras Abu Fontas combined power and desalination plant, near Doha, Qatar. The whole plant generates over 1600 MW of power using several gas turbines and produces about 100 ML per day of fresh water from sea water using the waste heat from the gas turbines.

Carbon capture and storage

Improved efficiencies and new technologies will have an impact on our fuel demand, and will certainly help to reduce carbon dioxide emissions. But the planet will continue to produce vast quantities of greenhouse gases, which will continue to build up in our atmosphere. Another solution, which is in the prototype stage in many countries, is to capture the carbon dioxide and bury it, safely underground, in rock formations such as coal seams, or in saline aquifers, where it cannot escape. This technique is known as carbon capture and storage or carbon sequestration (Figure 6).

?

Outline how the increased use of CHP electrical generation can reduce global warming.

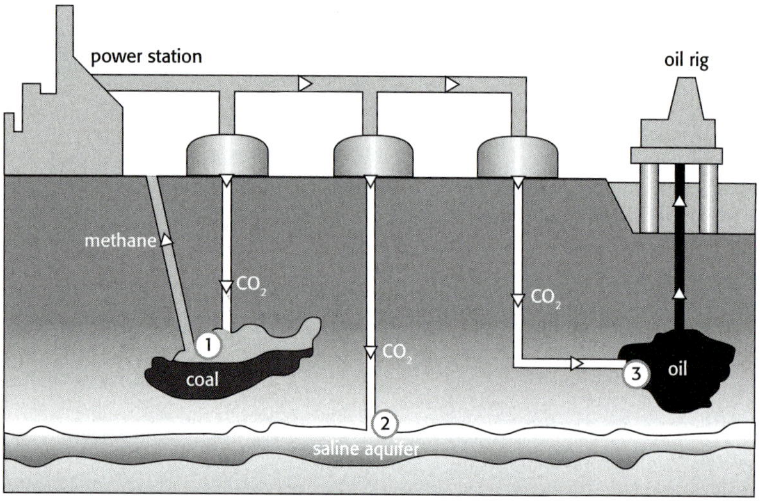

Figure 6 Carbon capture and storage

This is a particularly promising technique, because the necessary technology is already available, and it has the potential to reduce carbon dioxide emissions into the atmosphere significantly in a relatively short time. One added advantage is that the carbon dioxide can be pumped down into depleted oil reservoirs to help recover further oil.

One obstacle to the full-scale development of carbon capture and storage is that it is very difficult and the costs are prohibitive to "retro-fit" existing power plants with the technologies required to store carbon dioxide underground effectively. Only newly designed plants will be able to do this, and the design of a new power plant takes time.

The next few years

Critical decisions will be made in the next few years that may have a permanent impact on how our climate emerges from this current crisis. The adoption of these new emission-reducing technologies comes at a massive financial cost, but this is a very small cost compared with the potential impact of climate change on global society. A UK government report produced in late 2006 by Sir Nicholas Stern has suggested that an immediate increase in spending by governments of 1% annually to reduce emissions of carbon dioxide and other greenhouse gases could offset the potential costs of climate change, which were estimated to be as high as 20% of the world's total spending.

?

This question is based on the Ras Abu Fontas combined power and desalination plant near Doha, Qatar.

Data for sea water
density = 1.03×10^3 kg m^{-3}
boiling point = 101 °C
specific heat capacity = 3990 J kg^{-1} K^{-1}
1 litre = 10^3 cm^3

The whole plant generates over 1600 MW of power using several gas turbines.

(a) Outline some advantages of using gas rather than oil or coal in an effort to reduce global warming. [4]

As well as generating electricity, the Ras Abu Fontas plant produces fresh water from sea water using the waste heat from the gas turbines. The water is for the nearby city of Doha.

(b) Show that the amount of energy needed to raise the temperature of 1 litre of sea water from 25 °C to the boiling point is about 310 kJ. Use the data provided. [3]

(c) Show that the power needed to desalinate 100 ML day^{-1} of sea water is greater than 360 MW. [3]

To reduce the energy required, the pressure above the boiling sea water is reduced in a vacuum vessel so that the boiling point is lowered to 40 °C.

(d) Estimate the power needed to desalinate 100 ML day^{-1} at this lower boiling point. [2]

22 Renewable energy

Renewable sources of energy (see page 221 in Chapter 18) provide us with one of the greatest causes for optimism in the whole energy debate. With appropriate design, natural energy sources can be harnessed to produce a sustainable supply of electrical energy without making any significant demands on the Earth's dwindling resources. In many cases the energy source is effectively free, and, as an added bonus, some systems ensure that no pollutants or greenhouse gases are pumped back into the environment. It seems almost too good to be true: all we need do is to build enough renewable generating capacity to provide for our energy needs and we have solved both the current "fuel crisis" and the climate change crisis.

The combustion of biofuels and waste material to generate electricity is slightly different from the other renewable sources, because both release significant amounts of carbon dioxide into the atmosphere, and waste burning releases other pollutants as well. Neither source makes significant demands on the Earth's resources, but they are not necessarily environmentally friendly, and their use does not contribute to a reduction in greenhouse gas emissions.

Biofuels are produced from plants, which absorb carbon dioxide from the atmosphere through photosynthesis. They may then either be burnt directly (wood, for example) or converted into an energy-rich fuel such as ethanol, for example by fermentation, which is then burnt. In both cases the carbon dioxide absorbed in photosynthesis is released back into the atmosphere, so there is no net effect on atmospheric carbon dioxide concentrations.

Of the other renewable sources, hydroelectricity is the most used, but wind power is perhaps the most promising source of clean energy for the future. In 2003, 1% of all electricity globally was generated by wind turbines, and this is likely to increase dramatically. In some countries significant amounts of electricity are already generated in this way, and as technology improves its use is likely to expand further.

It is only comparatively recently that the focus in energy research has switched to renewables. The initial investment required to set up wind farms, dams, and tidal schemes can be prohibitive, and the efficiency of energy conversion into electrical energy is low. This means that large investments in capacity have produced only small returns in terms of electrical energy produced compared with nuclear power, for example. The location of renewable sources means that the energy often has to be generated in remote places, leading to additional costs in order to transport it to where it is most needed: the cities. These remote locations are often areas of outstanding natural beauty, and any human impact affects the environment with visual or noise pollution.

Hydroelectricity

Hydroelectricity is generated when water stored behind a dam is allowed to fall and drive a turbine. It may also be generated by pumped storage or from tidal energy in suitable river estuaries.

The Three Gorges Dam

One of the most ambitious hydroelectric schemes in history has been the construction of the Three Gorges Dam on the Yangtze River near Yichang in Hubei Province, Central China (Figure 1). The Yangtze River is the longest in Asia and the third longest in the world. It flows over 5000 km, from the edge of the Tibetan plateau to the East China Sea near the rapidly expanding city of Shanghai. The river itself has been a lifeline in central China for centuries, providing much-needed irrigation and a convenient means of transport for trade and travel in a region of mountainous and difficult terrain.

The dam itself will be completed by 2009, and will form part of the largest hydroelectric scheme in the world. It is designed to generate at least 22 GW at peak capacity in 2011, which is equivalent to approximately 22 conventional coal-fired power stations. It was originally intended to generate up to 10% of China's electricity, but because of China's rapid economic growth this figure has been revised down to 3%. This could still have a huge effect in reducing China's total carbon dioxide emissions, but it would be a mistake to assume that such hydroelectric schemes have zero impact on global warming. Reservoirs and similar bodies of water release small amounts of methane – another greenhouse gas – into the atmosphere produced in the biological decomposition of organic material.

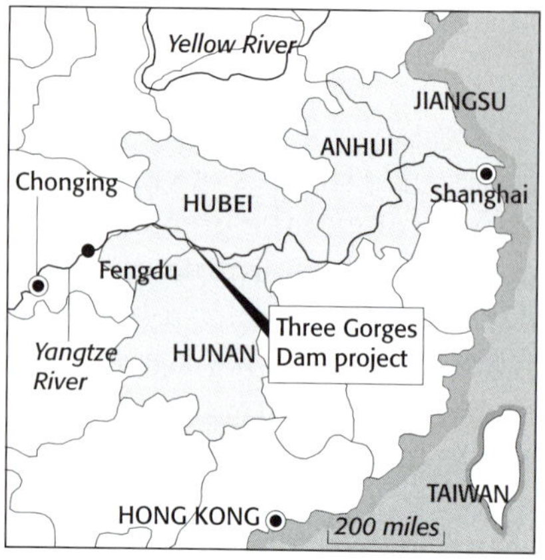

Figure 1 Location of the Three Gorges Dam

?

Use the data in Table 1 to answer the following questions.

Table 1 Data for the Three Gorges Dam

Economic	
Type	Concrete gravity dam
Construction dates	1993–2008 (original target date was 2009)
Official cost	US$22.5 billion (180 billion yuan at 2006 exchange rate)
Physical	
Height of dam	181 m
Length of dam	2.309 km
Power generation	
Left (north) bank	14 × 700 MW turbine generators (operating from 2003)
Right (south) bank	12 × 700 MW turbine generators (operating by 2008)
Underground	6 × 700 MW turbine generators (operating by 2011)
Reservoir data	
Length	660 km
Width	1.1 km (mean)
Area	1084 km^2
Land submerged	632 km^2
Water volume	39 billion m^3
Original river level	65 m above sea level
Final reservoir level	175 m above sea level

Inquiry: sources of renewable energy

Only hydroelectricity and wind power are covered in this book: hydroelectricity because, of all the renewable sources of energy, it currently generates the most electricity, and wind power because it is likely to be the leading renewable form of generation in the future.

Other promising renewable ways of generating electrical energy include:

- **Geothermal energy**. This uses thermal energy from "hot rocks" under the ground to produce steam, which can then drive turbines to generate electricity.
- **Solar energy**. Photovoltaic cells can convert solar energy directly into electrical energy.
- **Wave power**. An oscillating water column (OWC) ocean-wave energy converter changes the rising and falling motion of ocean waves into rotational motion, which is used to turn turbines. The total energy available within ocean waves is huge, but so is the cost of creating such schemes far out to sea.

?

Carry out some research to make sure you fully understand the basic principles of using each source, and their advantages and disadvantages.

1 (a) Estimate the mean drop of water in the reservoir, when it has risen to full height, as it falls through the turbines.

(b) State what assumptions you have made.

Hint: Fill in all the relevant heights above sea level on a sketch diagram like that shown in Figure 2.

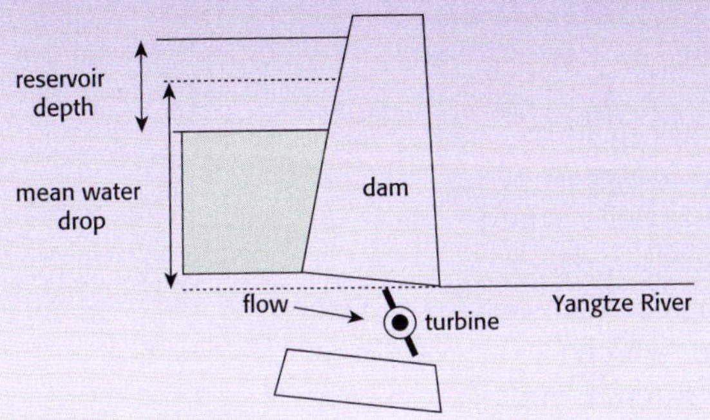

Figure 2 Sketch of dam cross-section

2 Show that the maximum power generated is approximately 22 GW.

3 A turbine converts kinetic energy in the water into electricity, which is distributed through China's electricity grid. If the turbines operate at 90% efficiency, calculate the kinetic energy per second of water flow required to generate power at 22 400 MW.
(If we ignore any resistive energy loses – a big assumption – it is reasonable to assume that all the potential energy stored in the water behind the dam is converted into kinetic energy as the water flows through the turbines.

4 Use the data to estimate the speed of the water as it flows through the turbines.

5 Estimate the total flow rate of the water through the turbines.

Data is a powerful tool of propaganda. It should always be viewed critically and never taken at face value. Information that is left out is just as important as the information that is included. The Chinese government would be keen to show how it is considering the energy needs of its fast-developing country with careful consideration of the environment and the need to reduce carbon dioxide emissions, and to control the regular flooding of the Yangtze. Environmental groups may be concerned about the effect on local people (it is estimated that more than 1.4 million will be displaced by the project), the changes in lifestyles caused by the loss of so much fertile farm land, the effects on the river downstream owing to increased siltation (deposition of sand and small rock particles) behind the dam caused by reduced flow, and the impact on flora and fauna in the Three Gorges area (which is designated by UNESCO as an area of outstanding natural value).

Tidal energy

Tidal energy schemes use a dam across a suitable river estuary. Sluice gates in the dam are opened as the tide rises and sea water fills the estuary. At high tide the gates are shut and the water is trapped at this high level behind the dam. At low tide, there is a significant height difference between the water level trapped behind the dam and the sea water outside. The trapped water is now allowed to run out through water turbines, generating electrical energy. Although the tide rises and falls only twice a day, a significant amount of power may be generated in this way owing to the huge volume of water involved. The environmental impact of such a scheme needs to be carefully considered. Currently the only tidal plant in Europe is on the Rance estuary in northern France. A useful website to start research on this is http://home.clara.net/darvill/altenerg/tidal.htm

Pumped storage

Pumped storage adds no overall electrical energy into the grid. In fact it takes energy. Electrical energy is very difficult to store, so pumped storage provides a useful way of producing electricity when it is needed. At peak times when electrical demand is high a water turbine may be switched into the national grid. Such water turbines may reach peak generating capacity in under a minute, which ensures that the electrical supply can adjust to rapidly fluctuating demand. At such peak times many electrical companies charge consumers more per unit of electricity consumed than at off-peak times, when demand is lower. During off-peak times the action of the water turbine is reversed, and excess capacity is used to pump water back up to the upper body of water ready to be used again at peak times. The Dinorwig power station in Wales has a good website for research: http://www.fhc.co.uk/dinorwig.htm

Wind power

Wind power has the potential to generate a significant proportion of the world's electrical energy needs. Recent improvements in technology and materials used have already led to the development of larger and cheaper wind turbines. The graph in Figure 3 indicates the growth of wind power in recent years.

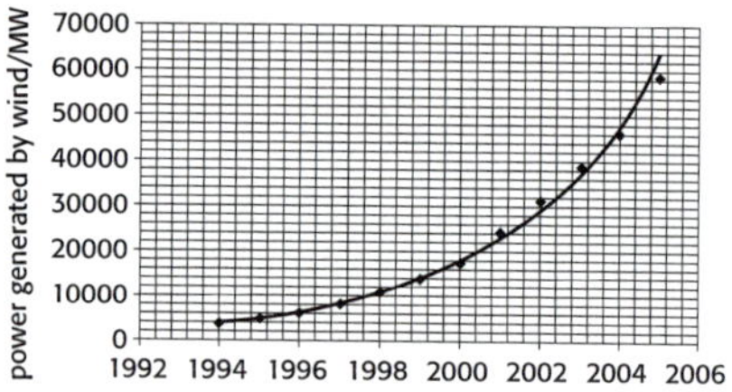

Figure 3 Recent growth in wind power. Source: Global Wind Energy Council (http://www.gwec.net)

Exponential increase

If the graph in Figure 3 can be described by an exponential function, then it should show a constant doubling time.

Time to grow from:

10 000 to 20 000 MW	~2.6 years
20 000 to 40000 MW	~2.6 years
30 000 to 60 000 MW	~2.5 years

It seems reasonable to assume that the growth in wind power generation is approximately exponential, with a mean doubling time of about 2.6 years.

We can use this data to project the wind power generated in 2020.

Power generated by the wind in 2004 = 48 000 MW

Number of doubling times from 2004 to 2020 = 16/2.6 = 6.2

$$\therefore \text{projected power in 2020} \approx 48000 \times 2^{6.2} = 3.5 \times 10^6 \text{ MW} = 3.5 \text{ TW}$$

?

1 Total world power demand is projected to be 5.4 TW in 2020. Discuss whether wind power can achieve the target of 34% of global electrical energy production, as suggested by the Global Wind Energy Council.
2 Use the data in Figure 3 to plot a graph of ln(power generated by wind) against time in years.
3 Does the graph plotted in question 2 suggest an exponential relationship?

Wind capacity

Wind turbines of different sizes can generate a range of powers, from 100 kW (enough for a village) up to 2 MW (adequate for a small town). The turbine blades may vary in size, from 30 m up to 80 m in length, and are designed to rotate at anything from 10–30 rev/min. Modern turbines can generate electricity at wind speeds as low as 4 m s^{-1}, reaching maximum capacity at about 15 m s^{-1}. Hurricane or typhoon winds can damage wind turbines, so a braking system is incorporated into the design so that they stop turning at speeds in excess of 30 m s^{-1}. With careful planning, wind power can be fed into any national grid system, backed up by conventional thermally generated power when the wind is low. Over a whole country it is often true that the wind is always blowing *somewhere*, so that wind power can be fed into the electricity grid fairly continuously.

?

1 Show that the maximum power P_0 that may be extracted from the wind per unit area is given by the formula $P_0 = \frac{1}{2}\rho v^3$, where v is the wind speed and p is the air density, assuming that the air stops as it hits the blades.
(Hint: Start by calculating the kinetic energy of air per second passing through the turbine blades.)

2 If the turbine blade rotates once every 3.0 seconds, compare the blade speed at the tip, 20 m from the axis,
with the speed at the other end, only 0.50 m away from the same axis.
(Hint: use $v = \omega r$.)

3 Show that the maximum power per m^2 in a wind of speed 7 m s^{-1} is 210.1 W m^{-2}. Assume that the density of air = 1.225 kg m^{-3}

4 Calculate the power per m^2 for an 8 m s^{-1} wind, and comment on why a small increase in wind speed causes a significantly larger increase in power.

23

Nuclear power

As an alternative to fossil fuel generation, nuclear power has had a controversial press. It has become a major source of energy in many countries across the world: the UK, USA, and France, for example, could not meet their current energy supply targets without it. The use of nuclear power to generate electrical energy is fairly widespread (Figure 1).

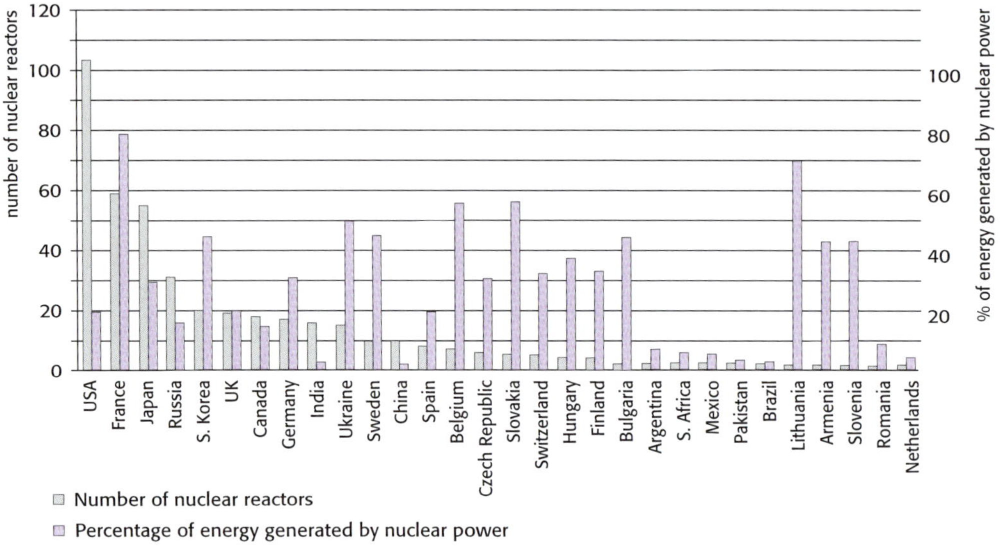

Figure 1 Use of nuclear power to generate electrical energy. Data from International Atomic Energy Authority, January 2007.

The thinking on the use of nuclear energy to generate electricity has changed since the 1950s, when it first became commercially available. The timeline in Figure 2 illustrates some trends in the development of nuclear power.

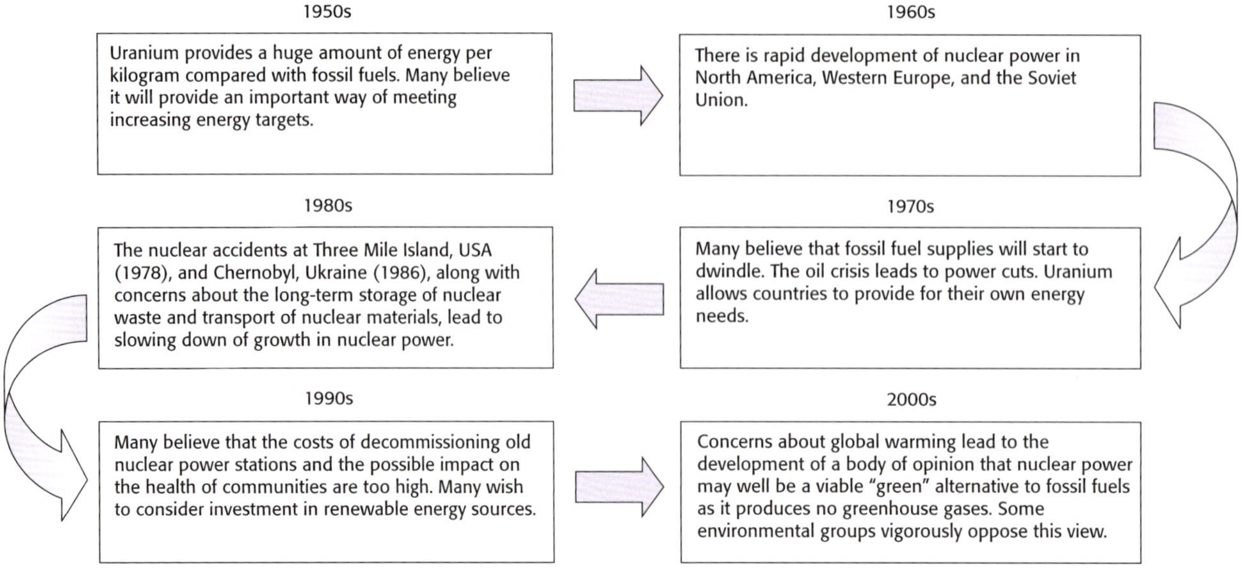

Figure 2 Timeline: the development of nuclear power

As an IB learner you will need to weigh up the evidence and form your own opinion on whether nuclear power is a safe, reliable, and efficient technology, and whether we should consider increasing investment in it. Alternatively, nuclear power may have such a negative impact on the environment in other ways that it could be wiser to invest more heavily in renewable technologies, such as wind, tidal, and wave power, to achieve our energy needs with minimal environmental impact.

The questions (right) aim to encourage you to think critically about nuclear power generation worldwide and to draw conclusions about the different circumstances and needs of individual countries.

Of the 442 operational nuclear reactors worldwide (2006), more than half of them are found in only four countries: the USA, France, Japan, and Russia. The first three of these have a very high demand for electricity, and rely heavily on nuclear power to achieve the required level of supply. France achieves a very large proportion of its electricity generating capacity through nuclear means. Table 1 enables you to compare the average rate of energy generation in these countries. The table also includes China and Lithuania for contrast.

?

1 Compare the number of nuclear reactors in OECD and non-OECD countries and outline your findings.
2 Outline some possible reasons that might explain why there are a relatively high number of reactors in Eastern Europe, previously the Communist bloc.
3 (a) State which three countries have the largest number of nuclear reactors.
  (b) Suggest reasons to explain this.
4 Outline what other data would be helpful in order to get a full picture of a country's nuclear power capacity.
5 Describe what the data tells us about nuclear power in the developing world.

Table 1 Comparison between six countries' nuclear power reliance

	Average **total** electricity power production* / MW	Number of nuclear reactors	Percentage of total electrical power produced from nuclear energy	Average electricity power production by nuclear power / MW
USA	460 889	103	19.3	
France	62 694	59	78.5	
Japan	109 250	55	29.3	
Russia	99 292	31	15.8	
China	282 500	10	2.0	
Lithuania	1689	1	69.6	

*Estimates.

Source: Data from www.iaea.org, October 2006.

?

1 Calculate the average rate of electrical power production from nuclear energy for each country in Table 1 and so complete the table.
2 Outline how the following factors may affect the growth of nuclear power in these countries.
  (a) Rising fossil fuel prices.
  (b) Protests and demonstrations by environmental groups at the source of uranium.
  (c) Environmental protests in the end-user country.
  (d) A small nuclear accident or leak of radioactive materials into the environment.
  (e) A transport strike in the generating country.

Inside the reactor

The enrichment process

Most naturally occurring uranium ores contain a mixture of two isotopes: uranium-235 and uranium-238. Most of the ore (99.3%) is uranium-238; the uranium-235 required for fission makes up only 0.7%. Modern reactors require enriched uranium, containing up to 3.5% of the isotope uranium-235.

In the **enrichment** of uranium the gas uranium fluoride is fed into a **centrifuge**. This contains a series of evacuated cylinders, each with a rotor between 1 and 2 m long with a diameter of 15–20 cm (Figure 3). The mixture of isotopes is spun round at rates of up to 70 000 rev/min. The heavier uranium-238 requires a larger **centripetal** force to stay in a circular path and thus moves to the outside of the centrifuge. The uranium-235 requires a smaller force and collects together, closer to the centre (Figure 3). The uranium-238 can then be removed, as **depleted uranium**, and the remaining mixture is then centrifuged several more times, in a cascade of centrifuges, before the concentration of uranium-235 has increased to the required 3.5%. The whole process is complicated, and involves some very technical and challenging engineering, partly because of the tremendous forces involved.

?

1 Estimate the mass of one molecule of uranium hexafluoride, UF_6, in kg. (Relative atomic mass U = 238, F = 19.0, 1 amu = 1.66×10^{-27} kg)

2 Assume the starting conditions are as follows:
Initial orbital radius r = 1.5 m
Rate of rotation = 70 000 rev/min.
Calculate:
(a) the number of revolutions per second.
(b) the period of rotation.
(c) the angular velocity.

Moderation and control

In order for a nuclear reactor to continue to release thermal energy at a steady rate, the reaction must be **self-sustaining**. For every one neutron that successfully collides with a uranium-235 atom, another one has to be produced that will definitely induce another fission. If too many neutrons are available for further fissions the rate of production of thermal energy in the reactor may increase in such a way that the reaction becomes uncontrollable, leading to an explosion. This uncontrolled fission, or **thermal runaway**, is the mechanism by which a nuclear bomb works. A great many precautions are used to ensure that this does not happen in a nuclear reactor.

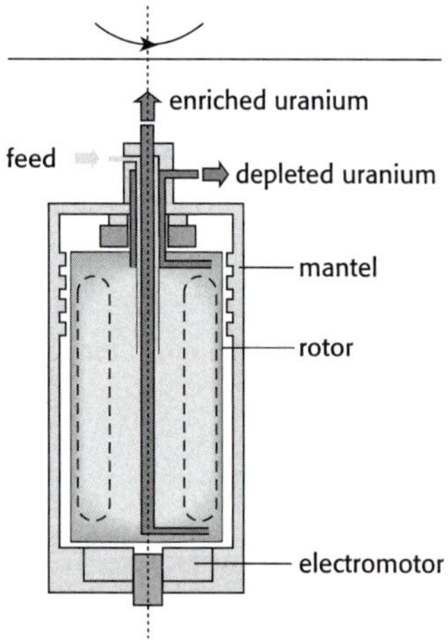

Figure 3 Structure of a uranium centrifuge. The rotor spins around a vertical axis. The uranium-235 collects towards the axis and the uranium-238 towards the outer wall of the rotor.

The nuclear energy spectrum in Figure 4 illustrates that most neutrons produced in a reactor are **fast neutrons** in the range 1–2 MeV. These have too much energy to cause further uranium-235 fission, and are far more likely to be **captured** by uranium-238. **Thermal neutrons** are more likely to cause further fission. These neutrons have energies of about 0.02 eV. **Moderators** are used to "slow" down these fast neutrons to thermal energies to increase the probability of further fission. **Control rods** are used to remove any excess neutrons to ensure the reaction continues safely.

Scattering cross-sections and probability

The fission equations show that for every inducing neutron that has a successful collision with a nucleus of uranium-235, two or three neutrons are produced in the resulting fission. In fact, the mean overall number of neutrons resulting from each fission is about 2.5. These neutrons may:

- leave the fuel rod through the external surface
- collide inelastically with a uranium-238 nucleus and be absorbed
- collide inelastically with a uranium-235 nucleus, inducing fission

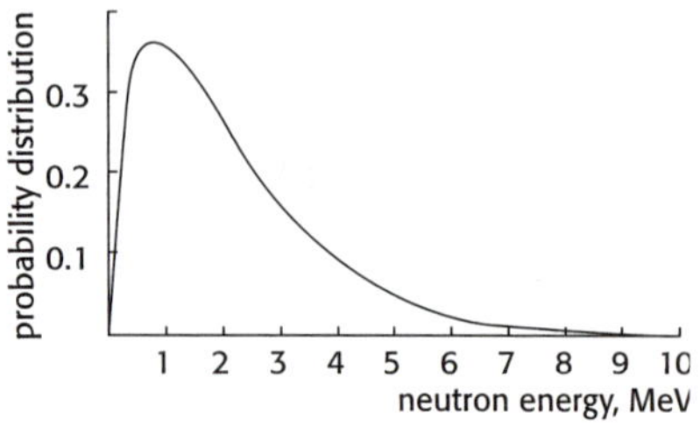

Figure 4 Neutron energy spectrum from a fission reactor core

- collide elastically with either a nuclide or the moderator and control rod molecules, transferring energy only.

The laws of probability are used to determine the chance of each outcome occurring. These calculations enable appropriate conditions to be set up within the reactor to ensure that criticality – a self-sustaining fission reaction – is maintained. As we are discussing many billions of events, the laws of probability work very well.

The concept of **scattering cross-sectional area** is used. A large scattering cross-sectional area implies a large probability of a particular event occurring (Figure 5).

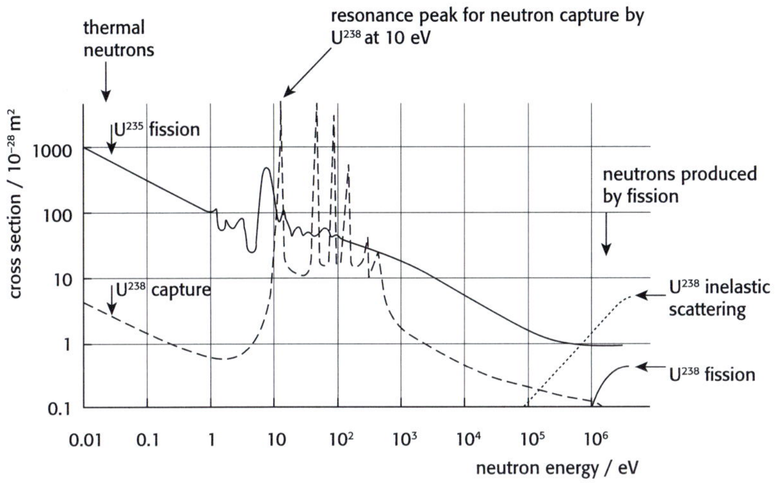

Figure 5 Scattering cross-sections for uranium nuclides

The curves are rather complicated, but show that
- the scattering cross-section and hence the probability of a fission for uranium-235 increases as the neutron energy is reduced
- the scattering cross-section for uranium-238 neutron capture is particularly large in the 10–100 eV energy range. The sharp peaks in this energy range are known as **resonances** (see Chapter 7, page 76).

The uranium-235 and uranium-238 nuclides present a similar geometric cross-sectional area to a neutron, but their scattering cross-sections depend on the neutron energy.

1 Consider a neutron moving along a tube of area *A* towards a nuclide of area a, as shown in Figure 6a. Write down the probability of the neutron colliding with the nuclide. (You can assume the neutron is a point object.)

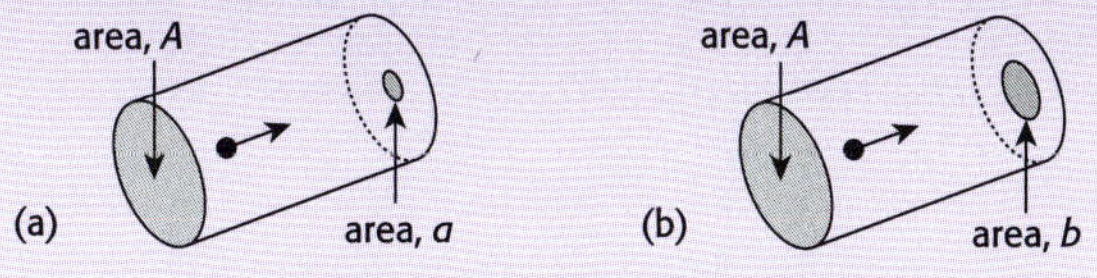

Figure 6 Calculating probabilities of nuclear collisions

2 Now assume that the tube contains a different nuclide of area *b* (Figure 6b). Write down the probability of a collision with the second nuclide.

3 (a) Write down the ratio of your answer to question 1 to your answer to question 2.
(b) Explain what this answer means in terms of probabilities.

4 Assume the neutron is a thermal neutron of energy 0.01 eV. Using the graph in Figure 5, determine the ratio of the probability of uranium-235 fission to uranium-238 capture.

5 Repeat question 4 for a fast neutron of energy 10eV.

6 Write a short paragraph to explain to a younger student how cross-sectional areas can be imagined to change in particle experiments.

Thinking about science: uranium enrichment

Nuclear power features frequently in the news. At the time of writing, both Iran and North Korea have featured regularly in the international news because of their nuclear programmes, but for different reasons. Iran has started the enrichment of uranium-235, and North Korea exploded a small test nuclear device in October 2006. It is interesting that so much attention should be focused on these two countries when many countries in the world have far more developed and advanced nuclear programmes. In addition, the USA, France, and Russia possess nuclear weapons, yet we rarely hear as much discussion or concern in the international press about these countries as we do about North Korea or Iran.

As well as relying on the use of radio, television, and newspapers to distribute their message, modern news agencies also use the Internet: websites, email, and podcasting. Many of these press agencies attempt to give a balanced view, and have no apparent political alignment. Some press agencies do, however, align themselves with political philosophies, parties, or organizations. In some countries newspapers have restrictions placed on what they can report. Other news organizations are part of profitable business concerns and can be obliged to represent the interests of their owners.

Often the press tends to present a viewpoint or opinion in order to make the news item more interesting or "newsworthy". Emotive or subjective language may be used to support an opinion, rather than relying on objective facts and evidence. Readers must remember to separate fact from opinion, and draw their own conclusions based on objective observation, just as should take place in an experiment.

As an IB learner you are encouraged to be thinkers, principled, and open-minded (see page 3). You should **think critically** about what you are reading, listening to, or watching. Ask yourself questions such as:

- What is the evidence to justify this opinion?
- What sources were used to produce the article?
- How does the type of language used distort or influence my viewpoint? (You will almost certainly have discussed this in your group 1 subject lessons.)
- What is the truth, and how do I know this?

In order to make a judgment, you may need to apply **principles** based upon your own moral and ethical standpoint in order to formulate your own personal opinions. By keeping **open-minded** you may be able to separate fact from misinformation, and even be able to use a blatantly distorted article to develop a more accurate picture of the truth because you are aware of the biases behind any article that you read.

Many news agencies report events from a prejudicial viewpoint. This may be the result of illogical argument or irrationality. It may be deliberate or unintentional. Prejudice consists of judging an issue or holding an opinion on it without rationally examining the evidence for it. Prejudice may well be based on preconceived false, or inaccurate, stereotypes.

?

Apply your critical thinking skills to the following situations. Try to give a balanced answer, considering all possible viewpoints. These questions are probably better to use in a class debate rather than as written answers. Discuss:

(a) whether developing countries have a moral responsibility to develop nuclear power.
(b) whether any country has the right to develop nuclear weapons for self-defence.
(c) whether it is better to isolate a country thought to be developing nuclear technology or to engage it with technological and logistical support.

Inquiry: types of nuclear reactor

Each reactor is unique, in that it is designed and engineered to fit a range of particular criteria. However, there are a few designs or templates that have popular appeal for reasons of cost, safety, and efficiency.

The most common type of reactor is the **thermal reactor**, so called because it is powered by the fission of uranium-235 with thermal neutrons. Although there are important differences between each type of thermal reactor, there are some common features. A block diagram for a nuclear power station has much in common with a fossil fuel power station (see page 224) – the key difference is that the fossil fuel furnace is replaced by the nuclear reactor (Figure 7).

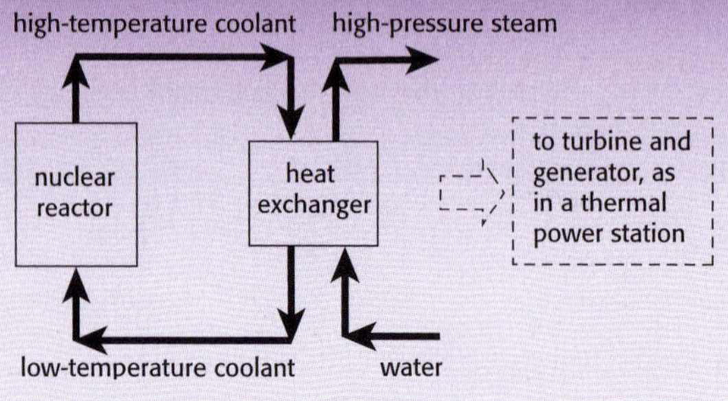

Figure 7 Block diagram for a nuclear reactor generating electricity. The source of thermal energy is illustrated.

The main types of thermal reactor in commercial use today are the:

- pressurized water reactor, PWR
- boiling water reactor, BWR
- pressurized heavy water reactor, PHWR
- gas-cooled reactor, GCR
- light water cooled, graphite moderated reactor, LWGM (also called RMBK in Russia)

There are important differences in design that effect the efficiency, the cost, and even the safety and risk involved. Some of the common features are described below.

- **Fuel rods.** All thermal reactors utilize the fission of uranium-235 to release thermal energy. The proportion of uranium-235 in the rods varies from 0.7% to 3.5% depending on the reactor.
- **Moderator.** Most neutrons released in fission are fast neutrons, so a moderator is used to reduce their energy down to thermal levels to ensure that the fission is self-sustaining. The moderators used vary: for example, solid graphite in the LWGM and steam in the BWR.
- **Coolant.** This is used to transfer the thermal energy released from the reactor core to raise steam. This steam then drives the turbines. The BWR is unique in that the coolant, steam, also drives the turbines directly. Both the BWR and PWR use water, in different states, as the coolant and moderator.
- **Control rods.** These are inserted among the fuel rod assembly and are used to control the reactor temperature, thus preventing it from overheating. If **criticality** is exceeded, neutron-absorbing rods made of cadmium or boron steel are lowered into the reactor core and any excess thermal neutrons are effectively absorbed.
- **Containment building.** The reactor core is usually surrounded by a steel vessel and a concrete structure, several metres thick, to absorb any excess radiation and help contain any potentially catastrophic event, such as an explosion, should all other safeguards fail.

The other main type of commercial reactor is the **fast breeder reactor**, FBR, which is quite different from thermal reactors.

To develop your **personal skills**, work in teams to research and present your findings on one of the four inquiries suggested below. It would be a useful exercise to work with students whom you do not normally partner so that team work can be practised ready for the **Group 4 project**, when these skills will be formally assessed. Strong team members are able to:

- offer ideas and advice to the group
- listen to the views of others
- agree and persevere on a specific task with a clear goal in mind
- work ethically and consider any environmental impacts, e.g. excess use of paper and the printer.

Some useful websites for the research include:
www.iaea.org
www.world-nuclear.org
www.uic.com.au

1 Compare and contrast the PWR and the BWR.
You should consider the following:
   - the proportion of uranium-235 in the fuel rods
   - the role of the coolant
   - the type of coolant and moderator used
   - operating efficiencies
   - safety features and possible risks
   - average power output of each reactor.

2 Find out about some of the advantages and disadvantages of the PHWR compared to the PWR and consider why many different countries have selected this design.

   In your research you could consider
   - the average power output
   - the efficiency of heavy water as a moderator compared with normal water
   - an advantage of using heavy water in terms of safety
   - why this reactor has been so successfully exported worldwide.

3 Find out about the FBR and how it works. In your research you should answer the following questions.
   - Is a moderator needed? Explain your answer.
   - Describe how this reactor is able to "breed" further plutonium-239 fuel.
   - Why are liquid metals suitable for use as coolants? Why would these coolants be inappropriate for a thermal reactor such as a PWR?
   - At the current rate of use, uranium may well run out in the next 60 years. How will FBRs affect this estimate?
   - Outline any ethical concerns involved when countries develop FBRs.

4 Find out what you can about the LWGM reactors used in Eastern Europe, and compare then with the PWR. The Chernobyl accident happened in a LWGM reactor. Try and find out what safety improvements have been put in place since this disaster took place in 1986.
You should consider how:
   - the LWGM is cooled and moderated
   - the accident happened
   - the design of subsequent generations of LWGM has improved safety.

The graph in Figure 9 illustrates the decay of the various components of nuclear waste. The graph has a log scale on both axes, and covers very large ranges of activity and time. This may make the graph appear difficult to interpret.

?

1 In Figure 11, identify which isotopes are fission products and which are transuranic elements.
2 Estimate the half-lives of caesium-137 and americium-241: a rough value is sufficient.
3 Explain why the isotopes thorium-229 and radium-226 only appear after about 5000 and 20 000 years respectively.
(Hint: Consider the possible decay chains in the waste fuel, starting with the alpha decay of uranium-238.)
4 Outline one useful purpose of the line labelled "Uranium ore equivalent to one tonne of fuel".
5 Does the waste fuel have an overall half-life? If so, is it possible to estimate its value?

Reactor accidents

In common with other sectors of the electricity generation industry, there have been many small-scale accidents at nuclear power stations. There have been very few that have had any serious effect on either the workers or nearby residents of the power station, although long-term environmental damage is much more difficult to gauge. The two most serious accidents were in 1979 at Three Mile Island, in Pennsylvania off the East Coast of the USA, and in 1986 at Chernobyl, in the Ukraine.

Three Mile Island was a PWR, now the most popular reactor worldwide, and Chernobyl was a LWGM, still widely used in Russia. Both had potentially disastrous consequences but only in Chernobyl was there a serious radioactive leak into the environment. In this case the leak was so serious that the effects were measured as far away as Finland to the North and Wales to the west. These tragic events led to major increases in safety precautions and any subsequent accidents in the last twenty years or so have been minor. Compared to other electricity generating sectors, nuclear power has a very safe record.

Development of nuclear weapons

Modern nuclear weapons use the uncontrolled fission of uranium-235 or plutonium-239 to generate tremendous amounts of energy. They require almost pure pieces of the fissile isotope, which involves a highly sophisticated and technical engineering capability to produce. They work on a very simple principle. The fissile material is converted into a critical mass by use of a chemical explosive. One method is to force two smaller pieces of uranium together, and another is to compress a piece of plutonium to increase its density (Figure 10). In both cases, the critical mass is exceeded and an uncontrolled fission chain reaction follows.

?

This question is about one of the challenges for nuclear engineers caused by the low number of thermal neutrons produced in fission. Without any further intervention, the fast neutrons will be captured by uranium-238 and the reaction will come quickly to a halt.

Moderators are used to slow down the fast neutrons to thermal energies in the process of moderation. The fast neutrons undergo a series of elastic collisions with the atoms or molecules making up the moderator. The formula below (which can be derived from conservation of momentum considerations) can be used to calculate the transfer of kinetic energy from the neutron of mass m and energy E_1 to the moderating particle of mass M and kinetic energy E_2 using:

$$\frac{E_2}{E_1} = \frac{4mM}{(M+m)^2}$$

In order to reduce the energy of fast neutrons down to thermal levels as quickly as possible a moderator is used whose constituent particles have a mass, M, as close as possible to the neutron mass, m.

For the questions below use the following relative masses:
neutron = 1, water molecule = 18, deuterium oxide molecule = 20

(a) Show that for
  (i) water moderation E_2/E_1 = 0.20, and
  (ii) deuterium oxide moderation E_2/E_1 = 0.18. [4]

(b) Calculate the proportion of its original kinetic energy retained by the neutron in each case. [3]

(c) Estimate the number of collisions a neutron of energy 1 MeV has to undergo in a water moderator to reach an energy of 0.02 eV. [2]

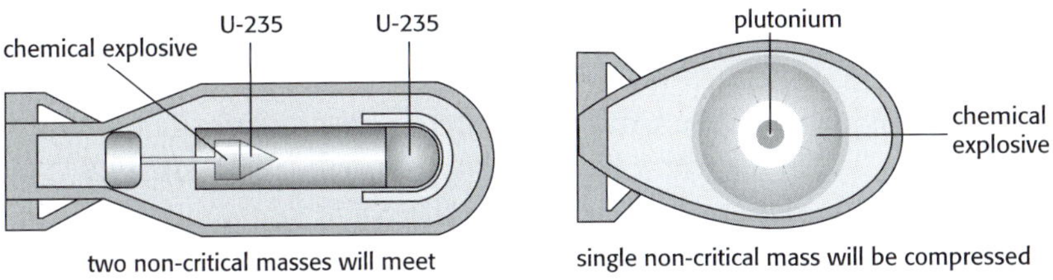

Figure 10 Two methods of achieving critical mass in a weapon

Pure uranium-235 does not occur in nature, and plutonium is not found naturally. Uranium-235 concentrations for conventional reactors are increased by centrifuging (see page 260), and this process may be continued to generate the higher concentrations needed for a weapon. Plutonium-239 is produced in the nuclear waste of most conventional reactors, and may be chemically separated out and purified. Fast breeder reactors use plutonium, but also "breed" further plutonium.

Nuclear reactors are therefore necessary in order to manufacture weapons. One of the risks of nuclear power is the possibility that weapons-grade fissile material may be produced as a by-product. Many governments have openly used a civilian nuclear programme to generate material for nuclear weapons; others have done it covertly. However, it is clear that nuclear reactors may be used for the totally ethical generation of electrical power without necessarily leading to the production of these weapons of mass destruction.

It is unlikely, but certainly a possibility, that terrorist groups could get hold of sufficiently pure material to make an effective fission weapon. Another fear is the use of so-called "dirty" bombs. These contain conventional, chemical explosives surrounded with highly radioactive waste. The explosive is used to disperse the waste over a large area, thus leading to large-scale radiation exposure and greatly increasing the number of fatalities.

Security at nuclear facilities is extremely high, at all stages of the nuclear cycle, in an attempt to reduce the risk of such awful weapons ever being produced.

The above section may make for depressing reading but there is much room for optimism. Despite continued fears, the nuclear generating industry has a very good overall safety record, and produces up to 15% of the world's much-needed electricity without producing greenhouse gases or other pollution from its generators. This is potentially very good news indeed.

The bigger picture: the nuclear debate

This final section in the chapter encourages you to weigh up the evidence for yourselves and decide whether the continued use of nuclear power is desirable and necessary or undesirable and avoidable. Much has been made of the potential for nuclear power to generate increased amounts of electricity without further adding to greenhouse gas emissions. However, there are strong arguments to increase the use of renewable energy sources, particularly wind power, in order to meet our electrical needs without compromising the climate.

In the nuclear debate you have to consider whether governments should:

- increase nuclear power generating capacity
- increase the use of renewable energy sources to generate electricity
- increase the use of both sources.

Which is going to be most beneficial option in the long term with the least costs? In order to get the debate started, two articles are presented below.

The first article comes from Greenpeace, a global environmental organization that campaigns against nuclear power, and argues as follows:

Greenpeace has always fought – and will continue to fight – vigorously against nuclear power because it is an unacceptable risk to the environment and to humanity. The only solution is to halt the expansion of all nuclear power, and for the shutdown of existing plants.

We need an energy system that can fight climate change, based on renewable energy *and energy efficiency. Nuclear power already delivers less energy globally than renewable energy, and the share will continue to decrease in the coming years.*

Despite what the nuclear industry tells us, building enough nuclear power stations to make a meaningful reduction in greenhouse gas emissions would cost trillions of dollars, create tens of thousands of tons of lethal high-level radioactive waste, contribute to further proliferation of nuclear weapons materials, and result in a Chernobyl-scale accident once every decade. Perhaps most significantly, it will squander the resources necessary to implement meaningful climate change solutions.

Things are moving slowly in the right direction. In November 2000 the world recognized nuclear power as a dirty, dangerous and unnecessary technology by refusing to give it greenhouse gas credits during the UN Climate Change talks in The Hague. Nuclear power was dealt a further blow when a UN Sustainable Development Conference refused to label nuclear a sustainable technology in April 2001.

The risks from nuclear energy are real, inherent and long-lasting.

Safety: *No reactor in the world is inherently safe. All operational reactors have inherent safety flaws, which cannot be eliminated by safety upgrading. Highly radioactive spent fuel requires constant cooling. If this fails, it could lead to a catastrophic release of radioactivity. They are also highly vulnerable to deliberate acts of sabotage, including terrorist attack.*

Waste: *From the moment uranium is mined, nuclear waste on a massive scale is produced. There is no secure, risk-free way to store nuclear waste. No country in the world has a solution for high-level waste that stays radioactive for hundreds of thousands of years. The least damaging option at this current time is for waste to be stored above ground, in dry storage at the site of origin, but this option also presents major challenges and threats.*

Weapons proliferation*: The possession of nuclear weapons by the US, Russia, France, the UK and China has encouraged the further proliferation of nuclear technology and materials. Every state that has a nuclear power capability has the means to obtain nuclear material usable in a nuclear weapon. Basically this means that the 44 nuclear power states could become 44 nuclear weapons states. Many nations that have active commercial nuclear power programs, began their research with two objectives – electricity generation and the option to develop nuclear weapons. Also, nuclear programs based on reprocessing plutonium from spent fuel have dramatically increased the risk of proliferation, as the creation of more plutonium means more nuclear waste, which in turn means more materials available for the creation of dirty bombs.*

Source: www.greenpeace.org/international/campaigns/nuclear

James Lovelock is a respected scientist and author, creator of the Gaia theory (see the biology course companion in this series of books), and a leading member of Environmentalists for Nuclear Energy. In an article published in the British newspaper *The Independent* in May 2004, entitled "Nuclear power is the only green solution", he makes the following points.

What makes global warming so serious and so urgent is that the great Earth system is trapped in a vicious circle of positive feedback. Extra heat from any source, whether from greenhouse gases, the disappearance of Arctic ice or the Amazon forest, is amplified, and its effects are more than additive. It is almost as if we had lit a fire to keep warm, and failed to notice, as we piled on fuel, that the fire was out of control and the furniture had ignited. When that happens, little time is left to put out the fire before it consumes the house. Global warming, like a fire, is accelerating and almost no time is left to act.

So what should we do? We can just continue to enjoy a warmer 21st century while it lasts, and make cosmetic attempts, such as the Kyoto Treaty, to hide the political embarrassment of global warming, and this is what I fear will happen in much of the world. When, in the 18th century, only one billion people lived on Earth, their impact was small enough for it not to matter what energy source they used.

But with six billion, and growing, few options remain; we cannot continue drawing energy from fossil fuels, and there is no chance that the renewables, wind, tide and water power can provide enough energy and in time. If we had 50 years or more we might make these our main sources. But we do not have 50 years; the Earth is already so disabled by the insidious poison of greenhouse gases that even if we stop all fossil fuel burning immediately, the consequences of what we have already done will last for 1000 years. Every year that we continue burning carbon makes it worse for our descendants and for civilization.

Opposition to nuclear energy is based on irrational fear fed by Hollywood-style fiction, the Green lobbies and the media. These fears are unjustified, and nuclear energy from its start in 1952 has proved to be the safest of all energy sources. We must stop fretting over the minute statistical risks of cancer from chemicals or radiation. Nearly one third of us will die of cancer anyway, mainly because we breathe air laden with that all pervasive carcinogen, oxygen. If we fail to concentrate our minds on the real danger, which is global warming, we may die even sooner, as did more than 20 000 unfortunates from overheating in Europe last summer.

Even if the greens are right about its dangers, and they are not, its worldwide use as our main source of energy would pose an insignificant threat compared with the dangers of intolerable and lethal heat waves and sea levels rising to drown every coastal city of the world. We have no time to experiment with visionary energy sources; civilization is in imminent danger and has to use nuclear – the one safe, available, energy source – now or suffer the pain soon to be inflicted by our outraged planet."

Source: James Lovelock, Nuclear Power is the only green solution, The Independent, May 24, 2006

Carefully consider the above two articles and write down the arguments for and against the use of nuclear power. Analyse the effect of the style of language used. Is the language objective or is it emotional? Do the articles just state facts or are they effective propaganda?

You should also consider the views of:

- the workers at nuclear power stations, and the trade unionists who represent them
- the local communities living near nuclear reactors and waste disposal units.

If possible, organize a formal debate in your teaching group or senior school on the nuclear issue.

24 Digital technology

HL Digital technology affects the way we work, learn, and use our leisure time. The Internet has proved to be an extremely useful tool for gaining knowledge, learning, and communicating. Its growth will continue, along with the development of new uses. However, although many of us take access to this technology for granted, there are many parts of the world where computers, mobile phones, and MP3 players are hardly known, and Internet access is still something for the future.

This chapter will consider the use of compact discs (CDs) and digital versatile discs (DVDs) to store data, and of charged-coupled devices (CCDs) to create digital images.

?

Make a list of electronic devices you have used in the last week, and sort the list into things used for work/learning, those used for entertainment, and those for any other uses.

Thinking about science: the Internet

The Internet has increased access to data, and hence to knowledge, in many parts of the world. The amount of knowledge available is difficult to quantify but it is estimated that in 2002 the Internet contained more than 530 000 terabytes (10^{12} bytes) of data – though much of this is not accessible or useful – and that between 2007 and 2010 the amount of digital data available is forecast to increase by 57% a year.

Knowledge is good

Education relies on the acquisition of knowledge and skills, and then on the application of these to help understand a range of different situations, contexts, or problems. Most, if not all, of this knowledge is now available from databases stored on servers all round the world, and is accessible from the Internet. This vast quantity of knowledge can cause difficulties. Not only does the sheer volume lead to inefficiencies, because time is wasted searching for what you need, but much information is replicated, and some is wrong. Accurate sources must be distinguished from misinformation and propaganda. It is important to be discerning when you are acquiring the knowledge you need to complete a task.

Ethics of the Internet

There is no one code of ethics governing the Internet. Essentially, anything goes. Websites that are considered immoral or unacceptable by some groups of people are seen as acceptable by others. For example, paedophiles, extremist political parties, criminals, and terrorists can post information on the Internet just as easily as transnational corporations and governments. Is it possible to set out a moral argument that correctly asserts that particular websites are immoral and others are not? Probably not. We each have our own moral code, based on our cultural upbringing and conscience; in other words, we all have our own set of standards. Public opinion also changes over time.

The Internet exemplifies the concept of freedom of speech, a principle closely upheld by many democratic organizations, including the United Nations. If this principle is acceptable on moral grounds, then it can be used to justify one's right to pervert, lie, and offend, which leads to an ethical dilemma, particularly in countries that have libel laws protecting the individual from false accusations. Many would argue that a private opinion voiced between adults is very different from a public statement posted on the Internet.

Material that can offend needs careful consideration. Many would argue, particularly educators, that a basic aim should be to ensure that children and young learners can use the Internet safely as a valuable educational resource. Parents and teachers trust that students are safe from the potential corrosive social effects and harm that many websites can cause. This can easily be achieved by the use of appropriate filtering software to remove any harmful or offensive material, which can therefore provide a safe refuge for learning.

Legal difficulties also arise when considering how to prosecute inappropriate web postings. Which country's laws apply? The country where the offensive article was posted, the country of the server where the website is stored, or the country where the material is accessed? Solutions have been achieved, however. International cooperation has been able to virtually outlaw the posting of child pornography for the use of paedophiles throughout the world. Uploading or downloading such material can lead to prosecution and lengthy prison terms by anyone involved.

The Internet in society

The Internet grew from developments at CERN, the European Centre for Nuclear Research based in Switzerland. CERN is an example of international cooperation between governments when no one country could afford the vast expenses of such research. A prototype of the Internet was developed to share data quickly with similar projects across the world. Nowadays email, chat sites, and Internet communications have greatly improved and reduced the cost of local and international conversations.

As well as bringing people together, the technology also has the potential to divide societies. Within the richer countries of the world, although most people have access, significant proportions of the population do not. See Table 1.

For example, nearly 70% of the population of the USA – the most Internet-connected country in the world – were estimated to have Internet access in January 2007. But this means that 30%, or over 90 million people, did not have access. The Internet can divide societies by creating an underclass of people without such access to information. Liberia in West Africa, for example, has the lowest Internet connectivity, with only 0.03% of its population being able to access the Internet. A lack of Internet access has the potential to further impoverish societies.

Analogue and digital signals

An analogue signal varies continuously between maximum and minimum values. A microphone, for example, produces an **analogue** signal that represents the sounds it detects. The **continuous** electrical signal generated contains all values of voltage between the maximum and minimum allowed values. The graph in Figure 1 represents an analogue signal that varies continuously between +6 and −6 V.

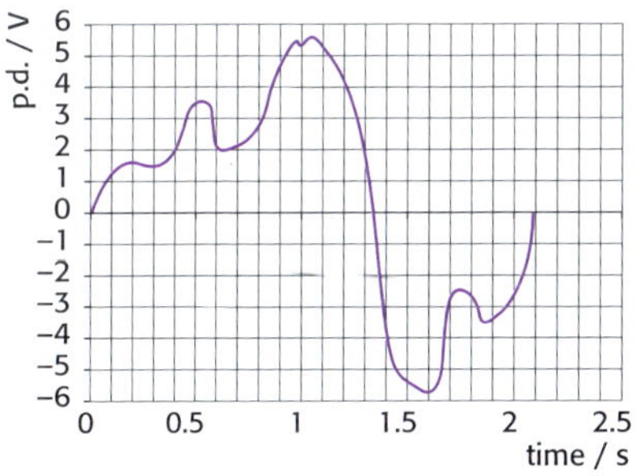

Figure 1 An analogue signal

A **digital** signal, by contrast, contains **discrete** (or quantized) values only. The signal in Figure 2 is a digital signal which has two values only; +6 or 0 V, representing a binary 1 or 0 (see page 272). Each 1 or 0 lasts for the same time: 0.2 seconds in this case.

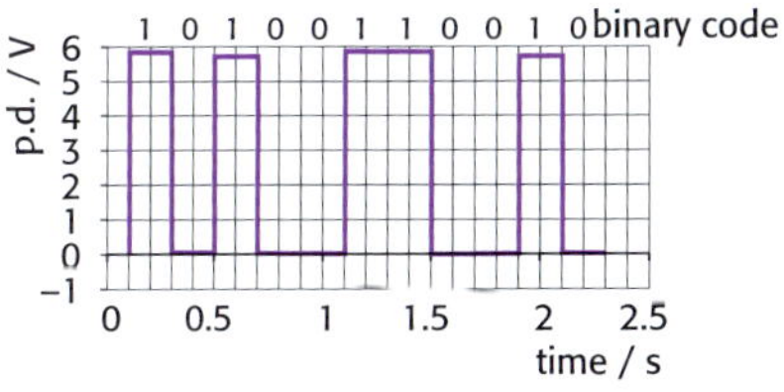

Figure 2 A digital signal

The photographs in Figure 3 show two ammeters reading the same value of current: Figure 3a is an analogue ammeter and Figure 3b is a digital ammeter. The analogue meter displays all values of current from the minimum to the maximum value allowable on the scale. Intermediate values can be achieved by interpolating (reading between) the gradations of the scale. If the position of the needle is in between two marks on the scale you can judge whether the current is closer to the lower or the higher value. In the picture below the needle is clearly between 0.20 A and 0.21 A and closer to 0.20 A. It is reasonable to estimate that the position of the needle in this case is about one quarter of the way between the two values: that is, 0.203 A ±0.003. The digital meter is accurate to ±0.01 A. The value displayed is somewhere between 0.20 A and 0.21 A. However, there is no way of knowing exactly what the reading is. In other words, the digital reading is quantized in steps of 0.01 A. The analogue meter can therefore be used to give a more precise reading than the digital meter.

Binary code

Computers use a binary code at their simplest, most basic level of operation.

The most common number system in general use is a base 10 system. It uses 10 digits represented by the characters, 0, 1, 2, 3, 4, 5, 6, 7, 8, 9. Numbers are written in columns, representing powers of 10:

$\mathbf{10^3} \quad \mathbf{10^2} \quad \mathbf{10^1} \quad \mathbf{10^0}$

So, for example, the number three thousand, seven hundred and forty-one is written 3741 and represents $(3 \times \mathbf{10^3}) + (7 \times \mathbf{10^2}) + (4 \times \mathbf{10^1}) + (1 \times \mathbf{10^0})$.

Numbers can be written to any base number (this is covered in the HL mathematics course), but base 10 emerged as the most useful system – presumably because humans have 10 fingers, and this was a convenient way to count.

Binary code is written in base 2, and so involves just two digits: 1 and 0. This is the simplest number system. Table 1 shows the first 16 numbers in base 10 and the corresponding binary numbers.

(a)

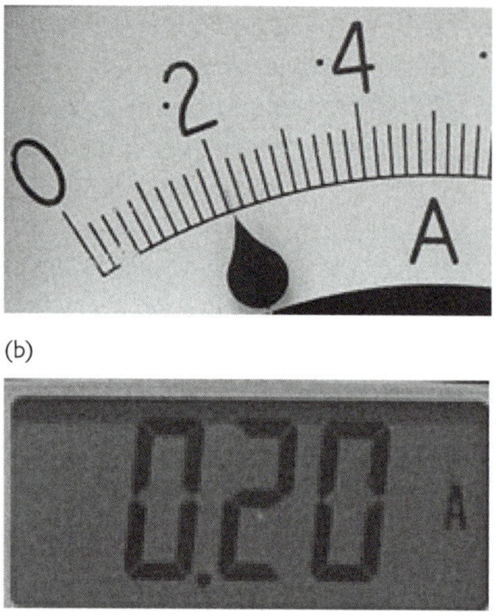

(b)

Figure 3 (a) An analogue ammeter, (b) a digital ammeter

Table 1 Conversion table for binary numbers

Base 10	2^3	2^2	2^1	2^0	Binary
0	0	0	0	0	0
1	0	0	0	1	1
2	0	0	1	0	10
3	0	0	1	1	11
4	0	1	0	0	100
5	0	1	0	1	101
6	0	1	1	0	110
7	0	1	1	1	111
8	1	0	0	0	1000
9	1	0	0	1	1001
10	1	0	1	0	1010
11	1	0	1	1	1011
12	1	1	0	0	1100
13	1	1	0	1	1101
14	1	1	1	0	1110
15	1	1	1	1	1111

?

Outline the differences between an analogue and a digital signal.

?

1 Write the following base 10 numbers in base 2:
16, 17, 45, 63, 100, 127.

2 Write the following binary numbers in base 10:
11110, 101010, 111000, 110111, 1010101.

One **binary** digi**t**, a 1 or a 0, occupies one **bit** of memory in a computer or any data storage device. The right-hand digit is called the **least significant bit**; when counting sequentially it changes most frequently, and represents the smallest value. The left-hand digit is called the **most significant bit**; this changes least often when counting, and represents the largest value. A binary signal containing a series of several 1s and 0s is called a **string**. An 8-bit number occupies 1 **byte** of memory.

?

Write the following hexadecimal numbers (a) in binary code and (b) as decimal numbers:
11, FF, A5, 5A, 97, 79, CD, DC.

The **hexadecimal** system, which is used in computing, uses a base of 16, with the digits 0–9 and the letters A–F. It can also be used to represent a byte. For example, the decimal numeral 93, whose binary representation is 01011101, is 5D in hexadecimal (5 = 0101, D = 1101). Each hexadecimal character represents a 4-bit string. The two strings are written down side by side to give an 8-bit string – a byte. Table 2 shows the conversion between decimal, binary and hexadecimal numbers.

8-bit numbers can be used to represent the basic word processing characters. The Notepad word processor in Windows uses an 8-bit string to represent each character. To check the size of a byte open Windows Notepad (or an equivalent program): in Windows XP the file path is *Start/All programs/Accessories*. Type the phrase "physics course companion". Save the file as *companion.txt* and close the file. The size of the file can be determined by right clicking on the file name in Windows Explorer and then clicking *Properties* (or by just holding the cursor over the file icon); it should show 24 bytes. The phrase uses 24 characters: 22 letters and two spaces, which are also considered characters. Windows Notepad stores text in its simplest form, so memory is used only to store the characters; there is no formatting (such as **bold** or *italic*) and hence no additional memory use, as is required from more sophisticated programs such as Microsoft Word.

The relationship between each character and its binary number is written in ASCII code, which may be researched further on http://www.asciitable.com/

Values stored on a CD are represented by a 2-byte number.

Table 2 Conversion table for hexadecimal numbers

Decimal	Binary	Hexadecimal
0	0000	0
1	0001	1
2	0010	2
3	0011	3
4	0100	4
5	0101	5
6	0110	6
7	0111	7
8	1000	8
9	1001	9
10	1010	A
11	1011	B
12	1100	C
13	1101	D
14	1110	E
15	1111	F

?

1 How many different numbers can be represented by
   (a) 1 byte
   (b) 2 bytes?
2 Computer memory is measured in kilobytes (kB), megabytes (MB), gigabytes (GB), terabytes (TB), and so on. Carry out some research to find out how many bytes there are in each of these values.
3 Suggest why the phrase "physics course companion" typed in Microsoft Word needs over 23 kB (kilobytes) of memory.

The bigger picture: analogue-to-digital conversion

The process of converting an analogue signal to a digital signal involves some complex electronic components; this is covered in more detail in the communications option chapter. The basic steps in digitization involve:

1 sampling the analogue signal at fixed time intervals
2 assigning a quantum level to the value of the sample
3 converting the quantum level into binary code

The key components are illustrated in Figure 4.

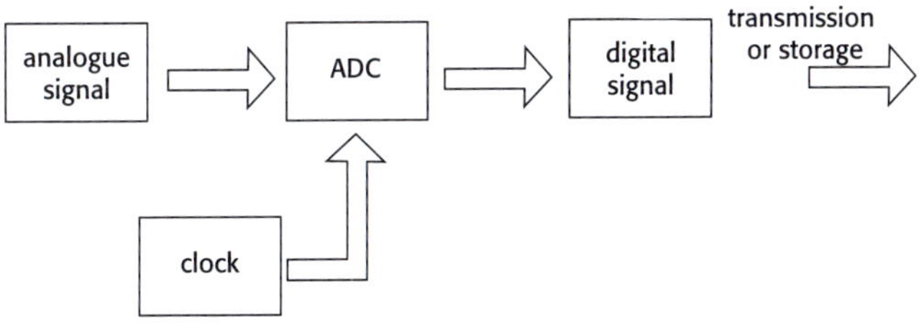

Figure 4 Analogue-to-digital conversion

The analogue signal is fed into the ADC (analogue-to-digital converter). This is a chip that samples the signal at a predetermined rate, assigns a quantum level to each sample, and converts these values into binary code. The clock produces a regular pulse to ensures that the 1s and 0s are produced at regular intervals. The digital signal produced is then either transmitted (e.g. a digital TV signal) or stored (e.g. on a CD).

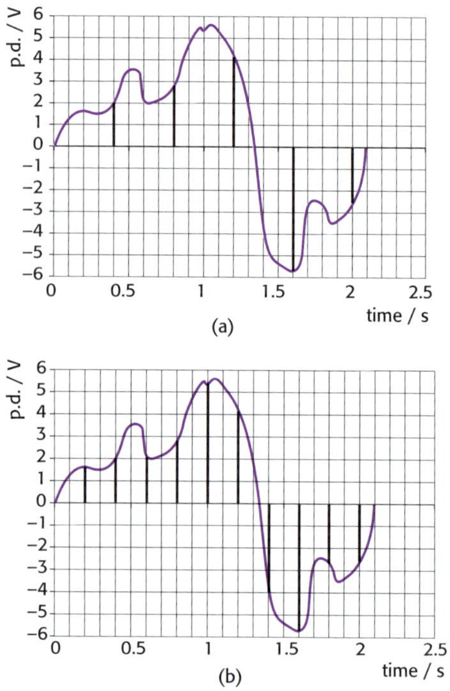

Figure 5 (a) Sampling every 0.4 s; (b) sampling every 0.2 s

Sampling

The analogue signal is read or sampled at specific time intervals, as illustrated in Figure 5.

The signal illustrated in Figure 5a is only sampled every 0.4 s. This means that significant amounts of information about the original signal will be lost in the conversion. To reduce this loss of information the number of samples each second – the **sampling frequency** – can be increased, as illustrated in Figure 5b. Sampling every 0.2 s allows more information to be retained, but increases the amount of data that has to be transmitted.

To ensure that enough data is transmitted, so that the original signal can be reproduced accurately, the signal must be sampled at a minimum rate of twice the maximum signal frequency. Humans are able to hear sounds in the frequency range of about 10 Hz to 20 kHz. The sine wave illustrated in Figure 6 has a frequency of 20 kHz. In general, the signal is sampled at least every half cycle for it to be fully reproduced after transmission.

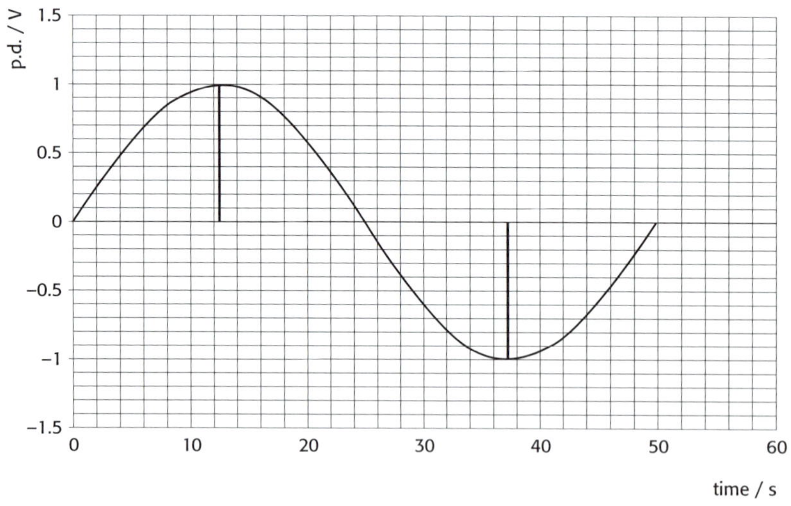

Figure 6 A 20 kHz sine wave sampled every half cycle

1. Record the sampled values and corresponding times in a table for:
   (a) the data in Figure 5a;
   (b) the data in Figure 5b.
2. Plot the data obtained in question 1 in two graphs, one for each table. Join up the points appropriately.
3. (a) Outline how you decided to join up the points in the graphs in question 2.
   (b) Discuss which graph represents the original signal more accurately.

Quantum levels and coding

The voltage values of the analogue signal samples are assigned quantum levels, as illustrated in Figure 7. These quantum levels are simply discrete numbers, and have nothing to do with the quantum levels assigned to electrons or nucleons in nuclear physics.

The sampling rate used to store music on CDs is 44 100 samples per second, written as 44.1 kHz. Explain why this rate is suitable.

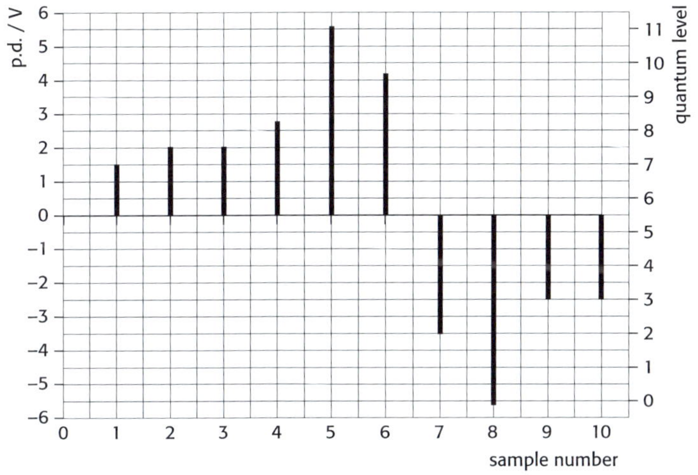

Figure 7 Quantum levels are assigned to samples from an analogue signal. The table illustrates the quantum level of each sample and the corresponding 4-bit binary code.

Sample no	Quantum level	Binary code
1	7	0111
2	7	0111
3	8	1000
4	8	1000
5	11	1011
6	10	1010
7	2	0010
8	0	0000
9	3	0011
10	3	0011

The signal illustrated in Figure 5b has been sampled every 0.2 s; each sample is shown with a thick line. The quantum levels of each sample are recorded in the table. Each quantum level is converted into the corresponding binary number. In the example above there are 12 quantum levels, labelled 0–11. These can be represented with a 4-bit binary code. The binary code is then transmitted as a series of 1s and 0s. To ensure that each sample is assigned an accurate quantum level, there need to be as many such levels available as possible. In the example illustrated in Figure 7, samples 1 and 2 are both given a quantum level of 7, yet their voltage values are clearly different. Level 7 represents all voltages from 1 V to 2 V. A 4-bit code does not allow the digital signal to accurately represent the original signal.

In order to re-create the original signal, the digital signal (taken from the storage medium that is being used) is fed into a DAC (digital-to-analogue converter). The binary signals are converted back into voltages. The DAC "estimates" the values of the signal in between the digitized values and reproduces an analogue signal.

?

1 (a) State how many quantum levels can be achieved with a 4-bit binary code.
  (b) Explain why this is not suitable for storing music on CDs.

2 Samples are stored as 16-bit (2-byte) binary numbers on CDs. Calculate how many quantum levels this represents.

Advantages of digital signals

Although analogue signals can accurately represent the source of the signal, they usually pick up noise in the transmission process. **Noise** is a form of interference that distorts the signal. It may be added to the signal during transmission, or when the signal is being played back. Depending on your own level of tolerance, noise can spoil the playback of a piece of music, for example.

The use of the term "noise" refers to modifications of the actual signal, not just to background noise, although this can be equally frustrating. Vinyl records or LPs (LP stands for long player) were a popular format for marketing music prior to the introduction of compact discs or CDs in the early 1980s. The musical signal in an LP is stored in a groove, physically cut into the record by a sharp stylus (Figure 8). The shape of the groove is an analogue representation of

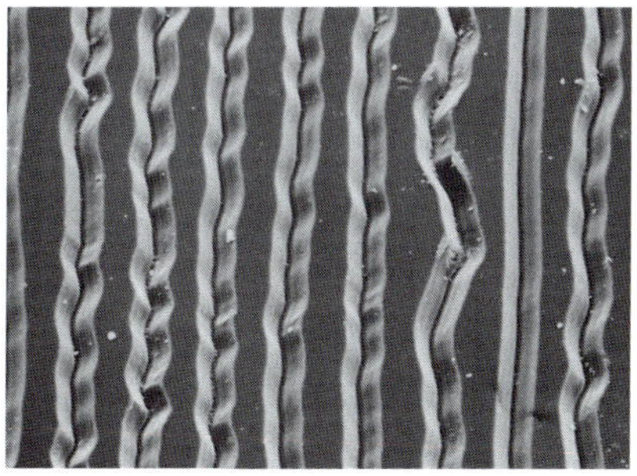

Source: www.audiography.com.au

Figure 8 Magnified view of grooves in an LP record.

the musical signal. The LP is played back by placing another stylus into the groove, and then rotating the record. The stylus vibrates in the groove and produces an electrical signal, which is played back through an amplifier and speakers so that it can be heard. Every time the LP is played, the stylus slightly damages and changes the shape of the groove. In addition, dust and other small particles settle in the groove. The stylus vibrates in this distorted and dust-filled groove, and so noise is added to the signal.

Digital signals, made up of binary strings (see below) of ON (1) and OFF (0) values may also pick up noise in transmission, but they are still recognizable as a series of 1s and 0s. Figure 5 on page 274 illustrates a binary signal where 1 = +6 V and 0 = 0 V. Music is stored on a compact disc in binary code. A digital receiver is programmed with a set cut-off voltage, often about 2 V. Any signal higher or equal to 2 V is identified as a 1 but anything below is read as a 0. A certain amount of distortion can therefore be tolerated by the system.

?

Use the Internet and other sources to find out how the following store data and, in each case, whether it is in an analogue or digital format.

(a) cassette tape
(b) floppy disk
(c) hard disk drive
(d) video (VHS) tape.

The choice, then, in deciding how best to store and transmit data is between an analogue signal, where the representation of the original variations is stored in a way that directly matches the original signal, and the use of an agreed code to translate between the original variations and a set of digital numbers. These digital numbers can be transmitted with effectively no corruption, but they then need to be translated back into the original analogue signal.

Compact discs

CDs have become one of the most popular formats for storing digital sound recordings. Large computer files can also be saved on CDs using CD writers (CD-R). Currently CDs can store about 0.75 GB (gigabytes) of data, and so are very convenient for transferring and storing resources.

Structure

CDs are just under 12 cm in diameter. The data is stored in a track of microscopic "pits" and "bumps", which are moulded into a thin layer of transparent plastic. These indentations are then covered with a layer of reflective aluminium (Figure 9).

The pits and bumps are arranged one after the other in a spiral track, starting from the centre of the disc, as illustrated in Figure 9b.

(a)

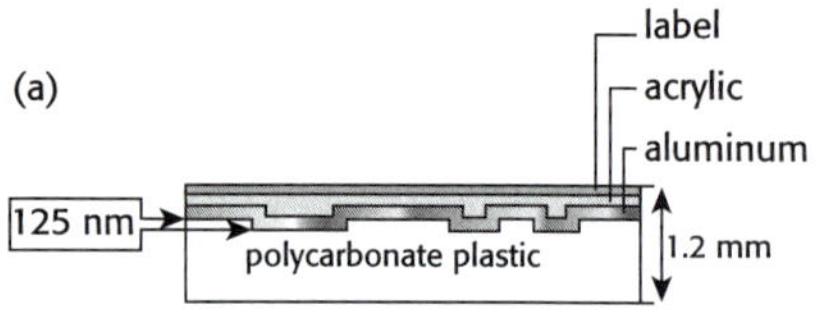

(b)

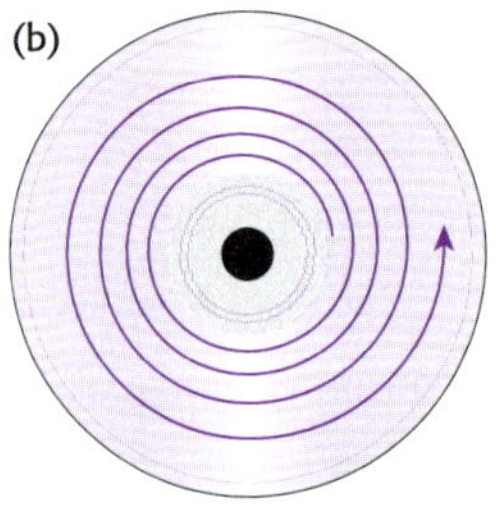

Source: *howstuffworks.com.*

Figure 9 (a) Cross-section through a CD; (b) the spiral.

(a)

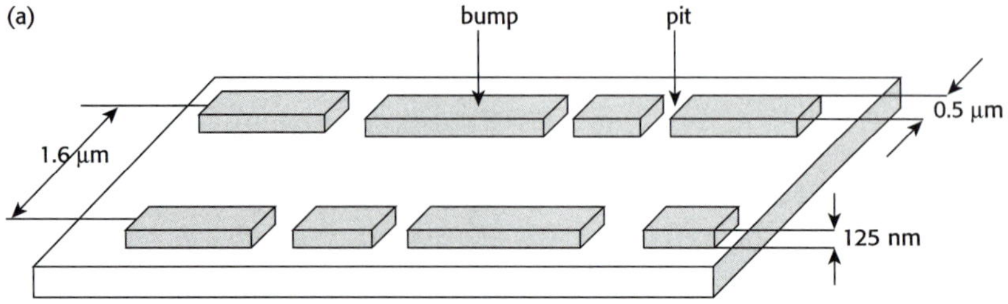

(b)

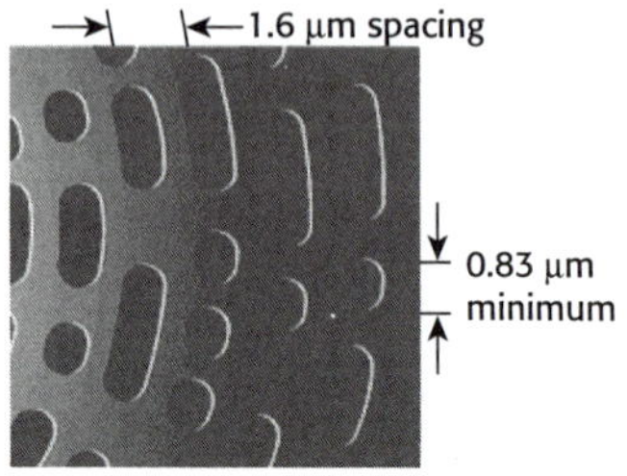

Source: *Computer Desktop Encyclopedia*

Figure 10 (a) Sketch showing the arrangement of "pits" and "bumps" in a track on a CD (not to scale). (b) Highly magnified plan view of a CD.

The diagram in Figure 10a illustrates the dimensions and arrangement of some "pits" and "bumps" in two consecutive arcs of the spiral.

The disk rotates at about 500 revolutions per second, and a focused laser beam "reads" the data by reflecting off the undulating track. The whole mechanism is very precise, and the laser is focused down to a spot of diameter 1.7 μm. The laser starts reading the data from the centre and moves outward along a radial line as the disk rotates rapidly next to it. One of the big advantages of a CD over an LP is that there is no mechanical contact between the laser and the disc, so it can be played and replayed without damage or the introduction of noise.

To replay the music perfectly from a CD, the laser must read the data at a constant rate. In order to do this the rotation rate of the disc must gradually change. Explain why this change is necessary, and outline how the speed changes while a CD is being played.

Reading the data

The laser beam is focused on the CD surface, and the reflected light is detected by a photodiode. The incident laser beam in reality is incident almost normal to the disc surface, but is shown in Figure 11 at an acute angle so that the incident and reflected beams are clearly visible. If the beam is entirely incident either on a bump or a pit then all parts of the reflected beam are in phase with each other. Constructive interference takes place, and a strong signal is detected by the photodiode.

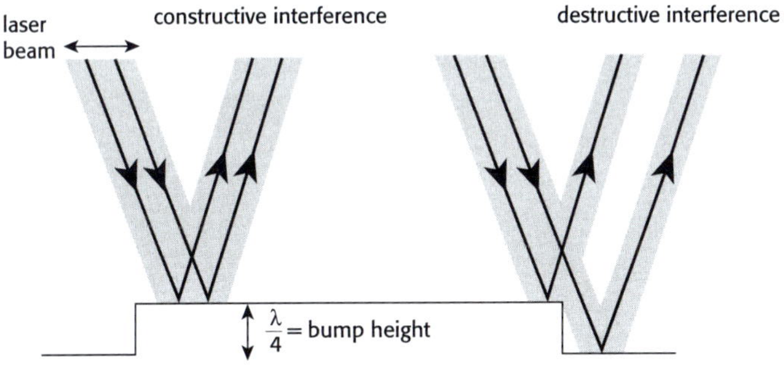

Figure 11 Reflection of the laser beam from the bump surface, and from the bump/pit transition.

If part of the beam is incident on a pit and part on a bump – that is, at the transition between the pit and the bump – then there is a path difference between the two parts of the reflected beam. If the pit height and laser beam wavelength, λ, are selected appropriately, this can lead to destructive interference at the photodiode. The signal is reduced in intensity. (See page 91 in Chapter 8 for more details of interference.) As the beam reflects off the spiral track, the destructive and constructive interference conditions mean that a digitally varying signal is detected at the photodiode. This signal, along with an accurate timing pulse, is converted into a series of 1s and 0s. Finally a digital-to-analogue converter (DAC) converts this digital signal back into an analogue signal. The width of the pits and bumps are varied to reproduce exactly the right binary code for conversion back into a recognizable analogue signal.

(a) The "bumps" for each digital bit on the surface of a CD are laid down along a spiral track around the disc. Dimensions are as follows:

Length of bump along track = 0.97 μm
Width of track = 2.1 μm

Show, using an estimate for the area for storage, that the disc can store about 4 gigabits (4×10^9 bits) of digital information. [2]

(b) Sound is stored in digital form by sampling the analogue waveform 44 000 times each second, and encoding each sample into a 16-bit number. Explain why 44 000 samples each second is an appropriate sampling rate. [2]

(c) Normally two stereo channels are recorded. Use this fact and the other information above to estimate the playing time available on a 4 gigabit music CD. [3]

Adapted from Edexcel Salters Physics, A2, paper 4, June 2001

Inquiry: DVDs and future technologies

Information technologies change rapidly as new techniques are discovered to store data on ever smaller devices and transfer data at ever increasing rates. At the time of writing (April 2007) DVDs are a very popular way of storing very large amounts of data – more than seven times the data capacity of a CD. This means that a DVD can hold the equivalent of about seven music albums or at least one full length movie. DVD writers (DVD-R) have been available for computers for a few years, and have now replaced the standard video (VHS) recorders used to record TV programmes and manage home movies.

In a small group, carry out some research to find out about DVDs and how they work. In particular, include:

- details of the structure and scale
- a comparison with a CD and explanation of why DVDs can store more data
- a brief outline of how a rewritable DVD-R works

Your findings should be presented in a format agreed with your teacher in advance (e.g. poster/article for school magazine/web page/podcast/essay/presentation). You should ensure that all ideas obtained in your research are properly cited.

Present your research to your class in a suitable way.

You may also want to include some details on the next possible technological breakthrough, which may change the way we store and access data.

?

1 Explain why the bump height is equal to ¼ of the incident beam's wavelength.
2 Calculate the wavelength of the laser beam for a bump height of 125 nm.
3 It has been suggested that a CD could be designed to store more data if a shorter-wavelength laser was used. Outline how the bump height would change.
4 A stereo music CD can store up to 74 minutes' worth of music. The sampling rate is 44 100 Hz, and each sample is stored as a 16-bit code. Show that a CD can store just under 800 MB of data. (Note: a stereo signal has two samples stored at the same time, one for each speaker.)

Data capture and the CCD

CCD (charge-coupled device) and CMOS (complementary metal oxide semiconductor) sensors have revolutionized the way images can be stored and captured. They are used in digital cameras, camcorders, fax machines, scanners, and astronomical telescopes. Both sensors operate on the same basic principle: light falling on the sensor results in an accumulation of electric charge. The value and position of this charge is then converted into a digital signal. CCDs produce better-quality images with less noise but are more expensive than CMOS sensors. Improvements in design, however, mean that at the time of writing (2007) it is not clear which type of sensor is going to be preferred for everyday applications. This section will focus on CCDs.

CCD basics

A CCD is composed of thin layers of light-sensitive semiconducting material arranged in small independent elements called **pixels** (**pic**ture **el**ements). A typical CCD is 2 cm × 2 cm in dimension and contains an array of 1000 × 1000 pixels (Figure 12).

The semiconducting material in each pixel is sandwiched between two thin electrodes. In a camera, light from the image is focused onto the CCD. Each pixel in a CCD behaves like a **solar** (photovoltaic) **cell** and converts light energy into electrical energy The incident photons free electrons from the semiconductor through the **photoelectric effect** (see page 207 for more details). The amount of charge that collects in each pixel is proportional to the number of photons or light intensity incident upon it. More charge collects in a pixel where the image is brighter than in a region where it is dim.

(a)

Source:NASA

(b)

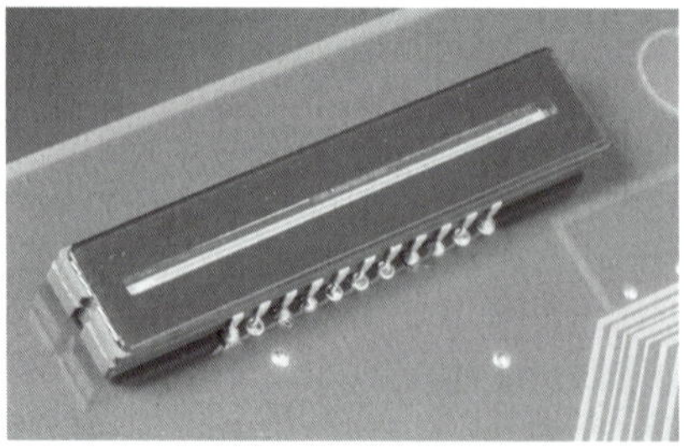

Source: http://www.betterlight.com/how_they_work.html

Figure 12 (a) A typical CCD, as used in a digital camera. (b) A linear CCD, as used in a scanner or fax machine

In this way the image is built up on the CCD: pixels with varying amounts of charge correspond to the different light intensities in parts of the image. The potential difference (p.d.) across each pixel increases with its charge. The value of the p.d and the position of the pixel are converted into a digital signal in binary code. The signal can be stored and then used later to reconstruct the image.

1 State what is meant by a semiconducting material.
2 A 2 cm × 2 cm CCD contains 1000 × 1000 pixels.
   (a) State the number of pixels in this CCD.
   (b) Calculate the length of one side of a pixel.
3 Briefly outline the photoelectric effect (revision exercise).
4 (a) Compare the amount of charge that collects in a pixel that is brightly illuminated with one that is not illuminated.
   (b) Describe how an image is represented on a CCD.
5 Explain why there is a difference in shape between a CCD used in a camera and one used in a scanner.

Capacitance

Each pixel in a CCD stores electric charge in a device called a **capacitor**. In its simplest form a capacitor consists of two parallel, conducting plates separated by a small insulating gap filled with a material called a **dielectric** (Figure 13). Real capacitors are generally quite small, about 1 cm or less in length, and generally have the plates and dielectric wrapped round in a cylindrical shape. Some typical capacitors are shown in Figure 14.

Capacitors may be charged either by contact with a charged object or by connecting them to a d.c. power supply. A charge, $-q$, builds up on one plate, inducing an equal but opposite charge of $+q$ on the other. If, for example, a capacitor is charged so that a charge of +250 nC collects on one plate and −250 nC on the other, it is conventional to say that the charge q on the capacitor is 250 nC.

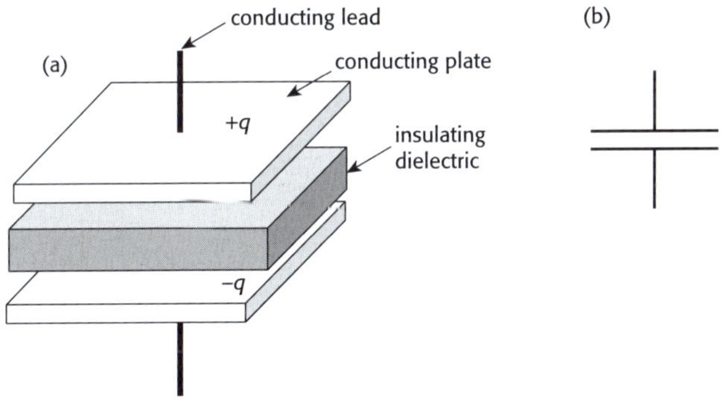

Figure 13 (a) Schematic illustration of a capacitor; (b) capacitor circuit symbol

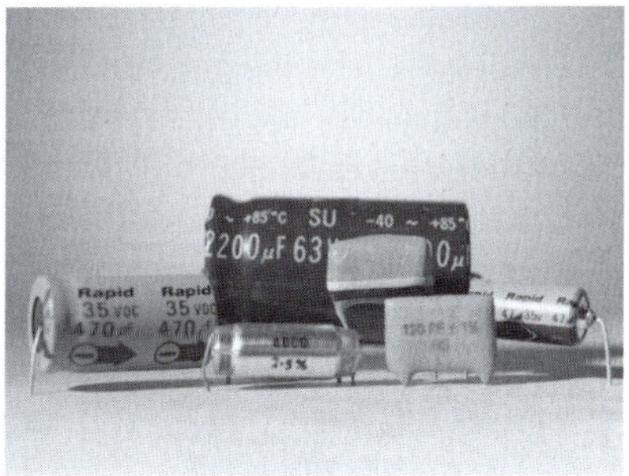

Figure 14 Miscellaneous capacitors

The circuit illustrated in Figure 15 can be used to charge a capacitor. Current flows for a short time until the potential difference across the capacitor is equal to the e.m.f. of the cell.

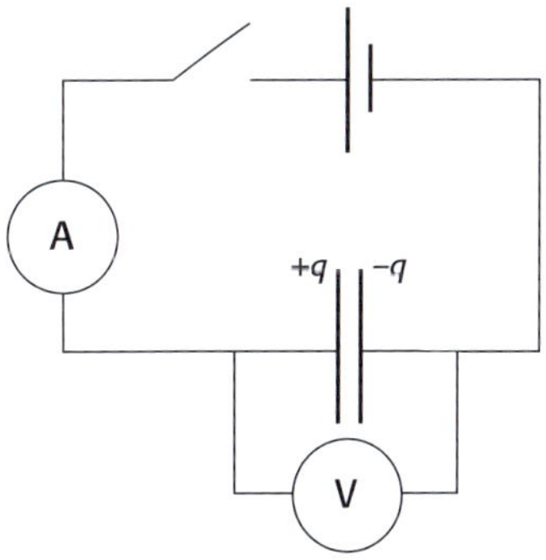

Figure 15 A circuit to charge a capacitor

Experiments show that the amount of charge stored on a capacitor, q, is proportional to the charging potential difference, V (Figure 16).

$$q \propto V$$

The constant of proportionality is defined as the **capacitance** of the capacitor.

$$q = CV$$

$$\Rightarrow C = \frac{q}{V}$$

1 Use a spreadsheet or otherwise to produce a graph of the data in Table 3, which shows the variation with voltage of the charge stored in a capacitor.
   (a) Add appropriate error bars to the graph.
   (b) Determine the capacitance of the capacitor.
   (c) Suggest an appropriate uncertainty for your answer in (b) using maximum and minimum gradients or otherwise.

Table 3

P.D. / mV ± 1 mV	200	190	180	170	160	150	140	130	120	110	100	90	80	70	60	50	40	30	20	10	0
Charge / nC ± 5 nC	200	190	180	170	160	149	140	129	120	110	100	90	80	70	60	50	40	30	20	10	0

2 A capacitor is charged with a 6 V d.c. supply. Calculate the capacitance of the capacitor when the maximum charge on the plates is
   (a) 30 nC
   (b) 300 μC.

3 A 20 μF capacitor is connected to a battery of e.m.f. 8 V. Calculate the charges on the plates.

4 Calculate the potential difference needed to charge the plates of a 10 μF capacitor to ±2.5 mC.

5 A constant 20 μA charging current is used to fully charge a capacitor in 20 s. During this time the potential difference across the capacitor rises from 0.0 V to 5.0 V. Calculate the capacitance of the capacitor.

6 If the circuit illustrated in Figure 15 is used to charge a capacitor, then the current falls from a maximum value to zero fairly quickly.
   (a) Outline in terms of electron flow how current can flow for this short time even though the capacitor is effectively an insulator.
   (b) Explain why the current falls from a maximum value to zero.

where
C is the capacitance
q is the charge stored in the capacitor
V is the p.d. across the capacitor

The unit of capacitance is the **farad**, F, where 1 F = 1 C V^{-1}. In reality the farad is an extremely large unit, so most commonly used capacitors have a capacitance measured in picofarads (pF), nanofarads (nF), or microfarads (μF).

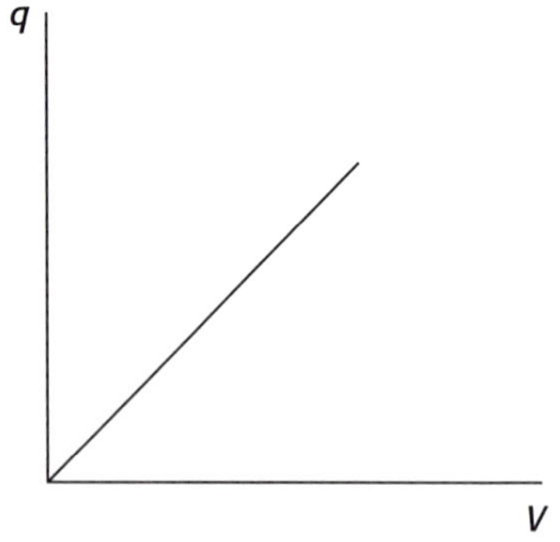

Figure 16 Sketch graph showing the variation of charge with p.d. across a capacitor

Quantum efficiency

The quantum efficiency of a CCD is defined as the ratio of the number of photoelectrons emitted in a pixel to the number of incident photons. That is:

$$QE = \frac{N_e}{N_p}$$

where
QE is the quantum efficiency
N_e is the number of photoelectrons emitted in a pixel
N_p is the number of photons incident on the pixel

In an ideal situation QE = 1: that is, each incident photon liberates an electron from the semiconductor. However, this is rarely obtainable.

In effect, the quantum efficiency is a measure of the CCD's sensitivity and can be related to the number of incident photons that the chip is actually able to detect. The quantum efficiency varies with the wavelength of the incident light, and is generally higher for back-lit CCDs than those that are illuminated from the front (see Figure 17).

Back-lit CCDs have relatively high quantum efficiencies, being greater than 70% for wavelengths of 450–800 nm, which corresponds to most of the visible waveband. Photographic film, by contrast, has a sensitivity of only about 10% – that is, only about 1 in 10 of the incident photons are recorded.

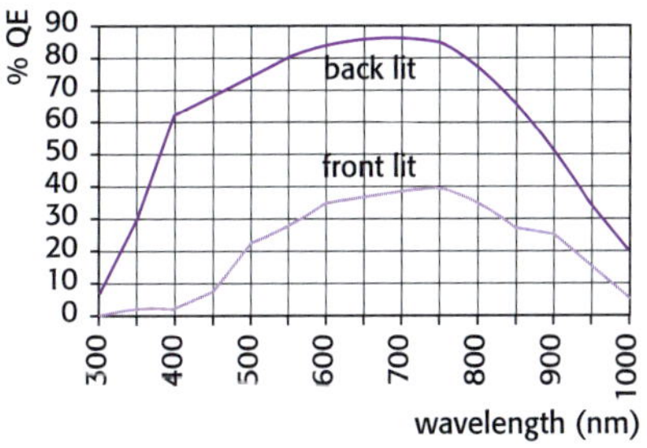

Source: http://www.ccd.com

Figure 17 Graphs illustrating the quantum efficiency of front-lit and back-lit CCDs over the range 200–1000 nm

Resolution

The resolution of a CCD is a measure of the smallest detail that the sensor is able to detect. The higher the resolution, the finer and more detailed the image becomes. The resolution of a CCD depends on the number of pixels. The more pixels there are in a given area, the greater the resolution. A typical chip with a 1024 × 1024 pixel array is said to have a size of 1 megapixel, although the actual number of pixels is slightly more than 1 million. At the time of writing, top-end professional cameras use CCDs with more than 8 megapixels.

Two adjacent points on an object are said to be just resolved by the CCD if the images of these points form on two pixels that are separated by at least one pixel in between. The Hubble Space Telescope was launched in April 1990 into orbit around the Earth. Being located above the atmosphere it is able to record much clearer, almost undistorted, images of distant astronomical objects, compared with telescopes on Earth. The image in Figure 18 is a globular cluster, NGC 2808, located in our Milky Way galaxy. This particular cluster contains more than a million stars.

The two adjacent stars, labelled A and B in the figure, are resolved. This means that they can be seen clearly as two separate objects. In order for this to happen the light from each star must fall on separate pixels on the CCD. The stars in region C are unresolved. They cannot be seen as separate objects, because light from several stars falls on one pixel.

In order to determine whether two separate objects in an image can be resolved or not, the magnification of the lens system that focuses light on the CCD also needs to be considered.

Magnification

When light from an object is focused onto a CCD, the image size is adjusted to fit appropriately onto the chip. Images of people, for example, are smaller than the people themselves, whereas pictures of, for example, animal cells are generally increased in size to allow detail to be studied. The **linear magnification** M of a lensing

?

This question is about a CCD.

(a) A digital camera is used to take a photograph of a freshwater crab. The CCD in the camera has 5.8×10^6 square pixels. Each pixel has an area of 5.8×10^{-10} m^2. A claw of the crab has an area of 1.5 cm^2. The image of the claw formed on the CCD is 0.10 cm^2. Two spots on the claw are separated by 0.80 mm. Deduce that the images of the two spots will be resolved. [4]

(b) Light is incident on the image collection area for a time of 50 ms. The number of photons incident on one pixel is 4.5×10^4. Each pixel has a quantum efficiency of 85% and a capacitance 20 pF.

(i) State what is meant by *quantum efficiency*. [1]

(ii) Estimate the change in potential difference across each pixel. [4]

(c) Outline how the variation in potential difference across individual pixels enables a black-and-white image to be produced by a digital camera. [2]

system is defined as

$$M = \frac{\text{image length}}{\text{object length}}$$

Digitizing the image

In order to store the image in digital format, at least two sets of data are required for each pixel:

- the value of the voltage
- the position of the pixel.

The voltage across each pixel is proportional to the intensity of the light falling on that pixel: the brighter the light, the higher the voltage. A high value represents a "white" part of the image, whereas a low value represents "dark grey" or "black". When the data for all pixels is combined, the image can be reconstructed.

A complex electronic process is used to record the value of voltage across each pixel and the pixel position. CCDs are designed so that these electronic components are mainly arranged around the edge of the chip. Some components do obscure the front of the CCD slightly, however – this is why backlit CCDs have a higher quantum efficiency. A sensor is not required for each pixel, so very little of the incident light is blocked out. CMOS chips, by contrast, detect the voltage directly at each pixel, which means that there are a lot more wires in front of the sensor. These block out some of the incident light, hence reducing the sensitivity of the chip.

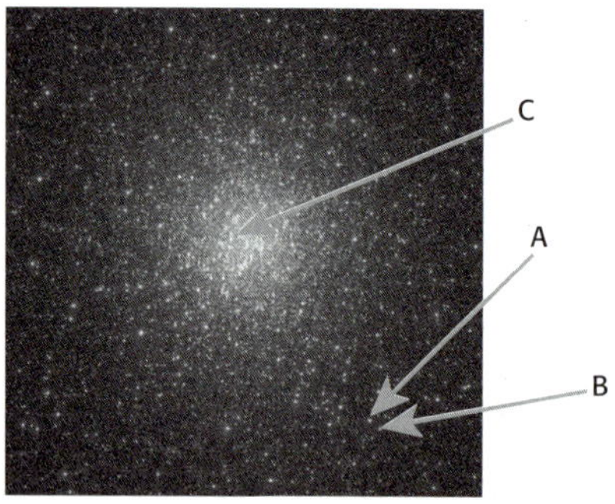

Source NASA, ESA. A. Sarajedini (University of Florida) and G. Piotto (University of Padua [Padova])

Figure 18 Globular cluster NGC 2808 taken by the Hubble Space Telescope in May 2005

A digital photograph is taken of a person whose eyes are 0.060 m apart. The distance between the eyes of the image incident on the CCD is 7.1×10^{-4} m. The CCD has an area of 4×10^{-4} m^2 and contains 5.0×10^6 pixels.

(a) Calculate the magnification of the camera.

(b) Calculate the length of one side of a pixel.

(c) By comparing the distance between two adjacent pixels and the distance between the eyes in the image on the CCD, determine whether the eyes are resolved.

Colour and bit depth

The CCD as described above produces grey scale images. Most chips use an 8-bit code, which means that there are 2^8 or 256 shades of grey available. A special filter, containing the primary colours of blue, red, and green, can be placed over the CCD so that the intensity of each specific colour can be detected. A special kind of filter called a **Bayer array filter** ensures that, in a group of four adjacent pixels, two detect green light, one red and the other blue (Figure 19); our eyes are more sensitive to green light, so this array simulates human vision reasonably accurately.

The p.d.s developed over each such group of pixels are combined to create a true colour image at each pixel.

If an 8-bit code is used to encode the intensity for each primary colour, this gives 256 different values. When all three primary colours are combined at each pixel, this gives $2^{(8 \times 3)}$ or 16 777 216 different colours. This is known as **true colour imaging**. This number of colours may seem a bit excessive, considering that the human eye can only distinguish about 10 million colours. The number of bits used to store the digital signal is called the **bit depth**.

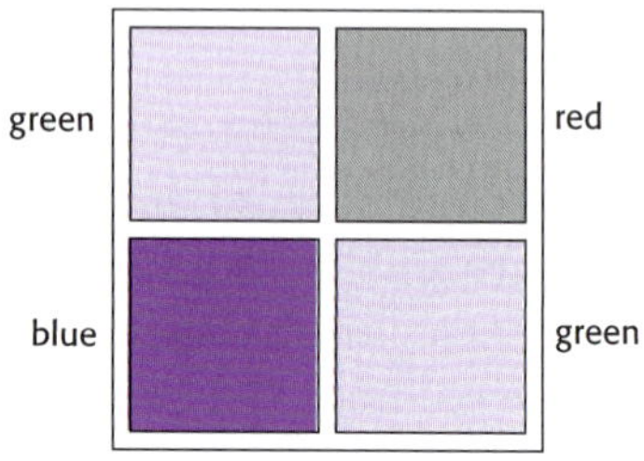

Figure 19 A Bayer array filter used with four adjacent pixels.

Carry out some research and make a list of some of the advantages and disadvantages of using CCDs to produce images compared with film.

5 The extended essay

One of the unique aspects of the IB Diploma Programme is the **extended essay**. Every student has to submit this formal piece of independent research or investigation, which is the result of approximately 40 hours of work. The essay (which should contain no more than 4000 words) is externally assessed, and, in combination with the theory of knowledge grades, can contribute up to three extra bonus points to the total IB Diploma score.

The essay needs to be associated with one specific Diploma Programme subject, and an appropriate teacher has to be available to act as a supervisor through the process in order to provide advice and guidance. Although many of the principles of this chapter apply to all extended essay subjects, it has been written with a physics extended essay in mind. Students familiar with the IB Middle Years Programme (MYP) will find that the extended essay is a natural progression from the MYP personal project.

A physics extended essay is a piece of **new research** with a clear emphasis on physics, and will be judged only on its physics content. New, in this context, means new to you, the student. Practical data needs to be linked to a physical model or hypothesis for any given situation. This also means that great care needs to be taken with an interdisciplinary area such as materials science, and ideally this should be avoided. Whatever topic is chosen, clearly any experimental work must be safe, and this may rule out some possible areas of focus.

Types of extended essay

There are several different approaches that can be used when writing a physics extended essay. The most successful essays are usually experimental in nature: the author designs and carries out an experiment and analyses the data appropriately.

In this situation, it is important to keep the initial experiments as straightforward and uncomplicated as possible. An essay that relies on the complicated construction of apparatus before any readings are available has the potential to waste a great deal of time. The effort expended in building the apparatus does not gain any credit in the assessment of the essay.

Other non-experiment approaches can still score highly but care must be taken to ensure that the essay is still a valid piece of *new research*, and that the analysis and the conclusions of the essay are the author's and nobody else's. For example, it would be inappropriate to attempt to use the essay to summarize current research on a particular area of physics. This does not constitute new research, and such an essay, if submitted, would score poorly. An essay on "black holes", for example, could be nothing more than a literature review based on the writing of other authors, unless you are able to collect and analyse your own data.

If secondary sources were used as the source of data, then an important aspect of the essay would be to assess the reliability of the sources and to challenge assumptions and statements with alternative interpretations. As a general rule, an essay that is based solely on material found in textbooks or on the Internet is unlikely to provide sufficient scope for critical analysis, and so is unlikely to score highly.

The key to a good extended essay is a well-focused, well-defined, and realistic research question that allows for an in-depth treatment within the limitations of your school science laboratories. It is recommended that the situation being studied can be understood without having to study new areas of physics beyond those already required by the IB Diploma Programme, though this does not mean that new areas cannot be considered. The idea for the essay should be yours, not your supervisor's. If you are unable to come up with a suitable research question in physics then it would be sensible to choose another subject for your essay.

Formal presentation of the extended essay

The vast majority of essays are word-processed. All extended essays must have the following:

- **Title page**. This provides a clear indication of the focus of the essay. It should be precise, and the title of the best essay is often written as a research question. The title "Force/extension characteristics", for example, is rather dull and extremely wide ranging. A much more focused title could be: "Do conditioners really soften your hair? A physical analysis of human hair involving force/extension graphs".
- **Abstract**. This is a separate page that consists of a couple of paragraphs not exceeding 300 words in total. It presents an overview or summary of the extended essay. It needs to include the precise research question, the process that was undertaken, and the conclusion of the essay. It is one of the last things to be written.
- **Contents**. All pages in the essay should be numbered, and this allows the different sections to be identified with subheadings that are easily located.
- **Introduction**. This is a summary section that places the topic of the essay into context, and links the research question to existing physics ideas. The research aim or question should be clearly and succinctly stated here, either as a direct question or as a statement or proposition for discussion. There is no need to "pad out" the essay with irrelevant information. Some poor essays involve the student writing all they can find out about their topic without any specific focus.
- **Body (development/methods/results)**. This is the core of the essay. The knowledge and understanding demonstrated should build on physics that you have already studied. All graphs, diagrams, and summary tables of data should be incorporated within the text and must be directly relevant to the essay. They should be numbered so that specific references can be made to them in the body of the essay. All personal views and conclusions

from the data need to be justified with supporting arguments. To improve the flow of the essay, it is sometimes convenient to put all your raw data into a suitably referenced appendix and include just a summary of the key data in the main body.

- Your essay needs to include an **evaluation** of the processes and/or experimental work that has been undertaken. It does not need to be a separate section, but it is an integral part of the essay.
- **Conclusion**. This summarizes the ideas that have been discussed in the body of the text. It should be based directly on your data, and could involve a quantitative statement such as a mathematical relationship between the variables under investigation. This is not the place to introduce new ideas.
- **References and bibliography**. All sources of quotations, ideas, and points of view used in the essay need to be cited properly at the appropriate point in the text, and accurately identified in a list of references in the bibliography. This is an alphabetical list of sources that were used in the production of the text. General sources of information that are not directly cited in the body of the essay should be cited in the introduction or a separate acknowledgment.
- **Appendices**. These contain additional information and data.

Citation

A very important aspect of the extended essay is the correct citation of all sources of information used. It does not matter what particular layout is used for this academic referencing, but if someone else's opinions or material is included in the essay without a proper citation then this would be **plagiarism** – an attempt to pass off other people's ideas as your own. Plagiarism is as serious as being caught cheating in an exam.

There are many possible documentation styles for being consistent with citations, and often your extended essay supervisor or coordinator will recommend a style for your school. The aim is to provide the reader with sufficient information to be able to locate the same source of information that the writer of the essay used. For a book the minimum information required is the author, the title, the publisher, and the date and place of publication. Web page citations need the home page title, the precise URL, and the date that it was accessed. The following website gives a useful overview of the different types of acceptable referencing system.

http://www.wisc.edu/writing/Handbook/Documentation.html

General knowledge and/or concepts such as Newton's law of gravitation or precise definitions do not require any specific reference.

The concluding interview **(viva voce)**

At the very end of the process, the supervisor gives the author a short interview, which is an opportunity to reflect on the whole process and on what has been learned. The supervisor writes a report, which is submitted with the essay to the external marker.

The length of the extended essay

The upper limit is 4000 words for all extended essays. This upper limit includes the introduction, the body, the conclusion, and any quotations, but does not include:

- the abstract
- acknowledgements
- the contents page
- maps, charts, diagrams, annotated illustrations, and tables
- equations, formulas, and calculations
- citations/references (whether parenthetical or numbered)
- footnotes or endnotes
- the bibliography
- appendices.

Essays containing more than 4000 words are subject to penalties, and examiners are not required to read material in excess of the word limit. It is, however, very difficult to be precise in physics, as most essays include tables. The following conversions apply if including computer code or if writing in Japanese or Chinese.

Japanese: 1 word = approximately 2 Japanese characters
Chinese: 1 word = approximately 1.2 Chinese characters
Computer code: Each line of code of a program fragment included in the body of the essay should count as two words towards the word limit.

Timescale

Although the task can initially seem daunting, the process becomes manageable if broken down into smaller individual tasks. For this approach to be successful it is important to set yourself deadlines for when these tasks are going to be completed. Different schools have different approaches.

Some schools allow students a great deal of flexibility, whereas others choose to control the whole process in great detail. Whatever process you are working under, it is extremely important to stick to deadlines. The list below identifies 15 important milestones. You need to know (or decide) when these tasks are going to be completed.

1. Choice of subject area and identification of supervisor
2. Choice of research question and general approach to be used
3. Preliminary experiments, pilot tests, and/or literature research
4. Main experiment and/or data collection
5. Analysis of data
6. Discussion of provisional findings with supervisor
7. First draft of the main body of the essay
8. First draft of abstract, references, bibliography, acknowledgments, etc.
9. Submission of complete first draft
10. Receipt of feedback from supervisor
11. Revision work on essay
12. Submission of second draft
13. Proofreading and final corrections
14. Submission of final draft
15. Concluding interview (viva voce)

Although it is impossible to plan for the unexpected, the timeline that you create should be flexible enough to be able to cope with delays and/or unforeseen problems.

Throughout the whole process it is very important to keep your supervisor up to date with your progress, and to discuss any difficulties, or any changes of direction you are contemplating.

Possible topics to consider

The IB Extended Essay guide presents some examples of research question essays that represent the range of approaches possible.

- Is it possible to determine the presence of a black hole at the centre of the Milky Way?
- Do wine bottles of different shapes behave as Helmholtz resonators?
- What will be the angular deflection of starlight by the Sun if Newton's universal law of gravitation is applied?
- Is the efficiency of electromagnetic damping of a moving glider a function of the initial kinetic energy of the glider?

Alternatively, an essay may focus on a statement or a hypothesis:

- The objective is to establish theoretically the proportionality existing between the terminal velocity of a cylindrical magnet falling down a metallic pipe and the resistivity of the metal of the pipe as well as the pipe's wall thickness.
- Water waves are observed in a long and narrow trough, and their speeds are measured. It is assumed that, for shallow water, the speed of the wave will be proportional to the square root of the depth of the water and independent of the wavelength.
- The objective is to establish the relationship between power and temperature for an incandescent lamp.
- A retractable ballpoint will be used to test the law of energy conservation.
- The objective is to establish an acoustic model of the concert flute.

26 Exam techniques

This chapter is devoted to improving your examination techniques. The two most important pieces of advice are:

- Don't panic.
- Read the question carefully.

Examination technique can be improved with practice, and the best students consolidate their knowledge and understanding at the end of a topic by attempting past examination questions. Initially these can be done without any time constraints, but as the real examinations come closer it is important to practice working within the same time as you will have in the real examination.

In every examination you have access to a clean copy of the physics Data Booklet. Apart from some basic pieces of information, such as the SI multipliers and the values of some fundamental constants, this document lists a large number of equations. It is important that you become familiar with these equations as you study the individual topics. Many schools give their students their own copy of the Data Booklet at the beginning of the course and allow them to add annotations and notes.

These notes cannot be brought into the final examinations, but they do allow students to ensure that they properly understand every formula that is contained in the booklet.

Always remember that the examiner can only give you credit if you are able to demonstrate knowledge and understanding. This means your answers must be understandable and legible. It is also important to show all the steps in your working. A wrong answer without any working must be worth zero marks. A wrong answer that allows the examiner to identify, for example, a simple mathematical slip during a calculation would lose only one mark from the total. Remember that, unlike your teacher, the examiner does not know you. In order to show excellence you must take care to craft your answers to the highest quality and not take short cuts or use abbreviations as you might do in internal tests.

Possible points to keep in mind include:

- If a question asks to you to explain how an observation is consistent with a named piece of theory, an important part of the answer would include an outline of the named theory.
- Sometimes there are several ways in which to approach a physics problem or explanation. If a question specifies the approach to take, it is very important to do just that. For example, if the question asks you to explain the expansion of a gas on heating *with reference to the kinetic theory of gases,* is very important to explain what this theory means and use it to answer the question.

Some calculations involve several steps, and these are often presented as different parts of one question: (a), (b) etc. In this case the answer in a later part of the question depends on an earlier answer. The principle of error carried forward (**ECF**) applies: if your working demonstrates that you have got the wrong answer to a later section because you have used an earlier wrong answer in the right way, you would be awarded the full marks for that section even if your answer was wrong.

Once you have arrived at an answer to a mathematical question, it is important to think about two additional things. First, the answer will almost certainly have units associated with it. If you forget to include these units, then your answer will not be complete and you will lose marks. Second, the number of significant digits that you quote in your answer should match the number of significant digits given in the original data.

For example, a calculator gives the result of dividing 8.3 by 1.7 as 4.882352941. The average speed of an object covering a distance of 8.3 m in a time of 1.7 s, however, is not 4.882352941 but should be correctly stated as 4.9 m s^{-1}. The units of the answer are derived from the units of the original data. The data was given to two significant digits, so the answer should be quoted to the same level of accuracy.

Command terms

The IB syllabuses for all group 4 subjects make use of the same limited number of **command terms** that define the level of detail required. The same command terms are used in the examination papers, so it is important to understand their precise meaning. The tables below lists the command terms classified by **objective level**.

The objective level is best imagined as a measure of the level of difficulty or detail of understanding required. There are three different objective levels: objective 1 is easiest and objective 3 the hardest. Every syllabus statement is categorized by objective level.

Objective 1

The syllabus statements and the questions that use these command terms require you to be able to remember information only, and so they represent the easiest tasks. You need to be able to demonstrate an understanding of facts, concepts, scientific methods and techniques, scientific terminology, and methods of presenting scientific information.

Define	Give the precise meaning of a word, phrase, or physical quantity.
Draw	Represent by means of pencil lines.
Label	Add labels to a diagram.
List	Give a sequence of names or other brief answers with no explanation.
Measure	Find a value for a quantity.
State	Give a specific name, value, or other brief answer without explanation or calculation.

Objective 2

The syllabus statements and the questions that use these command terms require you to be able to apply the information that you have learned in a given situation. They represent tasks that are probably reasonably straightforward but that involve some problem-solving ability. You need to be able to demonstrate an ability to apply and use scientific facts and concepts, scientific methods and techniques, scientific terminology to communicate effectively, and appropriate methods to present scientific information.

Annotate	Add brief notes to a diagram or graph.
Apply	Use an idea, equation, principle, theory, or law in a new situation.
Calculate	Find a numerical answer showing the relevant stages in the working (unless instructed not to do so).
Describe	Give a detailed account.
Distinguish	Give the differences between two or more different items.
Estimate	Find an approximate value for an unknown quantity.
Identify	Find an answer from a given number of possibilities.
Outline	Give a brief account or summary.

Objective 3

The syllabus statements and the questions that use these command terms require you to be able to apply the information that you have learned to a more complicated situation. They represent tasks and problems that involve several stages or steps before arriving at a possible solution. You need to be able to demonstrate an ability to construct, analyse, and evaluate hypotheses, research questions, predictions, scientific methods and techniques, and scientific explanations.

Analyse	Interpret data to reach conclusions.
Comment	Give a judgment based on a given statement or result of a calculation.
Compare	Give an account of similarities and differences between two (or more) items, referring to both (all) of them throughout.
Construct	Represent or develop in graphical form.
Deduce	Reach a conclusion from the information given.
Derive	Manipulate a mathematical relationship(s) to give a new equation or relationship.
Design	Produce a plan, object, simulation, or model.
Determine	Find the only possible answer.
Discuss	Give an account including, where possible, a range of arguments for and against the relative importance of various factors, comparisons, or alternative hypotheses.
Evaluate	Assess the implications and limitations.
Explain	Give a detailed account of causes, reasons, or mechanisms.
Predict	Give an expected result.
Show	Give the steps in a calculation or derivation.
Sketch	Represent by means of a graph showing a line and labelled but unscaled axes but with important features (for example, intercept) clearly indicated.
Solve	Obtain an answer using algebraic and/or numerical methods.
Suggest	Propose a hypothesis or other possible answer.

Structure of exams

The three papers in the physics exam are designed to test different topics, but also to test different skills and levels of understanding. A clean copy of the *Physics Data Booklet* is required for each paper at both SL and HL.

Paper 1

This is the multiple-choice paper and addresses objectives 1 and 2 (see above). Questions can be on any topic in the common core (topics 1 to 8 for standard level and 1 to 14 for higher level students). Time is very short: SL candidates have to answer 30 questions in 45 minutes and HL candidates have 40 questions in 60 minutes. This corresponds to 90 seconds per question for reading, solving, and entering your answer on the sheet provided.

You have access to the Data Booklet but are not allowed to use a calculator. This means that the examiners cannot ask questions that involve complicated calculations. Any sum must be able to be done quickly in one's head.

Paper 2

This is the longest paper and addresses all three objectives, and a calculator is required. Typically this paper is timetabled immediately after paper 1, so you need to be prepared for a long examination session. The paper requires answers that are written on the question paper. You can always ask for additional paper if you run out of space. As for paper 1, questions can be on any topic in the common core. The paper is divided into two sections.

In section A there is a data-based question that requires students to analyse a given set of data. The remainder of section A is made up of short-answer questions. This section is all compulsory. The total number of marks is 25 for SL candidates and 45 for HL candidates. A good rule of thumb is that there is about a minute available per mark awarded.

In section B students at SL are required to answer one question (worth 25 marks) from a choice of three, and students at HL are required to answer two questions (each also worth 25 marks) from a choice of four. These extended-response questions may involve writing a number of paragraphs, solving a substantial problem, or carrying out a substantial piece of analysis or evaluation.

At the beginning of the examination you will have 5 minutes' reading time. During this time it is important to read through the questions in section B and think about which question(s) you will attempt. One possible approach is to attempt the questions in the order in which you are most confident, attempting your "best" question first and your "worst" question last. With this approach you will tend to use time more effectively.

Paper 3

Paper 3 tests knowledge of the options and addresses objectives 1, 2, and 3; again, a calculator is required. You will also get 5 minutes' reading time. If you have managed to study more than the minimum number of options, which is not recommended, then you can use this time to decide which options you wish to attempt. If you have only studied the minimum number of options, as is generally the case, then you can use this time to start thinking about how you are going to tackle the questions.

It is very important that you do not attempt an option question in a topic that you have not studied. It often happens that the first part of an option question is fairly straightforward and can be attempted successfully by most students, whatever options they have studied. However, you will soon find that you will be unable to proceed further. Every year, it is obvious when students have attempted a question in an option they have not studied by the poor quality of their answers.

Students at SL are required to answer several short-answer questions in each of the two options studied. Each option is worth 20 marks. Students at HL are required to answer several short-answer questions and an extended-response question in each of the two options studied. The questions in each HL option are worth 30 marks.

The assessment specifications at SL and HL are summarized in the following assessment specifications sections.

Number of marks available

Each mark allocated to any given question corresponds to a step in a calculation or a point that is required in your answer. For example, an experimental description that is worth 4 points must be looking for four different pieces of information, and a calculation worth 3 marks will involve at least three identifiable steps before getting to the correct answer.

Many students take the number of lines available on the answer paper to be an indication of the amount of information required. For this reason the aim is to provide sufficient rather than excess space. If you need to write more information to complete your answer, make sure you ask for some continuation paper, which should be attached to the exam script. However, it is often the case that an answer that occupies a lot more space than is provided indicates that the student has not understood what is being tested.

Use of technical language

It is very important to use technical language in a precise way. For example, *speed* and *velocity* have the same meaning in everyday language, but they have different scientific definitions (speed is a scalar whereas velocity is a vector). It is correct to say that an object with constant velocity must have a zero resultant force acting on it, but it is not correct to say that an object with constant speed must have a zero resultant force. Confusions are also common with mass and weight.

Mathematical terms are often used inappropriately. See pages 20–1 for the difference between a proportional relationship and a linear relationship between two variables.

Describing graphs accurately can involve some complicated language. For example, consider an experiment in which an object is allowed to roll down a slope from rest. Measurements are taken of the velocity of the object at different times from the instant of release. The graph in Figure 1 is obtained.

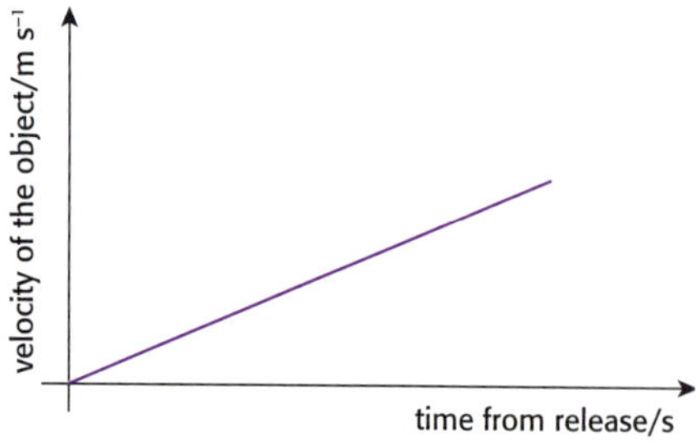

Figure 1 Is this relationship proportional or linear?

A precise way of describing this graph would be "the variation with time from release of the velocity of the object". The form that is used to describe the graph is "the variation *with* (the variable on the x-axis) *of* (the variable on the y-axis).

Use of drawings

There are many situations where a good, labelled diagram or sketch graph can be used demonstrate, quickly and precisely, knowledge and understanding. A poor diagram, however, can be confusing. Remember the following points:

- Precise labels are essential.
- Use a sharp pencil for drawing.
- Graph lines should be either a smooth curve or a straight line drawn with a ruler.
- Think about the need for precision in your drawings. For example, unless labelled otherwise, the relative lengths of the arrows in a free body diagram will be taken to represent the relative magnitudes of the forces involved.

In conclusion

If you are approaching your final examinations, you will have completed nearly two years of challenging study. So long as you have followed an extensive and thorough revision plan over several months there is no need for you to worry about the final exams. Consistent hard work throughout the course and a final dedicated revision programme are certain to get you successfully through the final examination.

Good luck with the exams and in your future studies!

27 Option E: Astrophysics

Have you ever been lucky enough to see stars on a clear night far away from the artificial lights of a city? Many of the pinpoints of light that you see whenever you look at the night sky may in fact be huge collections of stars, called galaxies. In fact, the observable universe is made up of more than 10^{20} stars at distances of up to 10^{27} m away from the Earth, making its truly huge dimensions hard to comprehend.

The Astrophysics option aims to give you a glimpse into the processes that have shaped the development of our universe into the one we now observe. This understanding can not only be used to predict the future development of the universe, but also often results in a sense of awe and wonder at our own place in space.

This chapter aims to provide the framework for studying the astrophysics option rather than all of the detailed information required. Many of the questions in the text require some independent research and there are many excellent sources of further information available on the Internet or in textbooks. Of particular note are the pages associated with space agencies that involve a great deal of international collaboration, for example NASA (the National Aeronautics and Space Administration, www.nasa.gov) and ESA (European Space Agency, http://sci.esa.int). The latter site has published teacher support material closely linked with this IB Diploma Programme Phyiscs option on their teacher pages, for example http://sci.esa.int/science-e/www/object/index.cfm?fobjectid=35713

Other possibilities include RKA (the Russian Federal Space Agency, www.federalspace.ru) or CNSA (China National Space Administration, www.cnsa.gov.cn).

Introduction to the universe

An observer at a given location on Earth sees a fixed and essentially unchanging pattern of stars. This pattern rotates from east to west through nearly 180° during the night. Although the pattern of stars remains essentially the same, slightly different sections of the pattern are visible at different times during a year. Different observers located around the world will see different sections of the whole pattern of stars and observers in the northern hemisphere see different stars from those in the southern hemisphere.

The fixed patterns of stars has been categorized into 88 **constellations**. The stars and galaxies in a constellation are all located in the same general direction from Earth but are not necessarily close to one another. Historical observers named the constellations after mythical figures and creatures which the shapes seemed to represent.

A change that is only noticeable over several nights is that during any given night the Moon and all the planets follow the same rotation as the stars but over time they move position with respect to the "fixed" background pattern of the stars.

It may not be surprising that ancient civilizations explained these observations by imagining the stars and planets to be fixed on transparent spheres that rotated around the Earth.

The same observations are better explained by the more modern view of the universe that says that the stars are fixed and the Earth rotates on its own axis (time period 24 hours) and revolves around the Sun (time period 1 year).

The distribution of stars

Stars are not uniformly distributed throughout space but are gathered together by their mutual gravitational attraction. Small groups of stars that gravitationally interact with one another are called **stellar clusters**. Very large numbers of stars bound together by gravity make up a **galaxy**.

To measure the large distances between stars, astronomers use a unit called the **light-year** (ly), which is the distance travelled by a beam of light in one year. An alternative unit is the **parsec** (pc) which is defined as 3.1×10^{16} m. The nearest star to the Earth (Proxima Centauri) is approximately 4.27 ly away.

The solar system

Our **solar system** consists of eight planets (Mercury, Venus, Earth, Mars, Jupiter, Saturn, Uranus, and Neptune) in elliptical orbits around the Sun. The gravitational attraction between the Sun and the planet provides the centripetal force necessary for the planet to stay in orbit. In addition to the planets, smaller masses called **planetoids** (e.g. Pluto), **asteroids**, and **comets** orbit the Sun. Many planets (including Earth) are also orbited by moons. The table below shows the approximate distances, comparative radii and masses of the planets.

Table 1

	Distance from Sun / 10^6 km	Radius / radius of Earth	Mass / mass of Earth
Mercury	57.9	0.38	0.06
Venus	108.2	0.95	0.86
Earth	149.4	1.00	1.00
Mars	227.9	0.53	0.15
Jupiter	778.3	11.19	1323
Saturn	1427	9.41	752
Uranus	2869.6	3.98	64
Neptune	4496.6	3.81	54

The average distance from the Earth to the Sun is called the **astronomical unit (AU)**. This unit is useful for talking about distances on the interplanetary scale.

?

1 Describe, using your own words and diagrams, the apparent motion of the stars over a period of a night and over a period of a year.
2 Explain the apparent motion of the stars over a period of a night and over a period of a year in terms of the rotation and revolution of the Earth.

?

1 Show that 1 light-year is approximately equal to 9.5×10^{15} m. How many light-years are there in a parsec? How far away is Proxima Centauri in parsecs?

Our own galaxy contains approximately 2×10^{11} stars and its diameter is approximately 10^5 light-years in diameter. Our Sun takes approximately 2×10^8 years to complete one orbit around the centre of the galaxy.

Galaxies also occur in **clusters**. There are approximately 20 other galaxies within 2.5×10^8 light-years of us including the Andromeda galaxy. Galaxy clusters also group together in **superclusters**. Despite the apparent abundance of stars, most of space is completely empty (a vacuum).

2 Compare, in terms of order of magnitude, the relative distances between stars within a galaxy and between galaxies.

?

1 Use Table 1 to compare the planets in terms of size and distance from the Sun.
2 Research the difference between an asteroid and a comet.
3 Research the nature and position of the asteroid belt.

Our Sun is a typical star. The only reason it appears to be bigger and brighter than all the other stars is that we are much closer to it than to the other stars.

Stellar radiation and stellar types

Stars are large nuclear fusion reactors, which convert light nuclei (hydrogen, helium, etc.) into heavier nuclei (carbon, iron, etc.). The process releases a great deal of energy as thermal energy and electromagnetic radiation. See page 302 for more details of the process.

Most of the atoms in a stable star are in the fourth state of matter – plasma. Many stars radiate energy whilst remaining stable for billions of years and from this we can infer that the processes going on in the star must be balanced. Firstly, the star must be in **thermal equilibrium**. As the star's temperature remains constant, the rate at which energy is being radiated must equal the rate at which nuclear energy is being converted into thermal energy.

Secondly, the star must be in **hydrostatic equilibrium** because it does not expand or contract. This is contrary to our expectations that a hot object in a vacuum would simply expand to fill the space. The only force keeping the constituent mass of the star together is gravity so at any given point in the star, the inward gravitational forces must balance the outward force. This is sometimes expressed in terms of pressure. We say that there must be equilibrium between the outward **radiation pressure** and the inward **gravitational pressure**.

?

Research the structure of our Sun with specific reference to the the following regions: core, radiative zone, convective zone, photosphere, chromosphere, and corona.

Given the huge distances involved, most, if not all, of the information that we can gain about any star comes from an analysis of the electromagnetic radiation that we receive from it. The total power radiated by a star is called its **luminosity *L***. The relative brightness of a star depends on the power received from it over a given area on the surface of Earth. This will depend on the luminosity and on how far away the star is. The **brightness *b*** of a star is defined as the power received per unit area:

$$L = \frac{b}{4\pi r^2}$$

where
L is the luminosity of the star, measured in W
b is the brightness of the star, measured in W m^{-2}
r is the distance between the star and the observer, measured in m

This relationship can be used to calculate the stellar distance once a star's luminosity and brightness have been determined. Comparisons of brightness of different stars are also often expressed in terms of a different measurement – the **magnitude**. See page 299–300 for more details.

The radiation emitted from stars corresponds closely to the theoretical black body model for radiation introduced on page 208. This allows us to utilize two further relationships to analyse radiation from stars.

The German physicist Wilhelm Wein (1864–1928) identified a relationship that allows us to calculate the surface temperature of stars from their colour. **Wien's law** links the absolute temperature of a black body and the wavelength at which the radiated energy reaches its maximum intensity.

$$\lambda_{max} = \frac{2.90 \times 10^{-3}}{T}$$

where
T is the absolute surface temperature of the star, measured in K
λ_{max} is the peak wavelength (where radiated energy reaches a maximum intensity), measured in m

The **Stefan–Boltzmann law** can be used to calculate a star's surface area (and hence size) if we know its luminosity and its surface temperature. For any black body radiator, the total power radiated is calculated as follows:

$$L = \sigma A T^4$$

where
L is the luminosity of the black body (i.e. the star), in W
σ is the Stefan-Boltzmann constant, 5.67×10^{-8} W m^{-2} K^{-4}
A is the surface area of the black body (i.e. the star), in m^2
T is the absolute temperature of the back body (i.e. the surface temperature of the star), in K

The surface area of a star is linked to its radius r:

$$A = 4\pi r^2$$

It should be noted that luminosity can only be calculated for a star from a measurement of its brightness providing we know the distance between the star and the observer. See page 299–300 for further details of how stellar distances as determined.

Finally, it is possible to deduce chemical and physical data from stellar radiation by analysing the wavelengths present in the light for evidence of atomic spectra. Further detail can be deduced from the relative intensities of each of the spectral wavelengths but details of this analysis are beyond the scope of this Course Companion.

Types of star

The analysis of spectral lines allow us to categorize the different types of star that exist into different spectral classes. Each star can be allocated into one of seven categories. Each category is designated by a letter which, in descending order of surface temperature, are: O, B, A, F, G, K, M. Each type is then further subdivided into ten groups designated by a number from 9 to 0. The Sun is in spectral class G2.

The majority of stars are, in fact, **binary stars**. This means that the points of light visible on Earth are not single stars but pairs of stars orbiting their common centre of mass. Some binary stars can be resolved either by eye or by telescope and are called **visual binaries** but the existence of the majority of binary stars has been deduced

?

1. Use Figure 1 to calculate the peak wavelength for a star with a surface temperature of (a) 12 000 K, (b) 6000 K, (c) 3000 K.
2. Use your answers to explain why relatively cool stars are orange/red in colour, medium temperature stars are yellow/white, and hot stars are bluish white.
3. Explain why green stars do not exist.

?

1. Calculate the luminosity of the Sun. Its radius is 6.96×10^8 m and the peak wavelength of its emitted radiation is 500 nm.
2. Use your answer to question 1 and the mass of the Sun (1.99×10^{30} kg) to calculate the power emitted per unit mass. Compare your answer with 1 W kg^{-1}, the average power emitted by a human being. What can be deduced from this information?
3. Explain whether starlight is an example of an absorption or an emission spectra. How are these spectra used to provide evidence for the chemical composition of stars?
4. Explain whether starlight allows us to deduce the elements in the interior of the star, in its outer layers, or both.

Table 2

Class	Surface Temperature / K	Colour
O	28000 – 50000	blue
B	9900 – 28000	blue
A	7400 – 9900	blue-white
F	6000 – 7400	white
G	4900 – 6000	yellow
K	3500 – 4900	orange
M	2000 – 3500	red

from the analysis of their stellar radiation. These systems are called **spectroscopic binaries**.

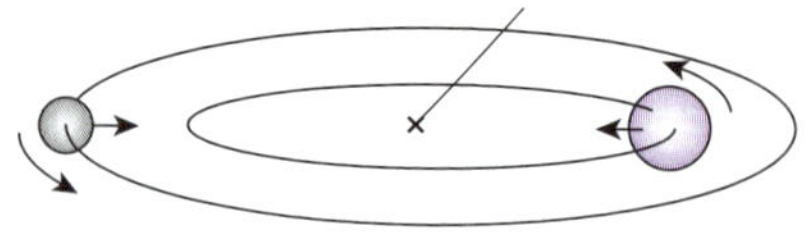

Figure 1 Binary stars – two stars in orbit around their common centre of mass

The existence of spectroscopic binaries is deduced from the periodic Doppler shift that can be seen in the spectra of the two stars that make up the binary system. If both stars are identical, over a period of time each spectral line can be seen to "split" into two separate wavelengths as a result of their orbital motion. During the "split" one of the stars in the binary pair has a component of its velocity moving towards the observer and thus has a Doppler blueshift whilst the other has a component moving away from the observer and has a Doppler redshift.

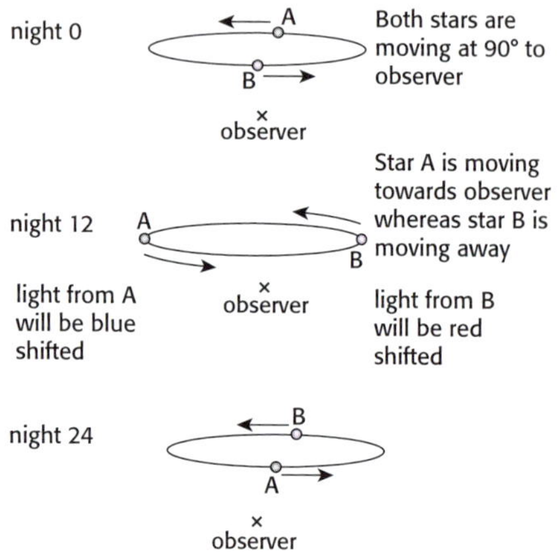

Figure 2 A spectroscopic binary system

Eclipsing binary stars are deduced from a periodic variation in the received brightness. In this situation, the orbits must be edge-on when viewed from Earth so that when one star passes in front of the other, the overall intensity received at Earth is reduced.

Sometimes individual stars are observed to have a regular variation in brightness. One class of this type of star are the **Cepheid variables**. They are named after the star delta Capheid in the Cephus constellation. Their brightness oscillations are believed to result from an oscillation in the star's physical size. As the size changes so does its surface temperature and its luminosity. See page 301 for information about the use of this type of star in calculating stellar distances.

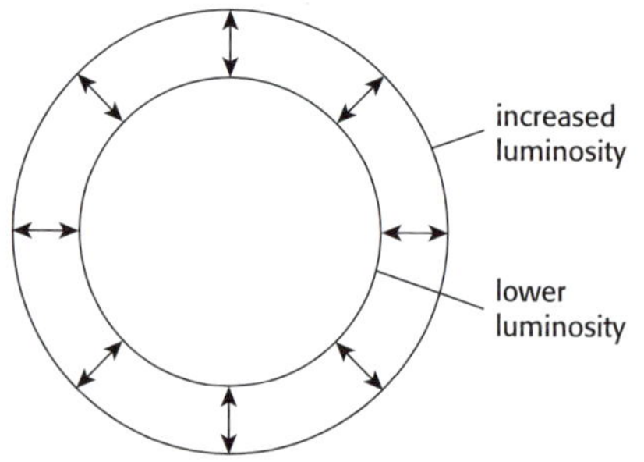

Figure 3 Periodic expansion and contraction of a Cepheid variable

Some stars are categorized as **giants** or even **supergiants**, depending on their size and mass. The **red giant** stars are in stellar class K and M and have low surface temperatures (in the range 3000 to 4000 K). At the other end of the size variation are the **dwarf** stars. **White dwarf** stars are known to have a high surface temperature (typically in stellar class B or A but sometimes at lower temperatures), their mass is of the same order of magnitude as our Sun but their volume is similar to that of the Earth. See page 302 for a description of how they can be formed.

Different stars are observed to have very different luminosities and surface temperatures. The range of surface temperature is shown by the stellar categories. Typically, star luminosities can range from ones that are a million times more luminous than our sun to ones that are ten thousand times less luminous. A very useful visual categorization of the properties of different stars is the **Hertzsprung–Russell diagram**.

On this diagram we plot increasing luminosity on the *y*-axis and decreasing temperature on the *x*-axis. Both scales are logarithmic. Most of the stars that we see in the night sky (approximately 90%) are located in the band that goes from the top left-hand corner of this diagram to the bottom right-hand end. This band is known as the **main sequence** and any star whose luminosity and surface temperature place it on this band is called a **main sequence star**. Stars that are not on the main sequence are more rare and most of these are either red giants or white dwarfs.

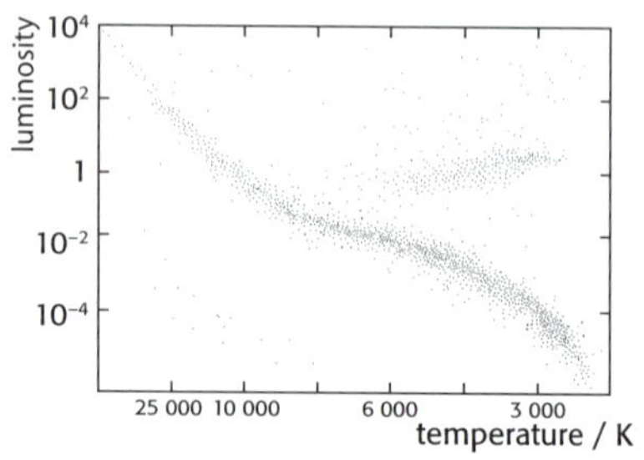

Figure 4 Hertzsprung – Russell diagram

Note that different versions of this diagram exist with different (but related) variables on each axis. An alternative variable for the y-axis is called the **absolute magnitude** of the star. This quantity is defined below and ranges from +15 (which represents a very low luminosity) to −10 (which represents a very high luminosity). An alternative variable for the x-axis is the spectral class of the star going from O to M.

The largest mass stars are towards the top of the diagram and this decreases down the main sequence from top left to bottom right. The variation with mass of the luminosity of main sequence stars is shown in Figure 5.

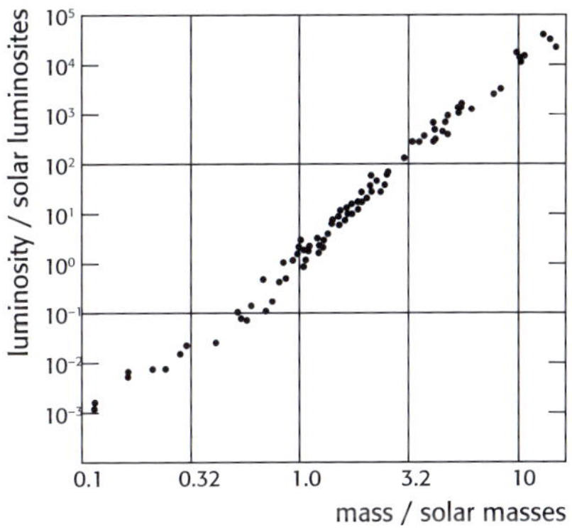

Figure 5 HR diagram (alternative axes)

Stellar distances

A key piece of information necessary for the analysis of the light from a star is the distance from observer to star. Unfortunately, this is not necessarily an easy measurement to achieve. Consider the difference between a distant star with high luminosity and a closer star with a lower luminosity. They will both appear to have the same brightness.

Astronomers use three different methods to find the distance to stars, each with an associated degree of uncertainty. They are: **stellar parallax**, **spectroscopic parallax**, and calculations involving **Cepheid variables**. Once the distance to a star has been found, measurements of brightness allow us to calculate its luminosity. The brightness of stars are usually compared using the magnitude scale.

Magnitude scale

For historical reasons, the brightness of stars is compared on the **magnitude scale**. The brightness of a star as seen by an observer on Earth is recorded as a magnitude, m. On this scale a bright star is magnitude +1 and a dim star (just visible with the naked eye) is magnitude +6. The bright star ($m = +1$) is roughly 100 times brighter than the dim star, corresponding to a difference of 5 magnitudes (i.e. $5 = 6 - 1$). This means that each step of the magnitude scale is equivalent to a change of brightness by a factor of $100^{0.2} = 2.512$.

If the distance to stars is known, then their luminosities can be compared directly using the concept of **absolute magnitude**, M. This is defined as the magnitude that the star would have if it were placed at a distance of 10 pc from Earth. When stars' absolute magnitudes are compared then we are directly comparing their luminosities. Each step on the absolute magnitude scale is equivalent to change of luminosity by a factor of 2.512.

The closest star to the Earth (apart from the Sun), Proxima Centauri, is 1.3 pc away and most stars are much further away. This means the majority of stars have a smaller numerical value for absolute magnitude than for apparent magnitude. Some stars would be so bright if placed 10 pc away that they have negative absolute magnitudes. A star of absolute magnitude −4.0 is five magnitudes brighter than a star with absolute magnitude +1.0 and thus its luminosity is 100 times greater.

?

Table 3 gives luminosity and surface temperature data for some stars. For each star, locate its position on the HR diagram and use this to categorize the stellar type.

Table 3

Star	Surface temperature / K	Luminosity / luminosity of the Sun
Aldebaran	4700	200
Betelgeuse	3000	20000
Regulus	14000	200
Rigel	13000	20000
Sirius B	20000	0.002

?

Star A and star B have the same brightness but star A is known to have 100 times the luminosity of star B. What is the distance of star A, d_A, in terms of the distance of star B, d_B?

HL

Stellar processes and stellar evolution

The nuclear fusion taking place in stars involves interactions between atomic nuclei. This can only happen at very high temperatures, so that the strong nuclear force (active at distances of 10^{-15} m or less) can overcome the Coulomb repulsion. The greater the density, the more likely a reaction is to take place. How did the stars become hot enough and dense enough for nuclear fusion reactions to start?

A star begins its life as a large cloud of gas. The mutual gravitational attraction causes the cloud to begin to collapse. The gravitational potential energy is converted into kinetic energy as the material accelerates, and the temperature of the gas cloud increases. If the initial mass of gas is sufficiently large it is possible for the gas cloud to get hot enough for nuclear reactions to start. Once they are underway, an equilibrium is created, with the energy released in the nuclear reactions balancing the energy radiated away from the star.

A main sequence star fuses hydrogen into helium. This is sometimes referred to as **hydrogen burning**, even though burning is a chemical reaction, rather than a nuclear reaction. The overall process can be viewed as the conversion of four protons into a helium nucleus, two positrons, and two neutrinos, as represented by the following equation:

$$4\,{}^{1}_{1}\mathrm{H} \rightarrow {}^{4}_{2}\mathrm{He} + 2\,{}^{0}_{1}\mathrm{e}^{+} + 2\,{}^{0}_{0}\upsilon$$

When there is not enough hydrogen left in a star's core to maintain the hydrogen burning required to support the star's equilibrium, the core collapses under gravity. This gravitational collapse causes the temperature of the core to rise still further and for most stars, the core is eventually hot enough for helium fusion reactions to take place.

When this process begins, typically helium nuclei are fusing in the core of the star to form beryllium, carbon, and oxygen in a process called the **triple alpha process**. There is still a thin shell of hydrogen fusion reactions around the helium burning core. The energy released by the collapse of the core causes a large expansion of the outer layer of the star, causing its structure to change. The radius greatly increases and most of the star is relatively low in temperature and density. It has become a **red giant** star.

The final fate of the star depends on its mass. Low mass stars, e.g. type G (up to about four times the mass of the Sun), are not massive enough to initiate any further fusion processes after helium burning. In the last stages of the process helium and hydrogen fusion takes place in shells around the contracting core. This causes the outer layers to be ejected as a spherical shell of cooling matter called a **planetary nebula**.

The remnant that is collapsing inward will keep on getting smaller until the matter is so densely packed together that **electron degeneracy pressure** prevents any further increase in density. In this condition it is known as a **white dwarf** star. The matter is extremely hot, extremely dense, and continues to radiate energy as it

?

1 Two possible mechanisms for the fusion of hydrogen into helium are the proton-proton chain (the p-p chain) and (for large mass stars) the carbon-nitrogen-oxygen (CNO) cycle. Research each of these series of reactions.

2 Use the masses provided in Table 4 to calculate the energy released in the production of one helium nucleus.

Table 4

Particle	Mass / kg
proton	1.673×10^{-27}
helium nucleus	6.6465×10^{-27}
positron	9.11×10^{-31}
neutrino	negligible

3 The fusion of four hydrogens into a helium nucleus releases about 3.9×10^{-12} J per helium nucleus. The luminosity of the Sun is 3.8×10^{26} W. Calculate:
(a) the number of reactions that are taking place each second,
(b) the mass of hydrogen converted into helium each second, and
(c) the mass converted into energy by the Sun each second.

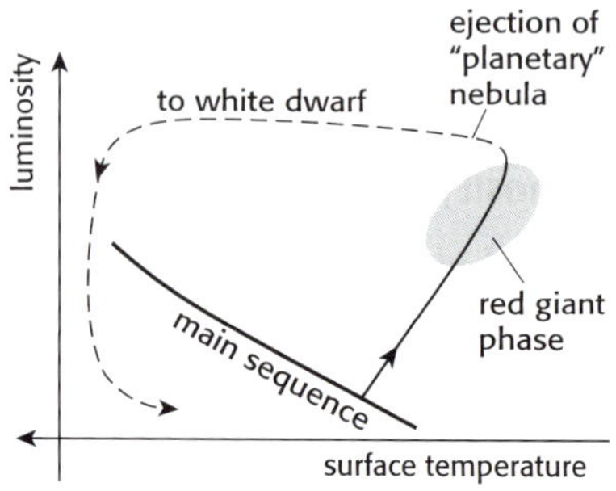

Figure 9 Evolution of a low-mass star

cools down but no further nuclear reactions are taking place so after billions of years it will cool and stop emitting energy.

The maximum possible mass of a white dwarf star is known as the **Chandrasekhar limit** and is calculated to be 1.4 solar masses. The original star that created this remnant can be up to about 4 solar masses.

Larger mass stars, e.g. type A, form **red supergiants** and are capable of sustaining further nuclear fusions. Carbon fusions can take place to produce neon, magnesium, and oxygen. Other reactions can take place in very large stars and even silicon can be involved in fusion processes and produce iron. Iron has the highest binding energy per nucleon (see page 204–5) so no more energy production stages are possible after this (Figure 10).

In the final stages of a red supergiant, the core of the star is composed of different layers each fusing different nuclei of increasing mass. Further fusion reactions would take in energy rather than release it and so do not happen inside the star. In the absence of any nuclear reactions occurring to sustain the equilibrium in the star, it collapses inwards.

If the mass of the core is greater than the Chandrasekhar limit, electron degeneracy pressure is insufficient to halt the collapse. Electrons and protons are crushed together to form neutrons and the density continues to rise. Eventually a process called **neutron degeneracy pressure** prevents any further collapse and the core becomes a **neutron star**. Any material falling inwards bounces off the inner core, creating a shockwave that causes the outer layers of the star to be expelled in a huge release of energy. This explosion is called a **supernova**.

Spinning neutron stars have been identified as the cause of **pulsars**. Pulsars are objects that are seen to emit pulses of radio waves with a regular time period (of about one second or less). The model astronomers have for the neutron star hypothesizes the creation of a very strong magnetic field. Protons and electrons on the surface of the neutron star are accelerated towards the magnetic poles and this creates a narrow beam of EM radiation. As the neutron star rotates, this beam of radiation sweeps around with the star. When the beam is pointing towards the Earth we receive the radiation. The period of the pulses then suggests that neutron stars have an extremely high rotational speed.

Cosmology

Given that the stars and galaxies that we observe at night do not appear to be moving it seems reasonable to assume, as Newton did, that the universe is both infinite and static.

However, the German astronomer Heinrick Olbers (1758–1840) identified a paradox that forced physicists to reconsider this model of the universe. The paradox states that if Newton's model of the universe were true then the night sky would be bright in all directions. This is not the case – so a new model was needed.

?

Use the binding energy per nucleon graph (page 204–5) to explain why red supergiants cannot continue to fuse elements heavier than iron in their cores.

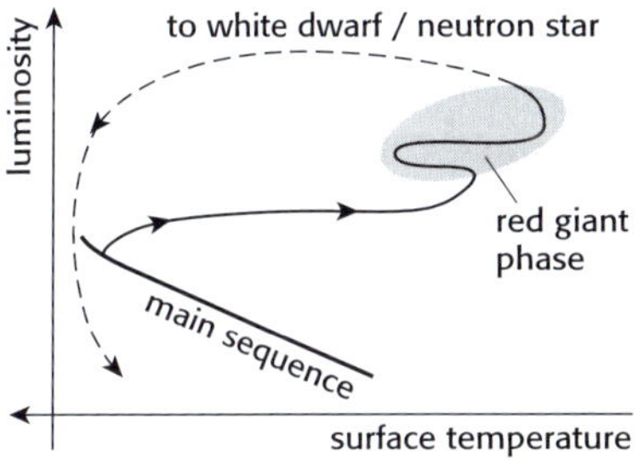

Figure 10 Evolution of a high-mass star

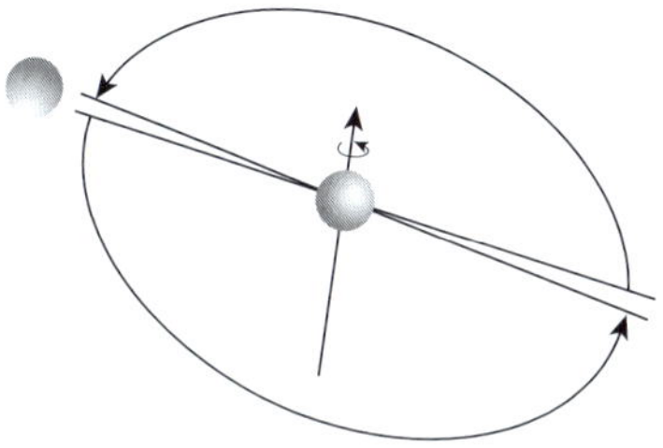

Figure 11 A spinning neutron star is a pulsar

?

1 Research the properties of a neutron star and pulsars.
2 Research how a physics student, Jocelyn Bell, first discovered pulsars.

Even neutron degeneracy pressure is not sufficient to halt the compression of the core due to gravity for very high mass stars. In this case the core is above the Oppenheimer–Volkoff limit for a neutron star of about 2.5 or 3.0 stellar masses. The core of such a star collapses and a black hole is formed. The density of a black hole is so high that even light is unable to escape.

3 Research what is meant by the following terms as applied to a black hole:
- singularity
- Schwarzchild radius
- event horizon.

There are three important statements that can be made about the observable universe. Firstly, the universe seems to be made up of the same kind of matter wherever we look; so we say that the universe is **homogeneous**. Secondly, it seems to be more or less the same in whatever direction we consider; the universe is **isotropic**. Finally, the physical laws that we observe to apply in our region of the universe seem to be the same throughout the universe; physical laws are **universal**.

To resolve Olbers' paradox physicists have hypothesized that the universe has not been around forever. The idea that the universe has a finite age can be linked to the expansion of the universe.

If a star is moving away from us, its light will be red shifted. If the star is moving towards us, it will be blue shifted (see page 98). When light is analysed from different galaxies, the vast majority of galaxies are red shifted and so we conclude that they must be moving away from us. In addition, the speed of recession is generally greater for the more distant galaxies. See page 306 for more details.

It is not correct to imagine the expansion of the universe as matter moving outwards from a point, filling up empty space. A better way is to imagine that space itself is expanding taking the matter with it. A good analogy is to imagine space as a sheet of rubber being pulled in all directions.

If the galaxies are all moving apart from one another (receding), they must have been closer together at some point in the past. The observations are consistent with a **Big Bang**. This theory says that all the observable matter in the universe was concentrated into one single point in space approximately 15 000 million years ago. In the Big Bang theory, space and time were created simultaneously at a single point, a singularity, of infinitely high temperature and density. After this creation, the matter expanded outwards and the matter has cooled down and eventually formed the universe we observe today.

Olbers' analysis assumes that no matter how distant a star is from Earth, its light will contribute to the brightness of the night sky but if the universe is only 15 000 million years old then we will only see light from stars less than 15 000 light years away. This is the solution to Olbers' paradox.

Einstein was able to show that space and time should not be considered as separate dimensions but rather considered together as an entity called the **space-time continuum**. The creation of space at the Big Bang also means the creation of time. From this viewpoint it does not make sense to ask what existed before the Big Bang – there is no such thing as "before".

The universe is expanding, but we do not know if this expansion will continue on forever. The rate of expansion will slow down due to gravitational attraction amongst the matter in the observable universe, but we do not know for certain if there is enough matter to eventually halt this expansion.

Research how the brightness of stars is expected to vary with distance, and how the number of stars is expected to vary with distance. Use your answers to show that the night sky might be expected to appear uniformly bright.

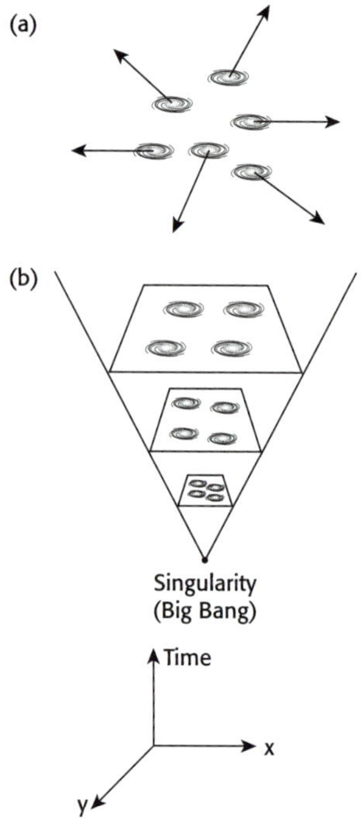

Figure 12 The expansion of the universe. (a) Physicists believe that the model of matter expanding in to empty space is incorrect. (b) Space itself is expanding from a singularity in this model, which is believed to be correct.

Another piece of evidence supporting the Big Bang model of creation is the discovery of cosmic microwave background (CMB) radiation by Penzias and Wilson. Research their discovery and explain how it is consistent with the Big Bang model. Your answer should compare the spectrum of microwave radiation with the expected characteristics of black body radiation and explain the significance of the CMB temperature of 3 K.

If the average density of the universe is sufficiently high, then the force of gravity will slow down the expansion to zero. The universe must then start to contract and all the matter will come back together (sometimes referred to as the **big crunch**). This is known as a **closed** universe.

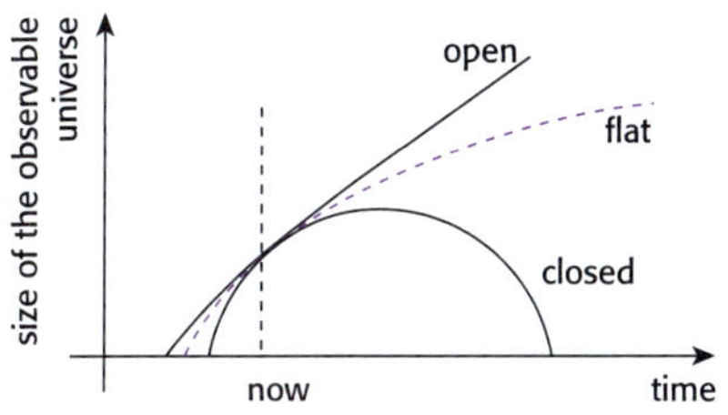

Figure 13 Different models for the development of the universe

If the average density of the universe is not sufficiently high enough for a closed universe, then, although the force of gravity will continue to slow down the expansion, it will never stop. The universe will continue to expand forever and this model is known as an **open** universe.

If the average density of the universe is at a certain value, **the critical density**, then the force of gravity will continue to slow down the rate of expansion but will take an infinite length of time to get to zero. This is known as a **flat** universe. The critical density works out to be approximately 1.9×10^{-26} kg m^{-3} or a few hydrogen atoms in every m^3.

Although current scientific evidence suggests that the universe is open, an accurate determination of the average density is extremely difficult. Observations of stars and galaxies can only identify matter that gives off light but the vast majority of the universe is already known to be **dark matter**, that is, matter which cannot be observed directly.

An estimation can be made of all matter present in a galaxy by estimating the number of stars it contains. This calculation is always several orders of magnitude smaller than the amount of mass calculated from its speed of rotation. Estimates that include all the known sources of mass (planets, asteroids, and so on) still fall short of explaining the matter that must be present in a galaxy. At the time of writing most of the mass of the universe is currently unexplained.

?

1 Research current possible theories that attempt to explain why so much dark matter exists. Possible explanations include MACHOs and WIMPs; find out the meaning of these terms.

2 Much of the current fundamental research that is being undertaken in astrophysics involves close international collaboration and the sharing of resources. Research one current example and present a summary of your findings. If possible you should try to include all of the following information:
- the title of the project
- an outline of the experimental procedures being followed
- the countries that are involved
- the timescale of the project
- the expected outcomes/benefits of the project.

3 All countries have a limited budget available for the scientific research that that country undertakes. Some would argue that research into the nature of the universe is one of the most fundamental, interesting, and important areas for mankind as a whole and it therefore deserves to be properly resourced. Others would argue that it is much more worthwhile to invest limited resources into medical research. This offers the immediate possibility of saving lives and improving the quality of life for some sufferers. List as many specific arguments as you can for and against investing significant resources into researching the nature of the universe. What is your view?

HL

Galaxies and the expanding universe

The observed redshift for different galaxies depends on their distance from Earth. In general, the further away a galaxy is, the greater its redshift and hence the greater its recessional velocity.

Hubble's law states that the recession velocity v for a galaxy is proportional to its distance away r.

$$v \propto r \qquad \text{or} \qquad v = H_0 r \qquad (5)$$

where
v is the recessional velocity of the galaxy, measured in km s^{-1}
r is the distance to the galaxy, measured in Mpc
H_0 is the Hubble constant, measured in km s^{-1} Mpc^{-1}

The value of the Hubble constant has a great deal of uncertainty associated with it, as the measurements we use in order to calculate it (distance and recessional velocity) suffer from a great deal of uncertainty.

A galaxy at a distance x then will have a recessional velocity of H_0x. If it has been travelling at a constant speed since the beginning of the universe (which is clearly not true), then the time taken will be:

$$\text{time} = \frac{\text{distance}}{\text{speed}} = \frac{x}{H_0 x} = \frac{1}{H_0}$$

The Big Bang model for the development of the universe imagines space expanding from a very hot, dense beginning to the present day. At the very high temperatures of the early universe (say up to 10^{-5} seconds after the Big Bang – the quark-lepton era), the universe was so hot that matter could not exist in the form that we see it today. Only elementary (fundamental) particles could exist. As the universe expanded it cooled, which allowed the formation of protons and neutrons to take place (from 10^{-5} to 10^{-2} seconds after the Big Bang – the hadron era).

The universe would have cooled sufficiently for helium nuclei to be formed from 10^{-2} to 10^{2} s after the Big Bang but for a significant time after this (10^{2} s to 300 000 years) the temperature of the universe was still so high that individual atoms could not exist. For these 300 000 years, all matter was in the form of a plasma with electrons, protons, neutrons, helium nuclei, and photons all interacting (the plasma/radiation era). Finally at about 300 000 years after the Big Bang, the universe was cool enough for the nuclei to capture electrons and for atoms to form.

At this point the universe was essentially 75% hydrogen and 25% helium. The elements that are more massive than helium were created when the first stars started to shine approximately 1000 million years after the Big Bang. A human being contains carbon, oxygen, and many other elements that are more massive than hydrogen and helium. Most of the atoms in your body must have been created as a result of fusion that took place in a star that shone many millions of years ago.

?

1 Use the information in Table 5 below to plot a graph to show the variation of recession velocity with distance. Comment on the shape of the graph.

Table 5

Name	Distance/ Mpc	Recessional velocity/km s^{-1}
NGC 55	1.99	129
NGC 157	21.5	1652
NGC 185	0.61	−202
SMC	0.06	158
Sculptor	0.08	243
M74	7.36	657
NGC 720	27.6	1745
NGC 891	9.81	528
NGC 908	18.4	1509
NGC 936	23	1430
LMC	0.05	278
UGC 4305	3.68	142
M82	3.68	203
LEO I	0.25	285
whirlpool	7.1	463
M 107	12.6	1048
NGC 1433	12.3	1075
NGC 2841	12.3	638
NGC 3198	14.4	663
M95	10.1	778
M96	10.1	897

2 Use the graph that you drew for question 1 to calculate a value for the Hubble constant.

3 Estimate the age of the universe from a value for the Hubble constant of 75 km s^{-1} Mpc^{-1}.

4 Explain why this must be an upper limit for the age of the universe.

5 Identify where the age of the universe as calculated from $\frac{1}{H_0}$ appears on Figure 13, which shows the possible future development of the universe.

8 Option F: Communications

We are all part of the digital revolution – a push for ever more efficient means of transmitting information. The 20th century saw improvements in the use of analogue signals, and in our ability to send these over large distances using cables and radio waves. With the advent of digitization we are now looking to develop better ways of getting such messages from one part of the world to another, whether this is over networks of cables and optical fibres, being carried by electromagnetic waves within our atmosphere, or by "bouncing" these signals off satellites in orbit around our planet.

Modern-day communications technology is a result of advances necessitated by the growth of civilization. Such developments have variously been driven by a historical desire to cooperate with, trade with, fight, and entertain each other better. This chapter aims to introduce you to some of the physics involved in communication.

Many things need to be considered in the building of an effective communications system. These include, but are not limited, to:

- **Speed**: How fast can you get your information from A to B?
- **Distance**: How far apart can A and B be?
- **Quality**: How similar is the received message to the original information that was sent?
- **Security**: In a world where personal privacy has become so important, how easy is it for others to access your information?
- **Mobility**: How important it is to be able to move during communications?
- **Cost**: Whether we are talking about hardware, about software, or in human terms, there are many economic and commercial demands that must be met.

It is impossible to optimize all these factors without excessive cost. We often have to balance demands in one area against needs in another.

?

Construct a timeline that summarizes some of the important steps in the history of communications. We suggest that you begin around 1800 (after electricity came under people's control), though going back further may allow you to appreciate the global nature of communications development. Along with the achievements of such well-known people as Edison, Marconi, and Bell, you should aim to include other milestones, including the invention of the first cathode ray tube, the first transistor, and the first magnetic recordings, as well as identifying when the first television signal was transmitted, what ARPAnet was, and when exactly the World Wide Web was legally born. Identify where a significant improvement was made to any of the factors listed above that contribute to a good communications system.

Channels of communication

A channel of communication is essentially any way of physically getting your message, coded or not, from one place to another. There have been some novel channels of communication developed over the ages.

Modern-day communications systems use one of three basic channels in order to transport data:

- electrically conducting wires
- optical fibre
- radio waves.

Digital signals

In most communications systems a message has to be **encoded**. A code is a way of transferring a message in a format that enables successful receipt at the other end. Some codes are used for preservation of privacy, but this is only one aspect of them. Encoding a signal often makes it easier to send it from one place to another, or assists in making communication more effective in another way.

Before continuing with this chapter it would be useful to review the material introduced in Chapter 24. You should be able to define or explain the following terms: digital, analogue, binary, bit, and byte.

As discussed on page 272, digital signals involve just two values – 0 (ON) and 1 (OFF). Digital signals can be more accurately conveyed in terms of avoiding **noise**. However, we have to accept a certain loss of information; what we receive will never be an exact replica of the original. Additional advantages of digital over analogue include the ease with which such information can be stored, and the ease with which signals can be transmitted and received. One of the biggest issues concerning modern communication is how we balance the quality of a transmitted digital signal against other factors such as speed and security.

Block diagrams

When discussing how any complicated system works, it is often easiest to use a block diagram approach and break the whole system down into its various stages. To do this with communications systems we need to refer to certain common devices. Examples of these include:

- **encoder**: converts the information you wish to send into a useful form
- **decoder**: extracts the information that has been sent
- **transmitter**: sends the communication signal
- **receiver**: receives the communication signal
- **input transducer**: converts the information to be sent into a form that can be processed (usually electrical)
- **output transducer**: converts the received signal back into a usable form
- **amplifiers**: designed to boost the strength of either signal or carrier
- **storage devices**: store information for use at a later time, e.g. disk drives, magnetic tape, CD, DVD.

Figure 1 shows a typical radio communications block diagram:

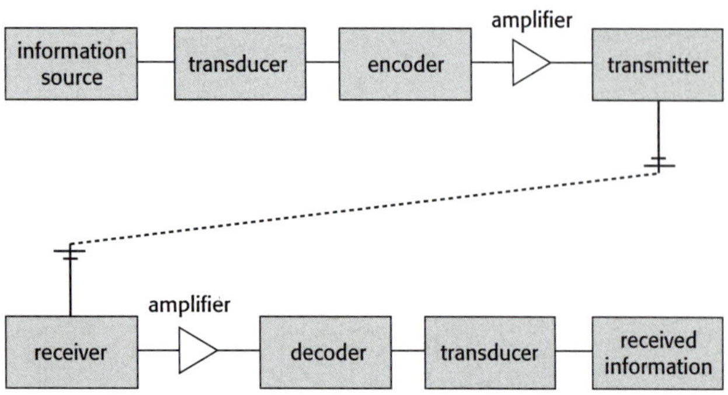

Figure 1 Block diagram of a typical communications system. This example uses radio waves, but the link could be by wire in a telephone system, for example.

A discussion of how this entire process works now follows.

Bits and bit rate

"640k ought to be enough for anybody." Bill Gates, 1981

Generally speaking, the more detail you need to send, the more bits you need to represent each part of the signal; but the more bits you have, the longer it will take to transmit the information. Another important factor is the channel of communication that you use to get your message across. These issues all involve the concept of **bandwidth**. In the digital world bandwidth means how many bits of information you can transmit every second. This is also called the **bit rate**.

- **Written text**: 8-bit binary is more than enough to cover the necessary range of letters, numbers, and symbols commonly found on a standard keyboard. This allows us 2^8 or 256 different values. Each symbol can be provided with a unique code in order to identify it. If all the text on this page were sent without any formatting in an email, then there would be roughly 3000 characters × 8 bits per character—in other words, 24 000 bits of information (or 3 kB).
- **Pictures**: Even with *compression* technology (the ability to save digital information at a lower memory than initially required – this is principally done using mathematical techniques) a good print-quality digital photograph is unlikely to be less than 2 MB or $2 \times 10^6 \times 8 = 1.6 \times 10^7$ bits of digital information.
- **Audio**: A standard 60-minute CD typically contains around 700 MB of digital information. This is considered to be high quality. Many audio formats, such as MP3, will be less detailed, and therefore require fewer bits.
- **Video**: A standard European TV screen refreshes 25 times per second, with a screen resolution in the region of 720 × 486 pixels, each of which requires information on colour as well as brightness. Add to this the desire for a cinema-quality surround sound audio track, and it starts to become clear why, at the time of writing, this level of communication is still being experimented with in terms of mass media availability.

There are, of course, many other things to consider. There is an obvious trade-off in terms of quality. An MP3 recording cannot compete with an original CD in terms of accuracy to the original sound, but users are happy with this, because it means they can carry more music around with them. You do not need to send *all* the information in a video stream 25 times every second, as much of the detail in any one frame is often similar to that in the previous one.

There are many tricks that can be employed to keep the amount of data down. However, that does not stop us from searching for faster ways of sending the information in the first place. And this is where we start to appreciate some of the physical as well as the economic limitations to our communications systems.

?

Research the following terms or phrases and explain what they mean in terms of digital communication. You may be asked to cover one or more of these, and you could share your research with the class in some way:

(a) sampling, quantization, pulse code modulation
(b) Internet telephony, radio and streaming audio
(c) shift registers
(d) companding
(e) aliasing
(f) Nyquist theorem.

Time division multiplexing

Whatever the information, a digital signal is only ever a very long series of 1's and 0's that need to be sent from one place to another. The simplest method would be to have a long piece of cable. Unfortunately, for global communications there will often be thousands of individual signals, which need to share one **channel of communication**. The physical limitations on how many signals we can send down one wire will be dealt with later. However, the process by which it is done is called **multiplexing**. With digital signals we shall deal with what is called **time division multiplexing**.

Suppose we have three signals from different sources (A, B, and C); each signal is 20 bits long, and they all have to travel along the same cable. You could send all of signal A, then all of signal B, and finally all of signal C. The people interested in signal A would receive the information quickly, but those looking for signal C would have to wait. The way round this is to allow each signal a share of the total time available to that channel.

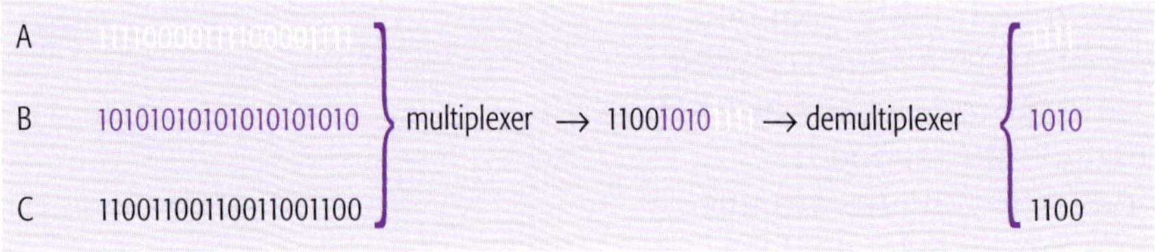

Figure 2 Principle of time division multiplexing

This is done by passing all three signals into a **multiplexer**, which cuts them up into equal-length slices of information (see Figure 2). For example, we might take the first 4 bits of signal A and then follow this with the first 4 bits of signal B and the first 4 bits of signal C after that. If we repeat this pattern, then everybody receives their signal at a constant rate, but at a lower bit rate than if one signal had the cable all to itself.

At the end of the communication channel there is a **demultiplexer**, which cuts the data stream into, in this example, 4-bit slices and then combines all the relevant signals back together again.

Any channel of communication can only cope with a limited number of messages at any one time, and the more signals the system needs to cope with, the slower the data transmission will be. The greater the capacity that a channel has, the more data it can accommodate – and the more expensive it will be! Internet traffic is making ever-increasing use of data in the form of images, audio and video, so there is a need for ever more efficient, high-volume channels of communication. This will continue to push communications engineering to the limit.

There are many consequences of this demand for better systems. These can be simple annoyances such as roads being dug up so that new cables can be laid but can lead us into some contentious issues.

?

Prepare some arguments for a debate entitled "The development of better communications systems is increasing the economic divide between rich western nations and poorer developing countries." You should research arguments for and against this statement and try to draw some conclusions.

?

A digital communications system includes some devices additional to those already mentioned. They are listed below. Construct a block diagram for a digital communications system using these new devices:

- **Sample-and-hold**: some way of measuring an input analogue voltage level and storing this temporarily
- **Clock**: controls the speed at which the system can process information
- **Analogue-to-digital converter (ADC)**: changes an analogue voltage to a digital code
- **Shift register (parallel to serial)**: takes many digital bits of information that are produced at once and puts them in a logical order ready for transmission
- **Shift register (serial to parallel)**: takes a series of data bits that come in one after another and then sends them all out at once down separate wires
- **Digital-to-analogue converter**: changes a digital signal into an analogue voltage.

Wires

Traditional wiring is increasingly being replaced for long-distance communications; it now tends to be used only at a very local level. Coaxial cable is superior to simple pairs of wires, but is still limited in how much information it can carry. The physics of how signals are transferred along cables is similar to that of the other two channels of communication covered in this section. Traditional wiring is increasingly being replaced by optical fibre, which we look at below.

Optical fibre

There are several types of optical fibre. They vary widely in terms of their cost and characteristics, but they all operate on the same basic principles of light transmission and **total internal reflection** (see page 87).

The basic construction of any optical channel is the same. A central core of extremely pure glass is surrounded by a **cladding** material of slightly lower refractive index. The whole structure is then protected, usually by a plastic outer layer. At one end of the fibre there will be a light emitting diode or laser that can be switched on and off. Digital signals are transmitted as pulses of light that indicate a binary level 1 when the light is on. When the light is not on a binary level 0 is assumed to exist.

Multimode fibre

Multimode fibres allow rays of light to follow many different paths. In this section we will look at **step indexed** fibres. These have a relatively large core of one material surrounded by a cladding material with a different refractive index. When a ray of light hits the boundary between core and cladding there is a distinct "step" or change in the refractive index. The rays that make up a beam of light will typically follow several different paths, as shown in Figure 3.

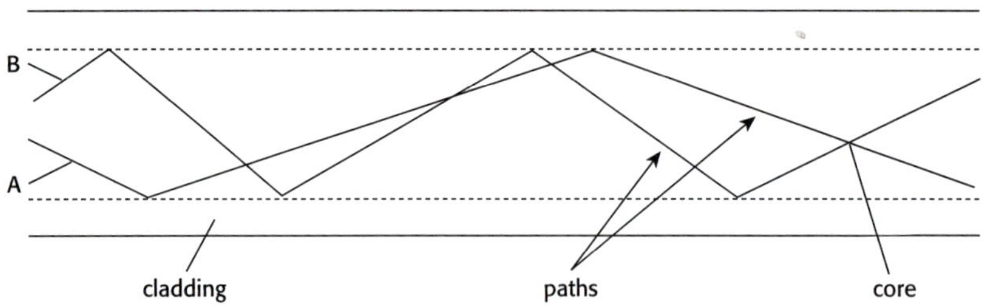

Figure 3 Two different ray paths along a multimode optical fibre

The ray labelled A will arrive at the end in a shorter space of time than B, because it has to travel a shorter overall distance. While the light travels more slowly in the glass than in a vacuum it still travels very quickly (~2×10^8 m s^{-1}). The high frequency switching of the transmission light source leads to a problem called **dispersion**, which is illustrated in Figure 8. An ON–OFF–ON sequence (binary 1–0–1) may start at one end of a cable, looking like Figure 4a.

But by the time it reaches the other end the information will have been spread out, as in Figure 4b, losing the middle OFF in the process. Dispersion effects can therefore limit the bit rate, especially over large distances.

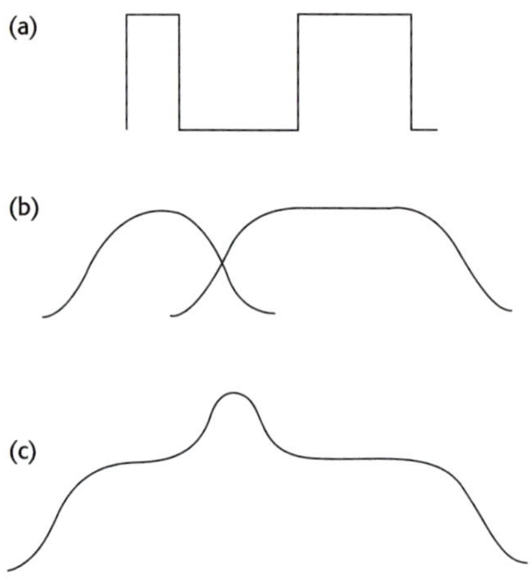

Figure 4 (a) Original pulses; (b) after dispersion; (c) combined effect

Monomode fibres

These have a much smaller diameter core. This means that there are significantly fewer possible paths for rays to follow, so dispersion more or less ceases to be a problem, and therefore higher bit rates can be attained (Figure 5).

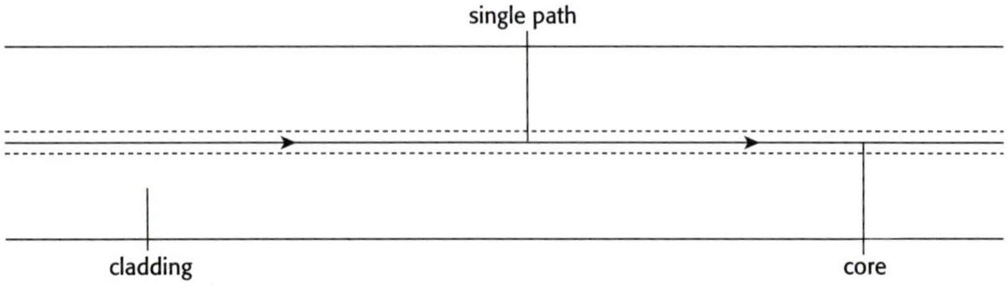

Figure 5 Monomode fibre

Attenuation

One problem for all channels of communication is the degradation of **signal strength**. Energy is being transferred from one place to another in the form of the communications signal, but some of that energy will be "lost" along the way. This process is called **attenuation**. Glass fibres are no exception. In a fibre optic channel the power of the signal decreases exponentially according to the equation

$$P = P_0 e^{-\alpha x}$$

where P is the instantaneous signal power, and P_0 is the original signal power; α is called the attenuation coefficient of the fibre – this is a measure of how good the fibre is at transmitting light; and x is the distance along the fibre (usually measured in km).

However well an optical fibre is manufactured there will always be some level of impurity, which causes scattering and absorption of light travelling along the cable. There are additional problems with aging of materials, because of impurities leaching into the glass over time, as well as with how individual lengths of fibre are joined together.

One of the big advantages of digital transmission is that, even though the signal attenuates and disperses over long distances it will still maintain the basic ON–OFF structure as originally sent. At some point the signal will degrade so much it will become impossible to decipher and so the signal must be tidied up periodically. This is done using an amplification device called a **regenerator** (which helps overcome attenuation problems) combined with something called a **reshaper** (which helps overcome dispersion issues) so that as the signal travels along the cable it will be kept as close to the original transmission as possible (Figure 6).

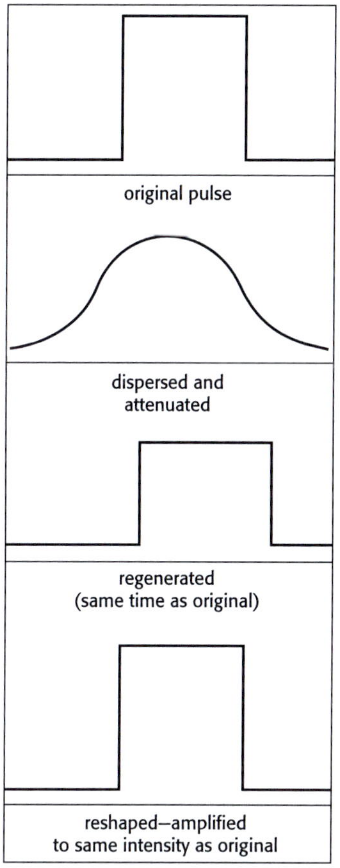

Figure 6 Regeneration and reshaping of an optical pulse

The various wavelengths of light are affected in different ways within a glass medium. Ideally, the signal will be carried by long wavelengths, because short wavelengths (higher-frequency light) will experience greater dispersion. Typically, red light or infrared is used. LEDs or tiny lasers transmit the pulses of light. Ideally the transmission will be from a monochromatic source of radiation in order to avoid **spectral spreading**, another dispersion effect caused by the fact that different wavelengths will travel at different speeds along the same channel (Figure 7).

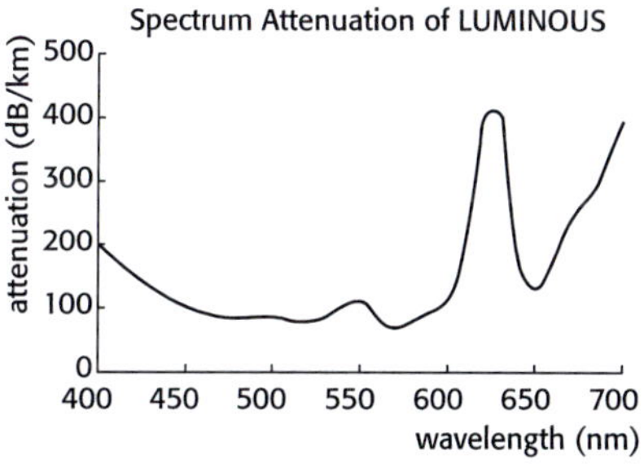

Figure 7 Typical spectral spreading in a glass fibre

Radio communication

One of the key trends in communication is mobility. Of course, the idea of being able to connect without the need for wires and cables is not new. The need to send messages to places where the use of a permanent channel is impossible has been with us for many centuries, made necessary by humankind's relentless exploration of the planet we inhabit. Military demands have often pushed communication technology, and the advent of naval warfare and – more recently – the use of airplanes demonstrated the importance of electromagnetic waves as a means for carrying information. Visual signals – flags, lamps and so on – have been superseded by the application of radio technology.

Modulation

The block diagram in Figure 8 shows the basic sequence of events in an amplitude-modulated radio transmission.

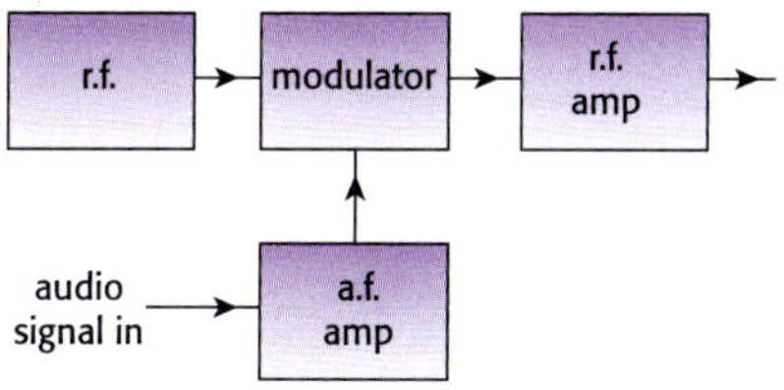

Figure 8 Sequence of events in amplitude-modulated radio transmission

The key to all this is the concept of **modulation**, which is a way of combining two electrical waves. One wave is the information that you wish to send – the signal; the other is the **carrier** wave. The key feature of the carrier is its frequency, which essentially does not change. This means that it is very easy to detect the carrier and "tune" in to it to facilitate retrieval of the information. The combination process takes place in a device called a **modulator**. At the other end after receipt of the transmission the information must be stripped from the carrier using a **demodulator**.

There are two forms of modulation. Both involve combining a signal at relatively low frequencies – what are termed **audio frequencies**, a.f. (human speech patterns vary between about 20 Hz and 20 000Hz) – with a significantly higher-frequency carrier oscillating at what is termed a **radio frequency** or r.f.

The frequencies used for carrier purposes are determined by the practicality of transmission and reception aerial size. Table 1 indicates how radio waves are classified.

Table 1 Frequencies and wavelengths for the classification of radio waves

Classification	Frequency range	Wavelength range
VLF (very low)	3–30 kHz	100–10 km
LF (low)	30–300 kHz	10–1 km
MF (medium)	300–3000 kHz	1000–100 m
HF (high)	3–30 MHz	100–10 m
VHF (very high)	30–300 MHz	10–1 m
UHF (ultrahigh)	300–3000 MHz	100–10 cm
SHF (superhigh)	3–30 GHz	10–1 cm
EHF (extremely high)	30–300 GHz	10–1 mm

Amplitude modulation (AM) is the more simple process. The amplitude of the carrier is altered depending upon the amplitude of the signal at different instances in time. If the signal amplitude is positive, then the amplitude of the carrier is increased; if the signal amplitude is negative then the carrier amplitude is decreased (Figure 9).

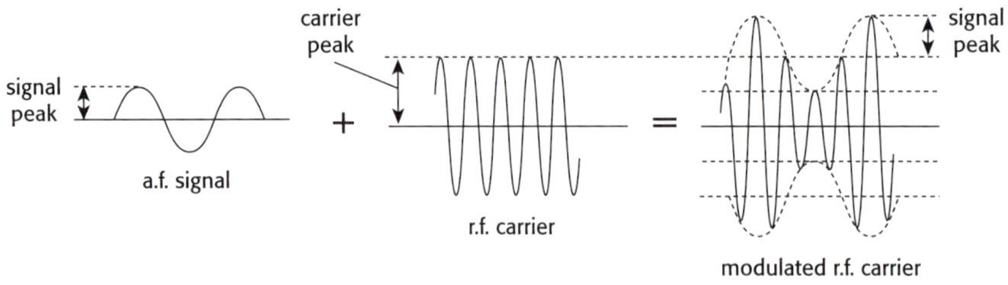

Figure 9 Amplitude modulation

The process is perfectly acceptable but does have a lot of problems in terms of quality being reduced due to the susceptibility of the transmission to noise. Also due to the need to increase **bandwidth** (covered later), the full range of audio frequencies is not transmitted, so this limits the accuracy of the received information.

Frequency modulation (FM) involves altering the frequency of the carrier. When the a.f. signal amplitude is positive then the carrier frequency experiences a small proportional increase. When the a.f. amplitude goes negative, then the r.f. frequency is decreased (Figure 10).

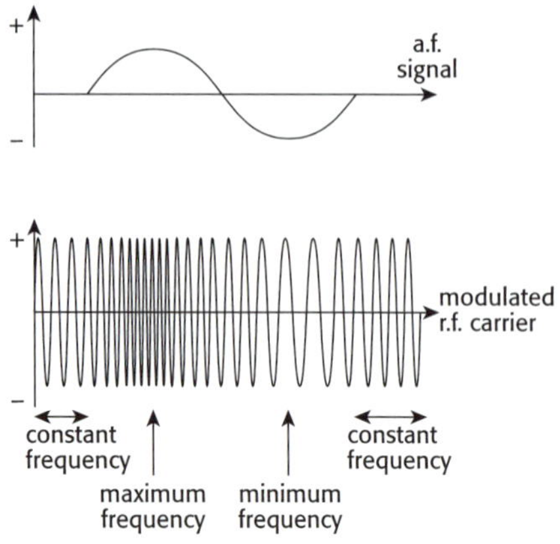

Figure 10 Frequency modulation

Tuning in

How can a receiver actually pick up the carrier wave – "out of the blue", as it were? A full understanding of this requires some knowledge of how aerials work but mainly an appreciation of what are known as **tuned circuits**. The way these work is touched on in a later section, but the basic principle is that what is called an **RC circuit** (R = resistance, C = capacitance) can cause electric charge to oscillate. Anything that is able to oscillate will, of course, have a natural or **resonant frequency** (see p xx).

By changing the values of either the resistance or the capacitance (usually the latter) in an RC circuit it is possible to select a particular resonant frequency. All oscillations of every frequency that an aerial is affected by will try to get the charge moving in the RC circuit, but only a very narrow range of frequencies will be truly effective in doing this. In Figure 11 anything below f_1 or above f_2 is not going to have much driving effect on the circuit.

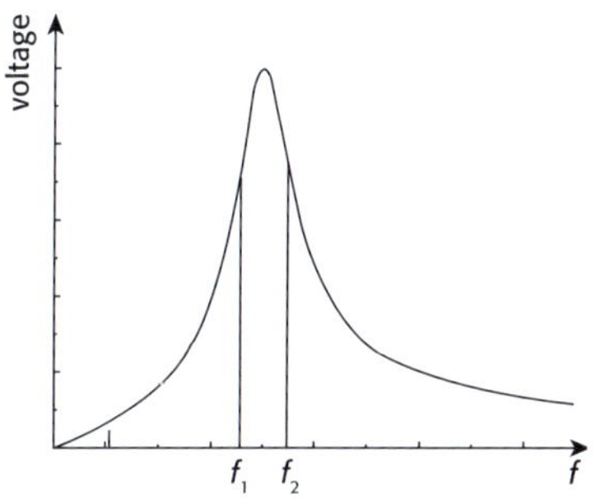

Figure 11 Resonance response of an RC circuit

Bandwidth

Certain queries should arise from all of this. For example, with frequency modulation, how much can the carrier frequency be allowed to change before we lose information because of poor tuning? Does the amount of data being transferred affect the modulation or tuning in any way?

As previously mentioned, it is quite easy to ascertain what is often called digital bandwidth but may more accurately be referred to as digital capacity. This is just a direct measure of how many bits per second can be transferred, and is probably best understood in relation to Internet connection "speeds" – a broadband connection typically offers nothing less than 100 megabits per second, for example.

Analogue signals, and the way the term "bandwidth" applies to them, may come across as being slightly older technology, but they are vital now as we move towards greater mobility with our communications.

AM bandwidth works like this. Because the r.f. carrier has been modulated, it is no longer sinusoidal in nature. From work covered in the waves units (Chapters 7–9) you will know that a wave such as that shown in Figure 12 can be made by superimposing several waves of different frequencies.

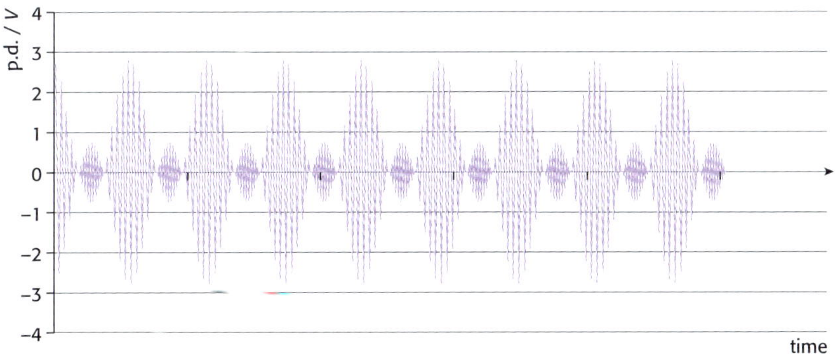

Figure 12 AM modulated wave

Power spectrums

It can be shown that, for an r.f. carrier of frequency f_0 being modulated by a single-frequency signal of frequency f', the overall signal can be split into three sinusoidal waveforms of the same amplitude with frequencies f_0, $f_0 + f'$ (upper side) and $f_0 - f'$ (lower side) (see Figure 13a). The difference between the upper side and lower side frequencies is the **bandwidth**.

For a normal signal that varies in frequency, as long as you know the range of a.f. frequencies, you can predict what are termed

sidebands. These are shown in Figure 13b, which is a **power spectrum** diagram.

If we wish to transmit digital information by radio, then we need as high a signal frequency as possible. This means that a large r.f. bandwidth also becomes big. CD-quality digitally transmitted sound requires a sampling (and therefore signal) frequency of 44.1 kHz, and therefore a bandwidth of around 88 kHz. Radio stations broadcasting on AM wavelengths are limited to a bandwidth of around 9 kHz by agreement.

FM bandwidth is different. In this case the r.f. carrier has its frequency altered in direct proportion to the **frequency** of the signal. A normal a.f. signal will produce many frequencies, which suggests that the bandwidth for a normal FM transmission would need to be extremely wide. However, if we calculate all the component frequencies we see a trend to lower amplitudes the greater the amount by which the component signal frequency differs from that of the carrier.

In practice, FM bandwidth is roughly calculated as $2 \times (\delta f_c + f_m)$ where δf_c is the deviation and f_m is the highest modulating frequency. By international agreement FM deviation must be no more than ±75 kHz, which provides plenty of bandwidth to deal with digital audio signals.

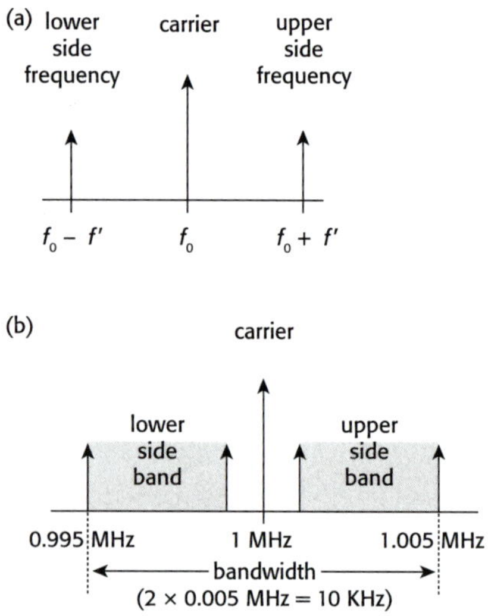

Figure 13 Power spectrum for (a) single frequency and (b) carrier wave of 1 MHz with audio frequencies

Bigger picture: Demodulator

Quite often at this level communications gets treated with a "black box" approach. A detailed understanding of the exact electronics or circuitry involved in doing a particular job is not normally considered necessary. This short section describes the basics of how a demodulator works, to show how some of the physics described in earlier sections "comes to life".

Consider the circuit in Figure 14.

As we mentioned on page 313, a demodulator is a circuit that is used to recover information from the carrier wave of a signal.

The diode rectifies the modulated carrier signal (essentially cutting out anything with negative potential). If the voltage through the diode is higher than the voltage stored on the capacitor, then the capacitor will charge up. At the same time current will flow through the resistor. When the diode is blocking reverse current in the cycle the capacitor discharges through the resistor. The resulting waveform is the original signal. The second capacitor simply prevents any d.c. passing on from this stage. This entire process is demodulation.

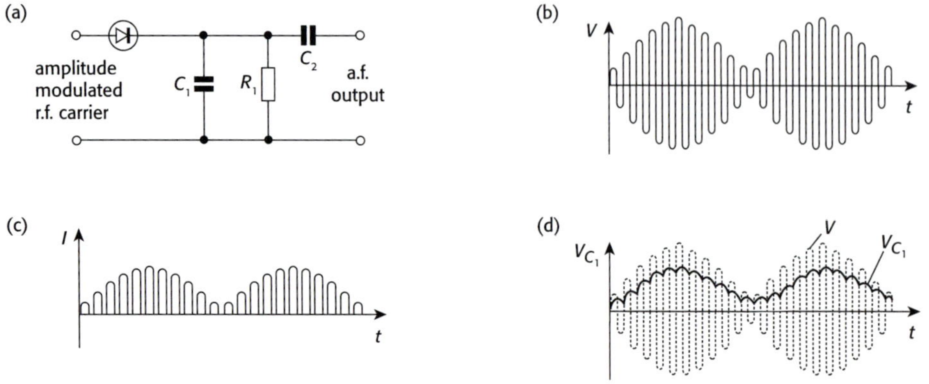

Figure 14 Demodulation

9 Option J: Particle physics

Particles and interactions

An atom is the smallest piece of an element that can exist and still be recognized as that element. Nearly all of these elements, such as the oxygen you breathe and the carbon in your skin, were made in stars about 5 billion years ago, at around the time that the Earth was forming. Hydrogen and helium are even older, most hydrogen having been made soon after the Big Bang.

A hundred years ago atoms were thought to be small impenetrable objects, like miniature versions of billiard balls perhaps. Today we know that each atom has a rich labyrinth of inner structure. At its centre is a dense, compact nucleus, which accounts for almost all of the atom's mass and carries positive electrical charge. It consists of protons and neutrons. In the outer regions of the atom there are tiny lightweight particles known as **electrons**.

Magnify a neutron or proton a thousand times and you will find that they too have a rich internal structure. They are clusters of smaller particles called **quarks**.

There is nothing smaller than these quarks, and so they are called **elementary particles**. There is another class of elementary particles, which includes electrons, and is called the **leptons**.

Elementary particles

Quarks and leptons are thought not to have any internal structure and hence are thought to be elementary particles. There are six known types (**flavours**) of each.

Antimatter

In addition, every particle has an antimatter partner with the same mass, but its other properties are opposite (electric charge, for example).

In 1932 Carl Anderson discovered the first evidence for antimatter. He observed a track on a cloud chamber photograph that had been left by a cosmic ray particle with the same mass as an electron, but a positive electric charge. He had discovered the **positron** – the antimatter partner of the electron.

In 1995 CERN announced that a small experiment using LEAR (Low-Energy Antiproton Ring) had managed to create atoms of anti-hydrogen (a negatively charged anti-proton with a positively charged positron in orbit).

Understanding antimatter is incredibly important in cosmology, because of the so-called asymmetry between matter and antimatter in the universe. See page 342 for more details.

Table 1 The six quarks

up (u)	charm (c)	top (t)
charge +2/3	charge +2/3	charge +2/3
mass 1/3m_p	mass 1.7m_p	mass 186m_p
down (d)	**strange (s)**	**bottom (b)**
charge −1/3	charge −1/3	charge −1/3
mass 1/3m_p	mass 0.5m_p	mass 4.9m_p

m_p is the mass of a proton. $m_p = 0.938$ GeV/c^2 = 1.67×10^{-27} kg.
Note that it is conventional in particle physics to fix the unit of charge as +1 so that the actual charge on an up quark is $2/3 \times 1.6 \times 10^{-19} = 1.07 \times 10^{-19}$C. This convention is followed from now on.

Table 2 The six leptons

electron (e)	muon (μ)	tau (τ)
charge −1	charge −1	charge −1
mass 0.0005m_p	mass 0.1m_p	mass 1.9m_p
electron neutrino (ν_e)	**muon neutrino (ν_μ)**	**electron neutrino (ν_τ)**
charge 0	charge 0	charge 0
mass *negligible*	mass *negligible*	mass *negligible*

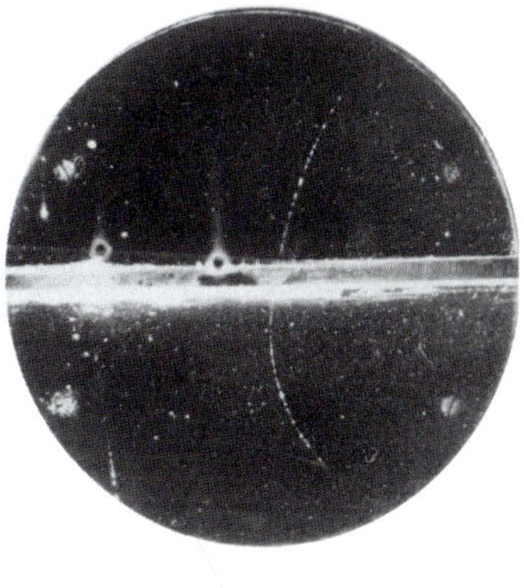

Figure 1 The first cloud chamber photograph of a positron. The line across the centre is a 6 mm lead sheet. The chamber was in a magnetic field which caused the track to be curved. The positron was moving upwards on the photograph.

Quantum numbers

All particles have intrinsic physical properties, for example, their rest mass (mass when they are stationary or moving at non-relativistic speeds). There is another set of intrinsic properties that all particles possess, called their quantum numbers. They are not necessarily directly related to physical properties but describe quantities that are conserved in interactions.

The Pauli exclusion principle

Different sets of quantum numbers are useful when we talk about different elementary particles because they do not all experience the same types of interactions.

The Pauli exclusion principle says that two fermions (the class of particles to which quarks and leptons both belong) in a closed system cannot have an identical set of quantum numbers.

This is very important in many areas of physics. For example, in astrophysics the maximum density of a neutron star is determined by the Pauli exclusion principle.

Atoms are also closed systems. The electrons are bound to the nucleus and have very similar quantum numbers to one another. However, they cannot be identical and this leads to them having slightly different energies. This is the origin of the different energy levels in atoms that give rise to the behaviour of the element.

The fundamental forces

A fundamental force is one that cannot be explained by the action of another force. For example, friction is actually due to the electromagnetic forces between the atoms in one object and those in another.

There are four known fundamental forces that are believed to govern all interactions in the universe. They are the electromagnetic, strong nuclear, weak nuclear, and gravitational forces. On the scale of the elementary particles the gravitational force is generally considered to be so weak its effects are negligible.

Table 3 The fundamental forces

Force	Felt by	Range	Relative strength*
Gravity	Any particle with mass	infinite	10^{-38}
Weak nuclear	Any particle	10^{-18} m	10^{-5}
Electromagnetic	Any charged particle	infinite	1
Strong nuclear	Only quarks†	10^{-15} m	100

* The strength of forces relative to the electromagnetic force has been estimated for two protons that are just touching.

† Technically, the strong force between nucleons is the remnant of the color force between quarks.

Thinking about science: conservation laws

Conservation laws place limits on the events that can take place in nature. The two most well known conservation laws, those of energy and momentum, state that whenever objects collide, explode, stick together, or repel each other, the total energy and momentum in the universe must remain the same.

The essence of such a conservation law is to compare the total amount of some quantity, such as energy, before an event has taken place with the total amount afterwards. If the total has not changed then the quantity is conserved. Conservation of energy and momentum are important in particle physics (always remembering to use the relativistic equations), but there are many other rules that introduce some new physics. By studying the quantities that are conserved in reactions and decays, particle physicists can learn more about the fundamental forces driving these processes.

Exchange particles

When an electron drops from an excited state to a lower energy in an atom, the energy difference is emitted in a quantum of electromagnetic energy called a **photon**. Photons play a very important role in the theory of the electromagnetic force at the level of elementary particles. Each of the fundamental forces is associated with a different quantum of energy which plays the role of the "carrier" of the force. It is the exchange of these particles (sometimes called **gauge bosons**) that give rise to the forces.

Table 4 Exchange particles

Force	Exchange particle	Charge	Mass
Electromagnetic	photon (γ)	0	0
Strong nuclear	gluon (g)	0	0
Weak nuclear	W^+	+1	$89m_p$
	W^-	−1	$89m_p$
	Z^0	0	$99m_p$

The Heisenberg uncertainty principle

"Anything that is not expressly forbidden by Nature is compulsory."

Richard Feynman

If you were in a car weighing 1 tonne and suddenly 89 tonnes were ejected, you would complain that something was wrong! But in the quantum world this kind of thing can happen if this violation of mass–energy balance is ephemeral, limited in time such that the product of the imbalance (ΔE) and the time it can last (Δt) cannot exceed Planck's quantum, h. In fact, it is usually written in terms of Planck's quantum, h, divided by 2π; as this combination occurs throughout atomic and particle physics it is denoted by the symbol $\hbar$.

Hence we write the Heisenberg uncertainty principle mathematically as

$$\Delta E \Delta t \geq \hbar / 2$$

This incredibly simple equation tells us that energy can, in a sense, be "borrowed", as long as it is "paid back" quickly enough. In this case, no physical laws are violated. Notice that the more energy is borrowed, the quicker it must be paid back. This has implications for the physical range of some of the forces.

?

What is the maximum energy that can be "borrowed" for one second?

Virtual particles

Table 4 lists the exchange particles for three of the four fundamental forces.

The particles that mediate the weak interaction, the W^+, W^-, and Z^0 were predicted in the late 1960s by Sheldon Glashow and Steven Weinberg in their theory of the electroweak force (unified theory of the electromagnetic and weak nuclear forces at high energies, see page 340).

These virtual particles are many hundreds of times more massive than the fermions whose reactions they mediate. They are not an intrinsic part of the quarks and leptons themselves, nor do they exist independently of them. In fact, they can only exist due to the Heisenberg uncertainly principle. The energy required for a single W boson (80 GeV) can occur for some 10^{-24} s, during which time not even light could travel more than about one-tenth of the distance across a proton.

The distance over which the W boson can transmit the weak force is limited by the fact that it is massive. This does not mean that it reaches a certain point and then turns off suddenly. Rather, its strength reduces with time, and by the time it has travelled a distance of the order of the size of a proton its strength has fallen away dramatically. Thus the force is known as the weak force. The weak force is responsible for the type of nuclear decay that produces beta radiation (beta decay).

The discovery of a massive W boson in CERN in 1983 is considered to be a major achievement of the Standard Model of particle physics (see page 341).

Feynman diagrams

In the 1950s the physicist Richard Feynman developed a method for dealing with the electromagnetic force between charged particles that has since been adapted for use with the other fundamental forces. The approach is very mathematical and requires an understanding of advanced quantum theory. However, the outline of his technique can be seen in his **Feynman diagrams**.

Consider a simple situation, such as two electrons scattering off each other. The event is completely described by stating the energy and momentum of each electron before and after. These are the only measurements that can be made without ruining the reaction.

In Figure 2 this is symbolized by the shaded "bubble of ignorance" in the middle. All we know for certain is what went in (*before*) and what came out (*after*).

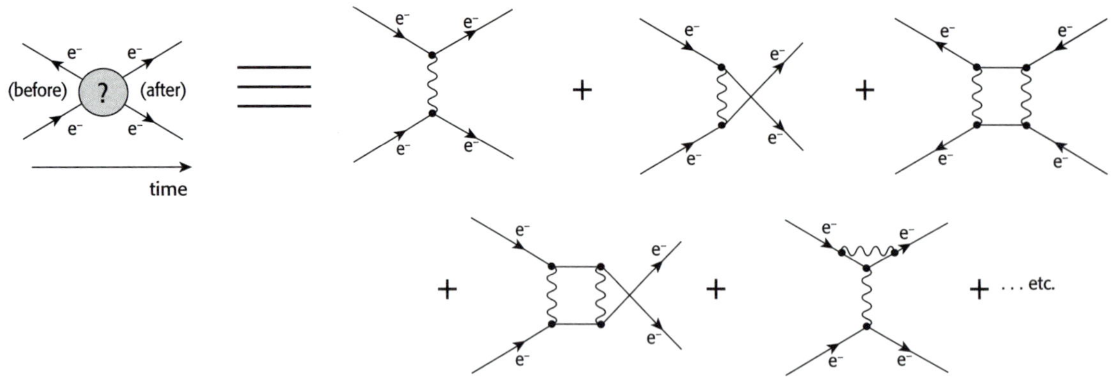

Figure 2 Feynman diagrams "fill in" the region between the start and end of an interaction. Each diagram represents a term in an infinite mathematical series; only the total of the series accurately represents the interaction.

Feynman worked out how to calculate the likelihood of a given initial situation leading to a given result. He did this by imagining all the possible things that might have happened inside the "bubble" to get from the initial to the final state. He represented each of the possible events as a "doodle", or Feynman diagram. Figure 2 shows the first five simplest Feynman diagrams that can be drawn for electron-electron scattering.

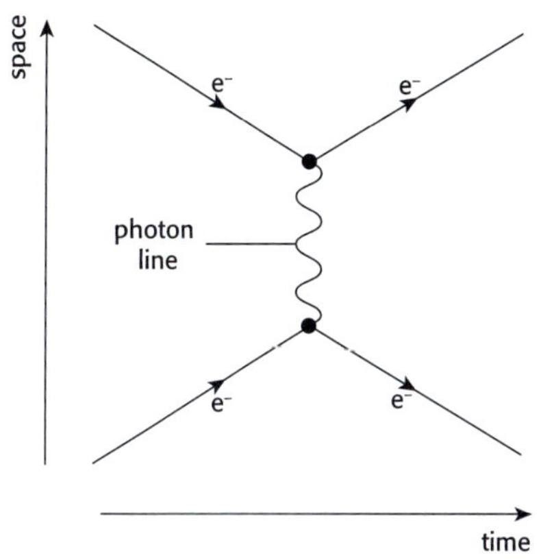

Figure 3 A Feynman diagram. In this diagram the photon appears to violate Einstein's theory of special relativity and travel at an infinite speed. However, remember that the diagram is just a representation of a mathematical expression and as such is not really physically meaningful.

In a Feynman diagram the straight lines represent the electrons and the wavy line is the energy and momentum being passed from one to the other in the reaction. The reaction is happening because of the electromagnetic force between the charged electrons so the energy and momentum is transferred by a photon. Each Feynman diagram represents a precise mathematical formula.

In fact, for any given interaction there is an infinite sum of possible Feynman diagrams. So why are they useful? It turns out that for each exchange particle involved, the probability of the reaction taking that particular pathway decreases by a known factor. For the electromagnetic force, each photon exchanged reduces the probability by a factor of $(137)^2$. The contribution to the sum of diagrams with increasing exchange particles decreases rapidly, and in this limit it is possible to use mathematical techniques to calculate the sum.

Annihilation and pair production

If a matter particle and an antimatter particle of the same type come into contact, then they spontaneously convert into energy. For example, an electron and a positron can annihilate into a high-energy photon, which is extremely unstable and will rapidly convert back into a particle/antiparticle pair. However, because the photon is a "clean slate" (in other words it has no charge, spin, etc.) any pair of particles can emerge:

$$e^- + e^+ \rightarrow \gamma \rightarrow e^- + e^+$$
$$e^- + e^+ \rightarrow \gamma \rightarrow \mu^- + \mu^+$$
$$e^- + e^+ \rightarrow \gamma \rightarrow \tau^- + \tau^+$$
$$e^- + e^+ \rightarrow \gamma \rightarrow q + \bar{q}$$

(in this reaction q can stand for any of the quarks).

Annihilations can take place between any matter particle and its antimatter partner. However, a photon is not always produced. Quarks react with each other far more readily via the strong interaction so if a quark and antiquark annihilate, a burst of strong-force energy (a gluon) is produced:

$$q_a + \bar{q}_a \rightarrow \text{gluon} \rightarrow q_b + \bar{q}_b$$

As in the case of a photon the gluon is a "clean slate" so the quarks in the final state need not be the same quarks as in the initial state.

Annihilation and pair production using Feynman diagrams

Figure 4 shows two annihilations and two pair production reactions. In Figure 4a, an electron and positron annihilate to form a photon, which then produces an electron/positron pair (as mentioned earlier, it could produce any particle/antiparticle pair).

Figure 4b shows a quark and antiquark annihilating to a gluon (strong force carrier), which then produces another quark/antiquark pair. A single quark can also produce a gluon which rapidly leads to

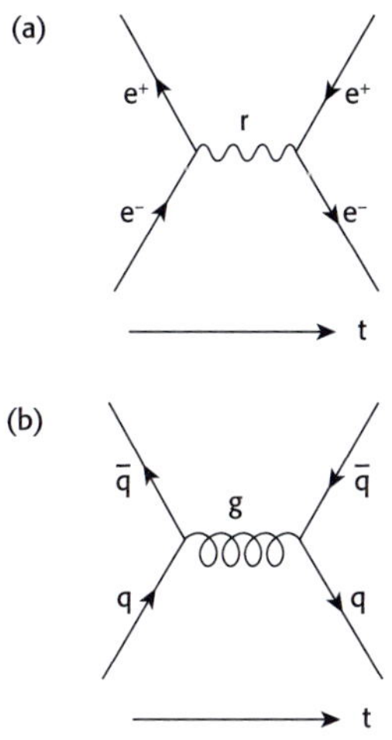

Figure 4 Some Feynman diagrams

pair production. This is the mechanism by which quarks remain in a bound state (nature does not allow quarks to exist as independent particles – see page 372).

The need for high energy particles

Energy requirement

In 1932 Carl Anderson discovered the positron (the common name for the anti-electron, e^+) in an event called a pair production:

$$\gamma \text{ (photon)} = e^+ + e^-$$

The rest mass of an electron or positron is 0.51 Mev c^{-2} so the minimum photon energy required (from the equation $E = hf$, see page 230) for such a reaction is 2×0.51 MeV $c^{-2} = 1.02$ MeV c^{-2}.

Dirac's theory predicted the existence of the antiproton (p^-). In this case, the pair production p^+/p^- would require a minimum photon energy of 2×938 MeV $c^{-2} = 1900$ MeV c^{-2}, almost 2000 times the energy required for the creation of an e^-/e^+ pair.

Actually, in order to create the proton/anti-proton pair, a beam of particles is used to strike a stationary (target) particle. For conservation of momentum the required initial kinetic energy of the incident particle must be about 5600 MeV c^{-2} (5.6 GeV).

Can you explain why the law of conservation of momentum means more energy is required?

Another reason to aim for high-energy particles is the requirement dictated by the need for high resolution.

Resolution of small targets

How close can two point sources of visible light be to each other and still be resolved with the human eye? Light diffraction is a limiting physical parameter. Rayleigh suggested the following criterion for circular apertures of diameter b:

$$\sin\theta = 1.22\frac{\lambda}{b} \qquad (1)$$

where
θ is the minimum angular separation of the two sources
λ is the wavelength of the visible light measured by the observer.

The shorter the wavelength, the smaller the minimum angle of resolution and so the greater the resolving power. The same principle applies to a telescope looking at stars or to a microscope looking at bacteria.

By decreasing the wavelength of the light observed, the resolution power increases proportionally. The smallest object that can be resolved in a visible light microscope is close to 1.2×10^{-7} m. In an ultraviolet microscope, targets of 5×10^{-10} m can be distinguished (the size of an atom!) since ultraviolet light has a wavelength of 10^{-7} m $- 10^{-9}$ m. As a rule of thumb, the size of a resolvable object is in the order of the wavelength of the light used.

Elementary particles are much smaller than an atom. For example, the cross-section of the proton is about 10^{-15} m. Gamma rays have a wavelength of the order 10^{-11} m, however, much shorter wavelengths are required to resolve individual protons. In order to study the structure of particles, the use of extremely high momentum/short wavelength particles is necessary. Quantum physics is coming to the rescue.

A 110 eV electron will have an associated wavelength of 1.2×10^{-10} m in the order of the wavelength of X-rays. This is also the order of magnitude of an atom. Electron diffraction patterns have been produced as predicted and they look like X-ray diffraction patterns.

To probe structures smaller than the size of an atom, electrons are accelerated so they have very large values of velocity and momentum.

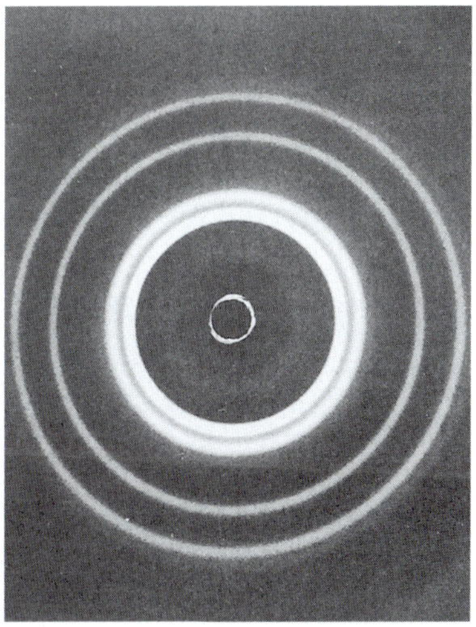

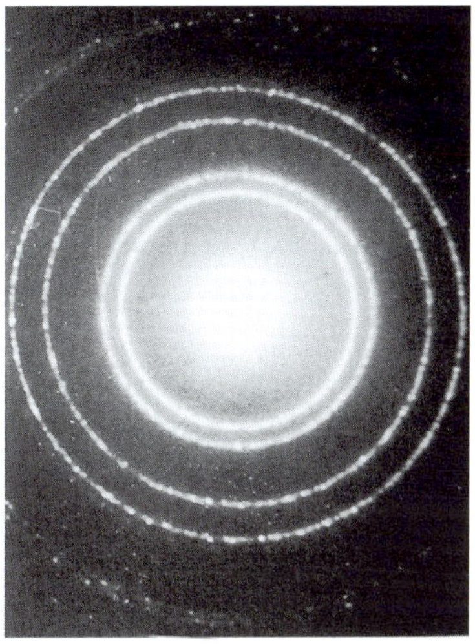

Figure 5 (a) diffraction pattern made by a beam of X-rays of wavelength 0.71×10^{-10} m from an aluminium foil target. (b) diffraction pattern made by 600 eV electrons from an aluminium foil target.

?

1. Estimate the wavelength associated with 30 keV electrons.
2. In a given accelerator, protons of 10 GeV are produced. Using the equation for the momentum of a particle $p = E/c$, estimate the wavelength associated with these protons.

The linear accelerator (linac)

A charged particle could acquire 20 MeV of kinetic energy by moving through a single potential difference of 20×10^6 V.

However, such high voltage cannot be used effectively in the laboratory. A linear accelerator can produce high-energy charged particles by accelerating them through successive application of low electric potential differences.

The operation of a linear accelerator

Figure 6 is a sketch of a typical linear accelerator.

Operation stages of a linac

Stage 1 Tubes 1, 3, and 5 are positively polarized by the oscillator, Tubes 2 and 4 are negatively polarized.
The protons are accelerated in the gap between tube 1 and 2, then drift inside tube 2 at a constant speed during half-period.

Stage 2 When the protons reach the end of the (now positive) tube 2, they are accelerated again towards tube 3 (now negative).

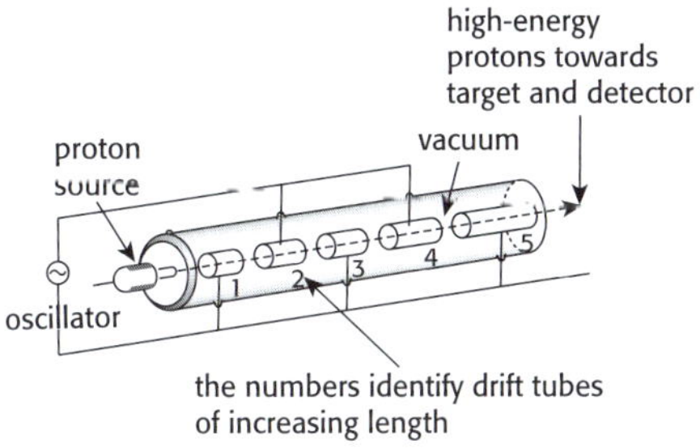

Figure 6 Linear acceleration

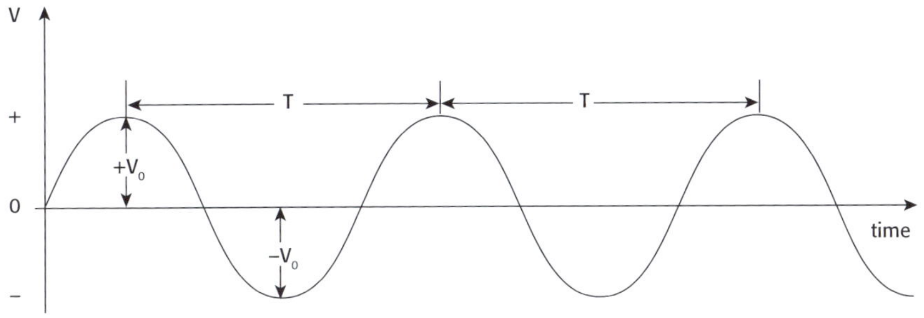

Figure 7 When the values of V are positive, the proton gains energy

Stage 3 The protons drift inside tube 3. The length of tube 3 is greater than the length of tube 2.

This process continues until the kinetic energy of the protons reaches the expected maximum value. The high-energy protons are then directed towards the target inside a detector.

?

1 State the reason why the main chamber is under vacuum.
2 Explain whether the time of flight of the protons inside a tube change from tube to tube.
3 Derive an expression giving the final kinetic energy of the proton at the end of the accelerator as a function of the total number of "gaps" n and of the *effective* voltage between successive tubes V_{eff}.
4 Determine whether such an accelerator can be used with electrons. What about heavy ions? If so, will all these charged particles have the same final kinetic energy? Will they reach the same final speed (assuming non-relativistic effect)?
5 (a) Discuss whether the oscillator should operate at low or high frequency to minimize the length of the tubes.
(b) A typical oscillator in a linac uses a radio frequency (RF) of about 2×10^8 Hz. What is the corresponding period of the signal?
6 Estimate the speed of the proton at the end of the first potential jump if the effective accelerating voltage is 25 keV. Assume its initial speed is effectively zero.

The special theory of relativity indicates that the speed of a particle can get very close to the speed of light c but will never reach or exceed it. So what would happen if energy were still given to the proton when the speed of the particle is very close to the speed of light?

The particle's speed will not change much (see Table 5), but its mass will increase by an amount Δm proportional to the given energy increment ΔE:

$$\Delta m = \frac{\Delta E}{c^2} \qquad (2)$$

There is equivalence between mass and energy. When a particle becomes relativistic[1] in a linear accelerator, its speed remains constant. Consequently, there is no need to change the length of the drift tubes so the design of a linear accelerator is much simpler for relativistic particles.

The constant length of these equal drift tubes is equal to c/2f, where f is the frequency of the oscillator.

Dependence of speed of particles on their kinetic energy

The cyclotron

In the cyclotron, a relatively low potential is used to accelerate a beam of particles in circular trajectories inside a homogeneous magnetic field.

[1] Generally, a particle moving at a speed v = 0.10 c or more is considered relativistic.

?

Table 5 Kinetic energy of different particles and their speed

Kinetic energy	Electron speed*	Proton speed*
0.1 MeV	0.553	0.0146
1 MeV	0.941	0.0462
10 MeV	0.9988	0.145
100 MeV	0.999987	0.429
1 GeV	0.99999987	0.876
10 GeV	0.9999999987	0.9963
100 GeV	0.999999999987	0.999957

* All speeds are expressed as fractions of the speed of light.

1 Using Table 5, plot a graph of the ratio v/c against E_k.
2 Explain why the accelerated electron reaches relativistic speeds at lesser kinetic energies than the accelerated proton.
3 Calculate the speed of an electron with a kinetic energy of 1.0 MeV, assuming Newtonian mechanics. Compare your results to the corresponding value in Table 5.

Trajectories of charged particles inside a homogeneous magnetic field

The behaviour of charged particles inside a homogeneous field can be deduced from knowledge of circular motion. Figure 8 illustrates the motion of a positive particle moving under a constant magnetic force $\mathbf{F}_m$ acting as the centripetal force. The field **B** is perpendicular and directed into the page. The radius of curvature is r and the velocity is **v**.

The vectors **B**, $\mathbf{F}_m$, and **v** are always perpendicular to each other. As a consequence, the direction of the velocity of the particle changes constantly but not its magnitude. No work is done by the magnetic field.

The magnetic force $\mathbf{F}_m$, acting on the charge q of mass m, is the centripetal force:

$$F_m = \frac{mv^2}{r} = Bqv \qquad (3)$$

$$r = \frac{mv}{qB} \qquad (4)$$

The product mv is the momentum p, so for constant q and B the radius of curvature r is a direct measure of the momentum p of the particle. The momentum is related to the kinetic energy E_k if the particle is non-relativistic:

$$E_k = \frac{p^2}{2m} \qquad (5)$$

Since p = rqB by equation 4:

$$E_k = \left[\frac{q^2B^2}{2m}\right]r^2 \qquad (6)$$

The kinetic energy of the particle is proportional to the square of the radius of curvature. If the radius of the cyclotron is increased by three, the energy of the particle increases by a factor of nine.

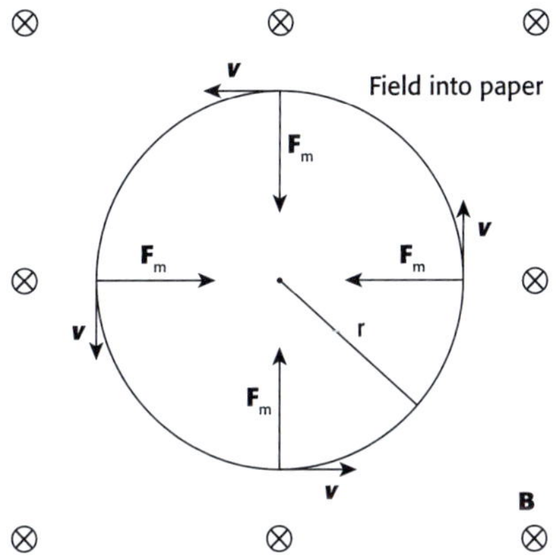

Figure 8 The field, velocity and force vectors are all perpendicular

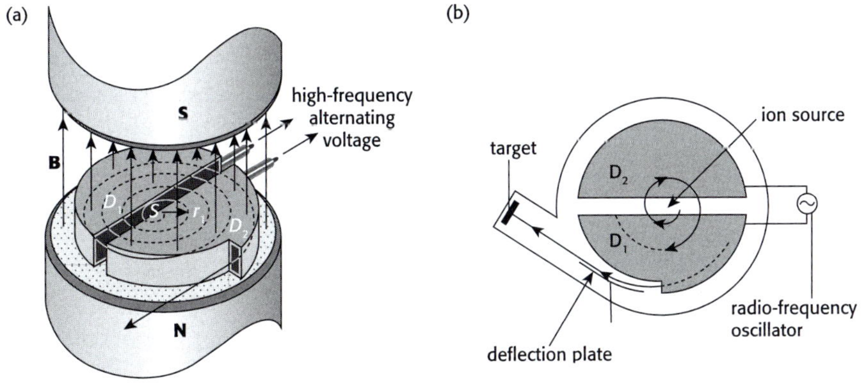

Figure 10 The design of a cyclotron

A homogeneous magnetic field B is created between the poles of an electromagnet. Two hollow semicircular electrodes are inside a vacuum chamber (because of their shapes, they are referred to as Dees). An ion source at the centre injects the particle between the two electrodes.

?

Deduce that the period T of one orbit is given by

$$T = \frac{2\pi m}{qB} \qquad (7)$$

Interestingly, the period is **neither** a function of the radius r **nor** of the speed v. How can this be? What happens to r when v increases?

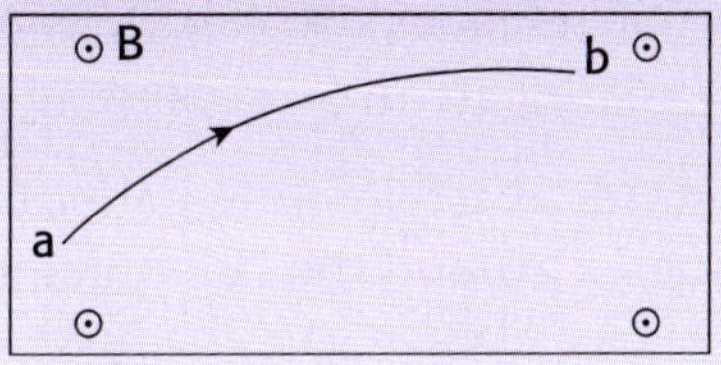

Figure 9 Analysis of a simple particle track

Figure 9 illustrates the trajectory of a proton moving from a to b in the plane of the page. The field (1.4 T) is perpendicular to the page and directed out of it.

1 State the sign of the charge.
2 Explain how you would go about measuring the radius of curvature of the trajectory.
3 The radius of curvature of the proton is 10 cm. Determine:
   (a) the momentum
   (b) the speed
   (c) the kinetic energy of this proton.

If D_2 is positive and D_1 negative, the particle will be accelerated towards D_1. Inside the Dee, the particle will not feel the electric field due to the charges on the surface of the electrode (see page 153), but will curve under the influence of the magnetic field. The time required to come back to the gap between the two Dees is not a function of the radius of the curvature or of the speed of the ion (see equation 7). During that time, the polarity of the two Dees reverses just in time for the particle to be accelerated again during its second passage through the gap.

After this second acceleration, the radius of curvature inside D_2 will be greater. If the particle remains non-relativistic, the time of travel inside the second Dee will be the same as it was inside the first. Under this condition, the frequency of the oscillator is constant and the accelerated particle remains in phase with the oscillator (accelerating field). The particle will follow a circular path with an increasing radius (spiral), until a deflecting plate sends it towards the target.

The frequency f of the oscillator is given by

$$f = \frac{1}{T} = \frac{qB}{2\pi m} \qquad (8)$$

The energy gained at each gap ΔE_k is

$$\Delta E_k = V_{eff}\, q \qquad (9)$$

where V_{eff} is the effective voltage of the oscillator.

The maximum value of the radius of the curvature is close to the actual radius of the Dee chambers.

Values for a typical cyclotron

Particle: proton
Radius: 0.40 m
Magnetic field: 1.5 T
Oscillator frequency: 22 MHz
Maximum kinetic energy: 17 MeV
Alternating voltage: V_{eff} = 200 kV

?

1 The proton moves at 19% of the speed of light. Assuming the proton is non-relativistic, determine how many times will it pass between the Dees before reaching its maximum energy. Assume that the initial kinetic energy of the injected proton is negligible.
2 If a deuteron ion is used, instead of a proton, state the new frequency of the oscillator.

The synchrotron

Theory

In the synchrotron, the radius of the orbit followed by the accelerated particles remains constant. Figure 11 shows a proton circulating in an evacuated tube in a magnetic field.

A series of accelerating stations (only one is shown in Figure 11) situated around the circumference of the synchrotron accelerate the particle. At each gap, the increase in kinetic energy of the particle is given by $V_{eff}\, q$.

As the speed increases, the magnetic field must be proportionally adjusted.

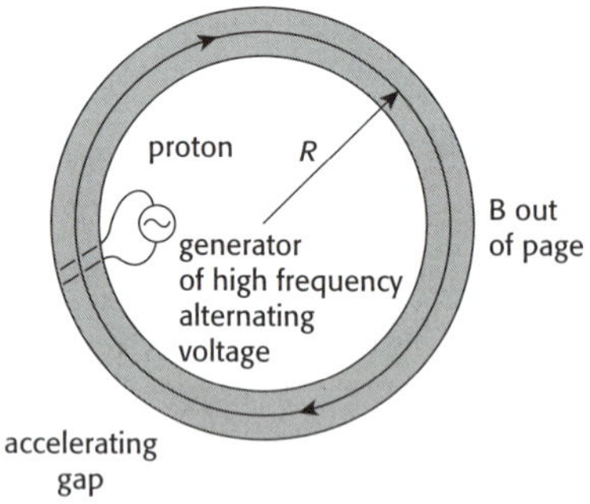

Figure 11

The equations $f = \frac{qB}{2\pi m_{rel}}$ and $f = \frac{v}{2\pi r}$

must both be satisfied during the acceleration of the particle around the ring. This is done by adjusting the magnetic field B and the resonance frequency f. As v tends to be the speed of light c, the frequency remains constant and B must be adjusted for increasing relativistic mass, m_{rel}.

The particle will stay in orbit of radius R if the magnetic field strength has the value given by

$$B = \frac{m_{rel} v}{qR} \qquad (10)$$

where the values of v and m_{rel} increase with time.

Mathematical physics: the energy required to create a set of particles at rest from the collision of an incoming particle with a stationary target

A relativistic incoming proton collides with a target proton. This scattering event produces a neutral pion π^0

$$p + p \rightarrow p + p + \pi^0$$

The rest mass of the pion is 135 MeV c^{-2}. More than 135 MeV c^{-2} is required since the incoming proton has momentum. Some of the particles must carry momentum after the collision in order to conserve momentum, and thus kinetic energy, away.

To simplify the mathematics we swap our viewpoint, or "frame of reference" to look at the event in the frame of reference of the centre of mass (CM) where, before the collision, the two protons move towards each other with opposite and equal velocities and the momentum of the system is zero. In the CM frame, the minimum energy required to create the neutral pion with the two protons, all at rest, equals the sum of the rest mass energy of these three particles. This sum must be equal to the total relativistic energy in the CM frame:

$$2(mc^2) = 2\,(m_p c^2) + m_\pi\, c^2$$

where m is the relativistic mass of each proton in the CM frame. The speed of the proton v in the CM frame can be determined by:

$$m = \frac{m_p}{\sqrt{(1 - v^2/c^2)}}$$

m_p: 938 MeV c^{-2}
m_π: 135 MeV c^{-2}
v: speed of each proton in the CM frame

Calculations yield v = 0.36 c.

However, in order to obtain a meaningful result we must look at the event from the laboratory frame of reference (where the target proton is at rest).

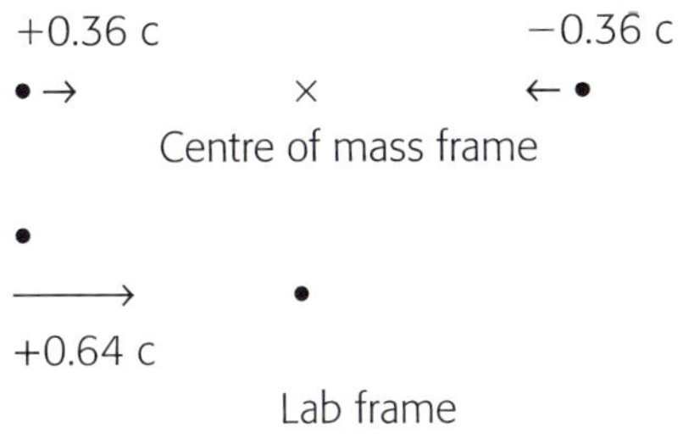

In the lab FR, V can be calculated by using the relativistic addition of velocities equation:

$$V_{tot} = \frac{V_1 + V_2}{1 + \frac{V_1 V_2}{c^2}}$$

The result is: V = 0.64 c (not 0.72 c!)

This is the velocity of the incoming proton. Its relativistic mass ($m = \gamma m_p$) is 1.3 m_p, and its kinetic energy ($E_k = \Delta m\, c^2$) is near to 280 MeV.

$$\text{The ratio } \frac{\text{energy of proton}}{\text{rest mass of pion}} = \frac{280}{135} \approx 2$$

The proton must have an energy of more than twice the rest mass of the pion to produce it. Collisions with a stationary target are not efficient, since a significant amount of kinetic energy is "used up" to conserve momentum.

If we generalize this approach we obtain

$$E_a^{\,2} = 2E\,Mc^2 + (Mc^2)^2 + (mc^2)^2$$

where E_a is the total rest mass energy of the product particles. This is the minimum energy that must be available for the event.

E: total relativistic energy (threshold energy) of the incoming particle (lab FR)
M: rest mass of the stationary target
m: rest mass of incoming particle

Experimentally the value of E_a is first determined, then its value is introduced into the equation to find the value of E.

If, in subsequent events, the total rest mass energy of the particles after collision is larger, does the efficiency of this collision (with a stationary target) increase?

For example, consider the event where an antiproton $\bar{p}$ and a proton p are produced in the collision of an incoming proton with a stationary proton:

$$p + p \rightarrow p + p + \bar{p} + p$$

Similar calculations show the incoming proton must have a kinetic energy of 6 $m_p c^2$ (total energy of 7 $m_p c^2$). Three times more energy than the rest mass energy is required. We can see that when higher energies are required, the efficiency drops.

The colliding-beam accelerator

The high-energy particles emerging from a synchrotron collide with a target to create new particles. Unfortunately, the total energy of the incident particle (kinetic energy + rest mass energy) is not entirely available when the target is fixed.

When colliding particles are moving towards each other at relativistic speeds much more energy is available for the creation of new particles.

Considering the amount of extra energy available for the creation of new particles it is clear that if both the "beam" and "target" protons are moving we are at a considerable advantage. However, aligning two protons so that they collide head-on (the radius of a proton is about 10^{-15} m.) is very challenging.

Comparative look at a relativistic collision between two moving protons

Two accelerators each produce a beam of protons (of rest mass m_0), moving at a speed v = 0.98 c relative to the laboratory frame of reference to which the accelerator tube is attached. The protons are aimed at each other.

In the lab frame, the gain in relative speed of approach is not great (0.0198 c). However, in the same frame, the relativistic energy is about 8 times greater than its Newtonian value. The energy gained in using colliding beams is significant. The principle of operation of the **colliding-beam** accelerator takes advantage of this fact. The probability of having two protons colliding head-on is not very high but the energy gain is so important that this practical approach represents the path for the future.

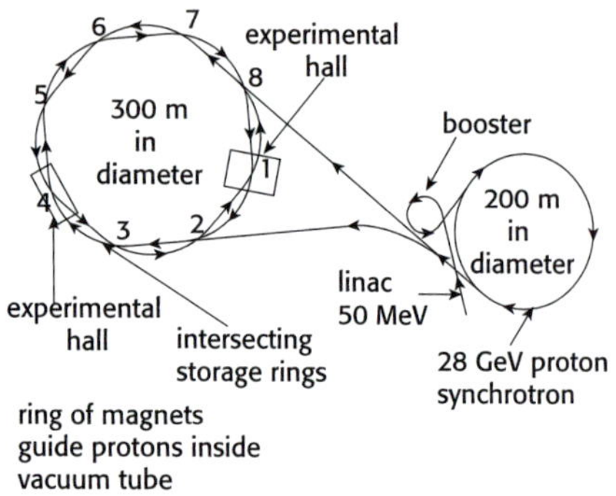

Figure 12

Figure 11 shows the proton synchrotron at CERN, which uses storage rings.

The proton beam from the synchrotron is divided in two separate beams moving in opposite directions, thus creating two beams intersecting at different points where the proton beams collide (almost head-on) with an energy equivalent to that produced by a single 1700 GeV accelerator.

The Large Hadron Collider will be operational in CERN in 2008. In this accelerator phyicists will be able to recreate the state of matter and energy a fraction of a second after the Big Bang.

> **Investigation**
> The Large Hadron Collider (LHC)
>
> Using the page found here: http://public.web.cern.ch/Public/Content/Chapters/AboutCERN/CERNFuture/WhatLHC/WhatLHC-en.html prepare a poster about the LHC accelerator. Include information about the type of accelerator, the detectors located around it, and some of its goals.

Bremsstrahlung (breaking radiation)

According to electromagnetic theory, an accelerated (or decelerated) electric charge emits electromagnetic waves. Figure 13 illustrates an electron moving in the Coulomb field of the nucleus of an atom.

The electric force changes the path of the charge. This change of path implies that the electron is accelerated and consequently will emit radiation or, in the context of quantum physics, a photon of energy hf:

$$\Delta E_k = hf \qquad (11)$$

where ΔE_k is the loss in kinetic energy of the electron.

This process is called slowing-down radiation or breaking radiation (Bremsstrahlung in German).

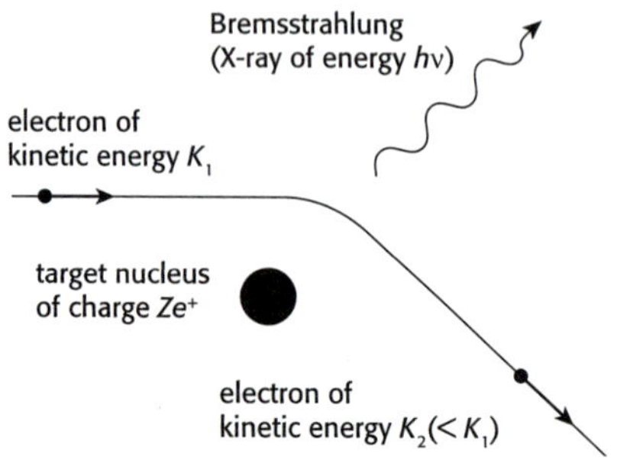

Figure 13

A succession of collisions will produce numerous decelerations as the electron collides with a series of atoms, producing a photon each time, so the kinetic energy of the electron is reduced by the radiation. Typically, electrons going through breaking radiation will produce a continuous X-ray spectrum. This is the case when electrons collide with a metallic target in an X-ray machine.

The breaking radiation effect

In a linear accelerator electrons can be accelerated to high energies without appreciable radiation in a cyclic accelerator, charged particles in circular motion are continuously accelerated towards the centre of rotation. As a consequence, they will radiate as electromagnetic waves in part in the visible region. Unfortunately, this process reduces the available kinetic energy of the particles. To reduce this effect in cyclic accelerators, the radius is increased, thus reducing the centripetal force a_c:

$$a_c = v^2/r \qquad (12)$$

In many accelerators, v is already near the speed of light ($v \rightarrow c$) so does not change appreciably as the particle circulates (the energy gained is reflected in the increase of its relativistic mass).

The bigger picture: can we justify the expense of particle physics research centres?

Consider the following points of view:

- In 1905, Einstein published three papers on light, special relativity, and the sizes and movements of molecules. These theories help our understanding of matter and the universe but, also, lay down the foundations for the technology and economy of today. The television, GPS, the mobile phone, and the electron microscope are examples of such applications.
- In the 18th century, Faraday, a brilliant scientist, was investigating magnetic fields and their effect on coils. Was his work then productive for his society? One day, Prime Minister Gladstone asked him what was the use of electricity "Some day you will tax it," answered Faraday. Today, electricity is an intrinsic part of our life thanks to his work and that of others.
- Colliding particles in accelerators produce new particles that are used for biological research and therapy. Synchrotron radiation has a large range of applications including biology, medicine, and technology.
- The sharing of large research cost by developing countries is a wise approach that permits reasonable scientific endeavour. The sharing of the design of accelerators and detectors as well as the sharing of research and data with developing countries encourage cooperation between people of different cultures and beliefs. It also feeds the progress of these nations.
- Without experimental physics, theoretical physics stagnates in pure speculation.
- Without fundamental research some new technology will not materialize.

Above all, curiosity is an essential part of the human mind.

Particle detectors

When the high-energy particles from accelerators strike a target, rapid reactions occur. These particles are invisible to the human eye so physicists design instruments that make it possible to "see" the path of charged particles in a medium. Three of these devices are described here: the bubble chamber, the spark chamber, and the photomultiplier. In each case, the ionizing power of charged particles moving in the medium is the dominating effect.

The bubble chamber

If pressure is exerted on a liquid, it can be superheated. By reducing the pressure suddenly, the liquid vaporizes. At that instant, incoming charged particles can ionize the liquid along its track, which creates gas bubbles around the ions. These bubbles reflect light and can be photographed by two cameras.

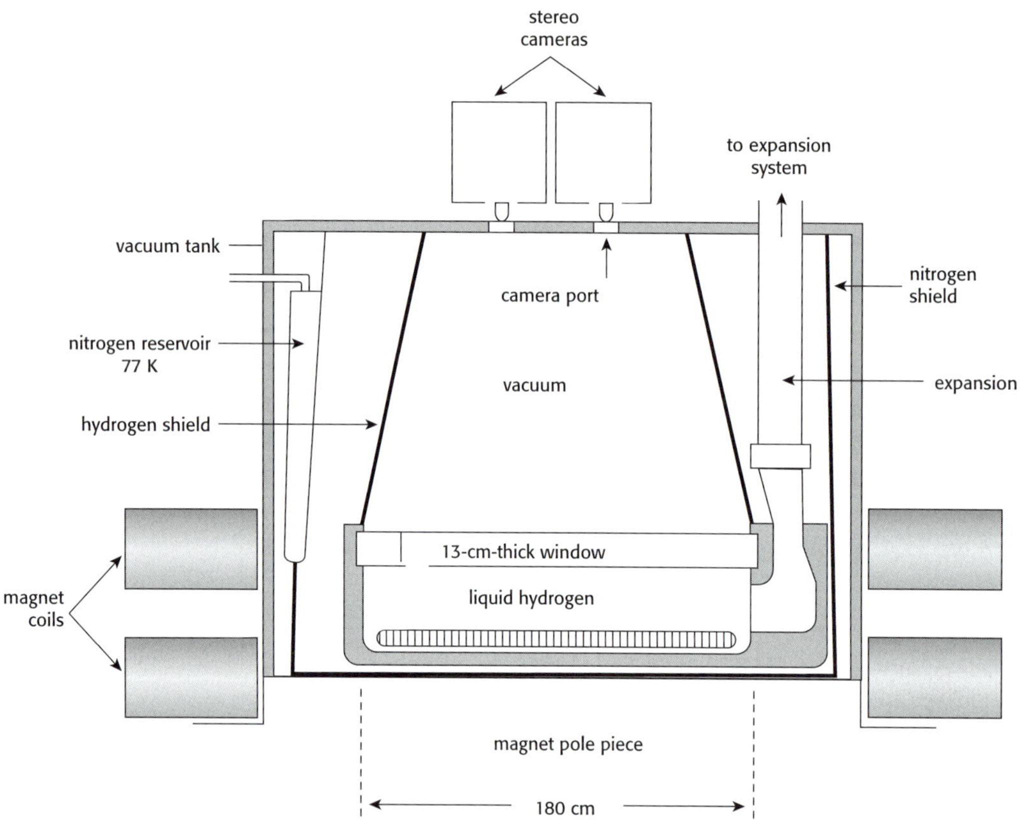

Figure 14 The bubble chamber

Liquid hydrogen is generally used in bubble chambers because it provides a dense target of protons. The analysis of events is also facilitated by the fact that the proton is the simplest nucleus available. However, the temperature at which the hydrogen is superheated is −246 °C, so a cooling system is required.

A magnetic field of about 1 T curves the tracks of charged particles.[2] Events unfold very rapidly so that the picture can reveal a series of cascading tracks across the chamber. Uncharged particles can be "seen" indirectly through their interaction with protons, which, after impact, will leave an ionic trail of their own.

Figure 15 A 4 MeV electron slows down in a liquid hydrogen bubble chamber immersed in a homogeneous magnetic field. The beaded trajectory is characteristic of electron tracks.

[2] As a comparison, a typical value of the field at the Earth's surface is 5.5×10^{-5} T.

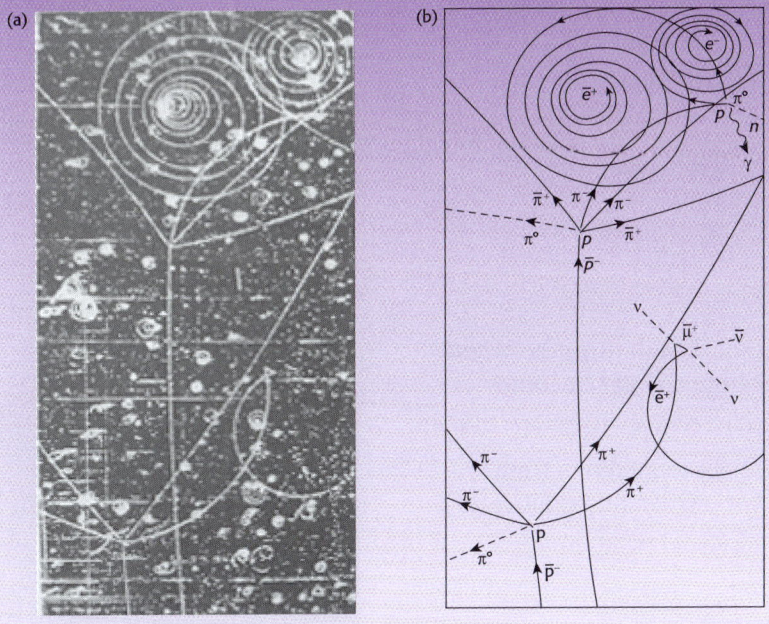

Figure 16 (a) Photograph of antiproton reactions in a hydrogen bubble chamber showing meson pairs in the kind of reaction that first led to the identification of the ω meson. (b) Legend diagram showing the unchanged trackless particles as dotted lines.

1 At the top of figure 16a, the spiralling trajectories of an electron and a positron are shown. Which particle carries less energy? What are your assumptions?

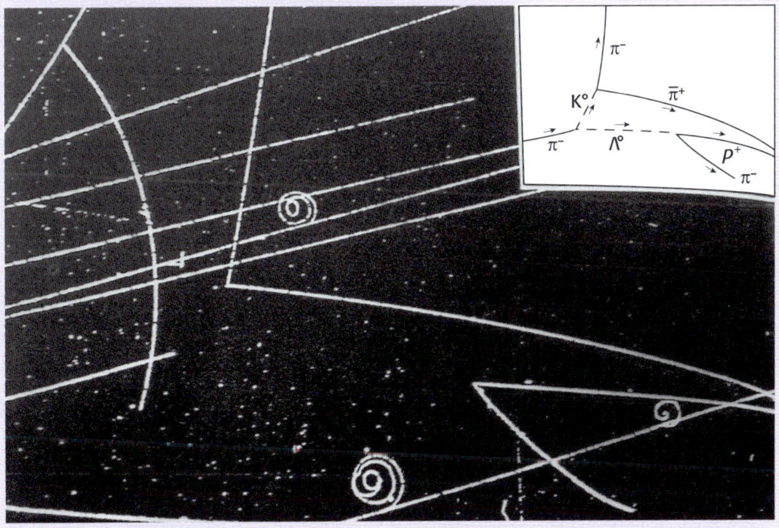

Figure 17 Bubble-chamber photograph with π^-, p, K°, λ°, $\overline{\pi}^+$, and π^- tracks, all in one related event. Legend diagram at upper right.

2 Neutral particles do not leave any tracks in the chamber. Explain why this is the case.

The spark chamber

Figure 18 shows two thin metallic plates immersed in a noble gas (neon or argon). A high voltage is applied to the plates. If the path of a charged particle runs between the plates, a trail of ion pairs will be produced in the space between the plates (Figure 18a).

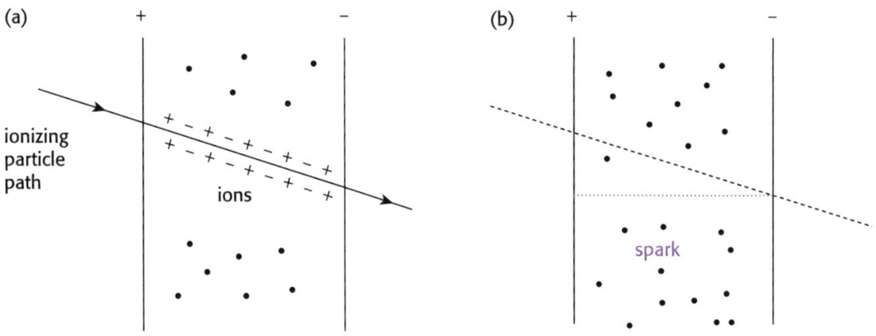

Figure 18

Because of the high potential difference, a spark will jump between the plates at the location of the ion pairs, where the gas becomes temporarily conductive (Figure 18b).

In a spark chamber, a series of electrodes are connected to a high-voltage source. A few millimetres separate successive foils (Figure 19).

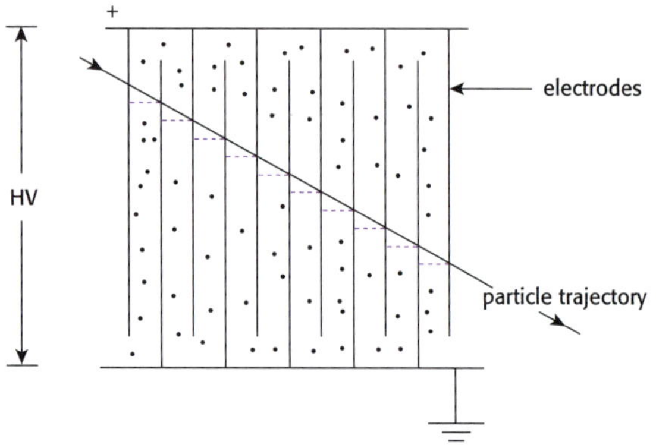

Figure 19

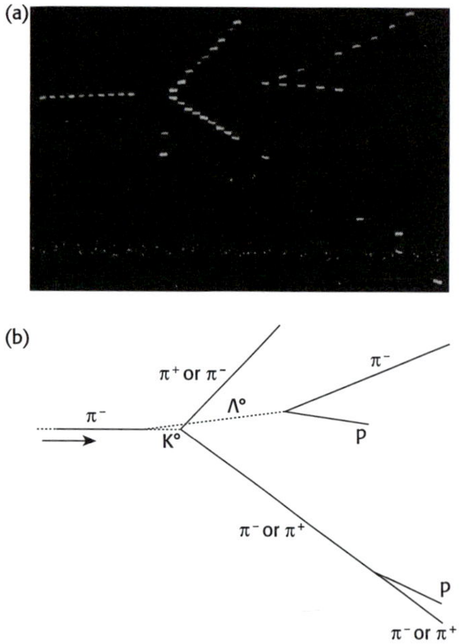

Figure 20 (a) shows appearance of particle tracks. No magnetic field is applied. (b) identifies the charged particles and their tracks as well as the implied tracks of the neutral particles.

When the potential difference is applied, a succession of sparks will reveal the trajectory of the particle through the chamber, since the charged particle will leave a trail of ions in the space between the electrodes. The series of sparks can be photographed or electronically counted (Figure 20).

The photomultiplier

Gamma rays produce very little ionization in a gas or liquid. They do not leave trails in a bubble chamber or in a spark chamber but a photomultiplier tube can be used to detect gamma rays and to measure their energy.

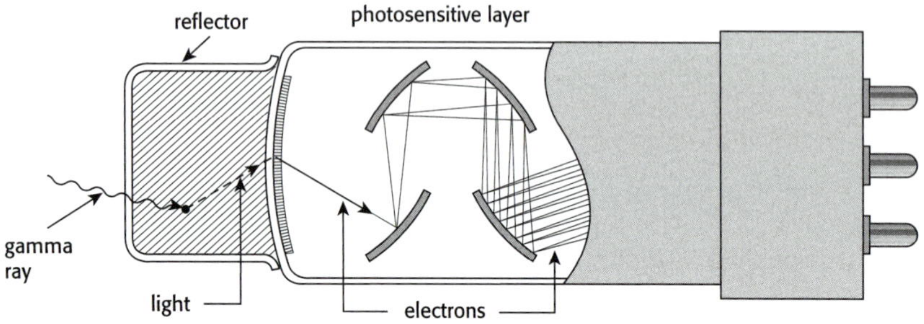

Figure 21 Scintillation detector coupled to a photomultiplier

A single gamma ray photon collides with an atom inside a transparent crystalline solid called a scintillator, producing light in the visible region. A reflector sends this light towards a photosensitive layer. The action of the light on this layer triggers the emission of a single photoelectron, i.e. an electron that has been liberated from its atom by a photon, which is accelerated towards an anode (+). Upon collision, a series of secondary electrons are liberated and accelerated towards a second anode, which is at a higher potential than the first one. At this anode, even more secondary electrons are ejected. This avalanche effect continues through a series of anodes called dynodes. The final signal is amplified electronically so the pulses can be counted. These pulses are also a measure of the energy of the incoming gamma ray. Ultimately, an energy spectrum of the photons entering the scintillation detector can be produced. The advantage of the photomultiplier is that individual photons of relatively low energy can be counted.

The particle physics international community

Particle research laboratories

In 1949, at the European Cultural Conference, in Lausanne, Switzerland, the French scientist Louis de Broglie proposed the idea for a European research laboratory where "greater resources than those available to the national laboratories and could then embark upon tasks whose magnitude and nature preclude them from being done by the latter on their own."[3] A series of meetings under the auspices of UNESCO brought about the foundation of CERN. Today, its membership comprises 20 European countries and more than half the world's particle physicists collaborate on the work done at CERN.

Scientists do fundamental research in the name of scientific endeavour, independently of their nationalities, politics, culture, and religion. This cultural melting pot is at the heart of the activities of these facilities. "... based in their own countries, members of collaborations not only provide most of the ambitious experimental apparatus, but they also contribute to a novel, global, powerful information and communication infrastructure using their own countries' industries and talents in a fair and constructive partnership. And the motivation for all this: cutting-edge physics."[4]

The Large Electron Positron collider (LEP) which ran from 1989–2000 contributed to the current Standard Model of particles and their interactions.

The new LHC project comprising the accelerator and four major detectors involves the efforts of more than 400 research laboratories from dozens of countries around the world. It is due to switch on in 2008 and physicists hope to detect evidence for the Standard Model.

Examples of some other big national laboratories:

- USA: Fermilab, SLAC
- Germany: DESY
- Japan: KEK
- Canada: TRIUMF

[3] *CERN Courier*, November 2004
[4] *CERN Courier*, March 2005

The International Linear Collider (ILC) is currently under development. It goes beyond the regional concept to a global one. The idea is to form a partnership between American, Asian, and European countries, which will share the design, construction, and management of the facility. Emphasis will be on the research and development of detectors. The goals of the ILC are:

- to create a research centre where strong international cooperation flourishes
- to ensure access to the facilities
- to include the broad community in the decision-making process: "access to materials, land use, ecology, and economic impacts due to the resources that are required."[5]
- to make the data available to the broader community
- to consider new forms of detector collaboration.

Scientists work together as colleagues and friends in a way that does not reflect the external tensions and hostilities existing between some countries and cultures. Even in times of conflicts, scientific collaboration continues to play a role in penetrating political barriers, as it was between Western and Soviet societies during the Cold War.

?

Research one of the following further examples of international collaboration in the particle physics community:

- JINR
- SESAME
- The Berlin Declaration
- The Pugwash conference
- ICSC
- The Erice statement
- ICTP

Quarks

Any particle that is made up of quarks is called a **hadron**.

According to the best theories at the moment, it is impossible for quarks to exist on their own; they must always be bound into particles by the strong force that exists between them. As it happens there are only three stable combinations that the strong force will allow.

Table 6 Combinations of quarks and antiquarks

Combination	Symbol	Name
three quarks	qqq	baryon
three antiquarks	$\bar{q}\bar{q}\bar{q}$	antibaryon
a quark and an antiquark	$q\bar{q}$	meson

?

Distinguish the category of hadrons into which the following particles can be placed.

1 uud (a proton)
2 udd (a neutron)
3 $u\bar{d}$ (a π^+ particle)
4 $\bar{u}d$ (a π^- particle)
5 $u\bar{u}$ (a π^0 particle)

The three π particles have very similar masses, and belong to a family known as the pions

Baryons, antibaryons, and mesons were discovered long before quarks and their properties were well known. The success of the quark idea was to draw together and explain their properties in terms of the properties of the quarks inside them.

Baryon number

Baryon number is a property of hadrons that was discovered by studying various reactions taking place in particle accelerators. It is

[5] *CERN Courier*, October 2005

possible to give every hadron a value of baryon number (B) based on a simple scheme:

Any baryon $B = 1$

Any antibaryon $B = -1$

Any meson $B = 0$

e.g. protons and neutrons have $B = 1$, an antiproton has $B = -1$, all pions have $B = 0$. In all reactions baryon number is found to be a conserved quantity. The baryon number of the various hadrons can be explained if the quarks are given values of baryon number shown in Table 7.

Table 7 Quarks and antiquarks

	Baryon number	Charge	Type of quark
Matter	1/3	+2/3	u c t
		−1/3	d s b
Antimatter	−1/3	−2/3	$\bar{u}$ $\bar{c}$ $\bar{t}$
		+1/3	$\bar{d}$ $\bar{s}$ $\bar{b}$

Antiquarks

Quarks all have antiquark partners. They have the same mass but opposite charge and **baryon numbers**.

Consider the baryon numbers of the quarks and antiquarks. You should be able to convince yourself that the sum of the baryon numbers of the quarks will produce the required baryon numbers of the hadrons.

Spin

An example of a quantum number (see page 318) is intrinsic angular momentum, or "spin".

The proton, neutron, and the quarks all have an intrinsic angular momentum of $\hbar/2$, or in the usual shorthand, "spin ½". Spin can be directed "up" (spin +½) or "down" (spin −½).

Quantum numbers are, in general, additive. This means that when elementary particles combine to form a new system, the quantum numbers of the particle that is formed are just the sum of the numbers of the constituent parts. Spin systems will tend to align so as to form pairs of spin "up" and spin "down" particles. When a proton is formed from three quarks, there will always be one spin "up" quark and one spin "down" quark. The third quark can be aligned in either of the two spin states so that the proton will have an overall spin of ±½.

example:

	u	+	u	+	d	=	uud	=	proton
spin	−½		+½		+½		−½+½+½		+½

?

Determine the possible values of spin of
(a) baryons
(b) antibaryons
(c) mesons

Conservation of baryon number

Baryon number was discovered when physicists noticed that some reactions that conserved electrical charge were not taking place:

$$p + p \nrightarrow p + \pi^+$$

$$p + n \nrightarrow p + \pi^0$$

$$p + p \nrightarrow \pi^+ + \pi^+$$

It seems that the total number of baryons must remain the same. Pions, which can appear in any number given enough energy, are mesons whilst protons and neutrons are baryons. This suggested a simple scheme in which the proton and neutron have $B = 1$ and the pions have $B = 0$. It is then easy to show that the reactions do not happen because they do not conserve baryon number:

$$\mathrm{p} + \mathrm{p} \nrightarrow \mathrm{p} + \pi^+$$
$$B \quad 1 + 1 \neq 1 + 0$$

$$\mathrm{p} + \mathrm{n} \nrightarrow \mathrm{p} + \pi^0$$
$$B \quad 1 + 1 \neq 1 + 0$$

$$\mathrm{p} + \mathrm{p} \nrightarrow \pi^+ + \pi^+$$
$$B \quad 1 + 1 \neq 0 + 0$$

Baryon number conservation indicates that the fundamental forces must all keep the net number of quarks (i.e. the number of quarks minus the number of antiquarks) constant.

Strangeness

In the early 1950s, particles with large masses and long lifetimes started to be discovered in cosmic ray experiments. In 1954 these particles were categorized and given a new property – **strangeness**. The odd lifetimes of the strange particles (heavy particles generally decay more rapidly than light ones) could then be explained by assuming that strangeness was conserved in strong force interactions but not conserved by the weak force.

Faced with the need for an arbitrary choice, physicists chose the K^+ meson to have strangeness $S = +1$, and categorized other particles relative to this. It is now understood that the strangeness of a baryon indicates the presence of **strange quarks** within it. An unfortunate consequence of the choice of the K^+ meson to have strangeness $S = +1$ is that the strange quark itself has $S = -1$, but the choice is arbitrary so it is not important. The antistrange quark has $S = +1$.

The other heavy quarks (charm, top, and bottom) are all defined to have their own associated quantum number in a similar fashion (see Table 8). These are very powerful concepts, but physicists do not yet fully understand their meaning.

Table 8 Quark properties

Quark	strangeness	charmness	bottomness	topness
u	0	0	0	0
d	0	0	0	0
s	−1	0	0	0
c	0	+1	0	0
b	0	0	−1	0
t	0	0	0	+1

Using conservation laws

Consider the following reaction in which X is a newly discovered particle:

$K^- + p \rightarrow K^0 + K^+ + X$

K particles are mesons. What can be deduced about the properties of X by applying the various conservation laws that we know?

1 Conservation of electric charge shows that X must carry a negative charge:

$$\begin{array}{llllllllll} & K^- & + & p & \rightarrow & K^0 & + & K^+ & + & X \\ Q & (-1) & + & (+1) & \rightarrow & (0) & + & (+1) & + & (-1) \end{array}$$

2 Conservation of baryon number shows that X must be a baryon:

$$\begin{array}{llllllllll} & K^- & + & p & \rightarrow & K^0 & + & K^+ & + & X \\ B & (0) & + & (+1) & \rightarrow & (0) & + & (0) & + & (+1) \end{array}$$

3 The next thing to decide is whether strangeness is conserved in this reaction. The reaction comes about due to the interaction of a K meson with a proton so it is highly likely that the strong force is responsible. Strangeness must then be conserved:

$$\begin{array}{llllllllll} & K^- & + & p & \rightarrow & K^0 & + & K^+ & + & X \\ S & (-1) & + & (0) & \rightarrow & (+1) & + & (+1) & + & (-3) \end{array}$$

This means that X is a very strange baryon indeed. In fact, X is the Ω^- particle (three strange quarks) whose existence was predicted by Murray Gell-Mann.

Color

The Ω^- is three apparently identical fermions bound in the same system, which suggests that the Pauli exclusion principle (see page 318) is being violated, so we must propose a new quantum number for quarks.

It turns out that there are three distinct varieties of strong charge and to distinguish among them we call them red (R), green (G), and blue (B). As such they are known as color charges, though this is nothing to do with colour in its familiar sense – it is just a naming convention.

Quarks cannot exist in isolation but must be bound in colorless combinations. This can only be achieved in one of two ways:

1 **Baryons**: 3 quarks each of which has a different color charge from the other two. Antibaryons are essentially the same but each antiquark possesses a different anticolor from the other two.
e.g. a proton u u d
(red) (green) (blue) = colorless

2 **Mesons**: a quark of one color and an antiquark of the corresponding anticolor.
e.g. a π^+ u $\bar{d}$
(green) (antigreen) = colorless

Gluons

Quarks interact via gluons. By studying the interactions of baryons and mesons we know that there are 8 distinct gluons, with complex color charges.

As well as mediating the interactions between quarks, gluons also bind the quarks together to form hadrons. You can imagine one of these particles as quarks surrounded by a "sea" of gluons.

Consider a π^+ meson. Energy must be put into the system to separate the quark and antiquark. If the energy in the system is high enough, one of the gluons in the "sea" can form a new quark/antiquark pair resulting in a new pair of mesons. See Figure 22.

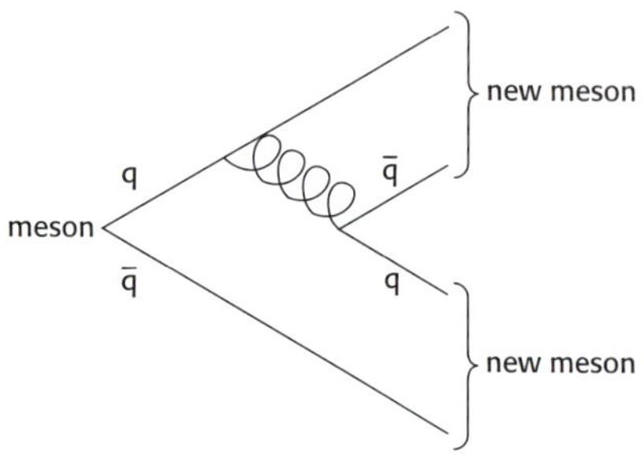

Figure 22 The formation of two new mesons by gluon emission

Conversely, as mutually attracting quarks are brought together, the strong force between them decreases so that the closer they are to each other, the smaller the force drawing them together. This is known as **asymptotic freedom**. Compare this feature of the strong nuclear force with, for example, the electromagnetic force between two oppositely charged particles (see Chapter 13).

Leptons and the standard model

In 1911–12 Victor Hess (1936 Nobel Prize winner) made a series of balloon flights to measure radiation in the atmosphere. He found that the intensity of radiation increased to five times the amount at sea level by about 5000 m altitude; hence the radiation must be coming from space. Robert Millikan coined the name **cosmic rays** for this penetrating and ionizing radiation, and believed it was electromagnetic in nature. However, but studying the intensity of radiation at different latitudes, Compton showed that the radiation was being deflected by the Earth's magnetic field and so must be charged.

Most of the cosmic rays striking the Earth are protons. Some are the nuclei of heavy elements (uranium nuclei have been detected) and there are also some very high-energy electrons. Typically they have energies between 1 and 10 GeV, but occasionally extraordinarily high-energy particles (up to 2×10^7 GeV) are detected, smashing into protons and producing hundreds of particles. The nature and origin of these cosmic rays is still a mystery.

The muon

Since its discovery the muon has been extensively studied. It is known to be unstable, and decays with a lifetime of 2.19703×10^{-6} seconds. This sort of time-scale is characteristic of a weak-force decay. Muons appear to be unstable "heavy electrons". It is possible to create atoms in which a muon replaces an electron in orbit around the nucleus. The biggest mystery that remains concerning the muon is why nature needs such a particle to exist.

The tau

In 1974 another "heavy electron", the **tau** (τ) particle, was discovered. This discovery came from the SPEAR collider (Stanford Positron-Electron Asymmetric Ring) built on a car park at SLAC, the Stanford Linear Accelerator. The particle was totally unexpected but it did not take long to confirm that it was another unstable lepton. However, the tau's enormous mass (3.5 million times that of the electron) and very short lifetime (3.3×10^{-13} seconds) means that it is very unlikely to be seen in cosmic rays.

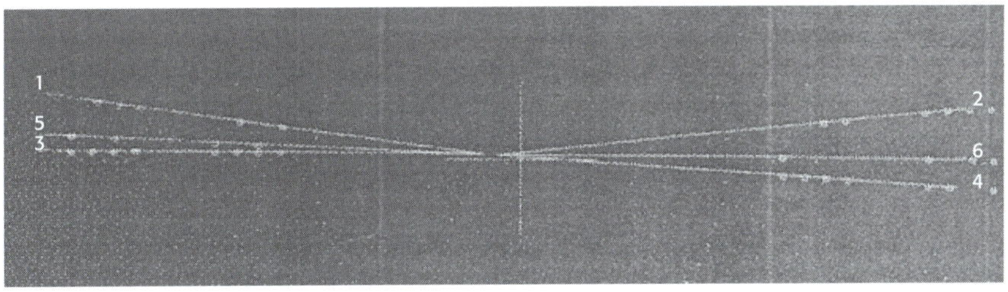

Figure 23 A typical tau decay. In this electronic track reconstruction a tau and antitau have been produced, travelled a short distance, and each decayed into three pions (numbered 1–6) and a neutrino. The small gap at the centre shows the distance travelled by the taus.

Lepton number and generations

There are three particles; the electron, the muon, and the tau, that appear to be versions of the same particle, but with increasing mass. They all carry the same electric charge and they all react via the weak and electromagnetic forces, but not the strong force. In each case there appears to be a partner: a neutral, almost massless object called a neutrino.

In order to categorize these particles physicists invented a labelling system called the **lepton number**, analogous to the baryon number of quarks.

Table 9 Lepton generations

First generation, $L_e = 1$	Second generation, $L_\mu = 1$	Third generation, $L_\tau = 1$
electron • found in atoms • involved in electrical and thermal conduction • produced in β^- radioactive decay	muon • produced in large numbers when cosmic rays collide with nuclei in the upper atmosphere–most cosmic rays that reach the surface are muons	tau • so far only observed in laboratories
electron neutrino • produced in β^- radioactive decay • produced in large numbers by nuclear reactors • produced in huge numbers by nuclear reactions in the sun	muon neutrino • produced by atomic reactors • produced in the upper atmosphere by cosmic rays • produced in the sun by nuclear reactions	tau neutrino • so far only observed in laboratories

Lepton number turns out to be more than simply a convenient way of tabulating the particles – it is also a conserved quantity.

Lepton number conservation

In any reaction involving leptons the total number in each generation must always remain the same.

?

State and explain which of the following reactions conserves lepton number.

(a) $\nu_e + \mu^- \rightarrow e^- + \nu_\mu$

(b) $\nu_e + n \rightarrow p + \mu^-$

Unification

Modern-day particle physicists are searching for beauty in the idea that the four fundamental forces can be united into one theory that shows them to be different aspects of the one force (just in the same way that electricity and magnetism can be brought together in electromagnetism). In 1979 Glashow, Salam, and Weinberg received the Nobel Prize for bringing this goal a step nearer with their combined theory of the weak and electromagnetic forces.

The electroweak force

In 1961 Sheldon Glashow showed how the weak force could be explained by Feynman diagrams and linked with the electromagnetic force. For 30 years physicists had known about the weak **charged current** reactions such as:

$\nu_e + n \rightarrow p + e^-$ W^+ exchange

$\bar{\nu}_e + p \rightarrow n + e^+$ W^- exchange

Glashow predicted the existence of another weak interaction (the **neutral current**) that involved the Z^0 exchange particle:

$\bar{\nu}_\mu + e^- \rightarrow \bar{\nu}_\mu + e^-$ Z^0 exchange

Evidence for such a reaction came from a giant CERN bubble chamber in 1973. Once the neutral current had been established the link with the electromagnetic force was more promising – in many ways the Z^0 and the photon are similar objects. The final problem was that the W and Z exchange particles are very massive and the photon is massless. Any theory seemed to require all the exchange particles to be massless.

The next step was taken in 1967 when Steven Weinberg and Abdus Salam applied an idea, first suggested by Peter Higgs, to the electroweak theory. Higgs's idea was to introduce another type of force field (now called the Higgs field) into the theory. By carefully selecting the right sort of field to add, Weinberg and Salam made the Higgs field interact with the W and Z in such a way as to give them mass.

The Higgs field is very important – without it you cannot explain why the W and Z exchange particles have mass. If the Higgs field exists then **Higgs bosons** should exist as well.

Higgs bosons

The Higgs field is currently the best explanation for the origin of the mass of not only the W and Z but all massive particles. So far there is no experimental evidence for Higgs particles (Higgs bosons). The next generation of particle accelerators (e.g. the Large Hadron Collider at CERN) have been designed specifically to find them. If they are not found, theoretical particle physicists will have a new challenge – to find a new explanation for the origin of mass.

Experimental evidence for the quark and standard models

Deep inelastic scattering

In the late 1960s, 20 GeV electrons were collided with a proton target. As a result of these collisions the electrons scattered off at some angle to the target, and the detectors used were able to measure the angle.

What the results showed was that at low energies the electrons collided "softly" with the protons and scattered away at quite small angles; the protons simply recoiled from the collision. However, once

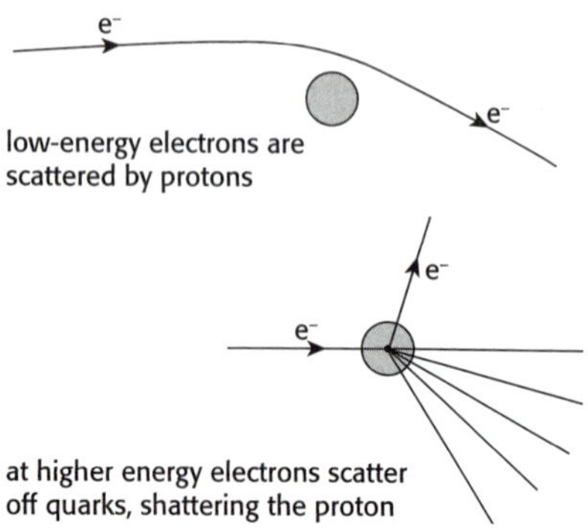

Figure 24 Deep inelastic scattering one particle to probe the structure of another

the energy of the electrons passed a certain minimum value they started to see much "harder" collisions – the electrons scattered off at greater angles and the protons shattered into a stream of hadrons rather than simply recoiling. This effect became known as **deep inelastic scattering**.

A careful study of the results, notably by Bjorken and Feynman in 1968, suggested that the protons contained three very small charged objects. At high energies they were able to "resolve" the very small objects inside the protons – in effect the electrons were starting to diffract off the the quarks inside the protons. As a result, one of the quarks was deflected away from the other two and shattered the proton into a stream of hadrons.

The idea is very similar to that of the Rutherford scattering experiment with α particles that revealed the presence of the nucleus inside the atom (Figure 25).

The SLAC results have since been refined using neutron targets and also neutrino beams. Physicists have been able to show that the three objects have the charges expected of the u and d quarks, and in the correct ratios.

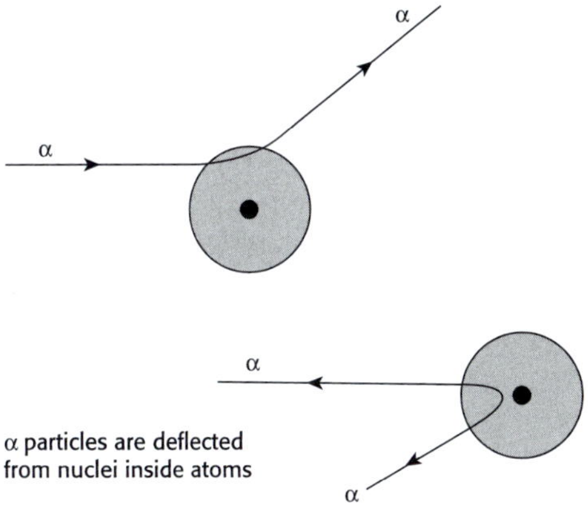

Figure 25 Rutherford scattering

The results of deep inelastic scattering experiments

When Gell-Mann and Zweig originally presented their ideas, only three types of quark were required to explain all the known hadrons. On 11 November 1974, two teams of physicists from America announced the discovery of a new meson. The team from SLAC names this new meson the ψ (psi) but the Brookhaven name was the J meson. To this day it is known as the J/ψ. At the time this was a remarkable discovery as the new meson had a mass of 3.1 GeV/c^2 (greater than any other known hadron) and a lifetime that was 2000 times longer than most other hadrons.

Within a very short time the new meson had been explained as the charm/anticharm meson ($c\bar{c}$) – a fourth quark had been discovered. Any lingering doubts about this interpretation went away when in 1975 and 1976 the charmed mesons D^+ ($c\bar{d}$) and D^0 ($c\bar{u}$) were discovered.

The Standard Model of particle physics

It seems that mass is the essential feature distinguishing the corresponding members of the three generations of basic particles (both quarks and leptons).

This "full set" of fundamental particles and the forces between them make up a complete description of nature, called the Standard Model of particle physics. There are many theories that propose further modifications but so far physicists have found no experimental evidence for any of them.

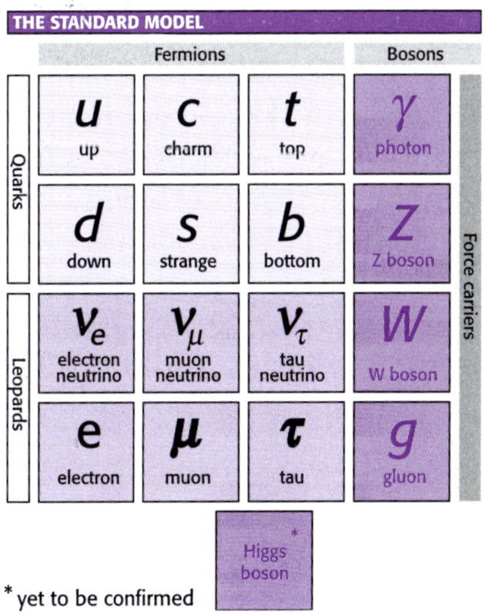

Figure 26 The Standard Model

The Big Bang

The universe as we know it began in a hot Big Bang approximately 14 billion years ago. The temperature of the universe immediately following the Big Bang is thought to have vastly exceeded 10^{30} K.

At this temperature particles had huge energies and so exotic particles would have existed that are no longer common today. As the universe cooled, these exotic particles would not have survived but would have annihilated or decayed to produce radiation and common fundamental particles (quarks and leptons). Within a millionth of a second, quarks had grouped to produce hadrons such as neutrons and protons and after about 3 minutes some light nuclei (Beryllium, Boron) had formed.

Some 300,000 years later, the ambient temperature had fallen below 10^4 degrees, that is, similar or cooler than the outer regions of our sun today. At these energies the negatively charged electrons were at last able to be held fast by the electrical attraction to the positively charged atomic nuclei whereby they combined to form neutral atoms. Electromagnetic radiation was set free and the universe became transparent as light could roam unhindered across space.

In today's universe, the once hot electromagnetic radiation now forms a black body spectrum with an effective temperature of 2.7 K (2.7 degrees above absolute zero). The discovery of this **cosmic microwave background** by Penzias and Wilson half a century ago is one of the great pieces of support for the Big Bang theory. Today precision measurements of the spectrum by instruments in satellites reveal small fluctuations in this radiation. These give hints of proto galaxies forming in the early universe.

Matter/antimatter asymmetry

The early universe contained equal quantities of matter and antimatter. If they had remained equally balanced we would expect that the particles and antiparticles would completely annihilate each other, leaving no matter. This is not the case, and we observe vastly more matter than antimatter in the universe. Physicists have observed a subtle difference in the behaviour of matter and antimatter that might explain this imbalance.

Figure 27 Two quests, one of the microcosmos deeper and deeper towards the very dimension of elementary particles, and the other of the macrocosmos, deeper and deeper to the edge of the universe and of time.

In the 1960s a type of weak decay was observed that did not conserve a quantity called "CP", which was previously thought to always be conserved in weak interactions. Since the Standard Model does not predict this **CP violation**, it is now thought that there might be gaps in the Standard Model, paving the way for new theories, known as **physics beyond the Standard Model**. Supersymmetry and string theory are two well-known examples of these types of theories. String theory perceives the fundamental particles as strings of energy, rather than points, using more than the four dimensions of space and time that are used in the Standard Model.

Particle physics and the quest for the infinitely small and infinitely large

Amazingly, these two quests are now converging towards the instant of creation, the Big Bang, an event particle physicists are aiming to reproduce in their laboratories!

Answers

Numerical answers only are provided.

CHAPTER 2

page 11

1 a 10^5
b 10^{-3}
c 10^9 or 10^{10}
d 10^{27}
e between 10^{-22} and 10^{-24}
f 10^{31}
2 B
3 C
4 B
5 C
6 C

page 23

1 B
2 A
3 B

CHAPTER 3

page 27

a i 18 t
ii 81 m
iii 27 m s^{-1}
iv 27(t − 6)
b t = 9.0 s

CHAPTER 4

page 34

D

page 36

1 B
2 A

page 39

2 c 0.3 N
d 1.5 kg

page 40

1 a 2.4 kN
b 19.2 kN
c 160 kN
2 a 2.3 kg m s^{-1}
b 4.6 N
3 a 14.5 squares = 2.9 N s
b 2.9 N s
c 48 m s^{-1}

page 41

5 a 2 m s^{-1}
b 24 kg m s^{-1}
c 120 N

CHAPTER 5

page 45

1 D
2 A
3 B

page 46

top
1 A
middle
a 2300 J
b 27 J
c 320 kJ
d 4.7 kJ
bottom
a 5.76 kJ
b 18 J
c 320 kJ
d 375 J

page 47

1 a 2.3 m s^{-1}
b 1 m

page 48

1 a 2.0 m s^{-1}
b 60 J
2 8 kg moving at −1 m s^{-1}, 12 kg moving at +4 m s^{-1}

page 49

bottom
1 3 kW
2 50 kW

page 50

top
1.5×10^6 J

CHAPTER 6

page 52

D

page 53

middle
a i constant vertical acceleration = −10 m s^{-2}
ii constant horizontal velocity
b i 0.4 s
ii 1.9 s
c i 6.9 m s^{-1} horizontally
ii 16.2 m s^{-1} at 28° to the vertical
d i 0.8 m
ii 13.0 m

page 54

a 9 m s^{-1} at 60° to the horizontal
b 10 m s^{-2}

CHAPTER 7

page 62

9.6 m s^{-2}; error range depends on graphs

CHAPTER 8

page 85

1 1.8×10^{20}
2 3.3×10^{15} Hz

CHAPTER 9

page 94

D

page 96

C

page 99

top
C

page 103

630 ± 30 nm

page 104

1 a No
b Yes
2 630 ± 60 nm

page 107

3 a 60°
b reduced to 0.25 of original
c 45°

CHAPTER 10

page 110

top
1 K s^{-1}; J s^{-1} or W
2 a 1°C
b 52 kJ
c 34°C
d 34°C
3 a 1.6°C s^{-1}
b 600 W

page 111

1 D
2 A

page 114

top
1 920 s
2 500g i.e. 0.5 litre
bottom
C

CHAPTER 11

page 121

a 5.25
b 27.6
c 34.3
d 5.86

page 125

b i CD and AB
ii DA
iii BC
d 210 J
e 525 J

page 126
273 K

CHAPTER 12

page 132
1 a together
b i 0 V ii 6 V iii 6 V

CHAPTER 13

page 156
a 6400 km
b 3.4×10^8 m
c 2.8×10^{-3} N kg^{-1}
d i field due to Earth
ii Yes

CHAPTER 14

page 160
C

page 163
top
1 V_{escape}(Earth) = 11.2 km s^{-1}; V_{escape}(Moon) = 2.37 km s^{-1}
2 a 11.2 km s^{-1}
b 11.2 km s^{-1}

CHAPTER 15

page 167
D

page 173
2 a 2 V
b 1.2 V
c 1.2 V

page 174
2 a 2.5 MW
b 0.5 MW
c 80%

page 175
b i 20 kW
ii 99.2%

CHAPTER 16

page 183
2 $\frac{E}{B}$

page 187
2 $v = \sqrt{\left(\frac{me^2}{4\pi\varepsilon_0 r}\right)}$

page 191
1 1.53×10^{-9} J; 6.85×10^{-13} J
2 1.8×10^{-30} kg

page 203
1 D
2 C
3 c 8.7×10^{-2} day^{-1}
d 29 days

page 206
e i 6; 12
ii 1.16613×10^{-11} J, or 7.3×10^7 eV
f iii 0.0213%

CHAPTER 17

page 210
2 3.97×10^{-24} kg m s^{-1}
3 1.67×10^{-10} m
4 a 1.65×10^{-10} m
5 $\lambda = \sqrt{\frac{h}{(2Vem)}}$
6 6.7×10^{-34} J s

page 217
1 $\lambda = 2l$; l; $2/3l$
2 $p = \lambda/2l$; λ/l; $3\lambda/2l$; $n\lambda/2l$
3 $\lambda^2/8ml$; $\lambda^2/2ml$; $9\lambda^2/8ml$
4 $n^2\lambda^2/8ml$

CHAPTER 18

page 222
2 coal: 155 years; oil: 41 years; gas: 65 years

page 224
3 2010: 84%; 2030: 86%

page 226
1 a 4.68×10^{-3} kg
b 3.21×10^{-3} kg
c 2.65×10^{-3} kg
5 a 8.1×10^6 kg
b 3.0×10^9 kg
6 a 890 MW
b 16.2 kg per day per household
7 a 5.0×10^6 kg
b 1.8×10^9 kg
9 100 000 tonnes

page 228
c i 3.6×10^6 J
ii 31%
e 1.4×10^{-3} m^3 s^{-1}

page 229
2 55%
3 46%
4 25 million
5 28%

CHAPTER 19

page 232
top
3 i 29 °C
ii 0.002 °C
bottom
1 i 520 nm
ii 820 nm
2 500 nm – visible
4 total energy emitted by the star in a given time
5 288 K; 10.1 µm; IR

page 233
1 3.82×10^{26} W
2 i 9.1×10^{-3} rad
ii 1.4×10^9 m
iii 6.2×10^{18} m^2
3 201×10^{15} W; 390 W m^{-2}

page 234
1 0.61
2 390 Wm^{-2}
3 Power flux = 240 Wm^{-2} at 255 K and difference between the two values is 63%

page 235
2 i 3
ii 9
4 i stretching (mode A)
ii stretching (mode A)

page 236
2 0.074°C decade^{-1}

page 238
b i 278 K
c i 1.51 kWm^{-2}
ii 4.76×10^{10} J
iii 1.07×10^{10} J
d i 92.3°C

page 239
1 D9: =(B9−C9)*$D$4/$B$4; E9: =E8+D9
2 i 291.5°C
ii 9.0 years
4 i 291.5°C ; just above 291.5°C
ii 10 years; 8 years

page 240
a 255 K
c 0.29

CHAPTER 20

page 245
3 a ±33%
b ±170%
7 K^{-1}
8 3.15 mm
9 a 3.9715 km
b 66.7 cm

CHAPTER 21

page 248
1 818 K
2 63%
3 a 273 K; 70%

page 252
d 71.3 MW; this is an underestimate as some energy would be needed to reduce the air pressure above the boiling water

CHAPTER 22

page 255
1 a 92 m
3 25 000 MW
4 42 m s^{-1}
5 28×10^6 kg s^{-1}

page 257
2 v = 42 m s^{-1} at 20 m and v = 1.1 m s^{-1} at 0.5 m
4 313.6 W m^{-2}

CHAPTER 23

page 259

bottom

1 88 952 MW; 49 215 MW; 32 010 MW; 15 688 MW;
5650 MW; 1176 MW

page 260

1 5.84×10^{-25} kg

2 a $1170\ s^{-1}$

b 8.55×10^{-4} s

c $14\ 700\ s^{-1}$

page 261

1 probability = a/A

2 probability = b/A

3 a a/b

4 200

5 0.01

page 266

top

2 Cs−137~ 30 − 40 years; Am−241 ~ 2000−3000 years

right

b 0.82

c 79

CHAPTER 24

page 272

middle

1 10 000, 10 001, 101 101, 111 111, 1100 100, 1111 111

2 30, 42, 56, 55, 85

bottom

a 00010001, 11111111, 10101010, 10010111, 01111001, 11001101, 11011100

page 273

1 a 256

b 65 536

2 1 kilobyte = 1 024 bytes

1 megabyte = 1 048 576 bytes

1 gigabyte = 1 073 741 824 bytes

1 terabyte = 1 099 511 627 776 bytes

page 275

1 a 16

2 65 536

page 276

a analogue

b digital

c digital

d analogue

page 277

c 58 minutes

page 278

2 600 nm

page 279

2 a 1 000 000

b 20 μm

4 a more charge collects on a brightly lit pixel than on a dim one

page 280

1 b 1 μF

2 a 5 nF

b 50 μF

3 160 μC

4 250 V

5 80 μF

page 281

a separation of spot images = 5.4×10^{-5} m; length of one pixel = 2.4×10^{-5} m

b ii 0.31 m V

page 282

a 2.53×10^{-3}

b 19.1 μm

c distance between image of the eyes on the CCD = 1.52×10^{-4} m; image/pixel length = 7.96; eyes are resolved as they are separated by 8 pixels

CHAPTER 27

page 295

middle

1 3.26 ly

page 297

top

1 a 2.4×10^{-7} m

b 4.8×10^{-7} m

c 9.7×10^{-7} m

middle

1 3.9×10^{26} W

2 2.0×10^{-2} W kg^{-1}

page 299

bottom

10 d_B

page 300

1 1.6

2 4450 L_{sun}

page 301

top

2 11 pc

bottom

a −4.5

b 3.6 Mpc

page 302

2 3.9×10^{12} J

3 a 9.7×10^{37}

b 6.5×10^{11} kg

c 4.2×10^{9} kg

page 306

2 73 km s^{-1} Mpc^{-1}

3 1.3×10^{10} years

CHAPTER 29

page 319

5.3×10^{-35} J

page 323

1 7.1×10^{-12} m

2 1.2×10^{-15} m

page 324

right

3 5.9×10^{8} m s^{-1}

page 325

3 a 2.2×10^{-20} kg m s^{-1}

b 1.3×10^{7} m s^{-1}

c 1.5×10^{-13} J

page 326

a 85 times

b 11 MHz

page 334

baryon; baryon; meson; meson; meson

page 335

a ±½

b ±½

c 0 or 1

page 339

a Does conserve lepton number

b Does not conserve lepton number

Index

Index entries are arranged in letter-by-letter sequence. Page numbers in bold refers to tables; page numbers in italics refers to figures.

absolute magnitude, 299–300
absolute scale of temperature, 120
absolute uncertainties, 15–16
absolute zero, 120
absorption spectra, 185–6, 214
  greenhouse gases, 236
abstracts, extended essays, 284
a.c. (alternating current), 172
acceleration, 26, 27
  average, 27
  calculating, 36
  centripetal, 43
  instantaneous, 27
acceleration–time graphs, 28–9
accuracy, 14
  improvement, 61–2
activation products, 266–7
ADCs (analogue-to-digital converters), 274, 311
adiabatic change, 123–4
air resistance, effects on projectiles, 53–4
albedo, 233
alchemy, 181
alpha radiation, 184
  alpha, 184
alternating current (a.c.), 172
AM (amplitude modulation), 313, 314
ammeters, 132
amperes, 127, 131
amplifiers, 308
amplitude, waves, 78–9
amplitude modulation (AM), 313, 314
analogue signals, 271
  coding, 274–5
analogue-to-digital conversion, 273–4
analogue-to-digital converters (ADCs), 274, 311
analysers, 106
Anderson, Carl (1905–91), 317, 322
Andromeda galaxy, 103
angular velocity, 42
  and speed, 42
annihilation, 321–2
anthropogenic effects, 231, 242
anti-neutrinos, 195
antibaryons, 334–5
antihydrogen, 197
antimatter, 196–7, 317
antiquarks, 321–2, 334, 335
antisound, 91
AP6 (Asia-Pacific Partnership on Clean Development and Climate), 246
apparent weightlessness, 164–5
Aristotle (c.384–322 BC), 181
artificial transmutations, 203–4
Asia-Pacific Partnership on Clean Development and Climate (AP6), 246
asteroids, 295
astronomical units (AUs), 233, 295
astrophysics, 294–306
asymptotic freedom, 338
atomic energy, 188, 206
atomic mass, 191–2
atomic model, 182–3
atomic physics, 180–206
atoms, 182
  internal structure, 183
  models, 180
attenuation, optical fibres, 312–13
audio, bit rate, 309
audio frequencies, 313
AUs (astronomical units), 233, 295
automated measurements, 138
average acceleration, 27
average global temperature, 234, 237
average speed, 27, 86
average values, 26
average velocity, 26
Avogadro constant, 115

background count, 198
background radiation, 198–201
bandwidth, 309
  radio communication, 314, 315
baryon numbers, 334–5
  conservation of, 335–6
baryons, 334–5, 337
batteries, 134–6
Bayer array filters, 282
Becker, Herbert, 185
Becquerel, Henri (1852–1908), 184
becquerels, 198
best-fit lines, 19
beta radiation, 184
Big Bang theory, 304–5, 306, 317, 328, 341–2
big crunch, 305
binary code, 272–3, 274
  and quantum levels, 274–5
binary digits, 272
binary stars, 103, 297–8
  eclipsing, 298
  spectroscopic, 298
  visual, 297–8
binding energy, 191
  nucleons, 204–5
biofuels, 253
biological effects, of radiation, 198–201
bit depth, charge-coupled devices, 282
bit rate, 309
bits, 309
black bodies, 209, 297
black body radiation, 208–9, 232
block diagrams, communication systems, 308
blood-flow measurements, 100–1
Bohr, Niels (1885–1962), 185, 194, 216
Bohr model, 185–8, 206
  problems, 187–8
  successes, 185–7
boiling, 114–15
boiling water reactors (BWRs), 263
bosons
  gauge, 319
  Higgs, 340
  W, 320
Bothe, Walther (1891–1957), 185
boundary conditions
  standing waves, 95–6
  wave functions, 217
Boyle's law, 120
breaking radiation, 328–9
Bremsstrahlung (breaking radiation), 328–9
Brewster's angle, 107
Brewster's law, 107
brightness, stars, 296
Brundtland Commission, 218
bubble chambers, 198, 330–1, 340
bubble of ignorance, 320
Bunsen, Robert (1811–99), 7
Bunsen flame, temperature estimation, 113
BWRs (boiling water reactors), 263
bytes, 272

calculus, 26
  and simple harmonic motion, 64
cancer
  and power lines, 177–8
  and radiation, 199, 200–1
candles, standard, 301
capacitors, 173, 279
carbon capture and storage, 251–2
carbon dating, 190
carbon dioxide, 234–5
  absorption spectra, 236
  atmospheric, 242, 253
  per capita emissions, 246
  vibrational modes, 235
Carbon Dioxide Information Analysis Centre, 242–3
carbon stores, 221
Carnot cycle, 125
Carnot engine, 125
carrier waves, 313
CAS (creativity, action, service), 229
cause, and correlation, 178–9
CCDs *see* charge-coupled devices (CCDs)
CDs *see* compact discs (CDs)
CE (conclusion and evaluation), 17, 23–4
centre of mass (CM), 25, 327
centripetal acceleration, 43
centripetal forces, 43, 260
Cepheid variables, 298, 299, 301
CERN *see* European Centre for Nuclear Research (CERN)
Chadwick, James (1891–1974), 185
Chandrasekhar limit, 303
channels, communication, 307
chaos
  and climate change, 244–5
  theory, 242
charge, 127
  definition, 130–1
  moving, 144
  nature of, 129–30
  negative, 127, 129–30
  neutral, 127, 129–30
  positive, 127, 129–30
  two-point, 154
charge-coupled devices (CCDs), 270, 278–82
  bit depth, 282
  capacitance, 279–80
  colours, 282
  image digitization, 282
  magnification, 281–2
  principles, 278–9
  quantum efficiency, 280–1
  resolution, 281
charged current reactions, 340
charged particles, trajectories, 325–6
Charles's law, 120
Chernobyl (Ukraine), nuclear accident, 203–4, 266
chi-square test, 13
China
  electrification, 228–9
  hydroelectricity, 254–5
CHP (combined heat and power), 251
circuits
  parallel, 129
  RF, 314–15
  series, 128
  short, 141
  tuned, 314
  see also electrical circuits
circular motion, 41–3
  and simple harmonic motion, 65–8
circularly polarized, 105
citations, extended essays, 285
cladding, optical fibres, 311
classical theory, vs. quantum theory, 207–9
climate change, 241–7
  balanced views, 243
  and chaos, 244–5
  effects, 244–5
  fossil records, 243
  future trends, 252
  global reduction efforts, 246–7
  long-term trends, 242–3
  predictions, 244
  reduction technologies, 248–52
  see also global warming
climate data, interpretation, 241–2
clocks, 311
closed universes, 305
cloud chambers, 317
CM (centre of mass), 25, 327
CMB (cosmic microwave background), 304, 342
CMOS (complementary metal oxide semiconductor) sensors, 278
coal, 224, 225
  efficiency improvements, 249
coherent light sources, 92
colliding-beam accelerators, 328–9
collisions, 48
  elastic, 48

explosive, 48
inelastic, 48
protons, 328
colours, and charge-coupled devices, 282
combined heat and power (CHP), 251
comets, 295
command terms, 289–90
communication channels, 307
communication systems
block diagrams, 308
components, 311
design criteria, 307
communications, 307–16
see also radio communication
compact discs (CDs), 270, 276–7
data reading, 277
structure, 276–7
see also digital versatile discs (DVDs)
complementary metal oxide semiconductor (CMOS) sensors, 278
completely inelastic collisions, 48
components, resolving, 34
compressions, 35, 82
Compton, Arthur (1892–1962), 338
concluding interview, extended essays, 285
conclusion and evaluation (CE), 17, 23–4
conduction, 112
conductors, 130
consequences, 84
conservation of energy, 44, 220–1
law of, 47
conservation laws, 318
using, 337
conservative fields, 158
constant angular speed, 65
constant ratio property, 201–2
constant-acceleration equations, 27–8
constellations, 294
containment buildings, 263
continuous electrical signals, 271
continuous spectra, 214
control rods, 260, 263
control variables, 18
convection, 112
coolants, 263
corpuscles, 207
correlation, and cause, 178–9
cosine rule, 33
cosmic microwave background (CMB), 304, 342
cosmic rays, 338
Cosmic Variance (website), 8
cosmology, 303–5
coulombs, 127, 129, 131
Coulomb's law, 150, 184, 186
CP violation, 342
creativity, action, service (CAS), 229
critical damping, 76
critical density, 305
cross product, 34
cumulative production, 222
Curie, Marie (1867–1934), 184, 185
Curie, Pierre (1859–1906), 184, 185
current, 127, 144
alternating, 172
definition, 130–1
direct, 172
eddy, 176
induced, 167
neutral, 340
phase, 173
virtual, 166–7
current density, 177
current-length, 145
cyclotrons, 324–6

DACs (digital-to-analogue converters), 277, 311
Dalton, John (1766–1844), 182
damping, 73–6
critical, 76
dark matter, 305
data
climate, 241–2
exponential, 201–3
processed, 23
see also raw data
data capture, and charge-coupled devices, 278–82
data collection and processing (DCP), 17, 19–23, 136–7
daughter nuclides, 189
Davisson, Clinton (1881–1958), 209
Davisson–Germer experiment, 209–10
d.c. (direct current), 172
DCP (data collection and processing), 17, 19–23, 136–7
de Broglie, Louis (1892–1987), 333
de Broglie hypothesis, 211
decay, rate of, 202–3
decoders, 308
deep inelastic scattering, 340–1
experiments, 341
degraded energy, 48, 111
degrees, 41
delocalization, 130
Democritus (c.460–370 BC), 181, 182
demodulators, 316
demultiplexers, 310
deoxyribonucleic acid (DNA), damage, 199
dependent variables, 18
depleted uranium, 260
derived units, 12–13
desalination plants, 251, 252
design, 17–19
detailed measurements, 138
detectors, particle, 329–33
deterministic laws, 211
diagrams
block, 308
Feynman, 320–2, 340
free-body, 37–8
Hertzsprung–Russell, 298–9
see also graphs
dielectrics, 279
diffraction
electron, 209, 323
Fraunhofer, 102
Fresnel, 102
waves, 78, 90, 101–3
diffraction gratings, 92, 213–14
digital communication systems, components, 311
digital signals, 271, 307–8
advantages, 275–6
digital technology, 270–82
digital versatile discs (DVDs), 270
advantages, 278
future trends, 278
see also compact discs (CDs)
digital-to-analogue converters (DACs), 277, 311
digitization, images, 282
Dirac's theory, 322
direct current (d.c.), 172
direction, 25
dirty bombs, 267
discovered reserves, 222
discrete values, 271
dispersion, 80, 312
displacement, 26, 63
particles, 86
sinusoidal, 55
displacement–time graphs, 28
dissipative energy change, 124
distance, 26
distribution, electricity, 175–6
DNA, damage, 199
doodles, 321
Doppler effect, 98–9
examples, 100–1
and spectral lines, 101
dose equivalent, 199–200
dot product, 34
drawings, in examinations, 292
driving forces, 76
DVDs *see* digital versatile discs (DVDs)
dwarf stars, 298, 302–3

earthquakes
epicentre, 82
explosions and, 83
focus, 82
hypocentre, 82
waves, 81–3, 88–90
ECF (error carried forward), 289
eclipsing binaries, 298
eddy currents, 176
edge effect, 155
efficiency, 49–51, 220–1
engines, 125
generators, 248–9
improvements, 248–9
quantum, 280–1
EIA (Energy Infomation Authority) (US), 227
Einstein, Albert (1879–1955), 208
Einstein's equation, 157
elastic collisions, 48
electric field lines, 148
electric field strength, 144
calculating, 150
electric fields
force direction, 146
patterns, 156
electric vehicles, 250
plug-in hybrid, 250, 251
electrical approximations, 130–1
electrical circuits, 127
parallel, 129
power dissipation, 133–4
series, 128
electrical definitions, 130–1
electrical meters
perfect, 132
real, 137, 140
electrical plugs, 141
electrical risks, 153–4
electrical sockets, 141
electrical splitters, 141
electricity, 127–41
distribution, 175–6
storage, 256
transmission, 175–6
electricity generation, 166
thermal, 224–6
electrification, China, 228–9
electromagnetic (EM) waves, 81, 207
polarization, 105
and risk assessment, 84–5, 177
and sound waves compared, 100
electromagnetic force, 142
electromagnetic induction, 166–79
electromagnetic spectrum, 84–5, 209
electromagnets, strength of, 156
electromotive force (e.m.f.), 135–6
generation, 166–9
transformer-induced, 171
electron degeneracy pressure, 302
electron diffraction, 209, 323
electron diffraction tubes, 213
electron orbits, modelling, 216–17
electron rest mass, 191
electrons, 130, 180, 317
early studies, 183
ground states, 215
kinetic energy, 217
orbitals, 212
potential energy, 217
electronvolts, 162, 188, 192
electrostatic fields, patterns, 153–5
electrostatic force, 142
electrostatic potential, 160
electroweak forces, 340
elementary particles, 317
elements, 181
elliptically polarized, 105
EM waves *see* electromagnetic (EM) waves
e.m.f. *see* electromotive force (e.m.f.)
emission spectra, 185–6, 213–14
emissions, simulated, 190
emissivity, 234
Empedocles (c.490–430 BC), 181
empirical falsifiability, 62–3
encoders, 308
endoscopes, 87–8
energy, 44–51
applications, 227
atomic scale, 188, 206
concept of, 192–3
conservation of, 220–1
degraded, 48, 111
dissipated, 48
Einstein's equation of, 157
internal, 109
lost, 111
in orbits, 164
packets, 186
and potential, 162–3
and poverty, 227–9
primary, 218, 221
renewable, 253–7, 268
secondary, 218
in simple harmonic motion, 70–2
thermal, 48, 109
tidal, 253, 255–6
transformations, 47–9
types of, 109
useage improvements, 248, 249–52
waves, 78–9
see also binding energy; conservation of energy; kinetic energy (KE); potential energy (PE)
energy absorption, modelling, 234–5
energy balance model, 233–4
energy change

dissipative, 124
ideal gases, 122–4
reversible, 123–4
energy conservation, in radioactive decay, 190–2
energy consumption, global, 218–19
energy crisis, 221
energy degradation, 220
energy density, fuels, 225–6
Energy Infomation Authority (EIA) (US), 227
energy sources, global, 218–30
energy-efficient vehicles, 250–1
engines
efficiency, 125
gas turbine, 248–9
enhanced greenhouse effect, 231, 236
modelling, 236–40
entropy, 125–6
epicentre, 82
equivalence, principle of, 165
error bars, 16
error carried forward (ECF), 289
errors
parallax, 14
random, 14
systematic, 14
zero, 14
escape speed, 162–3
essays, extended, 283–7
ethanol, 253
ethics, 219–20
European Centre for Nuclear Research (CERN), 197, 271
bubble chamber, 340
establishment, 333
Large Electron Positron collider, 333
Large Hadron Collider, 328, 333, 340
Low-Energy Antiproton Ring, 317
evaporation, 114–15
examinations
command terms, 289–90
drawings in, 292
marks available, 292
objective levels, 289–90
structure, 290–1
technical language, 292
techniques, 288–93
expanding universe, 306
experimental measurements, 10–27
experiments, thermal, 113
explosions, 48
and earthquakes, 83
explosive collisions, 48
exponential data, 201–3
exponential function, 256
extended essays, 283–7
abstracts, 284
appendices, 286
bibliography, 286
body, 285–6
citations, 285
concluding interview, 285
conclusions, 286
contents, 285
formal presentation, 284–5
introduction, 285
length, 286
process evaluation, 286
references, 286
timescale, 286–7
title page, 284
topics, 287
types of, 283–4
extension, 50
extrapolation, 120

fair tests, 18
falsification, principle of, 63
Faraday's law, 170
farads, 280
fast breeder reactors (FBRs), 263
fast neutrons, 260
FBRs (fast breeder reactors), 263
Feynman, Richard (1918–88), 7, 193, 196, 319, 320–1
Feynman diagrams, 320–2, 340
fibres
monomode, 312
multimode, 311–12
step indexed, 311
see also optical fibres
field lines, 147
field strength
definition, 143
electric, 144
gravitational, 143
magnetic, 144–5
fields, 158–65
conservative, 158
electrostatic, 153–5
radial, 148
see also electric fields; force fields; gravitational fields; magnetic fields
first harmonic, 96
first law of thermodynamics, 122
fission, mechanisms, 204
fission fragments, 266
flat circular coils, magnetic field patterns, 155
flat universes, 305
Fleming's left-hand rule, 146
flux, concept of, 169–70
flux lines, 147
flux linkage, 169–70
FM (frequency modulation), 314
focus, earthquakes, 82
force fields, 142–50
adding, 151
force direction, 145–7
representation, 147–9
forces
centripetal, 43, 260
direction of, 145–6
driving, 76
electromagnetic, 142
electrostatic, 142
electroweak, 340
fundamental, 318
magnetic, 142, 167
and momentum, 39–41
and motion, 32–43
nature of, 188
non-contact, 142–57
restoring, 55, 63
resultant, 63
strong nuclear, 188–9
types of, 34–5
see also electromotive force (e.m.f.)
force–extension graphs, 51
fossil fuels, 221–2
combustion, 231
dependence, 223–4
historical use of, 224
improved useage efficiencies, 248
reduced dependence, 248
see also coal
Fourier analysis, 72–3
Fourier series, 73
fractional uncertainties, 15–16
Fraunhofer diffraction, 102
free fall, 164
frequencies
audio, 313
classification, 314
natural, 76–7
radio, 313
resonant, 314
frequency modulation (FM), 314
Fresnel diffraction, 102
friction, 32, 35
fuel crisis, 221–2
fuel rods, 263
fuels
biofuels, 253
energy density, 225–6
see also fossil fuels
fundamental, 96
fundamental forces, 318
fusion, 114

Gaia theory, 268
galaxies, 103, 295
and expanding universe, 306
gamma radiation, 184, 185, 203, 322–3
decay, 189
gas turbines, 248–9
gas-cooled reactors (GCRs), 263
gas-fuelled generators, 248–9
gases, 113
analysis, 118–26
pressure, 116–17, 118–27
temperature, 118–27
volume, 118–27
see also greenhouse gases; ideal gases
gauge bosons, 319
GCRs (gas-cooled reactors), 263
GDCs (graphic display calculators), simple harmonic motion, 67–72
Geiger, Hans (1882–1945), 184
Geiger counters, 198
Gell-Mann, Murray (1929–), 341
generation
electromotive force, 166–9
thermal, 224–6
generators, 170
efficiency, 248–9
gas-fuelled, 248–9
Germer, Lester (1896–1971), 209
giant stars, 298, 302
Gigerenzer, Gerd (1947–), 201
Glashow, Sheldon (1932–), 319, 340
global energy sources, 218–30
global warming, 231, 236
causes, 241
and nuclear power, 268–9
see also greenhouse effect
gluons, 321–2, 337–8
GM tubes, 198
gPE (gravitational potential energy), 45, 46
gradients, 16
graphic display calculators (GDCs), simple harmonic motion, 67–72
graphs, 28–9
acceleration–time, 28–9
analysing, 40
displacement–time, 28
force–extension, 51
log-linear, 202
shapes, 20
velocity–time, 28
wave motion, 86–7
see also diagrams; straight-line graphs
gravitational field lines, 147–8
gravitational field strength, 143
calculating, 149–50
gravitational fields
combined, 152–3
force direction, 145
patterns, 151–3
planets, 151–2
gravitational potential, 158–9, 162–3
gravitational potential energy (gPE), 45, 46
gravitational pressure, 296
gravity, 32
falling under, 29–31
force of, 142, 157, 164
measurement, 58–62
universal, 89
greenhouse effect
mechanisms, 231
and solar radiation, 231–40
see also enhanced greenhouse effect; global warming
greenhouse gases, 231
absorption spectra, 236
greenhouses, 231–2
Greenpeace, 268
ground states, 215
group 4 project, 24, 263
growth, unsustainable, 218

hadrons, 334
harmonics, 73, 96
see also simple harmonic motion (SHM)
heat, 109
latent, 114
use of term, 109
see also specific heat capacity
heat engines, 124–5
Heisenberg, Werner Karl (1901–76), 212
Heisenberg's uncertainty principle, 212–13, 319, 320
Hertz, Heinrich Rudolf (1857–1894), 208
Hertzsprung–Russell diagram, 298–9
Hess, Victor (1883–1964), 338
hexadecimal system, 272–3
Higgs, Peter (1929), 340
Higgs bosons, 340
high energy particles, 322–3
homogeneous, 304
Hooke's law, 50, 57
Hubble's law, 306
Huygens principle, 102
hybrid vehicles, 250, 251–2
hydroelectricity, 253, 254–5
hydrogen, as fuel, 251
hydrogen burning, 302
hydrostatic equilibrium, stars, 296
hypocentre, 82
hysteresis, 176

ideal gases, 116
behaviour, 115–16
energy change, 122–4
microscopic explanation, 120–1

IEA (International Energy Agency), 227
ILC (International Linear Collider), 334
images, digitization, 282
imaging, true colour, 282
impedance, 173
impulse, 40
independent variables, 18
induced current, 167
inductors, 173
inelastic collisions, 48
inner product, 34
input transducers, 308
instantaneous acceleration, 27
instantaneous speed, 27
instantaneous values, 26
instantaneous velocity, 26
insulators, 130
intensity, waves, 78–9
Inter-Governmental Panel on Climate Change (IPCC), 236, 243, 246
interactions
  nature of, 188
  and particles, 317
  strong, 188–9
  weak, 189
intercept, 20
interference, waves, 78, 91
internal energy, 109
internal resistance, 135–6
International Energy Agency (IEA), 227
International Linear Collider (ILC), 334
Internet, 270–1
  ethics, 270–1
  in society, 271
inverse square relationships, 79, 148, 149, 151
inversely proportional, 21
inversely related, 21
IPCC (Inter-Governmental Panel on Climate Change), 236, 243, 246
isobaric change, 123
isochoric change, 123
isothermal change, 123
isotopes, 183
isotropic, 304
isovolumetric change, 123

Joliot-Curie, Frédéric (1900–1958), 185
Joliot-Curie, Irène (1897–1956), 185
Jonson, Ben (1572–1637), *The Alchemist* (1610), 181 2

KE *see* kinetic energy (KE)
kelvin scale, 120
Kepler, Johannes (1571–1630), *De Harmonices Mundi* (1619), 163
Kepler's law, 163–4
Keynes, John Maynard (1883–1946), 182
kinematics, 25–31
  see also motion
kinetic energy (KE), 45–6, 164
  electrons, 217
  organized, 111
  and particle speed, 324–9
  random, 111
  in simple harmonic motion, 70
kinetic theory, 116
Kuhn, Thomas (1922–96), *The Structure of Scientific Revolutions* (1962), 196
Kyoto Protocol, 246, 269

lamina, transformers, 176
Large Electron Positron (LEP) collider, 333
Large Hadron Collider (LHC), 328, 333, 340
Large Magellanic Cloud, 195–6
lasers, 103
latent heat, 114
LCDs (liquid crystal displays), 108
LDRs (light-dependent resistors), 138–40
LEAR (Low-Energy Antiproton Ring), 317
least significant bit, 272
Lenz's law, 168
LEP (Large Electron Positron) collider, 333
lepton generations, 339
lepton numbers, 339
  conservation, 339
leptons, 317, 338–9
  flavours, 317
lesions, 199
leukaemia clusters, 177–8
LHC (Large Hadron Collider), 328, 333, 340
lift, 35
light
  nature of, 207
  particle properties, 207
  as particles, 207
  redshifted, 101
  waves, 80–1
  as waves, 207
light water cooled, graphite moderated reactors (LWGMs), 263, 266
light-dependent resistors (LDRs), 138–40
light-years, 295
linacs (linear accelerators), 323–4
Lindzen, Richard (1940–), 243
line spectra, 214
  formation, 215–16
linear accelerators (linacs), 323–4
linear magnification, 281–2
lines, best-fit, 19
lines (power) see power lines
liquid crystal displays (LCDs), 108
liquids, 113
log-linear graphs, 202
long players (LPs), 275–6
longitudinal waves, 82
losses
  power lines, 175
  transformers, 176
lost energy, 111
loudness curves, 80
Love waves, 83
Lovelock, James (1919–), 268–9
Low-Energy Antiproton Ring (LEAR), 317
LPs (long players), 275–6
luminosity, 233
  stars, 296
LWGMs (light water cooled, graphite moderated reactors), 263, 266

macroscopic perspectives, 109
magnetic field lines, 148–9
magnetic field strength, 144–5
magnetic fields
  force direction, 146–7
  patterns, 155–6
  representation, 147
magnetic flux density, 169
magnetic force, 142, 167
magnetic leakage, 176
magnets, 144–5
magnification
  charge-coupled devices, 281–2
  linear, 281–2
magnitude, 25
  absolute, 299–300
  stars, 296
magnitude scale, 299–300
main sequence stars, 298
Malus's law, 107
manipulative skills (MS), 24
  summative, 17
Marsden, Ernest (1889–1970), 184
mass, 34
  atomic, 191–2
  attraction, 156–7
  centre of, 25, 327
  electron rest, 191
  molar, 115
  neutron rest, 191
  proton rest, 191
  relative molecular, 115
  units, 191–2
mass defect, 191, 194
mass–energy conservation, 191
matter
  amount of, 157
  dark, 305
  historical models of, 181–8
  wave properties, 207
matter waves, 211
matter–antimatter asymmetry, 342
Maxwell, James Clerk (1831–79), 207
Maxwell–Boltzmann distribution, 121
mean, 14
measurements, uncertainties in, 59
Mendeleyev, Dmitri Ivanovich (1834–1907), 182
mesons, 334–5, 337
microscopic perspectives, 109
Millikan, Robert Andrews (1868–1953), 183, 338
moderators, 260, 263
modulation
  amplitude, 313, 314
  frequency, 314
modulators, 313
molar heat capacity, 111
molar mass, 115
moles, 115
momentum
  conservation of, 41
  and force, 39–41
morality, 219–20
most significant bit, 272
motion
  circular, 41–3, 65–8
  electromotive force generation, 166–9
  and forces, 32–43
  kinematics, 25–31
  relative, 36
  wave, 86–7
  see also Newton's laws of motion; simple harmonic motion (SHM)
moving charge, 144
MS *see* manipulative skills (MS)
multimode fibres, 311–12
multiplexers, 310
multiplexing, 310
muons, 338
musical instruments, 96–7
  Fourier analysis, 72–3

natural frequency, 76–7, 235
natural oscillation, 76
nebulae, planetary, 302
negative charge, 127, 129–30
negative temperature coefficient (NTC) thermistors, 138
Neumann's equation, 170
neutral charge, 127, 129–30
neutral current, 340
neutrinos, 193–6
neutron degeneracy pressure, 303
neutron rest mass, 191
neutron stars, 303
neutrons, 130, 180
  discovery, 204
  early studies, 185
  fast, 260
  thermal, 260
Newton, Isaac (1643–1727), 32, 35, 89, 157, 182
  Opticks (1704), 207
  Philosophiae Naturalis Principia Mathematica (1687), 163
Newton's law of universal gravitation, 149, 151–2, 163–4
Newton's laws of motion, 35–8, 56–7, 116, 163–4
  first law, 35–6
  second law, 36, 38, 39, 63
  third law, 36–7
noise, 80, 275, 308
non-contact forces, 142–57
non-renewable resources, 221–2
normal distribution, 13–14
normal reaction, 34–5
NTC (negative temperature coefficient) thermistors, 138
nuclear accidents, 203–4, 263, 266
nuclear debate, 267–9
nuclear fusion, stars, 302
nuclear hazards, 264
nuclear physics, 180–206
nuclear power, 258–69
  development, 258
  and global warming, 268–9
  issues, 258–9, 262, 264, 267–9
  risks, 264–9
  safety issues, 268
  useage, 258
nuclear reactions, 189
nuclear reactors, 260–3
  accidents, 203–4, 263, 266
  control, 259, 263
  enrichment process, 259
  moderation, 259, 263
  scattering cross-sectional area, 260–1
  thermal, 262–3
  types of, 262–3
nuclear waste, 265–6, 268
  constituents, 266
nuclear weapons
  development, 266–7
  proliferation, 268
  and terrorism, 267
nucleons, 189
  binding energy, 204–5
nucleus, early studies, 184
nuclides
  daughter, 189

stability, 205
stable, 190
null points, 153

objective levels, 289–90
observationalist–inductivist perspective, 63
OECD (Organization for Economic Cooperation and Development), 218, 246
ohms, 131
Ohm's law, 143
oil, and fuel crisis, 221–2
Olbers, Heinrich (1758–1840), 303, 304
open universes, 305
oppositely charged parallel plates, 155
optical fibres, 311–13
attenuation, 312–13
cladding, 311
optically active solutions, 106, 108
orbitals, 212
orbits, 163–4
total energy in, 164
see also electron orbits
Organization for Economic Cooperation and Development (OECD), 218, 246
origin, 26
oscillations, 55
natural, 76
output transducers, 308

P waves, 82–3, 89–90
packets, energy, 186
pair production, 321–2
parabolas, 52
parallax
spectroscopic, 299, 301
stellar, 299, 300–1
parallax errors, 14
parallel circuits, 129
parallel plates, oppositely charged, 155
parallel resistance, 129
parallelogram rule, 33
parsecs, 295
particle detectors, 329–33
particle physics, 317–42
future research, 342
international community, 333–4
research cost issues, 329
research laboratories, 333–4
Standard Model, 341
particle speed, and kinetic energy, 324–9
particle tracks, 196, 197–8
particles
charged, 325–6
displacement, 86
elementary, 317
exchange, 319
high energy, 322–3
and interactions, 317
light as, 207
tau, 338
virtual, 319–20
Pauli, Wolfgang (1900–58), 194–5
Pauli exclusion principle, 318
p.d. *see* potential difference (p.d.)
PE *see* potential energy (PE)
pendulums
gravity measurement, 58–62
motion, 56–7, 58
Penzias, Arno (1933–), 342
percentage uncertainties, 15–16
period, 42
Periodic Table, 182
personal skills (PS), 17, 24
phase
current, 173
voltage, 173
phase angles, 65
phase change, 113–15
PHEVs (plug-in hybrid electric vehicles), 250, 251
photoelectric effect, 207–8, 278
photoelectrons, 208
photomultipliers, 332–3
photons, 85, 186, 208, 319, 321
photosynthesis, 221, 253
PHWRs (pressurized heavy water reactors), 263
physical laws, universal, 304
physics
approaches, 8–9
atomic, 180–206
learning in, 7–9
nature of, 7–8
nuclear, 180–206
quantum, 207–17
thermal, 109–17
see also particle physics
physics beyond the Standard Model, 342
pictures, bit rate, 309
pixels, 278
plagiarism, 285
Planck's constant, 85, 186, 208, 213
planetary nebulae, 302
planetoids, 295
planets, 295
gravitational fields, 151–2
orbits, 163–4
plasma, 113–14
plug-in hybrid electric vehicles (PHEVs), 250, 251
plutonium-239, 266–7
polarimeters, 106
polarization, 104–8
calculations, 107
examples, 107–8
polarizers, 106
Polaroid, 105–6
Polaroid dark glasses, 107–8
Popper, Karl (1902–94), 62–3
positive charge, 127, 129–30
positrons, 196, 317, 322
potential, 158–65
calculation of, 161–2
concept of, 158
electrostatic, 160
and energy, 162–3
gravitational, 158–9, 162–3
representation of, 160–1
potential difference (p.d.), 127, 131, 162, 164, 188
terminal, 135
see also voltage
potential dividers, 137–40
potential energy (PE), 158
electrons, 217
gravitational, 45, 46
in simple harmonic motion, 71–2
potential gradient, 161
potentiometers, 139–40
poverty, and energy, 227–9
power, 44, 49–51, 233
electrical, 133–4
wind, 253, 256–7
see also nuclear power
power cables, inappropriate use of, 141
power dissipation, electrical circuits, 133–4
power lines
and cancer, 177–8
dangers of, 177–8
losses, 175
power spectra, radio communication, 315–16
power supplies, 134–6
practical skills, 17
precision, 14
pressure, 116–17
electron degeneracy, 302
gases, 116–17, 118–27
gravitational, 296
radiation, 296
pressure law, 120
pressurized heavy water reactors (PHWRs), 263
pressurized water reactors (PWRs), 263, 266
primary coils, 173–4
primary energy, 218, 221
primary waves, 83
principle of superposition, 91
probabilistic laws, 211
probabilities, 84
and quantum states, 211–12
and scattering cross-sections, 260–1
and wave–particle duality, 211
probability waves, 211
procedures, evaluation, 23
processed data, presenting, 23
production, cumulative, 222
project planning cycle, 229–30
projectiles, 52–4
air resistance effects, 53–4
proportional, 20
proton rest mass, 191
proton synchrotrons, 328
protons, 130, 180
collisions, 328
early studies, 184–5
Proust, Joseph (1754–1826), 182
PS (personal skills), 17, 24
pulsars, 303
pumped storage, 256
PWRs (pressurized water reactors), 263, 266
Pythagoras's theorem, 33, 67

quanta, 85, 207, 208
quantum efficiency, charge-coupled devices, 280–1
quantum levels, and binary code, 274–5
quantum mechanics, 207
quantum numbers, 318, 335
quantum physics, 207–17
quantum states, and probabilities, 211–12
quantum theory, 186, 187, 207
vs. classical theory, 207–9
quantum tunnelling, 213
quarks, 317, 321–2, 334–8
colour, 337
flavours, 317
and standard models, 340–2
strange, 336

radial distance, 209
radial fields, 148
radians, 41
radiation, 112, 221
background, 198–201
beta, 184
biological effects, 198–201
black body, 208–9, 232
breaking, 328–9
and cancer, 199, 200–1
nonionizing, 185
stellar, 296–7
see also gamma radiation
radiation pressure, 296
radio communication, 313–16
bandwidth, 314, 315
modulation, 313–14
power spectra, 315–16
tuning, 314–15
radio frequency (RF), 313–14
radioactive decay, 189–91
energy conservation in, 190–2
rate of, 202–3
radioactivity, discovery, 184
radioisotopes, 201
random errors, 14
random processes, 203
rarefactions, 82
rate of change, 202
raw data, 18
processing, 22–3
recording, 22
Rayleigh, John Strutt (1842–1919), 322
Rayleigh criterion, 103–4
Rayleigh waves, 83
RC circuits, 314–15
reactions
normal, 34–5
self-sustaining, 260
receivers, 308
red giant stars, 298, 302
red supergiant stars, 303
redshift, 101, 306
Rees, Martin (1942–), 7–8
reflection
total internal, 87, 311
waves, 78, 87–8
refraction, waves, 78, 88–90
refractive index, 88
regenerators, 313
relative molecular mass, 115
relative motion, 36
relative velocity, 28
relativity, 187
remote measurements, 138
renewable energy, 253–7, 268
renewable resources, 221–2
research, and extended essays, 283
reserves, discovered, 222
reshapers, 313
resistance, 127–8
definition, 131
factors affecting, 132–3
internal, 135–6
parallel, 129
series, 128
resistivity, definition, 131
resolution
charge-coupled devices, 281
small targets, 322–3
resonance, 76–7, 235, 261
examples, 77
resonant frequencies, 314
resources
non-renewable, 221–2
renewable, 221–2
ultimately recoverable, 222
undiscovered, 222
restoring forces, 55, 63
resultant forces, 63

reversible energy change, 123–4
RF (radio frequency), 313–14
right-hand palm rule, 146
risk analysis, 201
risks
calculating, 84
electrical, 141
nuclear power, 264–9
radiation, 198–201
root mean squared (r.m.s.), 121
values, 172–3
rotating objects, 101
Royds, Thomas, 184
Rutherford, Ernest (1871–1937), 7, 184–5, 204
Rutherford scattering, 341

S waves, 82–3, 89–90
Salam, Abdus (1926–96), 340
sample-and-hold devices, 311
sampling, 274
sampling frequency, 274
satellites, orbits, 163–4
scalar addition, 33
scalar product, 34
scalars, 25, 32–4
scanning tunelling microscopy, 213
scattering, 235
Rutherford, 341
see also deep inelastic scattering
scattering cross-sectional area, 260–1
Schrödinger model, 206, 217
scientific method, 62–3
scientific notation, 10
SD (standard deviation), 14, 212
sea levels, rising, 244–5
seashore waves, 92–3
second law of thermodynamics, 125–6
secondary coils, 173–4
secondary energy, 218
secondary waves, 83
self-sustaining reactions, 260
sensors, 137–40
series circuits, 128
series resistance, 128
shift registers, 311
SHM *see* simple harmonic motion (SHM)
short circuits, 141
SI system (*Système international d'unités*), 11–13
sidebands, 316
sieverts, 199–200
signal strength, attenuation, 312–13
signals
continuous electrical, 271
see also analogue signals; digital signals
significant digits, 60
silence, 91
simple harmonic motion (SHM), 55–77
and circular motion, 65–7
damping, 73–6
driving equation for, 63–5
energy in, 70–2
everyday examples, 55–6
formulae, 56–7
on graphic display calculators, 67–72
heavy damping, 75
lightly damped, 74–5
restoring forces, 55
simulated emissions, 190
sine rule, 33
sinusoidal displacement, 55
sinusoidal sound waves, 91
size estimation, 10
skills
personal, 17, 24
practical, 17
see also manipulative skills (MS)
SLAC (Stanford Linear Accelerator), 338, 341
small targets, resolution, 322–3
solar cells, 278
solar constant, 233
solar radiation
definition, 231
and greenhouse effect, 231–40
mechanisms, 233
solar system, 295–6
temperatures, 234
solenoids, magnetic field patterns, 156
solids, 113
sound, speed of, 97
sound waves, 79–80
diffraction, 90
and electromagnetic waves compared, 100
Fourier analysis, 72–3
sinusoidal, 91
space–time continuum, 304
spark chambers, 331–2
SPEAR (Stanford Positron-Electron Asymmetric Ring), 338
specific heat capacity, 110
measurement, 113
specific latent heat, 114
spectra
continuous, 214
electromagnetic, 84–5, 209
emission, 185–6, 213–14
see also absorption spectra; line spectra
spectral lines, and Doppler effect, 101
spectral spreading, 313
spectrometers, 213–14
spectroscopic binaries, 298
spectroscopic parallax, 299, 301
speed, 26, 27
and angular velocity, 42
average, 27, 86
constant angular, 65
escape, 162–3
instantaneous, 27
measurement, 100
particle, 324–9
of sound, 97
see also velocity
spheres, charged, 153–4
spin, 335
spring constant, 50
springs, masses on, 56–7
stable nuclides, 190
standard candles, 301
standard deviation (SD), 14, 212
Standard Model, 338–9
physics beyond the, 342
standing waves, 94–8
boundary conditions, 95–6
and travelling waves compared, 95
Stanford Linear Accelerator (SLAC), 338, 341
Stanford Positron-Electron Asymmetric Ring (SPEAR), 338
stars
brightness, 296
constellations, 294
distribution, 295
dwarf, 298, 302–3
evolution, 302–3
giant, 298, 302
hydrostatic equilibrium, 296
luminosity, 296
magnitude, 296
main sequence, 298
neutron, 303
nuclear fusion, 302
red giant, 298, 302
red supergiant, 303
redshift, 101
thermal equilibrium, 296
types of, 297–9
white dwarf, 298, 302–3
see also binary stars; variable stars
stationary targets, 327
stationary waves *see* standing waves
steam engines, 125
Stefan–Boltzmann constant, 238, 297
Stefan–Boltzmann law, 233, 297
stellar clusters, 295
stellar distances, 299
stellar evolution, 302–3
stellar parallax, 299, 300–1
stellar processes, 302–3
stellar radiation, 296–7
step indexed fibres, 311
step-down transformers, 174
step-up transformers, 174
stochastic effects, 198–9
storage devices, 308
straight wires, magnetic field patterns, 155
straight-line graphs, 16, 20
log-linear plots, 202
strain gauges, 140
strange quarks, 336
strangeness, 336
stress analysis, 108
strings (binary), 272
strong interaction, 188–9
strong nuclear force, 188–9
substances, phase change, 113–15
Sun, luminosity, 233
superclusters, 295
supernovae, 195–6, 303
superposition, principle of, 91
sustainable development, 218
synchrotrons, 326
proton, 328
systematic errors, 14
Système international d'unités, units, 11–13

t-test, 13
targets
small, 322–3
stationary, 327
tau particles, 338
TDM (time division multiplexing), 310
technetium-99, risk factors, 200–1
technical language, in examinations, 292
temperature
absolute scale of, 120
average global, 234, 237
gases, 118–21
scales, 109–10, 120
solar system, 234
thermodynamic scale of, 120
use of term, 109
temperature change
macroscopic dimension, 109–11
microscopic dimension, 109, 111
tension, 35
terminal potential difference, 135
terrorism, and nuclear weapons, 267
text, bit rate, 309
thermal capacity, 110
thermal electricity generation, 224–6
thermal energy, 48, 109
thermal equilibrium, stars, 296
thermal experiments, 113
thermal neutrons, 260
thermal physics, 109–17
thermal reactors, 262–3
thermal runaway, 260
thermistors, negative temperature coefficient, 138
thermodynamic equilibrium, 124
thermodynamic scale of temperature, 120
thermodynamics, 118–26
first law of, 122
second law of, 125–6
thermometers, calibration, 113
Thomson, Sir Joseph John (1856–1940), 183, 184
Three Gorges Dam (China), 254–5
Three Mile Island (USA), nuclear accident, 266
threshold frequency, 208
tidal energy, 253, 255–6
timbre, 73, 96
time division multiplexing (TDM), 310
TIR (total internal reflection), 87, 311
toe (tonne of oil equivalent), 249
tonne of oil equivalent (toe), 249
total internal reflection (TIR), 87, 311
trajectories, charged particles, 325–6
transducers, 137
input, 308
output, 308
transformer cores, laminated, 176
transformer-induced electromotive force, 171
transformers, 173–5
applications, 175–6
losses, 176
step-down, 174
step-up, 174
symbols, 174
transmission, electricity, 175–6
transmitters, 308
transmutations, artificial, 203–4
transuranic elements, 266
transverse waves, 82
travelling waves, 94, 98
and standing waves compared, 95
triple alpha process, 302
true colour imaging, 282
tsunamis, 92–3
tuned circuits, 314
tuning, radio communication, 314–15
two-point charges, 154

ultimately recoverable resources, 222
uncertainties, 13–16
absolute, 15–16
assessment, 59–61
fractional, 15–16
graphical representations, 16
limits, 58–9
in measurements, 59
percentage, 15–16
uncertainty principle, Heisenberg's, 212–13, 319, 320
undiscovered resources, 222
UNFCC (United Nations Framework Convention on Climate), 246
unification, 339–40
United Nations Framework Convention on Climate (UNFCC), 246
units
derived, 12–13
SI, 11–13
universal gravitational constant, 149
universe
closed, 305
expanding, 306
flat, 305
homogeneity, 304
isotropic, 304
open, 305
overview, 294–6
unsustainable growth, 218
Updike, John (1932–), 195
upthrust, 35
uranium
depleted, 260
enrichment, 260, 262
uranium-235, 260, 261, 263, 266–7
uranium-238, 260, 261
decay chain, 189
utility, principle of, 220

vaporization, 114
variable stars, 298
Cepheid, 298, 299, 301
variables
control, 18
controlling, 19
dependent, 18
independent, 18
inversely proportional, 21
inversely related, 21
proportional, 20
selecting, 18
vector addition, 33, 151
vector multiplication, 34
vector product, 34
vector subtraction, 33
vector trigonometry, 33
vectors, 25, 26, 32–4, 145
decomposition of, 33
vehicles
energy-efficient, 250–1
hybrid, 250, 251–2
speed measurement, 100
see also electric vehicles
velocity, 26
average, 26
instantaneous, 26
relative, 28
see also angular velocity; speed
velocity–time graphs, 28
video, bit rate, 309
vinyl records, 275–6
virtual currents, 166–7
virtual particles, 319–20
visual binaries, 297–8
voltage, 127
phase, 173
see also potential difference (p.d.)
voltmeters, 132
real, 140
volume, gases, 118–21
Vostok (Antarctica), 242, 243

W bosons, 320
waste, nuclear, 265–6, 268
wave equation, 82–3
wave functions, 211, 212
boundary conditions, 217
wave mechanics, 216
wave motion, graphs, 86–7
wave phenomena, 94–108
waves, 78–93
amplitude, 78–9
carrier, 313
diffraction, 78, 90, 101–3
earthquakes, 81–3, 88–90
energy, 78–9
intensity, 78–9
interference, 78, 91
light, 80–1
light as, 207
longitudinal, 82
Love, 83
matter, 211
P, 82–3, 89–90
primary, 83
probability, 211
Rayleigh, 83
reflection, 78, 87–8
refraction, 78, 88–90
S, 82–3, 89–90
seashore, 92–3
secondary, 83
transverse, 82
see also electromagnetic (EM) waves; sound waves; standing waves; travelling waves
wave–particle duality, 211
weak interactions, 189
webers, 169
weight, 34
weightlessness, 38
apparent, 164–5
Weinberg, Steven (1933–), 7, 319, 340
white dwarf stars, 298, 302–3
Wien, Wilhelm (1864–1928), 297
Wien displacement law, 232
Wien's law, 297
Wilson, Robert Woodrow (1936–), 342
wind capacity, 257
wind power, 253, 256–7
wires, 311
work, 44–7
work done, 45
ideal gases, 122–3
with non-constant force, 50–1
work function, 208
written text, bit rate, 309

X-rays, 323

zero errors, 14
Zweig, George (1937–), 341

Physics

Course Companion

The ***Course Companion*** series has been designed to provide extra support for students of the International Baccalaureate Diploma Programme.

These resources aim to:

- support students throughout the course with content and activities that illustrate the values and principles of the IB and the IB learner profile
- encourage reflection, discussion and critical thinking
- enhance knowledge of a variety of cultural and ideological perspectives

The ***Physics Course Companion*** has been written for the 2007 syllabus, and includes material for all core units and selected option units at both standard and higher level. There are three key option units included: **Astrophysics**, **Communications** and **Particle physics**. Through activities and extra source material, this book emphasises wider issues and examines how the subject relates to international and real-world contexts.

This book:

- provides features such as *The bigger picture* and *Thinking about science* to enhance theory of knowledge (TOK), international-mindedness and discussion
- includes data-based and practice questions with answers and separate chapters on the internal and external assessment and the extended essay
- uses clear and concise language for those who have English as an additional language
- is written by experienced authors who were involved in writing the latest syllabus; a past chief examiner and current assistant examiner.

Authors
Tim Kirk
Neil Hodgson

OXFORD
UNIVERSITY PRESS

www.OxfordSecondary.co.uk

Orders and enquiries to Customer Services:

tel +44 (0)1536 741068
fax +44 (0)1536 454519

ISBN 978-0-19-91514